Names and Formulas of Some Common Ions

MONATOMIC IONS		POLYATOMIC IONS	
Group 1		*Cations*	
Lithium ion	Li^+	Ammonium ion	NH_4^+
Sodium ion	Na^+	Hydronium ion*	H_3O^+
Potassium ion	K^+	Mercury(I) ion (mercurous ion)	Hg_2^{2+}
Group 2		*Anions Single Charge*	
Magnesium ion	Mg^{2+}	Acetate ion	$CH_3CO_2^-$ or $C_2H_3O_2^-$
Calcium ion	Ca^{2+}	Chlorate ion	ClO_3^-
Barium ion	Ba^{2+}	Chlorite ion	ClO_2^-
Group 13		Cyanide ion	CN^-
		Formate ion	HCO_2^-
Aluminum ion	Al^{3+}	Hydrogen carbonate ion (bicarbonate ion)	HCO_3^-
Group 14		Hydrogen sulfate ion	HSO_4^-
Tin(II) ion	Sn^{2+}	Hydrogen sulfite ion	HSO_3^-
Lead(II) ion	Pb^{2+}	Hydroxide ion	OH^-
Transition Metals		Hypochlorite ion	ClO^-
		Nitrate ion	NO_3^-
Silver ion	Ag^+	Nitrite ion	NO_2^-
Zinc ion	Zn^{2+}	Perchlorate ion	ClO_4^-
Cadmium ion	Cd^{2+}	Permanganate ion	MnO_4^-
Copper(I) ion or cuprous ion	Cu^+	*Anions Double Charge*	
Copper(II) ion or cupric ion	Cu^{2+}		
Mercury(I) ion or mercurous ion	Hg_2^{2+}	Carbonate ion	CO_3^{2-}
Mecury(II) ion or mercuric ion	Hg^{2+}	Chromate ion	CrO_4^{2-}
Iron(II) ion or ferrous ion	Fe^{2+}	Dichromate ion	$Cr_2O_7^{2-}$
Iron(III) ion or ferric ion	Fe^{3+}	Oxalate ion	$C_2O_4^{2-}$
Chromium(III) ion	Cr^{3+}	Sulfate ion	SO_4^{2-}
Manganese(II) ion	Mn^{2+}	Sulfite ion	SO_3^{2-}
Nickel(II) ion	Ni^{2+}	*Anion Triple Charge*	
Cobalt(II) ion	Co^{2+}		
		Phosphate ion	PO_4^{3-}

*The hydronium ion exists only in water solutions, not in compounds.

Peroxide O_2^{2-}

hydrogen peroxide H_2O_2

hypochlorite

Sulfide – H_2S

ide – has no O

chloride Cl^-

INTRODUCTION

TO

CHEMISTRY

INTRODUCTION

TO

CHEMISTRY

EIGHTH EDITION

T. R. DICKSON

Cabrillo College

JOHN WILEY & SONS, INC.

New York   Chichester   Weinheim

Brisbane   Singapore   Toronto

Acquisitions Editors   Jennifer Yee, Cliff Mills
Marketing Manager   Karen Allman
Production Editor   Patricia McFadden
Designer   Kevin Murphy
Photo Editor   Nicole Horlacher
Illustration Editor   Edward Starr
Cover Photo   Tony & Daphne Hallas/Science Photo Library/Photo Researchers

This book was set in 10/12 Baskerville by UG
and printed and bound by Malloy Lithographing.
The cover was printed by Phoenix Color.

This book is printed on acid-free paper. ∞

The paper in this book was manufactured by a mill whose forest management programs include sustained yield harvesting of its timberlands. Sustained yield harvesting principles ensure that the numbers of trees cut each year does not exceed the amount of new growth.

Library of Congress Cataloging-in-Publication Data

Dickson, T. R. (Thomas R.), 1938-
    Introduction to chemistry / T. R. Dickson. — 8th ed.
        p.   cm.
    Includes index.
    ISBN 0-471-18042-4 (paper : alk. paper)
    1. Chemistry.   I. Title.
QD31.2.D53   1999
540—dc21   99-22865
                                                                 CIP

Printed in the United States of America

10 9 8 7 6 5 4 3

THIS BOOK IS DEDICATED TO
ALL CURIOUS AND ENTHUSIASTIC
CHEMISTRY STUDENTS

PREFACE

Introduction to Chemistry is designed for a one-term course in chemistry. It presents the foundations of chemistry to students who need preparation for further study as well as to those who wish to take only an introductory course. The subject of chemistry is introduced at a beginning level, and no previous background in chemistry is assumed.

A preparatory course should provide students with an honest view of chemistry as a central science and should offer a solid foundation and satisfactory preparation for future studies. This book is a product of more than 35 years of teaching, and it reflects my view that the introductory course is one of the most important chemistry courses offered. The overall purpose of this book is to teach chemistry by providing a dynamic, interesting, and relevant view of this vital science and its importance to society and our daily lives. This goal is accomplished through the interweaving of three main themes:

1. *Chemical thinking.* The ideas of chemistry and the words that underlie them are explained in detail. Natural processes and the behaviors of materials are vitalized from a chemical perspective using numerous examples and analogies.
2. *Chemical vision.* Images and models are used to clarify chemical concepts. Students are encouraged to use images to develop a chemical imagination. These images help in the visualization of dynamic chemical processes in nature and in the laboratory.
3. *Critical thinking and problem solving.* Methods of scientific thinking are emphasized to aid students in reading and understanding technical writings. Problem-solving methods of beginning chemistry are introduced, explained, and illustrated throughout the text. Approaches to solving chemical problems by reasoning and the unit-equation, factor-label, or dimensional-analysis methods are explained in detail. A variety of worked-out examples are carefully explained. A major goal is to instill problem-solving skills that will serve students in future courses.

I have worked hard to retain readability as a feature that has made this book useful through eight editions. To increase interest and to illustrate applications, several vignettes are dispersed throughout various chapters. These vignettes include topics related to health, technology, the environment, and society. Some examples are "Using a Calculator," "Seeing Atoms," "Nobel Prizes," "Food and Calories," "Conservation of Matter and Society," "Ethyl Alcohol," "A Hole in the Ozone Layer," and "Calcium in the Body." Several appendices with supplemental materials and a glossary of terms are provided at the end of the book.

● ORGANIZATION

This edition has been revised using comments and suggestions of teachers and students who have used the previous editions. A brief introductory chapter provides students with basic scientific ideas and gives an overview of coming attractions. Chapter 2 presents basic ideas involving measurements and problem solving. This material is vital to numerical and problem-solving topics covered in subsequent chapters.

As a science, chemistry is based on atoms and their behavior. Chapter 3 introduces the atomic theory and establishes a foundation for the development of a chemical imagination. Chemistry is an experimental science, and ideas of chemical formulas and reactions are fundamental. Chapter 4 presents formulas and their uses, and Chapter 5 presents reactions and equations. Chapter 6 emphasizes the visualization of ions and molecules and presents chemical nomenclature. Since problem solving involving stoichiometry is such an important skill, the topic of stoichiometry is covered in Chapter 7. An understanding of the mole and stoichiometric thinking is of great importance to a beginning chemistry student. Chapters 8 and 9 present the topics of atomic structure and chemical bonding. With a basis of (1) an atomic, ionic and molecular vision, (2) familiarity of chemical terms and nomenclature, and (3) calculation and stoichiometry skills, students are well prepared to continue exploring the chemistry presented in later chapters.

The remaining chapters have been rewritten and rearranged to vitalize the topics and to include interesting examples, analogies, and images. I resisted the temptation to add more material. The text retains its reasonable length. The topics covered in later chapters follow a purposeful order, but the topic and chapter order can be varied if desired.

● SUPPLEMENTS

A comprehensive package of supplements has been created to assist both the teacher and the student.

- *Study Guide* by **T. R. Dickson.** Each chapter contains objectives, glossaries, and summaries as well as further mathematical explanations as needed for text concepts. This valuable student guide also includes self-tests with solutions.

- *Laboratory Experiments* **by T. R. Dickson.** This manual contains a choice of experiments designed for students who have had no laboratory experience. Along with vitalizing important chemical concepts, the manual emphasizes laboratory techniques, procedures, and safety.

- *Instructor's Manual for Text and Laboratory Experiments* **by T. R. Dickson.** The *Instructor's Manual* includes solutions to all end-of-chapter problems and exercises in the text. Instructor's notes on materials and equipment as well as sample report sheets for the laboratory experiments are also provided.

- *Test Bank* **by T. R. Dickson.** The *Test Bank* contains approximately 1000 questions.

- *Computerized Test Bank.* IBM and Macintosh versions of the entire *Test Bank* are available with full editing features to help you customize tests.

- *A website for teachers and students.* URL: www.wiley.com/college/DicksonChem8E

ACKNOWLEDGMENTS

I am very grateful to all the students, teachers, and colleagues who have helped me in the preparation of this eighth edition. I appreciate the useful comments and suggestions of the following reviewers Rodney Oka, Monterey Peninsula College; Don Fenton, Kirtland Community College; Mary Ann Morrissey, Luzerne County Community College; Robert P. Metzger, San Diego State University; Margaret Benoit, San Diego State University; Carol Anderson, University of Northern Iowa; and Michael Schmidt, California State University, San Marcos. In addition, I thank the editorial staff at John Wiley for their many years of support, advice, and outstanding professional service.

T. R. Dickson

TABLE OF CONTENTS

CHEMISTRY

1-1 A TALE OF TWO TIMES

In 1991, hikers discovered the freeze-dried body of a Stone Age man preserved in the ice of a mountain high in the Italian Alps. The man apparently had met a natural death about 5300 years ago. His is the oldest body ever discovered fully dressed in the clothing of the times. He wore well-tailored deerskin pants and top complemented with leather boots and a parka-like outer garment made of reeds. His belongings included a pouch containing small tools and some fungus that may have had a medical use. He wore a small jewelry-like item most likely of artistic or religious significance. A wood-framed leather backpack and a copper ax were found nearby.

Let's compare his clothing and possessions with a modern-day hiker. A typical hiker might wear well-tailored wool pants, an Orlon™ sweater, and a parka made of Gore-Tex™, two common synthetic fibers. She might have leather boots with synthetic rubber soles and carry a first aid kit containing synthetic medicines. Around her neck might be a chain with a piece of jewelry of artistic or religious significance. Her backpack could be nylon with an aluminum frame, and she could have a stainless steel ice ax.

Some things have not changed much in more than 5000 years. What has changed dramatically is the availability of materials that humans can put to a variety of uses. Furthermore, it appears that the Stone Age man was very resourceful and aware of his environment but he did not have the opportunity to learn chemistry.

1-2 USEFULNESS AND CURIOSITY CREATE CHEMISTRY

Chemistry is the science of the properties, composition, and behavior of matter. Matter refers to the varieties of materials or substances that comprise the universe. It is interesting that our word *matter* is derived from *Mater*, the Mother Earth, whom the ancients viewed as the source of all material things. Chemistry as a science evolved from human curiosity about the natural world. One human activity that influenced chemistry was the

Chemistry:
(kem ə strē), noun chemistries, plural 1. the study of substances that compose the universe and the processes by which they act on one another. 2. the science dealing with characteristics of elements or substances, the changes that take place when substances combine to form other substances, and the laws of the behavior of substances under various conditions. 3. the application of the science of chemistry to certain subjects: the chemistry of iron, the chemistry of life. 4. Slang, the existence of some factor: the chemistry between us is good, you make my molecules move.

1

Chem-, Chemi-, and Chemo-:

prefixes referring to chemistry:

chemotherapy: *the treatment of a disease by use of specific chemicals.*

chemiluminescence: *the production of light during a chemical process without a rise in temperature.*

search for and the isolation of useful materials: dyes for cloth, ores for metals, materials for construction, natural products for medicines. This field became applied chemistry, or chemical technology, that today involves the isolation of hundreds of useful materials from the environment and the manufacture of thousands of different consumer products. Obvious examples of useful materials are iron ore and aluminum ore, sources of iron and aluminum metals and crude oil, the raw material for fuels, petroleum products, plastics, and many synthetic medicines. Applied chemistry also deals with processes and methods involving recycling of materials and pollution control.

Another part of chemistry is philosophical and involves speculation about the nature of materials. How are materials similar and different, and what are things made of? This area is called theoretical or pure chemistry, and it provides a chemical view of nature and explanations of varied natural processes. What are materials such as iron ore, aluminum ore, and petroleum made of? What are the natures of metals, medicines, and plastics? Pure chemistry provides answers to these kinds of questions. Modern chemistry includes academic and industrial research and investigations of practical applications of materials. In your study of chemistry, you'll learn many interesting applications and you'll gain a useful chemical vision of the world.

Chemistry is one of the four basic branches of modern natural science:

Chemistry is the science of the properties, composition, and behavior of matter.

Life science or biology is the science of living systems.

Earth science or geology is the science of the earth and earth processes.

Physics is the science of matter, energy, forces, and motions.

Chemistry contributes ideas and information to the other sciences and, in turn, borrows from them. Furthermore, a close relation exists between the evolution of chemistry and the development of medical sciences. This is why pharmacists are often called chemists in Britain. Modern medicine depends on a chemical understanding of life processes, and some important chemical ideas have come from the curiosities and imaginations of medical practitioners. Knowledge of chemistry is vital to the large variety of scientific disciplines ranging from agriculture to zoology.

Chemical terminology is part of modern, everyday communications:

Olive oil contains monounsaturated fats.

I'm on a low-sodium diet.

Watch your calories.

You need more iron and calcium in your diet.

Write a few sentences using some chemical terms that you know.

Your study of chemistry will give you an interesting view of nature and help you understand chemical methods and terminology. What you learn about chemistry in this book will serve you in future chemistry courses and other science courses.

A variety of materials make up the environment. By definition, **matter** is anything that has mass and occupies space. Mass refers to how much material is in a sample of a substance. Mass is a property of matter that gives it inertia, the resistance to being set in motion or resistance to any change in motion. The amount of mass in an object is directly related to its weight. A baseball is much easier to throw than a more massive shot-put ball and obviously easier to catch. Typically matter occurs in the physical forms of solid, liquid, and gas. Matter is a collective term used to refer to all natural materials. Specific kinds of matter are called chemicals.

Practitioners of chemistry are, of course, chemists. Chemists observe matter in the environment and they collect, separate, and isolate samples of interest. In the laboratory, chemists examine, test, measure, analyze, purify, combine, and thoroughly study samples. In times past, chemists isolated and classified different chemicals from the environment and developed methods to make a variety of "synthetic" chemicals. Modern-day chemical research still involves such activities. Each year, chemists isolate many new chemicals from natural products and develop methods to make new synthetic chemicals. Some have practical use, whereas others are analyzed and studied to satisfy intellectual curiosity and add to the scientific understanding of nature. As of the beginning of 1999 more than 19.3 million chemicals were cataloged and registered in the Chemical Abstracts Service database.

Chemical Abstracts Service

1-3 SCIENTIFIC METHOD

In reference to development of natural science, Albert Einstein, one of the most famous twentieth-century scientists, said, "The most incomprehensible thing about the world is that it is comprehensible." **Natural science** refers to the accumulation of knowledge about nature and the attempt to understand the workings of the physical world. Science, however, is more than just an accumulation of knowledge. It represents an extremely useful way of thinking and provides a vision of nature and natural occurrences.

The scientific method includes observations, measurements, experimentation, replication of results, development of hypotheses, analysis of connections of cause and effect, factual documentation, and development of scientific theories. Let's take a closer look at the important parts of the scientific process.

Making Observations

Careful observations are crucial in science. Noting details and looking for patterns requires focused attention. Some observations are qualitative descriptions: the material is black and a solid. Some observations are

Find a new and shiny penny. Inspect and observe the penny as closely as you can. Record all your observations about the penny.

quantitative: the solid weighs 2.3 pounds. A quantitative observation is expressed as a measurement having a number, like 2.3, and a unit, like pounds. Statements of scientific facts about materials and measurements of these materials are part of the story of chemistry. The study of materials involves descriptions and measurements.

Stating a Hypothesis

A **hypothesis** is an educated guess that serves as a possible explanation of certain observations. Such an explanation needs to make sense within the framework of science. An interesting or useful hypothesis is subjected to careful scientific scrutiny. Extensive testing and experimentation may suggest alterations to a hypothesis or even its complete rejection as a valid explanation.

Devising Experiments

The term *experiment* relates to the word *experience*. Experiments are carefully controlled observations. Often experiments are designed to try to confirm or support a hypothesis or to disprove it. One valid experiment that disproves a hypothesis may be enough to reject it, or at least require its alteration. Thomas Henry Huxley, an English biologist, in 1870 gave eloquent testimony to this idea when he said, "The great tragedy of Science [is] the slaying of a beautiful hypothesis by an ugly fact."

Chemistry is an experimental science. Often when new and significant experimental results are reported in scientific literature, other scientists repeat the experiments to see if they get results that are the same or at least very similar. In 1989, two scientists claimed that they could carry out a process called nuclear fusion using simple laboratory equipment at room temperature. If their hypothesis proved correct, cold fusion had the potential of providing a vast source of inexpensive energy. Scientists around the world tried to repeat the cold fusion experiments. After months of effort, no scientist could confirm the results as originally reported. Because the original experimental results could not be reproduced, the hypothesis put forth by the two scientists was rejected. Nevertheless, research on the topic of cold fusion is continuing. Experiments are important in chemistry. That's why the chemistry laboratory is an integral part of chemistry. The laboratory serves as the crucible for putting hypotheses to the acid test.

Establishing Scientific Laws

A hypothesis is supported and confirmed by experimental observations carried out by many scientists over time. In this way, a hypothesis is accepted. A few ideas that began as hypotheses are of general importance and have become **scientific laws** or **natural laws**. A law is an expression of some notable and consistent pattern in nature. A common example is the law of gravity first stated in the late 1600s by Isaac Newton, the famous English scientist, as, "Every particle of matter attracts every other particle with a force proportional to the product of the masses and inversely proportional to the square of the distances between them." You experimentally confirm this law every time you use a bathroom scale to weigh yourself. A scientific law may be changed if experimental evidence warrants a change. In other words, scientific laws are not necessarily fixed and unchangeable. Many laws, however, have survived the test of time and the skepticism of scientists; they are considered unlikely to change.

An example of a law important in chemistry is the law that describes forces between electrical charges. Under certain circumstances, objects can carry positive or negative electrical charges. Coulomb's law relates to the behavior of charged objects. A statement of the law is:

Like charges repel and unlike charges attract. The force of attraction or repulsion that exists between two charged objects is proportional to the product of the sizes of the two charges, divided by the square of the distance between them.

This law is called **Coulomb's law**, since it was first stated by the French scientist, Charles Coulomb, in the late 1700s.

In later chapters, you'll learn about other laws that are important to chemistry. Learning chemistry does not mean that you have to learn about numerous laws. Actually, only a few laws are vital to your understanding of chemistry.

Express the first part of Coulomb's law by using pictures to illustrate the idea.

Developing Theories or Models

A law reveals what to expect in nature, and a **scientific theory** provides a grand explanation of natural processes. Theories are elegant treasures of science and provide a framework to envision physical reality and make sense of mystifying natural occurrences. They provide general explanations of observations and allow the prediction of events yet to occur. Often a theory is stated as a broad generalization about nature. Some theories are called **scientific models.** An example is the scientific model of the solar system. This model provides a beautiful and comprehensible view of the planets in motion about the sun; it explains their relative positions and motions. The model is used to predict future positions of planets and other phenomena, such as solar and lunar eclipses or phases of the moon. Furthermore, the model gives us a mental image of the solar system. The image is made into a drawing or a three-dimensional representation. (You have never seen the solar system but you probably can draw a sketch of it.) Sketches and three-dimensional constructions are simplified visions of the larger reality. As you continue your study of chemistry, you'll learn about a few very important chemical theories.

Ockham's Razor

William of Ockham was a medieval scholar of the early 1300s. He stated a rule that has become very important to scientific thinking. The rule is called Ockham's razor since it is used to shave away fuzzy thinking. Simply stated, the rule is, "Do not multiply entities beyond necessity." From a more contemporary view this rule is stated in other ways, such as:

Do not read more into facts than the facts allow.

If two hypotheses seem to explain the observed facts, the simpler hypothesis is more acceptable.

"Freedom is the oxygen without which science cannot breathe."
David Sarnoff,
1954

The development of scientific theories does not always happen easily, quickly, or smoothly. Evolution of thought takes time. The modern view of the solar system, for example, took thousands of years and countless astronomical observations to develop. At times, new ideas meet significant resistance. The famous Italian scientist Galileo Galilei (1564–1642) was forced by church authorities to retract his views that Earth moved around the sun. Furthermore, he was put under house arrest for the last few years of his life so that he could not communicate his ideas to others. There are many cases involving the suppression of scientists by political, social, governmental pressures, and prejudices. Such suppression not only occurred in the medieval times of Galileo but also happens in modern times. In the early 1900s, Marie Curie, a Polish-born French scientist, was a pioneer in the newly discovered field of radioactivity. Despite her many honors, including two Nobel Prizes, she was never elected to the French Academy of Sciences. Apparently she was slighted because she was Polish born and a woman. In the 1950s, Linus Pauling, an American chemist, had his passport restricted by the government and was not allowed to travel out of the United States. In the 1970s and 1980s, Andrei Sakharov, a Russian physicist, was exiled to a small Russian city and not allowed to talk with other scientists. Both Pauling and Sakharov were punished for speaking against the development of nuclear weapons. Despite his harassment by the government, Pauling was awarded the 1963 Nobel Peace Prize for his work to ban the testing of nuclear weapons in the atmosphere. Sakharov won the Peace Prize in 1975 for his opposition to nuclear weapons. Before his death, Sakharov was recognized for his scientific integrity and opposition to the government of the former Soviet Union. Recently, the Catholic Church admitted that Galileo was treated unfairly in the 1600s, and Marie Curie's remains were moved to an honorary grave in the Pantheon of Paris 60 years after her death.

Scientists are only human and subject to personal value judgments, biases, prejudices, misinterpretations, self-serving behaviors, and peer pressures. Our discussion of the evolution of scientific ideas is based on hindsight. Here the results are given and descriptions of the hard work,

Nobel Prizes

Alfred Nobel, a Swedish scientist, invented dynamite in 1866. In his search for a way to make the explosive nitroglycerin safer to handle, he accidentally discovered that nitroglycerin absorbed on diatomaceous earth (a readily available mineral) was quite stable and could be safely handled before use. He also invented blasting gelatins and a smokeless gun powder. Profits from his inventions exploded, and he became quite wealthy. He thought that his inventions would be used for peaceful engineering purposes but, unfortunately, humans used them as weapons of war. Apparently out of guilt over the destructive uses of dynamite, he established a foundation to awards prizes (now called the Nobel Prizes) in five fields: peace, literature, physics, chemistry, and physiology/medicine. Later a prize for economics was added. These prizes, awarded annually since 1901, are among the highest honors scholars or scientists can earn. In 1958, Alfred Nobel was honored by scientists when a chemical element, nobelium, was named for him.

Two of the many Nobel Prize winners serve as interesting examples. Marie Curie and her husband, Pierre, won the physics prize in 1903 for their work on radioactivity. She later was awarded the 1911 Nobel Prize in chemistry for the discovery of the element radium. In 1935, her daughter Irène shared the chemistry prize with her husband. Linus Pauling was one of the most outstanding American chemists of modern times. He is unique because he won prizes in two different categories. In 1954, he won the chemistry prize for his brilliant work on the nature of the chemical bond and, as mentioned above, he won the Nobel Peace Prize in 1963.

---

mistakes, invalid experiments, rejected hypotheses, and personal foibles are left out.

Open and international communication and exchange of knowledge are essential to the development of science. Results of scientific studies and research are reported in numerous scientific publications around the world. One organization of chemists is the **International Union of Pure and Applied Chemistry (IUPAC).** The name refers to the two aspects of chemistry. Some chemical research deals with **pure chemistry** in which materials and phenomena are investigated because they are scientifically interesting. **Applied chemistry** concentrates on practical uses of materials and the development of useful processes to manipulate materials. The applied and pure parts of chemistry often overlap. Some applied research reveals new ideas about nature; some pure research generates practical applications.

The **American Chemical Society (ACS)** is an organization of research chemists, industrial chemists, and chemists involved primarily in chemical education. The ACS sponsors periodic conferences and meetings and publishes specialized books and various scientific journals. Similar organizations exist in other countries.

Nobel Prizes

International Union of Pure and Applied Chemistry

American Chemical Society

Fact or Fiction?

Beware of certain so-called facts. A scientific fact is a statement of a verifiable observation. A fact should be unambiguous, and reproducing it or at least confirming it should be possible. Sometimes apparent facts are superfluous or are used to mislead or deceive the reader. Sometimes opinions or points of view masquerade as facts. A scientist reads "facts" from a skeptical point of view and applies Ockham's razor. The use of facts can vary from quite reliable or quantitative to outright distortions or deceptions. Some examples are:

Fact: The major components of air are nitrogen and oxygen.

Superfluous fact: "Contains zero calories," from the nutritional label of bottled water. This statement is superfluous and unnecessary since water contains no food energy.

A vague or questionable fact: "Low-calorie dessert," a claim on some packaged desserts. This statement is vague since there is no basis of comparison for the term *low*.

A distortion of fact: "Contains no chemicals," a claim on some packaged foods and bottled water. This statement is a distortion since all materials are chemicals. Possibly the unstated message is that the contents contain no added synthetic chemicals.

An outright distortion of fact: "All chemicals are dangerous."

A nonfact: "Little green aliens landed in New Mexico." This statement is an old fabrication with no basis in fact.

ACTIVITY
1-4

Find a "fact" in an advertisement, newspaper or book and analyze it.

I celebrate myself,
And what I assume
you shall assume,
For every atom belonging to me as
good belongs to
you.

Walt Whitman,
"Song of Myself,"
1855

1-4 CHEMICAL VISION

Chemistry is sometimes called the conceptual science. Chemistry is conceptual because it requires the use of mental visions of the particles that structure matter.

> If in some cataclysm, all scientific knowledge were to be destroyed, and only one sentence passed on to the next generation of creatures, what statement would contain the most information in the fewest words? I believe it is the *atomic hypothesis* that *all things are made of atoms—little particles that move around in perpetual motion, attracting each other when they are a little distance apart, but repelling upon being squeezed into one another.*

These are the words of Richard Feynman, a modern American scientist, who won the Nobel Prize for physics in 1965.

This atomic or particle view of matter relates to the amazing success of chemistry as a science. All material objects are composed of very tiny, characteristic particles. Chemical particles are far too small to see with typical microscopes so you cannot observe them. Part of learning chemistry involves learning about and visualizing these important particles.

The fundamental particles that structure matter are atoms. As we will see, various kinds of matter have characteristic atoms. Simply, atoms are viewed as very tiny spherical objects. For example, atoms of hydrogen, carbon, nitrogen, oxygen, and argon are pictured as:

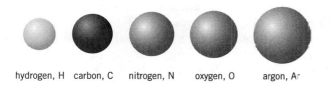

    hydrogen, H   carbon, C   nitrogen, N   oxygen, O   argon, Ar

Matter, whether in the form of a gas, a liquid or a solid, is viewed as collection of particles. Idealized (stop motion) pictures of a solid, liquid, and gas are shown in Figure 1-1. A **gas** sample is pictured as a diffuse and dynamic collection of high-speed particles with a relatively large amount of space between them. The particles are bouncing about, colliding with one another and any objects in their environment. The particles in a gas are in continuous, random, and high-speed motion. The motion of the gas particles "fills" the entire container. Such incessant motion also accounts for the ways in which gases behave. Imagine a **liquid** as a teeming collection of particles in which there is relatively little space between them. Individual particles can

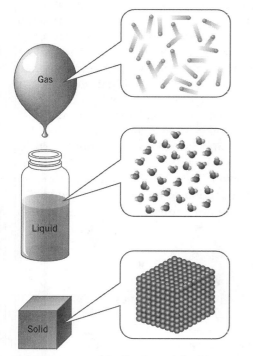

FIGURE 1-1   Idealized particle views of a solid, a liquid, and a gas.

move about in the collection, and the entire collection is quite flexible and conforms to the shape of the container. Picture a **solid** as a collection of particles with fixed positions in space arranged in some three-dimensional pattern. The particles can move or vibrate about their positions but cannot easily migrate like particles of liquids or gases. Solids have fixed shapes.

A few chemical substances are composed of individual atoms. Many substances, however, contain particles made of combined or bonded atoms. The nature of atomic bonding is discussed in Chapter 9. Until then, it is useful to imagine that atoms can attach or bond to form characteristic atomic groups. Particles composed of groups of combined atoms are called molecules. By definition, a molecule is a particle composed of two or more chemically combined atoms. For example, a sample of water is a collection of a very large number of molecules each containing two hydrogen atoms combined with one oxygen atom.

Molecules of various substances range from those involving only two atoms to those having many atoms. Molecules composed of two bonded atoms are called diatomic molecules. Molecules of most substances include more than two atoms. Notice that water molecules are triatomic. Some naturally occurring biological materials and synthetic plastics are composed of molecules having hundreds or thousands of atoms. Such relatively "giant" molecules are called macromolecules. Although such molecules contain hundreds or thousands of atoms, individual molecules are still quite small and it is not possible to see individual molecules.

In contrast to those materials that contain molecules, some substances are composed of collections of electrically charged atoms called ions. Ions occur as positively charged particles and as negatively charged particles. Substances that contain ions have positive ions balanced with negative ions. For example, a sample of table salt (also called sodium chloride, represented as NaCl) is pictured as a collection of positively charged sodium ions balanced by negatively charged chloride ions.

sodium
ion

chloride
ion

As you learn chemistry, you will learn more detail about chemical particles. Part of chemistry involves explaining the behavior of matter in terms of these characteristic particles. Take a moment to look closely at the cubic foot of air in front of you. Move any people or objects out of the way so that you can focus on the air. What do you see? You don't see anything. How is such a sample of air visualized from a chemical point of view? As it turns out, a typical cubic foot of air on average contains about 800,000,000,000,000,000,000,000 particles. We say "on average" because the particles in air are in rapid, random motion, so various particles move in and out of the cubic foot of air in front of you. Such particles are moving at speeds of hundreds of miles per hour and collide billions of times per second. Wave your hand in front of you; you can feel some of them and knock them about. How are the particles that compose air pictured? Air is a mixture of several different chemicals. Let's consider the major components since their particles are the most populous in a sample of air. If the air is totally dry, about 78% of the particles are diatomic nitrogen molecules.

N_2
nitrogen

About 21% of the particles are diatomic oxygen molecules.

O_2
oxygen

One percent of the particles are atoms of argon.

Ar
argon

Typical air, however, is not dry but contains variable amounts of water vapor made up of molecules of water. The remaining chemical components of air are present in very low concentrations. One important but minor component of air is carbon dioxide, which occurs in triatomic molecules pictured as:

CO_2
carbon dioxide

Using these images gives us a way to visualize air from a particle view. Again focus on the cubic foot of air in front of you and imagine the particles that you find moving about.

1-5 LEARNING CHEMISTRY

This chapter is a preview of coming attractions and is designed to help you in your study. The ideas and examples presented are intended to give you a general view of scientific thinking and what you'll learn in this text. You are going to embark on an exploration of the basic ideas and techniques of chemistry. This text and your teacher are your guides.

For this first course in chemistry, Chapter 1 serves as a general introduction. The next chapter introduces basic ideas about chemistry and gives you tools to use throughout the remainder of the book. We'll begin with fundamentals and build on them as you explore the science of chemistry. You'll find the exploration interesting and useful to your future studies in science.

Chemistry has its own language. Many new words and terms are introduced along with the "grammar" for using them. Learning the language of chemistry gives you a new way to describe the world. In learning chemistry, you'll gain a new vocabulary and new visions. Studying chemistry involves learning new ideas, new words, and new ways of looking at the world. You need to develop your own approach to learning chemistry. The following sections will aid you in your adventure.

Tools

Your textbook is the main source of information for your study of chemistry. Look at the table of contents and note that there are some useful appendixes at the end of the book and a glossary of important terms. Your instructor will assign specific chapters in the book. Some students find it useful to scan a chapter first to see what topics and information are presented. Each chapter has a focus section that reviews or summarizes important ideas.

As you carefully read assigned chapters, make notes on your reading. Underlining important ideas is useful. Many examples of problem solving are given in the text. Study the examples carefully for better understanding. Important new terms are printed in the text in boldface.

Each chapter has a set of activities and end-of-chapter questions. Use them to test your knowledge and to practice problem solving. A variety of questions and problems are given to provide a range of choice. Material related to specific sections in the chapter are set off by section number. You are not expected to answer all the end-of-chapter exercises, and your instructor may choose to assign selected questions. Those problems marked with an asterisk (*) have answers listed in the answer section at the back of the book.

Some students find the *Student Study Guide for an Introduction to Chemistry* a useful supplement. Your science department or school library has various chemistry reference books. Consult references or computer Web sources for additional information about chemistry. A calculator is an invaluable tool to help you in calculations. A scientific calculator is the most useful type.

Resources

Your teacher is your resource for guidance and information. Be creative and impress your teacher by asking good questions and participating in groups and discussions. Lecture notes are very useful, so practice taking careful notes and use them for study purposes. Homework questions and problems are assigned to allow you to practice what you are learning. Some students find study groups very useful. Two or more students gathered together make a group. Group sessions are a good place to do homework, talk about ideas, or study for exams. Your school may have tutoring services available. Use the chemical visions you learn to explore the world. Practice telling others about what you are learning. Visit the Web site at www.wiley.com/college/Dickson/Chem8E for interesting supplemental information. Some other interesting Web sites are noted in various chapters. When you see the icon shown in the margin visit the Wiley Web site to find links to the indicated topic.

Terms and Vocabulary

Notice that chemistry has its own jargon and unique terminology. Terms introduced in this chapter include matter, molecule, and atom. Some commonly used terms have special meanings in chemistry. For example, the term *matter* is derived from the Latin word for stuff (*materia*) and has several definitions. Look it up in a dictionary. The study of chemistry involves learning new terms. You'll have to memorize the definitions of some important terms. Be selective about the terms you decide to memorize, however. If you are not sure about a term, ask your instructor or your study group. Many students use flash cards to aid memorization. Some students record tapes as study aids. Writing definitions and terms by hand allows you to picture them and tune into them.

Give definitions for the following terms: *mass, hypothesis, experiment.*

Problem Solving

Part of learning chemistry includes problem solving and calculations. Techniques of problem solving and calculation are introduced in Chapter 2 and are emphasized throughout the remaining chapters. To learn useful methods of problem solving, carefully study examples given in the text and work problems for practice. Some students find practicing problem solving with groups essential to their learning process.

Concepts and Visions

Three important goals of your study of chemistry are to practice critical thinking, to develop chemical visions, and to achieve an understanding of fundamental chemical concepts and problem-solving methods. The accomplishment of these goals requires practice. Consider some examples.

1. A scientific fact should make sense, not just appear to make sense. Get used to thinking critically when you see a stated fact or a supposed fact. Try to interpret, expand, or explain such statements. As an example, consider the statement: "Smoking causes cancer." This is not a clear statement. Of course, it does not mean that the act of smoking causes cancer. A more precise or complete statement is that certain chemicals in cigarette smoke are known to cause cancer or that statistical studies reveal that certain types of cancer are common among long-term smokers.
   To practice, comment on the following "facts":

   "This product contains 25% fewer calories."
   "I believe that the positions of the planets influence our lives."
   "One in fifty adult Americans—3.7 million—may have been abducted by UFOs."

2. Water can exist as a gas (water vapor or steam), a liquid, and a solid (ice). Sketch simple pictures to show the differences between these three physical forms of water. Water molecules are triatomic particles, but for convenience use small circles in your sketches.

3. Write a definition for matter. Write a definition for chemistry. Give definitions that are not exact copies of definitions given in the text.

QUESTIONS

Each chapter has a set of problems and questions. Use them to test your knowledge and to practice problem solving. A variety of questions and problems are given to provide a range of choice. Material related to specific sections in the chapter are set off by section number. You are not expected to answer all the end-of-chapter exercises, and your instructor may choose to assign selected questions. Those problems marked with an asterisk (*) have answers listed in the answer section at the back of the book.

Sections 1-1 to 1-3

1. Give a definition for the term *theory* that uses terms that are different from those used in the definition given in the text.

2. Do you think that the Stone Age man "knew" some applied chemistry? Why?

3. What two human activities lead to the development of the science of chemistry?

4. What is natural science?

5. What is the difference between a qualitative observation and a quantitative observation? Give some observations about yourself to illustrate the difference.

6. What are the general parts or aspects of scientific thinking?

7. Describe each of the following.
   (a) hypothesis
   (b) experiment
   (c) scientific law

8. Give a statement of the law of gravity.

9. Give a statement of Coulomb's law.

10. Explain the second part of Coulomb's law as given in Section 1-3.

11. What is a theory or a model?

12. Why are theories useful in science?

13. Give a statement of Ockham's razor in your own words.

14. Critically analyze the following "facts."

   (a) 50% less salt

   (b) Eating sweets makes you fat.

   (c) All natural

   (d) Contains no cholesterol

   (e) An apple a day keeps the doctor away.

   (f) A monkey is reported to have given birth to twin humans.

   (g) 95% no-fat potato chips

   (h) Steam is invisible.

   (i) low in sodium

Section 1-4

15. Use sketches to describe the differences between the distribution of particles that make up solids, liquids, and gases.

16. Describe and use a sketch to explain what particles you imagine populate the cubic foot of air in front of you.

17. Here is a question involving scientific ethics. The tissue of the Stone Age man found frozen in the ice was essentially freeze-dried. Scientists analyzed the DNA in the cells of a sample of his tissue. What are your opinions about such an analysis, or even the use of this DNA from an ethical perspective?

MEASUREMENTS

IN CHEMISTRY

2-1 TAKING MEASURE

If asked to describe yourself in physical terms, you might give a qualitative observation like your hair color or the color of your eyes. Other quantitative descriptions include measures of your height and weight. (What is your height in centimeters and your weight in kilograms?) Through time, human societies developed different units of measurement and various measuring devices. Modern supermarkets have electronic balances to weigh produce, whereas merchants in rural China accomplish the same thing using handheld balances. Historically, commerce, trade, and taxation led to the development of units of measure along with appropriate measuring instruments and calculating devices.

A few examples of the origins of units show how closely they relate to humans. The cubit, a biblical measure, was the length from the elbow to the tip of the middle finger. The foot originated from the length of the human foot, and the inch came from the width of a human thumb. A mile was first defined by the Romans as 1000 double paces and later was more precisely defined as 5280 feet by the English. The pound became the name of a commercial unit of weight first used in Europe. The term *pound* comes from the Latin word for pound weight (*libra pondo*). The word *libra* is part of the Latin name for a pound so that is why our abbreviation for pound is lb. A unit of weight still used in the pharmaceutical industry is the grain, which originated from the weight of a grain of wheat.

Until the beginning of the 1800s, various city-states, countries, and regions had local systems of measuring and weighing. These systems often used units that differed from those used by others, which caused confusion and inconvenience in commercial communication. Scientists of the time also used various commercial units of measure. The lack of standardized units not only made commerce and trade difficult, but it complicated communications between scientists. In 1791, French scientists began to establish a new system of measurement that has become the modern **metric**

system. Usually, convincing the general populace to adopt a new system of measure is quite difficult. The late 1700s and early 1800s, however, included the years of the French Revolution, a time of significant social change in Europe. In those times, the metric system came into use as a commercial and scientific system of measure. The metric system uses units like the meter and the gram instead of miles, yards, feet, inches, tons, pounds, and ounces.

How is it possible to describe objects and events in the physical environment? From a subjective view, poets and writers of literature have done this throughout the ages. To measure is to find the extent, size, quantity, or capacity of something. To describe the environment, we measure objects within it. To express measurements, we need fundamental units for length, mass, time, temperature, and volume.

2-2 THE METRIC SYSTEM

SI and Metric Units

The metric system is the common system of reference units used in science. It has evolved into the **International System of Units**, sometimes called **SI units**, from the French, *Système International d'Unités*. Over the years, units of the metric system have been refined, and today the system consists of a set of specifically defined units. Nearly all countries of the world use metric or SI units for commercial and scientific applications. These units provide for worldwide agreement in measurement.

The basic units or base units of the metric system are listed below. Each unit has an abbreviated symbol used for convenience.

OBSERVATION	UNIT	SYMBOL
mass	**gram**	**g**
length	**meter**	**m**
time	**second**	**s or sec**
temperature	**kelvin**	**K**
	or	
	degree Celsius	**°C**
volume	**liter**	**L**

A variety of measuring instruments are available that allow the precise measurements in these metric units. For example, balances are used to measure mass. The process of measuring the mass of an object on a balance is called weighing. Thermometers and thermistors are used to measure temperature. Precise clocks or chronometers are used to measure time. Calibrated wheels or rulers are used to measure distances. Some surveying instruments make distance measurements using laser beams. Laser measurements have been used to measure the distance between Earth and the moon. During one Apollo journey, astronauts placed a mirror on the moon. Later, a laser beam from Earth was reflected from the mirror. The time needed for the beam to travel to the moon and back was used to measure the distance from Earth to the moon.

2-3 THE LARGE AND THE SMALL

Objects and the distances between them range from very large to very small. The invention of the telescope allowed scientists to look at planets in the solar system, various stars, and galaxies. The word *telescope* comes from the Greek *tele* (distant) and *skopos* (watcher). The invention of the microscope allowed scientists to observe very small objects. The word *microscope* comes from the Greek *mikros* (small) and *skopos* (watcher). Objects that are visible only with the aid of a microscope are microscopic in size. Things that are too small to be seen with typical microscopes are submicroscopic. Atoms, molecules, and ions are submicroscopic particles.

A **meter** is slightly longer than a yard, and it is a good unit to measure people-sized objects. The height of a typical person is between 1 and 2 meters. The diameter of the sun is about 1,400,000,000 meters, and the diameter of Earth is about 13,000,000 meters. In contrast, the average diameter of a red blood cell is 0.000009 meter, and the length of an oxygen molecule is about 0.0000000001 meter.

In science, large and small numbers are often written in a form called **exponential notation** or **scientific notation.** Exponential notation is a method by which magnitudes of numbers are expressed in powers of 10 rather than the normal form. Various multiples of 10 are expressed as 10 raised to a power.

$$100 = (10)(10) = 10^2 \qquad \text{one hundred}$$
$$1,000 = (10)(10)(10) = 10^3 \qquad \text{one thousand}$$
$$10,000 = (10)(10)(10)(10) = 10^4 \qquad \text{ten thousand}$$
$$100,000 = (10)(10)(10)(10)(10) = 10^5 \qquad \text{one hundred thousand}$$
$$1,000,000 = (10)(10)(10)(10)(10)(10) = 10^6 \qquad \text{one million}$$
$$1,000,000,000 = (10)(10)(10)(10)(10)(10)(10)(10) = 10^9 \qquad \text{one billion}$$

A fraction of 10 is expressed as 10 raised to a negative power.

$$\tfrac{1}{10} = 10^{-1} \qquad \text{one tenth}$$
$$\tfrac{1}{100} = 1/(10)(10) = 10^{-2} \qquad \text{one hundredth}$$
$$\tfrac{1}{1,000} = 1/(10)(10)(10) = 10^{-3} \qquad \text{one thousandth}$$
$$\tfrac{1}{1,000,000} = 1/(10)(10)(10)(10)(10)(10) = 10^{-6} \qquad \text{one millionth}$$

To change a number greater than one to exponential notation, move the decimal point to the left until it is between the last digit and the next-to-last digit. Then, count the number of places the decimal was moved; use the count as an exponent of 10. Finally, drop any extra zeros and write the number as the remaining digits multiplied by 10 raised to this exponent.

$$1,400,000,000 \text{ becomes } 1.4 \times 10^9$$

To change a number less than one to exponential notation, move the decimal point to the right until it is on the right of the first nonzero digit. Then, count the number of places the decimal point is moved and use the count as the negative exponent of 10. Finally, drop the leading zeros, then write the number as the remaining digits multiplied by 10 raised to this negative exponent.

$$0.000\ 009\ 2 \text{ becomes } 9.2 \times 10^{-6}$$

Exponential notation is a convenient way to write large and small numbers to avoid the use of numerous zeros. A few examples of measurements in exponential notation illustrate its convenience. The diameter of the sun is about 1.4×10^9 meters. Earth has a diameter of about 1.3×10^7 meters. The diameter of a typical red blood cell is about 9×10^{-6} m. An oxygen molecule has a length of about 1×10^{-10} m. Figure 2-1 shows the range of sizes of a variety of objects using exponential notation.

EXAMPLE 2-1

The distance from our solar system to the nearest star is about 100,000,000,000,000,000 meters. What is this distance in exponential notation?

$$100,000,000,000,000,000 \text{ m becomes } 1 \times 10^{17} \text{ m}$$

EXAMPLE 2-2

A certain type of virus measures 0.0000008 meter in length. Express this length in exponential form.

$$0.0000008 \text{ m becomes } 8 \times 10^{-7} \text{ m}$$

In the year 2000, the population of Earth is about 6.1 billion people. Write this number in exponential form and as a normal number using zeros.

Changing exponential numbers to normal form just requires moving the decimal point and adding zeros as needed. For example, to change the number 3.4×10^6 to normal form, we move the decimal six places to the right and add zeros; thus, $3.4 \times 10^6 = 3,400,000$. When an exponential number has a negative exponent, change to the normal form by moving the decimal point to the left. For example, to change 2.8×10^{-9} to normal form, move the decimal nine places to the left; thus, $2.8 \times 10^{-9} = 0.0000000028$.

Distance to nearest galaxy 10^{21} meters

Diameter of galaxy 10^{19} meters

Distance to nearest star 10^{17} meters

Diameter of the sun 10^9 meters

Diameter of Earth 10^7 meters

Mountains 10^3 meters
                                    (1 kilometer)

Height of human   1 meter

Diameter of head of 10^{-3} meters
straight pin   (1 millimeter)

Diameter of red blood cell 10^{-5} meters

Submicroscopic

Diameter of a hydrogen atom 10^{-10} meters

Diameter of atomic nucleus 10^{-15} meters

FIGURE 2-1 Variations in sizes and distances.

2-4 METRIC PREFIXES

Prefixes are sometimes used in language to refer to size; a microcomputer, is a small computer, and megabuck is slang for a million dollars. Expressing very long and very short distances in the metric system is possible using a set of prefixes that denote large and small multiples of

Metric Prefixes

PREFIX	SYM-BOL	VALUE
yotta	Y	10^{24}
zetta	Z	10^{21}
exa	E	10^{18}
peta	P	10^{15}
tera	T	10^{12}
giga	G	10^{9}
mega	M	10^{6}
kilo	k	10^{3}
hecto	h	10^{2}
deka	dk	10^{1}
deci	d	10^{-1}
centi	c	10^{-2}
milli	m	10^{-3}
micro	μ	10^{-6}
nano	n	10^{-9}
pico	p	10^{-12}
femto	f	10^{-15}
atto	a	10^{-18}
zepto	z	10^{-21}
yocto	y	10^{-24}

the basic units. The prefixes are names for various multiples or fractions of 10.

To illustrate how prefixes work in the metric system, suppose we want to measure a distance somewhat less than 1 meter. By dividing 1 meter into 100 equal parts, we can measure lengths in units of $\frac{1}{100}$ of a meter. This fraction is given a unique prefix name, and appropriate lengths are expressed using the prefix. In the metric system, the prefix **centi-** means $\frac{1}{100}$. Some measurements of length are expressed in units of centimeters (cm). We know by definition that 1 centimeter is $\frac{1}{100}$ of a meter and that there are 100 centimeters in 1 meter.

As another example, suppose you wanted to express your weight in metric units. You weigh several thousand grams, so the gram is not a convenient unit to express your weight. For example, 130 pounds is 59,000 grams. The metric prefix **kilo** means 1000, so you could express your weight in units of kilograms (kg). Doing this, we know that 1 kilogram is 1000 grams. Therefore, 130 pounds is 59 kilograms. Note that kilo means 10^3, so 59 kilograms is 59×10^3 grams.

Prefixes are used to express multiples or fractions of any metric base unit. The three most commonly used prefixes are listed below. (See the margin for less common prefixes.)

PREFIX	ABBREVIATION	MEANING
kilo-	k	1000 or 10^3 times unit
centi-	c	$\frac{1}{100}$ or 10^{-2} times unit
milli-	m	$\frac{1}{1000}$ or 10^{-3} times unit

Prefixes may be used with any metric base unit. A prefix is placed in front of the base unit to represent the indicated multiple or fraction. Lengths and distances are sometimes expressed in km (read as kilometers), cm (read as centimeters), or mm (read as millimeters). The distance 2000 m is expressed as 2 km, and the mass 0.0023 g is expressed as 2.3 mg (read as milligrams). Note that the prefixes centi- and milli- have the same meaning they have in our monetary system. For example, 1 cent is $\frac{1}{100}$ of a dollar, and 1 mill is $\frac{1}{1000}$ of a dollar.

2-5 CONVERSION FACTORS

Experimental chemistry involves making measurements. Suppose the mass of an object is 288 g. The unit "gram" reveals that this is a measure of mass. Take note that all measurements must include both a number and a unit. A measurement without a unit is meaningless. Computations with measurements are quite common in chemistry, so let's consider some typical computations using some simple examples.

Imagine a wooden board is 78 inches long. How long is it in units of feet? To change inches to feet, we need the relation between them: (1 ft/12 in.). Note that the abbreviation in. is used for the inch and a slash (/) is read as per, so we read this as 1 foot per 12 inches. Such a relation is a conversion factor because we use it to change or convert from one unit to the other. The pattern used in conversion is important, so let's look closely at how a conversion factor works in terms of the units involved. Computation with the inch-to-foot conversion factor is represented as

$$78 \text{ in.} \left(\frac{1 \text{ ft}}{12 \text{ in.}} \right) = 6.5 \text{ ft}$$

A **conversion factor** of this type always has a numerator involving one unit and a denominator involving another unit. Notice how the units divide out or cancel, as shown by the cancel marks. That is, the inches cancel and the desired units of feet remain. Canceling like units in the numerator and denominator follows the rules of algebra. A term in the numerator cancels a like term in the denominator. The numerical part of the calculation involves dividing 78 by 12.

Since the use of conversion factors is important in many computations, let's see how their use is justified. Conversion factors express a relation between two units. For example, we know that the relation between inches and feet is

$$1 \text{ ft} = 12 \text{ in.}$$

From this relation, we can obtain a conversion factor to convert from inches to feet and a factor to convert from feet to inches.

$$\left(\frac{1 \text{ ft}}{12 \text{ in.}} \right) \quad \text{or} \quad \left(\frac{12 \text{ in.}}{1 \text{ ft}} \right)$$

Which factor is used depends on whether we want to convert from feet to inches or vice versa. Conversion factors like these are sometimes called unit factors. They relate units of measure. Using conversion factors in problem solving is sometimes called unit analysis or dimensional analysis since the units or dimensions are used to aid the conversion process. Remember that the units are as important as the numbers. As you will soon see, conversion factors are very useful in chemical computations. A few more examples illustrate the use of such factors.

EXAMPLE 2-3

A person is 6.0 ft tall. What is this height in inches? Although you can do this calculation easily, consider the formal or logical "setup" involving this problem. The calculation involves multiplying the measurement by the unit factor relating inches to feet.

$$X \text{ in.} = 6.0 \text{ ft} \left(\frac{12 \text{ in.}}{1 \text{ ft}} \right) = 72 \text{ in.}$$

We start by deciding what unit is needed in the answer, then choose the factor that allows the unwanted units to cancel, which gives the answer in the desired unit. The numerical parts of the factors are multiplied to give the numerical part of the answer.

EXAMPLE 2-4

Express 1 hr in units of seconds. In other words, find the number of seconds in 1 hour. One approach to this problem is to convert the hours to minutes, then the minutes to seconds. This process requires the unit factor relating minutes to hours (60 min/hr), and the factor relating seconds to minutes (60 s/min).

To solve the problem using these factors, first multiply 1 hr by the minutes-per hour factor, then multiply this product by the seconds-per-minute factor. Start by noting the unit needed in the answer.

$$X\,s = 1\,\cancel{hr}\left(\frac{60\,\cancel{min}}{1\,\cancel{hr}}\right)\left(\frac{60\,s}{1\,\cancel{min}}\right) = 3600\,s$$

Notice that the desired units result when we algebraically cancel the hour and minute units. If we use the wrong factors, the units do not work out correctly. Including the proper units is essential and serves as a double check on the calculation process.

Sometimes a measurement or a quantity must be multiplied by a series of factors to give a desired result. When a quantity is multiplied by a series of factors, its units and numerical value change but it still represents the same measurement; 1.5 yards, 4.5 feet, and 54 inches are three ways of expressing the same distance.

FOCUS
2

Problem Solving

An important part of your study of chemistry involves learning problem-solving methods. Each problem is unique, but it is useful to follow a general pattern in solving problems. Problems are stated in words, so you need to read carefully and interpret them. Typically, unit factors are used in the solutions of problems. Consequently, the solution pattern is called unit factor analysis or dimensional analysis. Let's consider the pattern by using an example. A fast racehorse runs a distance of 7.0 furlongs in 81 seconds. What is the average speed of the horse in miles per hour?

1. Search the problem to find what is to be solved for, the **unknown.** Look for a sentence with a question mark and look for key words such as "what," "what is," "how many," or "how much." Express

the sought unit or units of the unknown quantity. In this example, the unknown is the speed of the horse in units of miles per hour (mi/hr).

2. Search the question for any given information, the **givens.** Record the givens with their units. In this example, it is given that the horse runs 7.0 furlongs in 81 seconds. Make a summary of the givens.

$$\text{Given: distance 7.0 furlong} \qquad \text{time 81 s}$$

3. Decide what **concept** or **principle** is involved in the problem. If possible, express the concept as a definition or a formula. In this example, the furlong is a unit of distance and the second is a unit of time. Thus, the concept is speed. Speed is the ratio of distance to time, $s = d/t$.

4. Use the givens as they relate to the concept. Write them down and follow them by the unknown, which is the **goal** of the problem. In this example, the given speed is 7.0 furlongs per 81 seconds.

Given: Goal: Find the speed.

$$\left(\frac{7.0 \text{ furlongs}}{81 \text{ s}}\right) \qquad X \text{ mi/hr} = ? \text{ (Note that we represent the unknown as } X \text{ having the desired units.)}$$

5. Decide what conversion factors or other information is needed in the solution of the problem. In this example, you need to change seconds to hours and furlongs to miles. Since there are 60 seconds in 1 minute and 60 minutes in 1 hour, to convert seconds to hours use the factor

$$\left(\frac{3600 \text{ s}}{1 \text{ hr}}\right)$$

To change furlongs to miles, we need to know the factor that relates these units. If you do not know a specific factor, you need to look it up. The dictionary defines a furlong as a length used in horse racing that is equivalent to one-eighth of a mile. The factor, found by dividing 1 by 8, is

$$\left(\frac{0.125 \text{ mi}}{1 \text{ furlong}}\right)$$

6. Use the concept, the givens, and the factors to set up the calculation sequence or do a sequence of intermediate calculations. Include units in the setup. In the example, the setup is

$$X \text{ mi/hr} = \left(\frac{7.0 \text{ furlongs}}{81 \text{ s}}\right)\left(\frac{3600 \text{ s}}{1 \text{ hr}}\right)\left(\frac{0.125 \text{ mi}}{1 \text{ furlong}}\right)$$

Notice that the setup begins with an X having the units we want in the answer, which reminds us that we need to end with these units.

7. Use the setup to calculate the numerical answer, and be sure that the units cancel to give the desired units for the answer. Canceling the units serves as a double check of the setup. In this example,

$$X \text{ mi/hr} = \left(\frac{7.0 \text{ furlongs}}{81 \text{ s}}\right)\left(\frac{3600 \text{ s}}{1 \text{ hr}}\right)\left(\frac{0.125 \text{ mi}}{1 \text{ furlong}}\right) = \frac{39 \text{ mi}}{1 \text{ hr}} \text{ or } 39 \text{ mi/hr}$$

This example illustrates a very useful method used in problem solving. Note that the general pattern involves multiplying the givens by one or more conversion factors that relate to the concept involved and any factors that change the units or dimensions to obtain the desired units in the answer. In general, the pattern is:

$$X \text{ desired units} = \text{givens (factor or factors related to concept or units)} = \text{answer with units}$$

EXAMPLE 2-5

Commercially, concrete is sold by the cubic yard, which is referred to as "yards." Suppose you plan to make a concrete slab measuring 30 ft by 18 ft by 0.50 ft. How many cubic yards of concrete do you need for the job?

GIVENS	GOAL	CONCEPT	FACTORS NEEDED
slab dimensions 30 ft by 18 ft by 0.50 ft	cubic yards $X\,yd^3 = ?$	volume of a rectangular solid $V = $ length $\times$ height $\times$ width	cubic feet to cubic yards

The factor relating cubic feet to cubic yards is found by cubing the basic relation between yards and feet.

$$\left(\frac{1\ yd}{3\ ft}\right)^3 = \left(\frac{1\ yd^3}{3^3\ ft^3}\right)^3 = \left(\frac{1\ yd^3}{27\ ft^3}\right)$$

First, find the volume of the slab in cubic feet, and then use the factor to change the units to cubic yards.

$$X\,yd^3 = (30\ ft)(18\ ft)(0.50\ ft)\left(\frac{1\ yd^3}{27\ ft^3}\right) = 10\ yd^3$$

Double check the setup of the problem by making sure that the units cancel to give the desired unit.

2-6 METRIC UNIT CONVERSION

Occasionally, changing from a base metric unit to a prefixed unit or from a prefixed unit to a base unit is necessary. For instance, we may change a measurement from meters to kilometers or from kilometers to meters. The conversion factors come from the definitions of the prefixes. Consider the relations between the meter as a base unit and the common prefixes used with the meter. (Similar relations are possible for any metric unit.)

$$1000\ mm = 1\ m, \qquad 100\ cm = 1\ m, \qquad and\ 1000\ m = 1\ km$$

Conversion factors obtained from these relations are

$$\left(\frac{1000\ mm}{1\ m}\right) \quad or \quad \left(\frac{1\ m}{1000\ mm}\right)$$

$$\left(\frac{100\ cm}{1\ m}\right) \quad or \quad \left(\frac{1\ m}{100\ cm}\right)$$

ACTIVITY 2-2

When you are traveling 55 miles per hour, how fast are you going in feet per second? Hint: The speed can be expressed as (55 mi/1 hr).

and

$$\left(\frac{1000 \text{ m}}{1 \text{ km}}\right) \quad \text{or} \quad \left(\frac{1 \text{ km}}{1000 \text{ m}}\right)$$

To convert a measurement in a prefixed unit to the base unit, just multiply by the factor corresponding to the meaning of the prefix. The convenience of the metric system is that these computations involve simple multiplication or division by a multiple of 10.

EXAMPLE 2-6

A quarter-mile running track is 440 yards or 402 meters around. What is this distance in kilometers? Multiplying by the factor (1 km/1000 m) gives the distance in kilometers.

$$X \text{ km} = 402 \text{ m} \left(\frac{1 \text{ km}}{1000 \text{ m}}\right) = 0.402 \text{ km} \text{ or } 4.02 \times 10^{-1} \text{ km}$$

Division by 1000 involves moving the decimal point three places to the left.

EXAMPLE 2-7

An inch is equivalent to 2.54 cm. How many meters is 2.54 cm?

The factor (1 m/100 cm) is used in the conversion.

$$X \text{ m} = 2.54 \text{ cm} \left(\frac{1 \text{ m}}{100 \text{ cm}}\right) = 0.0254 \text{ m} = 2.54 \times 10^{-2} \text{ m}$$

Division by 100 involves moving the decimal point two places to the left.

A conversion factor is chosen so that the numerator has the desired unit and the unit in denominator cancels. Consequently, including the correct units in a conversion factor is important and allows us to be sure that the units cancel.

EXAMPLE 2-8

One pound is equivalent to 0.454 kg. How many grams is 0.454 kg?

The factor relating grams to kilograms is (1000 g/kg). This factor is used to find the number of grams corresponding to 0.454 kg.

$$X \text{ g} = 0.454 \text{ kg} \left(\frac{1000 \text{ g}}{1 \text{ kg}}\right) = 454 \text{ g}$$

Multiplication by 1000 involves moving the decimal point three places to the right.

ACTIVITY 2-3

Carry out the following metric conversions.

(a) One metric ton is 1000 kg. How many grams are there in 1000 kg? What is the common name for this number?

(b) A 10 K race is 10 kilometers long. How many meters is this?

(c) How many kiloseconds are there in 30 minutes?

EXAMPLE 2-9

A hand calculator can add two numbers in about 40 ms (milliseconds). How many seconds is this?

We want an answer in seconds and need cancel milliseconds, so we use the factor (1s/1000 ms).

$$X\,s = 40 \; \cancel{ms} \left(\frac{1 \text{ s}}{1000 \; \cancel{ms}} \right) = 0.040 \text{ s} = 4.0 \times 10^{-2} \text{ s}$$

2-7 SIGNIFICANT DIGITS

Typically, we work with three different kinds of numbers: counted, defined, and measured. Counted items are expressed as exact whole numbers. If we count several people, for instance, the number is always an exact whole number, never a fraction. Defined relations also involve exact numbers but not always whole numbers. By definition, there are exactly 12 inches per foot, there are exactly 60 seconds per minute, and 1 inch is exactly 2.54 centimeters. Such numbers come from definitions, not measurements. In contrast, measured numbers come from reading measuring devices; they are never exact.

Imagine measuring a height, as shown in Figure 2-2. The measuring device shown in Figure 2-2(a) is calibrated to measure lengths to the nearest tenth of a meter. The first digit of the measurement comes from the calibrations of the scale, and the final digit of the measurement comes from interpolation. **Interpolation** means to read a measurement more closely than the calibration lines on a scale by estimating the distance between the lines. The first measurement in the figure has two digits and a certainty of

FIGURE 2-2 Measuring a height with rules of varying calibration.

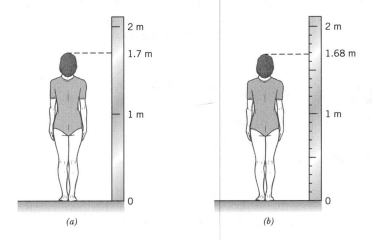

(a)   (b)

one tenth of a meter. If the measuring device is more precisely calibrated, we can give more digits in the measurement, as shown in Figure 2-2(b). Here the height measurement includes three digits and is certain to the nearest hundredth of a meter.

Some measuring devices, such as rulers, require us to estimate the last digit by interpolation. Other devices, such as the odometers on cars, are digital; they give the numbers of the measurement as a series of digits. The odometer in an automobile is a measuring device for distance. Imagine taking a trip in which the distance measured by the odometer is 370.2 miles. The odometer measures to the nearest 0.1 mile. This is its limit of precision or limit of certainty as a measuring device. The certainty means that the device measures precisely to the nearest tenth of a mile, but not to any smaller fraction of a mile. All measuring devices have specific limits of precision or certainty, which means they are readable to a specific number of digits.

Suppose each of five people carefully measures the length of a metal rod using a ruler calibrated to the nearest tenth of a centimeter. The group records the set of measurements shown below.

25.0 cm   25.1 cm   25.0 cm   24.9 cm   25.0 cm

Why are some of these measurements different from the others? A calibrated measuring instrument allows measurements to a specific certainty. Here the ruler is calibrated so that it can be read to a certainty of 0.1 cm. Such a certainty means that the ruler is used to make measurements known to plus or minus one tenth of a centimeter. The average of the above measurements could be expressed as 25.0 ± 0.1 cm; all the measurements agree within the 0.1 cm certainty of the measuring instrument.

Because measuring instruments have limits of certainty, measurements are never exact. Counting numbers and defined numbers are always exact, but measurements are never exact numbers. All measuring instruments are calibrated to a specific certainty. They can be, and normally should be, read to that certainty. If calibrated correctly, instruments allow accurate measurements to a specific number of digits. The ruler referred to above is read to a certainty of plus or minus one tenth of a centimeter (± 0.1 cm) but not to the nearest hundredth of a centimeter, which would exceed its limit of certainty.

A measurement made with an instrument of known certainty should be expressed to this certainty. For example, if you measure the mass of an object using a balance calibrated to read to the nearest 0.01 gram, you can express the measurement to ± 0.01 g. Suppose, using such a balance, you find that the mass of an object is 32.26 g. When a measurement is expressed this way, you can assume that it is known to a certainty of plus or minus one digit in the last digit of the measurement (e.g., 32.26 g means 32.26 ± 0.01 g). Whenever you see a measurement in this book, you can assume it is known to a certainty of plus or minus one digit in the last digit. Normally, including the $\pm$ notation in a measurement is not necessary.

A measurement is expressed as a sequence of numbers containing a specific number of digits followed by a unit. Measuring instruments are designed to give specific numbers of digits. All the digits that are a necessary part of a measurement are called **significant digits** or **significant figures.** The measurement 370.2 mi has four significant digits; 25.0 cm has three significant digits. Note that significant digits are any of the nonzero digits (1, 2, 3, 4, 5, 6, 7, 8, or 9). Zero may be a significant digit or simply function as a placeholder. As a result, the zeros that are part of a measurement are sometimes ambiguous. When dealing with a number obtained from a measurement, the following guidelines apply.

1. Count significant digits by starting with the first nonzero digit on the left and continue to the right. For example, 0.057 m has two significant digits; 12.67 cm has four significant digits.
2. Leading zeros occur at the beginning left-hand side of a number. They are never significant; they are merely decimal placeholders to show the position of the decimal point. For example, 0.0056 g has two significant digits. Notice that 0.0056 g and 5.6 mg are two ways of expressing the same measurement; both have two significant digits.
3. Zeros between nonzero digits are always significant. For example, 10,001 kg has five significant digits; 0.00003003 m has four significant digits. (Here the zeros on the left are not significant but the two zeros between the nonzero digits are significant.)
4. Zeros that follow nonzero digits and are on the right of the decimal point are significant digits. For example, 125.0 m has four significant digits; 0.250 s has three significant digits. (Here the zero on the left is not significant but the zero on the right is.)
5. Trailing zeros are ambiguous; they may or may not be significant. Trailing zeros are zeros to the right of nonzero digits in numbers that contain no decimal point. For example, 25,000 m has five significant digits if it is measured to the nearest 1 m or two significant digits if it is measured only to the nearest 1000 m. The ambiguity occurs because zeros may be part of a measurement or they may be placeholders to show the position of the decimal point.

Sometimes a trailing zero that is significant is written with a line above it. Following this pattern, 25$\bar{0}$ g has three significant digits and 3$\bar{0}$00 m has two significant digits. The use of lines, however, is not a common practice. Furthermore, if a decimal point is added to a number, it is assumed that the zeros to the left of the point are significant. Thus, 20. g has two significant digits and 2500. miles has four significant digits. Decimal points, however, are not always used in this way.

In this book, you can assume zeros that follow nonzero digits are not significant unless otherwise suggested by the context. The context relates to how reasonable the measurement is and the precision of instruments typically used to make such measurements. For example, a measurement expressed as 20 g is assumed to have two digits since typical balances are accurate to at least the nearest gram. Thus, if we know that a typical measuring instrument is accurate to at least the first digit before the decimal point, a zero in this position is a significant digit. Of course, to avoid ambiguity, it is best to express the measurement in exponential notation showing all the significant digits in the nonexponential part. Thus, 20 g written as 2.0×10^1 g shows that it has two significant digits. If numerous zeros are included in a measurement, we do not consider them as significant digits unless otherwise indicated. For instance, the measurement 150,000 °C is assumed to have two significant digits since it is unlikely that a temperature measurement has six significant digits. Stating that the height of Mount McKinley is 20,300 ft is ambiguous. The zeros on the right may or may not be significant. Writing a number in exponential notation is the most effective way to show whether or not zeros are significant. If the height of Mount McKinley is written as 2.03×10^4 ft, we know the measurement is expressed to three significant digits.

EXAMPLE 2-10

How many significant digits are in each of the following measurements?

(a) 5.07 s (three significant digits; zeros between nonzero digits are significant)

(b) 0.0007 g (one significant digit; leading zeros are never significant)

(c) 203 km (three significant digits; zeros between nonzero digits are significant)

(d) 1.005 m (four significant digits; zeros between nonzero digits are significant)

(e) 200. g (three significant digits; zeros are significant when the decimal point is given)

(f) 2500 m (We assume four significant digits, but it would be best expressed as 2.500×10^3 m.)

(g) 10.0 s (three significant digits; otherwise, there is no reason to give the zero following the decimal point)

(h) 18,000,000°C (We assume two significant digits, but the measurement is best written as 1.8×10^7 °C to avoid ambiguity.)

(i) 2.80×10^3 m (three significant digits, as shown in the nonexponential part)

(j) 12 in./ft (This is a definition and is exact.)

(k) 10 cm (We assume 2 significant digits since instruments normally measure to ±1 cm.)

(l) 2346 (This is a trick question; significant digits only apply to measurements. This number has no unit and is not a measurement, so it has nothing to do with significant digits.)

How much do you weigh? To one significant digit? To two significant digits? To three significant digits? (Scientific observation involves making and reporting honest measurement. You can, however, use an imaginary weight in this exercise.)

2-8 SIGNIFICANT DIGITS IN CALCULATIONS

After driving a car from Mexico City to Guadalajara, suppose the odometer shows a driving distance of 558 km. Expressing this in meters, the distance is 558,000 m. Does this mean that we know the distance in meters to six significant figures? No! The number of significant digits in a number is not changed by multiplying or dividing it by a unit conversion factor. When a measurement is used in a calculation, the number of significant digits in the measurement dictates the number of significant digits in the answer. Simply put, a calculated result cannot have greater certainty than the certainty of the measurement used in the calculation.

An important fact about significant digits in measurements is that the number of significant digits in the measurement cannot be increased or decreased by multiplying or dividing by a defined (exact) factor or a counting number.

Significant Digits in Multiplication or Division

The rule for finding the number of significant digits in a multiplication or a division calculation differs from that used in an addition or subtraction calculation. First, consider the case of multiplication or division. Suppose you drove a car 627.2 miles as measured by the odometer. When filling the gasoline tank before and after, you find that the car used 14.1 gallons of gasoline. To calculate the mileage in miles per gallon (gal), simply divide the miles by the gallons. A calculator reads

$$\left(\frac{627.2 \text{ mi}}{14.1 \text{ gal}}\right) = 44.48227 \text{ mi/gal}$$

Has the calculator created seven significant digits for us? No! Calculators cannot understand significant digits, but humans can. The calculated result can have no more significant digits than the measurement having the fewest significant digits. In other words, a calculated result is no more certain than the data from which it is obtained. We know the gallons used to only three significant digits, so the mileage must be expressed to three significant digits. We can do the calculation to as many digits as we like, or to as many as the calculator produces, but we must round the final answer to the correct number of significant digits. Thus, the answer for the mileage is rounded to three significant digits to give 44.5 mi/gal. To round numbers, follow these simple rules.

1. If the number following the digits you want to keep in the answer is less than 5, the digits kept are not altered. For example, 2.034 rounded to three digits is 2.03.

2. If the number following the digits you want in the answer is 5 or greater, add 1 to the last digit kept. For example, 1.999517 rounded to four digits is 2.000.

EXAMPLE 2-11

Round the following numbers to the number of digits indicated.

(a) 253.507 to three digits (add 1 to the last digit kept: 254)

(b) 0.0727 to two digits (add 1 to the last digit kept: 0.073)

(c) 1.0039 to three digits (the three digits kept are not changed: 1.00)

(d) 15.9994 to four digits (add 1 to the last digit kept: 16.00)

Using a Calculator

An electronic calculator is quite convenient for most computations you need to do when solving problems. It does not solve problems for you, but it does find the numerical answer once a problem has been set up. The example given here illustrates the use of typical algebraic calculators.

A baseball pitch is timed at 92.1 miles per hour. What is this speed in units of feet per second? First the question is summarized as

Given	Goal
speed of 92.1 mi/hr	X ft/s = ? or change mi/hr to units of ft/s

To change the units, the given speed is multiplied sequentially by the factors relating hours to minutes, minutes to seconds, and feet to miles. The setup for the calculation is

$$X \text{ ft/s} = \left(\frac{92.1 \text{ mi}}{1 \text{ hr}}\right)\left(\frac{1 \text{ hr}}{60 \text{ min}}\right)\left(\frac{1 \text{ min}}{60 \text{ s}}\right)\left(\frac{5280 \text{ ft}}{1 \text{ mi}}\right)$$

To do the multiplications and divisions, the numbers are keyed into the calculator starting from the left and moving in sequence to the right. Each number is followed by the division or multiplication key according to the pattern given in the setup. You do not have to key the number 1 because multiplication or division by 1 does not change the result. For this example, the key sequence using a typical calculator is

$$92.1 \div 60 \div 60 \times 5280 = 135.07982$$

Try the calculation on your calculator. Some calculators have an enter key instead of an = key. The calculator gives too many digits, so rounding to three digits gives 135 ft/s. A calculator can save you time and effort in most of the numerical calculations in this book. To practice, use your calculator to confirm answers given in examples throughout the book. See Appendix 2 for more about the use of calculators.

If you prefer, you can do the above calculation one step at a time, writing down the answer at each step, but be sure not to round the intermediate answers. Keep more digits than you need in any intermediate answers. Round only when you have the final answer. Rounding too soon can make your final answer incorrect.

When calculating with measurements, be aware of the number of significant digits allowed in the calculated result. The pattern is simple. When two or more measurements are used in multiplication and division calculations, the answer can have no more significant digits than the measurement with the fewest number of significant digits. So just count the digits in the measurements and use the least number of significant digits in the answer. Remember, factors that include exact numbers or counting numbers are not considered in the determination of significant digits in an answer. Simply put, in a multiplication or division calculation, the answer can have no more digits than the measurement that has the fewest number of digits. Remember too that in a multiplication or division calculation, the position of the decimal point floats and has nothing to do with the number of significant digits in the answer. With practice, you'll get used to expressing calculated answers to the correct number of digits. If you are not sure how many that is, just ask your instructor or study group for advice. When in doubt, ask.

EXAMPLE 2-12

How many yards are in 237 inches?

Given	**Goal**
237 in.	X yd = ? or how many yards

We need the factors relating feet to inches (1 ft/12 in.) and yards to feet (1 yd/3 ft).

$$X \text{ yd} = 237 \text{ in.} \left(\frac{1 \text{ ft}}{12 \text{ in.}}\right)\left(\frac{1 \text{ yd}}{3 \text{ ft}}\right) = 6.5833333 \text{ yd rounds to } 6.58 \text{ yd}$$

The answer has three significant digits because the measurement has three significant digits.

You may question why the one and two digits in the conversion factors in Example 2-12 do not dictate the result. These exact factors come from defined relations and thus have no influence on the significant digits. Remember, whenever a factor is used that comes from a defined equivalence, it does not influence the number of significant digits. Significant digits only apply to measurements or factors derived from measurements.

EXAMPLE 2-13

If a car travels 15.2 miles in 21 minutes, what is the speed in miles per hour?

Given	**Goal**
15.2 miles	speed in miles per hour
21 minutes	X mi/hr = ?

We need to change the ratio of miles to minutes to miles per hour using the factor relating minutes and hours (60 min/hr).

$$X \text{ mi/hr} = \left(\frac{15.2 \text{ mi}}{21 \text{ min}}\right)\left(\frac{60 \text{ min}}{1 \text{ hr}}\right) = \frac{43.428571 \text{ mi}}{1 \text{ hr}}$$

The mile measurement has three significant digits. The time measurement, however, has only two, so the answer needs two significant digits; thus, it is rounded to 43 mi/hr. Again, the defined conversion factor relating minutes and hours has no influence on the number of digits in the answer. Round a calculated result as you read it from the calculator display or write it down from the display and then round it to the desired number of digits.

A gasoline pump records volumes of gasoline. Check a pump at a gas station. How many significant digits are given in the digital read-out of volume?

Significant Digits in Addition or Subtraction

When a calculation involves addition or subtraction, the number of significant digits is determined in a way that differs from multiplication and division calculations. The number of digits in a result calculated by addition or subtraction depends on the position of the decimal point. A calculated result can have no more significant digits with respect to the decimal point than the measurement with the fewest digits with respect to the decimal point; the least certain measurement dictates the certainty of the result. We cannot increase the certainty with respect to the decimal point by just adding a more certain quantity. Be aware that addition and subtraction of numbers can change the total number of significant digits in the result.

Suppose you get on an elevator with a group of people. First, you want to find the sum of the weight of each passenger. Each person knows his or her weight to the nearest pound except one person, who claims to weigh about 190 pounds, which we'll assume is known to only two significant digits. When we add the weights, the quantity having the least certainty dictates the answer.

$$
\begin{array}{r}
132 \text{ lb} \\
122 \text{ lb} \\
183 \text{ lb} \\
+ \ 190 \text{ lb} \\
\hline
627 \text{ lb}
\end{array}
$$

This calculation suggests we know the sum to the nearest 1 pound. We cannot be that precise, however, because the last measurement is known only to the nearest 10 pounds. The rule is that the sum or difference of measurements cannot be expressed more precisely than the measurement that is least precise with respect to the decimal point. Here the answer must be rounded to 630 lb or 6.3×10^2 lb because the least certain measurement limits the answer to the nearest 10 pounds.

In most calculations, it is a good idea to carry along any extra digits until the final answer, then round the result to the correct number of digits. Some examples are

$$\begin{array}{ll}
52.50 \text{ m} & \text{(two digits past the decimal)} \\
\underline{+\ \ 2.017 \text{ m}} & \text{(three digits past the decimal)} \\
54.517 \text{ m} & \text{(answer is rounded to 52.52 m because the} \\
& \text{result requires only two digits past the decimal)}
\end{array}$$

$$\begin{array}{ll}
12.2 \text{ g} & \text{(one digit past the decimal)} \\
\underline{+\ \ 4.032 \text{ g}} & \text{(three digits past the decimal)} \\
16.232 \text{ g} & \text{(answer is rounded to 16.2 g because the result} \\
& \text{requires only one digit past the decimal)}
\end{array}$$

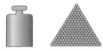

1 in = 2.54 cm

1 pound = 453.6 grams

1 kilogram
II
2.205 pounds

1 gallon
II
3.785 liters

EXAMPLE 2-14

Express the correct number of significant digits in the following cases.

(a)
$$\begin{array}{ll}
63.4 \text{ g} & \text{(one digit past the decimal)} \\
\underline{+\ 59.8 \text{ g}} & \text{(one digit past the decimal)} \\
123.2 \text{ g} & \text{(answer should have one digit past the decimal point, but} \\
& \text{notice that this addition has increased the number of signifi-} \\
& \text{cant digits to four)}
\end{array}$$

(b)
$$\begin{array}{ll}
63.4 \text{ g} & \text{(one digit past decimal)} \\
\underline{-\ 59.8 \text{ g}} & \text{(one digit past decimal)} \\
3.6 \text{ g} & \text{(The answer should have one digit past the decimal point,} \\
& \text{but notice that this subtraction has decreased the number} \\
& \text{of significant digits to two.)}
\end{array}$$

2-9 ENGLISH TO METRIC CONVERSIONS

Speed limits in Mexico are given in kilometers per hour and distances in kilometers or meters. In Italy, you buy grapes by the kilo (kilogram) and talk about the weather in degrees Celsius. Nearly every major country of the world uses the metric system for business and commerce. In the United States, we use a system of measure inherited from the past, known as the English system of measure. The United Kingdom, however, has changed to the metric system, so these units are now primarily used in the United States. Metric units do not have simple relations to English units. For rough estimations of how some metric units compare with our typical commercial units, consider the following. One inch is about 2.5 cm, so an inchworm is a 2.5 cm worm in metric terms. One kilogram is about 2.2 lb, so a kilo of Italian grapes is about 2.2 lb. There are about 0.6 mi/km; thus, if you noticed a sign with a speed limit of 80 km/hr in Mexico, it would mean about 50 mi/hr, about 0.6 of 80. One meter is about 1.1 yd, so the 100 m dash is about 10 percent longer than the 100 yd dash.

The metric system is commonly used for technical and scientific purposes. Typically, you'll see metric units of mass used on over-the-counter medicines and in nutrition data. Converting precise measurements made in English units to metric units or from metric to English is sometimes necessary. Appropriate conversion factors that relate the two systems are used in such calculations. Appendix 1 lists the most common conversion factors, and a few comparisons of English and metric units are given in the margin on page 36. For rough estimates between the two systems, you can use the information given in the previous paragraph.

To convert a precise measurement made in one system to another, just look up the appropriate conversion factor. These factors are used in forms having the desired unit in the numerator and the unit to cancel in the denominator. To get precise answers, do not round a factor before using it.

ACTIVITY 2-6

Do you think that the United States should change commercial measuring units to the metric system? Give reasons for your answer.

Unit Conversion

EXAMPLE 2-15

A person is 60.0 inches tall. What is this height in centimeters?

$$X \text{ cm} = 60.0 \text{ in.} \left(\frac{2.54 \text{ cm}}{1 \text{ in.}} \right) = 152 \text{ cm}$$

EXAMPLE 2-16

If somebody weighs 80.7 kg, how much is this in pounds?

$$X \text{ lb} = 80.7 \text{ kg} \left(\frac{2.205 \text{ lb}}{1 \text{ kg}} \right) = 178 \text{ lb}$$

EXAMPLE 2-17

How many meters long is a 100 yd football field? (Assume that one hundred yards is known to three significant digits.)

$$X \text{ m} = 100 \text{ yd} \left(\frac{0.9144 \text{ m}}{1 \text{ yd}} \right) = 91.4 \text{ m}$$

EXAMPLE 2-18

A speed limit sign reads 80 km/hr. What is the speed limit in miles per hour?

We need to convert the kilometers to miles.

$$X \text{ mi/hr} = \left(\frac{80 \text{ km}}{1 \text{ hr}} \right) \left(\frac{0.6214 \text{ mi}}{1 \text{ km}} \right) = 50 \text{ mi/hr or } 5.0 \times 10^1 \text{ mi/hr}$$

EXAMPLE 2-19

A high jumper leaps 6.25 ft. How high is this in centimeters?

The given measurement is in feet. We want the answer in centimeters. It is convenient to first convert the feet to inches, then the inches to centimeters. First, multiply the 6.25 feet by the feet-to-inches factor.

$$X \text{ cm} = 6.25 \text{ ft} \left(\frac{12 \text{ in.}}{1 \text{ ft}} \right)$$

Next, multiply this product by the inches-to-centimeters factor. The two steps are included in a single setup as

$$X \text{ cm} = 6.25 \text{ ft} \left(\frac{12 \text{ in.}}{1 \text{ ft}} \right) \left(\frac{2.54 \text{ cm}}{1 \text{ in.}} \right) = 191 \text{ cm}$$

Using several factors in any conversion is allowed if they are needed.

EXAMPLE 2-20

One Olympic event is the 800 m run. How many miles is this?

The calculation sequence can involve the conversion of the meters to kilometers, then the kilometers to miles. The setup includes multiplying by both the meters-to-kilometers factor and the kilometer-to-miles factor.

$$X \text{ mi} = 800 \text{ m} \left(\frac{1 \text{ km}}{1000 \text{ m}} \right) \left(\frac{1 \text{ mi}}{1.609 \text{ km}} \right) = 0.497 \text{ mi}$$

Notice the way the miles-to-kilometers factor is used. If the alternate form is needed, a conversion factor is inverted or "turned over" by switching the numerator and denominator. The new factor is the reciprocal of the original factor. In this case

$$\left(\frac{1.609 \text{ km}}{1 \text{ mi}} \right) \quad \text{is inverted to give} \quad \left(\frac{1 \text{ mi}}{1.609 \text{ km}} \right)$$

2-10 VOLUME MEASUREMENTS

Matter is anything that has mass and occupies space. The term *volume* is used to describe how much space is occupied; volumes represent amounts of three-dimensional space. In the English system, volumes are sometimes expressed in cubic inches or cubic feet. In the metric system, cubic centimeters or cubic meters are used as units of volume.

Imagine a piece of clay in the form of a cube measuring 1 cm on an edge. The volume of the cube is 1 cubic centimeter or 1 cm³. Notice this volume unit corresponds to a cubic length (1 cm $\times$ 1 cm $\times$ 1 cm = 1 cm^3) and is read as cubic centimeter. Even if the cube of clay is rolled into a ball, flattened into a sheet, or distorted in another way, the volume remains 1 cm^3.

When working with volumes of liquids in the English system, we do not usually use units of cubic inches. Normally, special units such as fluid ounces, pints, quarts, or gallons are used instead. A special metric unit is used in chemistry to measure and express volumes. This unit is the liter, L. By definition, one liter is the volume equal to 1000 cm^3 (see Figure 2-3). In this text, L is used as the symbol for the liter.

Metric prefixes are used with the liter. If 1 L of volume is divided into 1000 equal parts, each part is 1 milliliter (mL). There are 1000 mL in 1 L. One liter is also equivalent to 1000 cm^3, so the milliliter and cubic centimeter are the same amounts of volume.

$$1 \text{ mL} = 1 \text{ cm}^3$$

Since the milliliter and cubic centimeter have the same meaning, they are often used interchangeably. From the definition of the liter, we can write the factors

$$\left(\frac{1 \text{ L}}{1000 \text{ cm}^3}\right) \quad \left(\frac{1000 \text{ cm}^3}{1 \text{ L}}\right) \quad \left(\frac{1 \text{ L}}{1000 \text{ mL}}\right) \quad \left(\frac{1000 \text{ mL}}{1 \text{ L}}\right)$$

These factors are used to change volumes from cubic centimeters or milliliters to liters or vice versa.

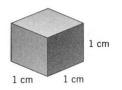

One Liter:

The volume equal to one thousand cubic centimeters

1 L = 1000 cm^3

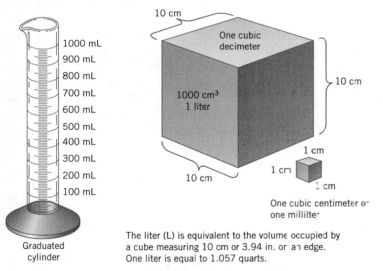

The liter (L) is equivalent to the volume occupied by a cube measuring 10 cm or 3.94 in. or an edge. One liter is equal to 1.057 quarts.

FIGURE 2-3   The liter. (This figure is not to scale.)

ACTIVITY

2-7

A beverage can lists its contents as 355 mL. What is this volume in liters? What is this volume in centiliters?

EXAMPLE 2-21

Express 0.233 L in units of milliliters. The volume is multiplied by the factor (1000 mL/1 L) to obtain the desired result.

$$0.233 \; \cancel{L} \left(\frac{1000 \text{ mL}}{1 \; \cancel{L}} \right) = 233 \text{ mL}$$

2-11 DENSITY

If you hold a rock in one hand and a piece of wood of comparable size in your other hand, the rock feels heavier. In fact, the piece of wood is likely to float on water, whereas the rock is likely to sink. How can we describe this difference between these two forms of matter? Remember, matter is anything that occupies space and has mass. In your immediate environment, substances might include wood, plastic, metals, rocks, various liquids, and even gases. One way to compare samples of matter is to find the masses of a fixed unit of volume. To do this, we carefully measure the mass and volume of a sample of a substance. The characteristic property of that substance is the ratio of its mass to its volume; this ratio is called the density. By definition, **density** is the mass per unit volume of a material.

Suppose a sample of iron metal has a mass of 18.8 g and occupies a volume of 2.39 cm^3. We find the density (d) by dividing the mass by the volume.

$$d = \left(\frac{18.8 \text{ g}}{2.39 \text{ cm}^3} \right) = \left(\frac{7.87 \text{ g}}{1 \text{ cm}^3} \right) \quad \text{or} \quad 7.87 \text{ g/cm}^3$$

Notice that the numerical part of the calculation involves dividing 18.8 by 2.39. The units remain with grams in the numerator and cubic centimeters in the denominator. One cubic centimeter represents one unit of volume. Once a density is calculated, it is often expressed as a certain number of grams per cubic centimeter or g/cm^3. Thus, the above density is expressed as 7.87 g/cm^3, which shows that the density is known to three significant digits. Since a density is found by experiment, the number of significant digits depends on the number of digits in the mass and volume. Density tells us the mass of one cubic centimeter or a unit volume of a substance. Each cubic centimeter of iron has a mass of 7.87 g. Ratios used to express the amount of one quantity per unit amount of another are quite common and useful in chemistry. We often use ratios like these every day (e.g., miles per hour, cents per pound, or dollars per gallon).

Since density is a ratio of mass to volume, it is represented by the algebraic expression

$$d = m/V$$

$$d = \frac{m}{V} = \left(\frac{\textbf{mass in grams}}{\textbf{volume in cm}^3 \textbf{ or mL}} \right)$$

An algebraic expression of this type serves as a succinct definition of density and suggests how the density of a substance is found by experiment. Typical units for mass are grams, and volumes are most often measured in cubic centimeters or milliliters; thus, density is the number of grams of a substance per cubic centimeter or milliliter. Densities are found by experiment and are then tabulated in reference books. For example, *The Handbook of Chemistry and Physics* is a reference book that contains a variety of information, including the densities of many chemicals. When densities are recorded, the temperature at which the density was determined is usually given because volume, and therefore density, varies with temperature. Some reference books give densities as specific gravities. For solids and liquids, the specific gravity is the ratio of the mass of a given volume of material to the mass of an equal volume of water at 4°C. Since the density of water is 1.00 g/cm^3 at 4°C, the specific gravity of a material is the same as its density. Thus, if we find that the specific gravity of ethyl alcohol is 0.789 at 20°C, we know that its density at this temperature is 0.789 g/cm^3. In other words, for liquids and solids, specific gravities are numerically equal to densities.

Table 2-1 lists the densities of a few representative substances. Gases typically have low densities, and some metals have relatively high densities. For example, the density of gold is 19.3 g/mL, whereas the density of helium gas is 0.000178 g/mL or 0.178 g/L. This means that gold is more than 100,000

Table 2-1 *Densities of Some Common Chemicals*

Liquids	Density at 20°C (g/mL)
Water	1.00
Ethyl alcohol	0.79
Carbon tetrachloride	1.59
Mercury	13.55

Solids	Density at 20°C (g/mL)
Magnesium	1.74
Aluminum	2.70
Iron	7.87
Gold	19.3
Sulfur	2.07
Salt	2.16
Sugar	1.59
Ice (0°C)	0.92

Gases	Density at 0°C (g/L)
Air (dry)	1.29
Oxygen	1.43
Nitrogen	1.25
Carbon dioxide	1.96
Chlorine	3.17
Helium	0.178

Why are gases less dense than liquids or solids? Draw pictures to illustrate your answer.

times denser than helium. Notice that the relatively low densities of gases are typically expressed in grams per liter rather than grams per milliliter.

Originally, the gram, as a unit of mass, was defined as the mass of 1 cubic centimeter of water. Consequently, water has the easily remembered density of 1 g/cm^3. Since the density of a substance varies with temperature, water does not have a density that is exactly 1 g/cm^3 at all temperatures. The density of water, however, is 1.00 g/cm^3 at room temperature and is very close to this value for liquid water between 0°C and 100°C.

Table 2-1 shows that helium gas has a density about seven times less than the density of air. A balloon or even a blimp filled with helium gas floats in the air. Helium has such a low density that it floats in air and even allows its lightweight container to float. The average density of the helium and the container is less than the density of air. The densities of gases decrease with increasing temperatures. Air at 40°C or 104°F, for instance, has a density of 1.12 g/L compared with air at 20°C or 68°F with a density of 1.20 g/L.

An object placed in water floats if it is less dense than the water. This is true if the object does not dissolve in water. When we pour two liquids together and they do not mix, the liquid of lower density forms a layer on top of the liquid of higher density. The oil in an oil and vinegar salad dressing floats on the vinegar. The oil must be less dense than the vinegar. Vegetable oils and petroleum oils are less dense than water, so they float on water. When an oil spill occurs in the ocean, the oil floats on the water.

There are some practical implications to oils and petroleum products floating on water. One method used to obtain crude oil from nearly depleted wells involves pumping water into the well. The oil accumulates and floats on the water, and it is then pumped out easily. Never try to put out a gasoline fire with water. The burning gasoline floats on the water and spreads the fire. A carbon dioxide fire extinguisher should be used on a gasoline fire, or the flames should be smothered using a blanket or dirt. For a similar reason, never try to put out a grease fire in the kitchen using water. Instead, use a carbon dioxide fire extinguisher, dump baking soda on the flames, or cut off the air supply by covering the flames.

A 325 mL sample of carbon dioxide gas has a mass of 0.637 g. What is the density of the gas sample in grams per liter? Would you expect a balloon filled with carbon dioxide gas to float in air? Why?

EXAMPLE 2-22

The density of water is 1.00 g/mL. Ice floats on liquid water, so it must be less dense than liquid water. If a 3.3 g sample of ice has a volume of 3.6 mL, what is the density of ice?

GIVEN	GOAL	CONCEPT
mass 3.3 g	density in g/mL	$d = m/V$
volume 3.6 mL		

Dividing the mass by the volume gives the density.

$$d = \left(\frac{3.3\ g}{3.6\ mL}\right) = \left(\frac{0.92\ g}{1\ mL}\right) \quad \text{or} \quad 0.92\ g/mL$$

EXAMPLE 2-23

Ebony is a dark, hard wood. A rectangular piece of ebony has a mass of 522 g and a volume of 435 cm³. Find the density of ebony and decide whether it is likely to float on water. Dividing the mass by the volume gives the density.

$$d = \left(\frac{522 \text{ g}}{435 \text{ cm}^3}\right) = \left(\frac{1.20 \text{ g}}{1 \text{ cm}^3}\right) \quad \text{or} \quad 1.20 \text{ g/cm}^3$$

A calculator may read 1.2 for this calculation. The extra zero is added because it is a significant digit. Again, calculators cannot understand significant digits, but humans can. The piece of ebony is denser than water, so it won't float.

Would you expect margarine to float or sink in water? Why?

Dense Humans

You might have some friends who are denser than you are. If you place some butter or margarine in water, it floats because fats are less dense than water. Body fat is also less dense than water. That is why a fatter person finds it easier to float on water than a leaner person. If you want to win a competition with a muscular athlete, challenge him or her to a floating contest. The most common chemicals in the human body are water, protein, bones, and fat. Proteins are the main components of muscles. The portion of the body consisting of bones, protein, and water is known as the lean body mass. Proteins and bones are denser than fats and do not typically float on water. People differ in their amount of body fat and lean body mass, which means that they have different densities. Overweight people are less dense than leaner people. (Why?)

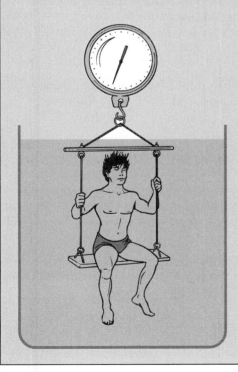

One method used to measure the percentage of body fat is based upon how dense a person is. A large tank of water into which a person is totally immersed is equipped with a seat attached to a weighing scale. This apparatus is used to weigh a person under water. See Figure 2-4. Everyone weighs less under water compared with the weight out of water. The greater the percentage of lean body mass a person has, the more he or she weighs under water compared with a person having a higher percent of body fat. The more fat a person has,

FIGURE 2-4   Determination of body fat content by immersion.

the more he or she floats and the less he or she weighs under water compared with a person having a higher percent of lean body mass. By comparing the actual weight of a person out of water to his or her weight under water, it is possible to find the precise lean body mass and percent of body fat.

You can use an approximation method to estimate your percent of body fat. The desirable body fat percent for a person relates to their body mass index, or BMI. The BMI is calculated using body measurements. By definition, the BMI is the mass of a person in kilograms divided by the square of his or her height in meters. Normally, a BMI value between 20 and 25 is desirable and a value over 30 is often undesirable and potentially unhealthy. Of course, remember that individuals vary in their body fat percent as a result of diet, body type, and heredity factors. Your BMI is not an exclusive indicator of health and is not a substitute for professional medical advise.

Scientific American note on BMI

Calculate your personal BMI.

2-12 DENSITY AS A RATIO OF UNITS

Densities are useful in chemistry because they express the relation between the mass and volume of substances. The density as a ratio of units is used as a conversion factor to find the mass of a given volume or the volume of a given mass. Let's see how this works. Gold has a density of 19.3 g/cm^3. A troy ounce of gold is 31.1 g. What volume does an ounce of gold occupy? The density is a ratio expressed as a factor relating mass and volume or is inverted to relate volume and mass.

$$\left(\frac{19.3\ \text{g}}{1\ \text{cm}^3}\right) \quad \text{or} \quad \left(\frac{1\ \text{cm}^3}{19.3\ \text{g}}\right)$$

To find the volume, the mass is multiplied by the second form of the factor, which is the reciprocal or inverse of the density. Remember that any factor is usable in its normal form or in its reciprocal form.

$$X\ \text{cm}^3 = 31.1\ \cancel{\text{g}}\left(\frac{1\ \text{cm}^3}{19.3\ \cancel{\text{g}}}\right) = 1.61\ \text{cm}^3$$

Density is an example of an experimentally derived ratio of units used to relate one property to another. In a sense, density is a property factor used to convert from one property to another. For any substance, the density relates mass and volume.

$$d = m/V$$

To use a factor like density, be aware of the units of the desired answer. If you want units of mass, density is used directly as a factor. To find the mass corresponding to a given volume of a substance, multiply the volume by the density:

$$m = Vd$$

EXAMPLE 2-24

A sample of gasoline has a density of 0.68 g/mL. What is the mass of 90.5 mL of this gasoline?

GIVEN	GOAL	CONCEPT
volume 90.5 mL	find the mass	$d = m/V$
density 0.68 g/mL	$X\,g = ?$	

Density is the factor. To get mass, the volume is multiplied by the density.

$$X\,g = 90.5\ \cancel{mL} \left(\frac{0.68\ g}{1\ \cancel{mL}}\right) = 62\ g$$

Remember that density is an experimental value that is based on measurement. The number of significant digits in a density must therefore be considered when deducing the answer. In this example, the volume has three significant digits but the density has only two, so the answer has two significant digits.

If we want to obtain the volume of a substance from its mass, the inverse or reciprocal of the density is used as a factor. That is, the density is "turned over," switching the numerator and denominator. This gives a factor relating volume to mass.

$$V = m\frac{1}{d}$$

EXAMPLE 2-25

A 10-carat diamond has a mass of 2.0 g. If the density of the diamond is 3.2 g/cm³, what is the volume of this diamond?

GIVEN	GOAL	CONCEPT
mass 2.0 g	volume in cm³	$d = m/V$
density 3.2 g/cm³	$X\,cm^3 = ?$	

The density expressed as a factor is

$$\left(\frac{3.2\ g}{1\ cm^3}\right)$$

Since we want volume from mass, however, we invert the density to give

$$\left(\frac{1\ cm^3}{3.2\ g}\right)$$

2-12

The volume of a drop of mercury is 0.0500 cm³. If the density of mercury is 13.55 g/cm³, what is the mass of a drop of mercury? How does this mass compare with the mass of a drop of water having the same volume?

The mass is multiplied by the inverse of the density to give the volume.

$$X \text{ cm}^3 = 2.0 \text{ g} \left(\frac{1 \text{ cm}^3}{3.2 \text{ g}} \right) = 0.63 \text{ cm}^3$$

2-13 TEMPERATURE

Temperature is used to express relative hotness and coldness. We use temperature to convey information about heat, but temperature and heat are not the same thing. Temperature and heat are both related to the motions of the particles of matter. When a mercury thermometer is placed in contact with a material, the chemical particles collide with the glass and, thus, indirectly with mercury particles composing the thermometer. Figure 2-5 illustrates this concept. These collisions cause the motions of the particles in the thermometer to increase or decrease, since they exchange energy with the particles in the material. As the liquid in a thermometer exchanges heat, it expands or contracts to reveal the temperature. In an electronic thermometer, a solid-state device called a thermistor changes its electrical properties as it is heated or cooled by collisions with particles.

Any moving object is said to have kinetic energy or energy of motion. Kinetic energy is proportional to the speed of an object and its mass. Chemical particles that compose gases, liquids, and solids are in various states of motion and, therefore, have kinetic energy. When a material is heated, the average kinetic energy of the particles increases as the temperature increases. Cooling a sample decreases the average kinetic energy and, therefore, the temperature decreases. The motion of the particles

FIGURE 2-5 Idealized view of a mercury thermometer being heated.

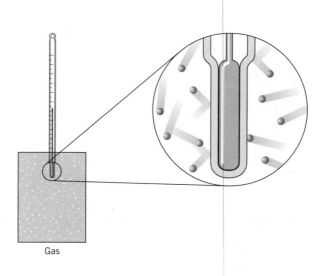

Gas

gives an object its measurable temperature, and **temperature** is a reflection of the average kinetic energy of the particles. Higher average kinetic energy corresponds to higher temperature; lower average kinetic energy corresponds to lower temperature. A direct relation exists between the average kinetic energy of the particles and the temperature.

Heat also relates to the motion of particles. Heat is thermal energy and is a medium of energy exchange between objects. When two objects at different temperatures are placed in contact, their particles collide and exchange energy, somewhat like pool balls that exchange energy when they collide. The hotter object cools down because the average kinetic energy of its particles decreases, while the cooler object warms up because the average kinetic energy of its particles increases. In time, the objects in contact come to "thermal equilibrium," which means that the average kinetic energies of their particles are the same, and, therefore, the objects have the same temperature.

Heat and temperature are related but are not the same. Two objects can have the same degree of hotness or coldness—that is, the same temperature—but one has a greater capacity to transfer heat than the other. For example, a burning match and a bonfire may have the same temperature, but the fire has a much greater capacity to transfer heat. The temperature can be measured anywhere within a sample and does not depend on the size of an object, but the ability to transfer heat does depend on size. Temperature reflects the average kinetic energy of the particles, whereas the ability to transfer heat depends on the number of particles and their motions. Temperature is an example of an **intensive property** of matter, a property that does not depend on size. Heat is an **extensive property**. The amount of heat transferred depends on the extent or size of a sample. (Is density an intensive or extensive property of matter?)

Lower temperature

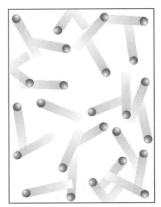

Higher temperature

Temperatures are measured using relative scales of comparison. The Celsius scale (in the past known as the centigrade scale) was first defined using the freezing point of water as zero degrees Celsius and the normal boiling point of water as 100 degrees Celsius. The Celsius degree was defined as $\frac{1}{100}$ of the temperature difference between the freezing point and boiling point of water. Temperature-measuring devices are calibrated to read temperatures on the Celsius scale.

In the late 1800s, Lord Kelvin, a Scottish scientist, established a fundamental temperature scale that used the lowest possible temperature as the beginning reference point. The lowest possible temperature, sometimes called absolute zero, is defined as zero Kelvin (0 K). This temperature corresponds to $-460°F$ or $-273°C$. The size of the degree Kelvin is chosen to be the same as a Celsius degree so that a simple relation exists between the two scales. Note, by convention, that the degree sign (°) is not used when giving temperatures on the Kelvin scale. The Celsius scale and the Kelvin scale are used in science, and the Fahrenheit scale is used for everyday applications in the United States.

The relation between the Kelvin and Celsius scales is quite simple. Zero degrees on the Celsius and Kelvin scales differ by 273 (or more precisely, 273.15). Thus, to change a Celsius temperature to a Kelvin temperature, just add 273: $T(K) = T(°C) + 273$. To change a Kelvin temperature to the Celsius scale, just subtract 273: $T(°C) = T(K) - 273$. As discussed in Chapter 10, Kelvin temperatures are useful when working with gases.

Sometimes a Celsius temperature needs to be converted to the Fahrenheit scale or vice versa. As shown in Figure 2-6, the Fahrenheit degree is smaller than the Celsius degree. Between the freezing point and the boiling point of water are 100°C and 180°F. The relation between the degrees on these scales is

$$\left(\frac{180°F}{100°C}\right) = \left(\frac{1.8°F}{1°C}\right) = \left(\frac{9°F}{5°C}\right)$$

This relation is used to convert temperatures on the two scales. As seen in Figure 2-6, however, the zero points on the scales are different. The zero points differ by 32°F. Consequently, the difference in zero points need adjustment. To convert a Fahrenheit temperature to the Celsius scale, first subtract 32°F to adjust the zero point and then multiply by $\frac{5}{9}$ (5°C/9°F):

$$T(°C) = [T(°F) - 32°F] \frac{5°C}{9°F}$$

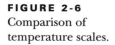

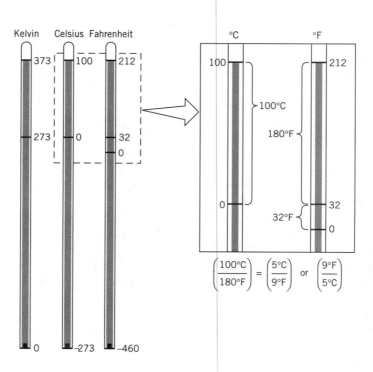

EXAMPLE 2-26

Express 68°F in the Celsius scale.

First, subtract 32, then multiply 5 and divide by 9.

$$T(°C) = (68 - 32)\frac{5}{9} = 20°C$$

To convert a Celsius temperature to the Fahrenheit scale, first multiply by $\frac{9}{5}$ (9°F/5°C) and then add 32°F to adjust the zero point:

$$T(°F) = [T(°C)]\frac{9°F}{5°C} + 32°F$$

EXAMPLE 2-27

Express 20°C in the Fahrenheit scale.

$$T(°F) = (20)\frac{9}{5} + 32 = 68°F$$

EXAMPLE 2-28

What is −40°F in degrees Celsius?

$$T(°C) = (-40 - 32)\frac{5}{9} = -40°C$$

Notice that, by coincidence, −40°F equals −40°C. Since these temperature scales happen to have the same value at this temperature, the following alternative equations are used to convert temperatures:

$$T(°C) = [T(°F) + 40]\frac{5}{9} - 40$$

$$T(°F) = [T(°C) + 40]\frac{9}{5} - 40$$

To convert from Fahrenheit to Celsius, add 40, multiply by $\frac{5}{9}$, and subtract 40. To convert from Celsius to Fahrenheit, add 40, multiply by $\frac{9}{5}$, and subtract 40.

How Hot Is It?

For temperatures in the range of about 10°C to 60°C, it is possible to estimate the corresponding Fahrenheit temperature. To do this, double the Celsius temperature and add 25 to give the approximate Fahrenheit temperature. For example, 40°C is about 105°F (40 × 2 + 25 = 105). Actually, it's 104°F. To estimate Celsius temperatures that correspond to Fahrenheit temperatures in the range of about 40°F to 140°F reverse the process. That is, subtract 25 from the Fahrenheit temperature and divide by 2. For example, 85°F is about 30°C (85 − 25 = 60, $\frac{60}{2}$ = 30). It's actually 29°C. Remember that this estimation method only works within the ranges indicated.

QUESTIONS

Each chapter has a set of problems and questions. Use them to test your knowledge and to practice problem solving. A variety of questions and problems are given to provide a range of choice. Material related to specific sections in the chapter are set off by section number. You are not expected to answer all of the end-of-chapter exercises, and your instructor may choose to assign selected questions. Those problems marked with an asterisk (*) have answers listed in the answer section at the back of the book.

1. Match the following terms with the appropriate lettered definitions.

metric system	a. mass per unit volume
gram	b. 1000 cm^3
meter	c. $\frac{1}{100}$ or 10^{-2}
second	d. $\frac{1}{1000}$ or 10^{-3}
K	e. 1000 or 10^3
centi-	f. scientific system of units
milli-	g. metric unit of mass
kilo-	h. metric unit of length
significant digits	i. standard unit of time
liter	j. Kelvin temperature degree
density	k. measure of the average kinetic energy of an object particle
temperature	l. the digits that are part of a measurement

Section 2-1

2. List the four qualities of the physical world that are used as the basis for defining units of measure.

3. Imagine a system of measure based on the following definitions for units of mass and length (the unit for time is the normal unit). The unit of mass is the stone (an old English unit corresponding to 14 pounds), and the unit of length is the league (an old English unit corresponding to 3 miles). How does a system of measure based on these units compare with the American and metric systems? Express your height, weight, and a 60 mile per hour speed limit in terms of this new system of measure (use any prefix as needed).

4. Create your own system of measure by using references for units of mass and length (use the normal units for time). Determine how your system of measure compares with the typical units of the American and metric systems. Express your height, weight, typical distances, and speed limits in your units.

Sections 2-2 to 2-4

5. List the names and abbreviations of the basic metric units.

6. List the three common metric prefixes along with their numerical meanings.

7. *Convert the following to exponential notation or standard form.
   (a) 695,950.0 km (the average radius of the sun)
   (b) 5.684 × 10^{26} kg (the mass of Saturn)
   (c) 2.998 × 10^{10} cm/s (the speed of light)
   (d) 0.0000000000000000000000001672 g (mass of a proton)
   (e) 3.1536 × 10^7 s (the number of seconds in a non–leap year)
   (f) 0.000000028841 cm (the distance between the centers of two gold atoms in solid gold)

(g) 125,200 cm/s (the average speed of a helium atom in the atmosphere to four digits)

(h) 3.45×10^5 cm/s (the speed of sound at sea level)

8. Convert the following to exponential notation or standard form.

(a) 36,000,000°F (the estimated temperature at the center of the sun)

(b) 6.022×10^{23} atoms (the number of atoms in 12 grams of carbon)

(c) 1.67×10^{-24} g (the mass of a typical hydrogen atom)

(d) 0.00000000000000000000001675 g (the mass of a neutron)

(e) 4.7×10^3 ft/s (the speed of sound in water)

(f) 0.0000000186 cm (the radius of a typical sodium atom)

(g) 299,800,000 m/s (the speed of light)

(h) 1.86×10^5 mi/s (the speed of light)

(i) 6×10^{12} miles (the number of miles in a light-year to one digit)

(j) 92,950,000 miles (the average distance of Earth from the sun to four digits)

Sections 2-5 and 2-6

9. How many millimeters are there in 1 cm? How many centigrams are there in 1 mg?

10. *Which is longer, 0.017 m or 71 mm? Which is a greater mass, 75 mg or 0.5 g?

11. The RDA (recommended dietary allowance) for vitamin B12 in adults is 0.0031 mg/kg of body weight. If a woman weighs 55 kg, how many grams of vitamin B12 are needed each day?

12. *The factor relating centimeters to meters is (100 cm/1 m). Give the factors that relate the following units.

(a) liters to milliliters

(b) grams to kilograms

(c) milliliters to deciliters

(d) meters to centimeters

(e) milligrams to grams

(f) centimeters to millimeters

(g) milliseconds to seconds

(h) meters to kilometers

13. *Convert the following measurements to the units indicated.

(a) 4.37 cm to m

(b) 0.024 kg to mg

(c) 2.8 cm to nm

(d) 30 s to ms

(e) 537 mm to m

(f) 750 mL to L (three digits)

(g) 16,890 mg to kg

(h) 0.0821 L to mL

14. Convert the following measurements to the units indicated.

(a) 0.914 km to cm

(b) 375 mg to g

(c) 263 ms to s

(d) 550 mL to L

(e) 0.647 g to mg

(f) 43.9 cm to mm

(g) 0.0000001 L to mL

(h) 219 mm to cm

Section 2-7

15. *Give the number of significant digits in each of the following measurements.

(a) 0.050 m

(b) 28.30 g

(c) 0.046 L

(d) 0.0007 kg

(e) 974.32 g

(f) 5.080 s

(g) 33.6 L

(h) 103,000 cm (measured to the nearest 1000 cm)

16. Give the number of significant digits in each of the following measurements.

(a) 20.8 km (the diameter of Phobos, a moon of Mars)

(b) 1500 g (the approximate mass of a typical adult liver)

(c) 0.000072 cm (the average length of a human taste bud)

(d) 1769 mL (the measured volume of urine excreted in one day)

(e) 24,901.47 miles (the circumference of Earth at the equator)

(f) 0.360 s (the duration of a heartbeat)

(g) 103 m (the measured height of a redwood tree)

(h) 0.08 mm (the average thickness of human skin)

Section 2-8

17. *Indicate the number of significant digits that should be expressed in the following calculations.

(a) $973 \, \text{ft} \left(\dfrac{12 \, \text{in.}}{1 \, \text{ft}} \right) \left(\dfrac{2.54 \, \text{cm}}{1 \, \text{in.}} \right)$

(b) $26 \, \text{mi} \left(\dfrac{1760 \, \text{yd}}{1 \, \text{mi}} \right) \left(\dfrac{3 \, \text{ft}}{1 \, \text{yd}} \right)$

(c) $0.500 \, \text{m} \left(\dfrac{1 \, \text{km}}{1000 \, \text{m}} \right)$

(d) $4.70 \, \text{mL} \left(\dfrac{2.7 \, \text{g}}{1 \, \text{mL}} \right)$

(e) $\quad 2.37 \, \text{g}$
$\underline{+ \; 23.06 \, \text{g}}$

(f) $\quad 0.776 \, \text{cm}$
$\underline{+ \; 133.54 \, \text{cm}}$

(g) $\quad 300.47 \, \text{kg}$
$\underline{- \; 189.290 \, \text{kg}}$

(h) $\quad 7 \, \text{s}$
$\underline{+ \; 4 \, \text{s}}$

18. Indicate the number of significant digits that should be expressed in the following calculations.

(a) $3179 \, \text{s} \left(\dfrac{1 \, \text{min}}{60 \, \text{s}} \right) \left(\dfrac{1 \, \text{hr}}{60 \, \text{min}} \right)$

(b) $6.28 \, \text{hr} \left(\dfrac{95 \, \text{mi}}{1 \, \text{hr}} \right) \left(\dfrac{1.609 \, \text{km}}{1 \, \text{mi}} \right)$

(c) $25 \, \text{mL} \left(\dfrac{1 \, \text{L}}{1000 \, \text{mL}} \right) \left(\dfrac{1 \, \text{gal}}{3.785 \, \text{L}} \right)$

(d) $5.0 \, \text{g} \left(\dfrac{1 \, \text{oz}}{28.35 \, \text{g}} \right) \left(\dfrac{1 \, \text{lb}}{16 \, \text{oz}} \right)$

(e) $\quad 803 \, \text{g}$
$\underline{+ \; 1647 \, \text{g}}$

(f) $\quad 0.911 \, \text{m}$
$\underline{+ \; 237.48 \, \text{m}}$

(g) $\quad 562.1 \, \text{kg}$
$\underline{- \; 526.98 \, \text{kg}}$

(h) $\quad 6 \, \text{cm}$
$\underline{+ \; 8 \, \text{cm}}$

(i) $\quad 750 \, \text{m}$
$\underline{+ \quad 65 \, \text{m}}$

19. A bag of nuts is labeled "16 oz (453.59 g)." Comment on the validity of the numbers used on the label in terms of the number of significant digits.

Section 2-9

20. *To three significant digits, how many millimeters are there in 1 yard?

21. A blood capillary has the same diameter as a red blood cell, 0.001 mm. How many inches is this?

22. *A home run hit over the back wall of a baseball field travels 403 ft. How many meters is this?

23. A 10 km race is how many miles?

24. Mary's little lamb weighs 127 lb. What is this in kilograms?

25. A thoroughbred racehorse has a mass of 549 kg. What is this mass in pounds?

26. Mount Olympus, Greece, is recorded as 9573 ft high. What is the height in meters?

27. *Express the speed 55 mi/hr in terms of kilometers per hour.

28. A baseball is pitched at 98 mi/hr. What is this speed in kilometers per hour?

29. *An average bee carries 0.0018 oz of nectar per trip to the hive. What is this mass in milligrams? How many trips does it take to transport 1.00 kg of nectar?

30. The speed of light is 3.00×10^8 m/s. What is this speed in miles per hour?

Sections 2-10 to 2-12

31. What is a liter and how is it defined?

32. *Determine the number of gallons per 1.00 L, using 1 L = 1.057 qt and 4 qt = 1 gal.

33. What is the volume of a cup of liquid in milliliters expressed to three digits? (Note: 4 cups = 1 qt.)

34. Some melons at their most active phase grow more than 5 cubic inches a day. What volume is this in liters? (Hint: Change cubic inches to cubic centimeters first.)

35. A winery ferments cabernet sauvignon in 250-gallon barrels. What is this volume in liters to three digits? How many 750 mL bottles can be filled with a barrel of wine?

36. The volume of blood flowing through an adult's liver is 25 mL per second. How many gallons per minute is this?

37. *If your economy car averages 41.5 mi/gal and the fuel tank holds 34.0 L of gasoline, how many miles can you drive using one tank of gasoline?

38. A can of soda contains 12 fluid oz. What is this volume in mL expressed to three digits?

Sections 2-11 and 2-12

39. What is density and in what units is it usually expressed?

40. *Calculate the densities of the materials listed below using the experimental measurements given.

    (a) a 34.1 mL sample of water with a mass of 34.2 g

    (b) a 133 mL sample of gold with a mass of 2.568 kg

    (c) a 47 mL sample of ice with a mass of 43.24 g

    (d) a 61.9 mL sample of ethyl alcohol with a mass of 49.1 g

    (e) a 58.7 cm^3 sample of aluminum with a mass of 158.5 g

    (f) a 1.06 L sample of air with a mass of 1.35 g

41. Calculate the densities of the materials listed below using the experimental measurements given.

    (a) a 42.43 cm^3 silver sample with a mass of 449.8 g

    (b) a 95.5 mL sample of water with a mass of 96.0 g

    (c) an 18.0 g sample of salt with a volume of 8.27 cm^3

    (d) a 0.279 g sample of copper with a volume of 0.0314 cm^3

    (e) a 63.1 L sample of ice at $-10°C$ with a mass of 62.8 kg

    (f) a 275 mL sample of ethane gas with a mass of 0.368 g

42. Explain how it is possible for two substances to have the same density.

43. *Determine the mass or volume for each of the following, given the density and the measured mass or volume.

    (a) What is the mass of 10.0 L of water if its density is 1.00 g/cm^3?

    (b) What is the volume of a 6.72 g piece of sulfur if its density is 2.07 g/cm^3?

    (c) What is the volume of a 31.3 g sample of aluminum if its density is 2.70 g/cm^3?

    (d) What is the mass of a 456 cm^3 sample of iron if its density is 7.87 g/mL?

    (e) A cup contains 237 mL of sugar. What is the mass of the sugar if its density is 1.59 g/mL?

    (f) The total lung capacity of a typical male is 5.9 × 10^3 mL. What is the mass of air in the lungs if the density of air is 1.29 g/L?

    (g) There are 3.785 L per gallon. What is the mass in grams of 15 gallons of gasoline if the density of gasoline is 0.70 g/mL? What is the mass in pounds?

44. Determine the mass or volume for each of the following, given the density and the measured mass or volume.

    (a) What is the mass of 118 mL of milk if its density is 1.035 g/mL?

    (b) What is the volume of 0.250 kg of water if its density is 1.00 g/cm^3?

    (c) What is the mass of a piece of paper with a volume of 0.747 cm^3 if its density is 0.279 g/cm^3?

    (d) A quintal is an esoteric unit of mass equal to 100 kg. What is the volume of 2.50 quintals of salt if its density is 2.16 g/cm^3?

    (e) What is the mass of a 425 cm^3 sample of paraffin wax if its density is 0.88 g/cm^3?

    (f) What volume does a 0.639 g breath of air occupy if its density is 1.29 g/L?

    (g) A croissant recipe calls for 700 g of flour to three digits. According to Jacque Pepin's data, the density of flour is 0.620 g/mL. How many mL of flour are needed for this recipe?

45. *The largest very pure gold nugget ever found had a volume of 3619 cm^3. If the density of gold is 19.3 g/cm^3, what was the mass of this nugget in grams? How many pounds is that?

46. The largest topaz ever found came from Brazil and had a volume of 11,026 cm^3. If the density of topaz is 3.25 g/cm^3, what is the mass of this topaz in grams? What is the mass in pounds?

47. *The density of water is 1.00 g/cm^3. What is the density of water in pounds per cubic foot? (Hint: There are 28,317 $cm^3/1 ft^3$.)

48. If a very large hailstone weighs 2.0 lb and the density of ice is 0.92 g/cm^3, what is the volume of the hailstone in units of quarts?

49. A bucket of molasses has a volume of 0.321 cubic feet and weighs 6.95 lb. Calculate the density of molasses in pounds per cubic feet and in grams per cubic centimeter. (See Question 47.)

50. When vegetable oil is poured into water, it forms a layer on top of the water. What can you deduce concerning the density of vegetable oil compared with the density of water?

51. When grenadine syrup is poured into orange juice, it sinks to the bottom. What can you deduce concerning the density of grenadine syrup relative to orange juice?

52. Ice floats on liquid water. What can you deduce concerning the density of ice compared with the density of liquid water? (See Question 40(a) and (c) to compare the actual densities.)

53. *A gold-colored ring has a mass of 31.6 g and a volume of 1.59 mL. Is the ring pure gold? The density of gold is 19.3 g/mL.

54. Why can a person float in salt water more easily than in fresh water?

55. Assuming a density of 1.0 g/cm^3, what is the volume of a 139 lb person in cubic centimeters? What is this volume in liters?

56. *In certain types of stars, matter is considered highly compressed. If a handful of this matter, about 72 mL, has a mass of 1,440,000 kg, what is the density to two significant digits?

57. Which has a greater mass, 1 L of water or 1 L of gasoline? The density of water is 1.00 g/mL, and the density of gasoline is about 0.68 g/mL.

58. A small box is filled with liquid mercury. The box measures 3.44 cm wide, 6.88 cm long, and 1.56 cm high. If the density of mercury is 13.6 g/mL, what is the mass of mercury in the box?

59. *Pumice is a volcanic rock that contains many trapped air bubbles. A 550 g sample occupies 578.3 mL. What is the density of pumice to three digits? Does it float on water?

60. A reflecting pool is in the shape of a rectangular solid measuring 75 m by 22.5 m by 0.275 m. Using a density of 1.00 g/cm^3 for water, determine the number of grams of water the pool contains. Also determine the number of pounds of water in the pool.

61. *A steel ball bearing in the form of a sphere has a mass of 0.216 g. The diameter of the bearing is 0.3742 cm. Calculate the density of the ball bearing. The volume of a sphere is given by the formula $V = (\frac{4}{3})(\pi r^3)$ or $(\pi d^3)/6$, where r is the radius, d is the diameter, and π is 3.142.

62. A sample of aluminum metal is in the form of a cylinder of 0.17 cm radius and 1.18 cm height. If the mass of the sample is 0.289 g, calculate the density of aluminum. The volume of a cylinder is given by the formula $V = (\pi r^2)h$, where r is the radius, h is the height, and π is 3.142.

Section 2-13

63. What is temperature?

64. What is heat? How do heat and temperature differ?

65. How is heat exchanged between objects?

66. What is the Celsius temperature scale and how is it defined?

67. What is the Kelvin temperature scale and how is it defined?

68. *Convert the following temperatures.
   (a) 98.6°F to °C
   (b) 25°C to °F
   (c) 212°F to K

69. Use the approximation method and the precise calculation method to change the following Celsius temperatures to the Fahrenheit scale. Express the answers to two digits.

10 °C   20 °C   30 °C   40 °C   50 °C   60 °C

Questions to Ponder

70. Which is heavier, a pound of feathers or a pound of lead?

71. Is helium lighter than air? Explain your answer.

72. What are some advantages and problems associated with the United States switching to the metric system?

73. How would you explain why a helium balloon rises in air using your knowledge of density and gases?

74. An old bromide about water based liquids says "a pint's a pound the world around." Use the density of water to confirm this saying.

75. Change the following to metric units.
   (a) Using romantic enticements, he inched his way into her life.
   (b) Give them an inch and they take a mile.
   (c) An ounce of prevention is worth a pound of cure.
   (d) A centipede and a millipede.
   (e) You have reached a milestone in your life.

ATOMS

3-1 THE EVOLUTION OF AN IDEA

Matter is the interest of chemistry. The materials surrounding you, the ground beneath your feet, the air you breathe, and even the stars of the universe are forms of matter. As mentioned in Chapter 1, one of the most important ideas in chemistry is that all matter consists of very tiny, characteristic particles called **atoms.** The atomic view of matter is not a new idea. Around 500 B.C.E., the Greek philosopher Democritus, extending the ideas he learned from his teacher who likely borrowed from ancient Hindu ideas, incorporated the particle nature of matter into his philosophy. Democritus suggested the term *atom* for particles of matter. Atom comes from the Greek *atomos* (that which cannot be divided or broken down). The idea of the atom came from philosophical questions about matter. Democritus asked a hypothetical question: What happens if a sample of matter is divided into smaller and smaller bits? Would such subdivision reach an ultimate particle, or could the subdivision occur indefinitely and still be characteristic of the sample? Democritus favored the particle view and developed an atomic vision of matter as part of his philosophy. Dominant Greek philosophers such as Plato and Aristotle, however, considered matter as continuous and not particulate.

The atomic view survived as a philosophical idea and was adopted by a later Greek philosopher, Epicurus (about 300 B.C.E.). Atoms as particles of matter became part of Epicurean philosophy, which later influenced Roman thinkers. In 50 B.C.E., Lucretius, a Roman philosopher and poet, wrote *De Rerum Natura* (The Nature of Things), an epic poem describing the atomic view of matter. A few hand-transcribed copies of this poem survived over the centuries. Gutenberg developed the printing press in the 1400s, and the epic Lucretian poem became one of the many works of literature first set in print. Copies of the poem helped carry the atomic hypothesis to modern times. The idea survived among many scientists, or natural philosophers as they were called in past times.

The philosophical idea of atoms was not based on reproducible experimental evidence and measurements. Experimental science developed in the 1600s and 1700s. One result of this development was that experimental evidence for an atomic theory began to accumulate. In 1662, Robert

Boyle, an English scientist, experimented with samples of air. To explain why air is easily compressed, he imagined an air sample as composed of tiny, invisible particles separated by space. In other words, Boyle used the hypothesis of atoms to explain the behavior of air.

More experimental evidence was collected by Antoine Lavoisier, a French chemist of the late 1700s. Lavosier once said: "Three things distinguish every science: the series of facts that constitute the science, the ideas that call the facts to mind and the words that express them. The word should give birth to the idea; the idea should depict the fact." He is known as the father of modern chemistry because he developed careful methods of observation and measurement, and emphasized the use of precise balances to weigh chemicals used in experiments. His emphasis on research and careful measurement established chemistry as a fundamental science. To fund his laboratory and scientific research, he invested in a tax collection agency in prerevolutionary France. During the French Revolution he was arrested and was executed in 1794. His death, at the peak of his scientific career, was a great loss for France and the world of science.

Lucetius Information
Poem by Lucretius

Lucretius: On the Nature of Things

The epic poem by Titus Lucretius Carus, a Roman philosopher and poet, served as the repository of the ideas of the Ancient Greek atomists. Let's consider some portions of the poem, written in 50 B.C.E., as translated from the original Latin.

Lest you yet

Should in any way to doubt my words
Because the primal particles of things
Can never be distinguished by the eyes,
Consider now these further instances
Of Bodies which you must yourself admit
Are real things, and yet cannot be seen.
First the wind's violent force scourges the sea,
Whelming huge ships and scattering the clouds. . . .
Winds therefore must be invisible substances
Beyond all doubt, since in their works and ways
We find that they resemble mighty rivers
Which are of visible substance. Then again
We can perceive the various scents of things,
Yet never see them coming to our nostrils:
Heats too, we see not nor can observe
Cold with our eyes nor ever behold sounds:
Yet must all of these be of bodily nature,
Since they can act on our senses.

3-2 STATES OF MATTER

As you know, matter typically occurs in one of three states or forms: the solid state, the liquid state, or the gas or vapor state. Solids include such materials as rocks, soils, minerals, ores, crystals, metals, glasses, plastics, and fibers. A **solid** has a definite volume and fixed shape. Solids are characteristically rigid and their shapes are changed by crunching them or re-forming them. A **liquid** has a definite volume but flexible shape. Some familiar liquids are water, oils, ethyl alcohol, and gasoline. A liquid takes on the shape of the container in which it is placed, and it easily pours and spreads out on a flat surface. Some common gases are air, nitrogen, oxygen, helium, and carbon dioxide. A sample of a **gas** has neither a definite shape nor volume and occupies the entire container in which it is placed. When you fill a balloon with air, the air sample fills the entire volume of the balloon. This happens, of course, because of the incessant and random motion of the particles making up a gas.

Upon heating, some materials burn or decompose. Many chemicals, however, simply change their physical state when heated or cooled. For instance, upon heating, ice melts to give liquid water. Further heating causes the liquid water to boil; this produces steam, which is water vapor. The various changes of state for materials are shown here. Note that heating supplies energy and cooling decreases energy.

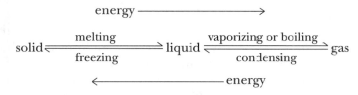

3-3 PHYSICAL AND CHEMICAL PROPERTIES

Chemists are interested in finding out what things are made of. To do this they probe and manipulate samples of matter and observe their various properties. **Physical properties** are related to the physical nature of matter. Examples of physical properties are appearance, color, physical state, density, heat conductivity, electrical conductivity, and melting and boiling points. Water, for example, is a clear, colorless liquid with a density of about 1 g/mL, a melting point of 0°C, and a boiling point of 100°C at typical atmospheric pressure. Pure liquid water does not conduct electricity and is not a great conductor of heat. Once heated, however, a sample of water retains its temperature and cools slowly. Water is said to have a relatively high heat capacity. Metals that are good conductors of heat do not retain their temperatures and have relatively low heat capacities.

Chemical properties refer to the chemical behavior of a form of matter. Chemical properties include how a chemical behaves when mixed with other chemicals. Does it burn in air? Does it decompose upon heating? Does it rust or corrode? Does it change chemically when mixed with water or other chemicals? A **chemical property** corresponds to a chemical either decomposing into other chemicals or combining with other chemicals to make new chemicals. These characteristics, or properties, serve to distinguish one type of matter from another. Table sugar, for example, readily dissolves in water, and a sample readily melts but easily burns when heated in air. In contrast, a chunk of iron does not burn when heated and does not dissolve or change when placed in water. The iron does, however, slowly corrode or rust when exposed to air or a moist environment. The ability to burn in air and the ability to corrode or rust are chemical properties. Normally, we express physical properties as measurable facts, whereas chemical properties are "verbs" expressing chemical behavior or chemical changes.

3-4 MIXTURES AND SOLUTIONS

Many rocks appear as mixtures of various materials, whereas a nickel coin appears uniform and, even when it is cut into bits, the pieces have the same uniform appearance. Some materials occur in the form of heterogeneous mixtures. **Heterogeneous** means made of different parts. A mixture of materials can often be separated into various homogeneous forms of matter. **Homogeneous** matter is uniform throughout, and any portion of a sample has the same properties and composition.

A **solution** is defined as a homogeneous mixture of two or more components. A solution is homogeneous since it is uniform throughout. In other words, every portion of the solution is the same as any other portion; it has the same density and the same appearance, even under microscopic examination. A solution consists of a **solvent** in which one or more substances are dissolved. The dissolved substances are called **solutes.** When salt is dissolved in water, the salt is the solute and the water is the solvent.

Separating the components of a solution by simple filtration is not possible. The components can, however, be separated in other ways. The salt in a saltwater solution is isolated by allowing the water to evaporate. Pure water is obtained by distilling it from water solutions. Distillation involves heating a water solution until it boils then condensing the vapors. Distilled water is prepared this way.

3-5 PURE SUBSTANCES OR CHEMICALS

With some effort, it is possible to separate most heterogeneous mixtures and solutions into pure substances. **Pure substances** or **pure chemicals** are forms of matter that have the same properties throughout and that have definite and unchanging chemical composition. Pure substances are homogeneous. Water is an example of a pure chemical. No matter where a sample of pure water comes from, it has the same properties as any other sample of pure water. The key property of a pure substance is that it has fixed chemical composition. Water from any source is always the same from a chemical perspective. In contrast, although solutions are homogeneous forms of matter, they do not have fixed unchanging chemical composition. Solutions of salt, for instance, are all homogeneous and contain salt mixed with water, but such solutions can range from slightly salty to very salty. They don't have fixed composition so they are not pure chemicals. In contrast, salt and water are pure chemicals with invariant composition. A **pure chemical** means one kind of chemical with very little or no other chemicals mixed in.

Realizing that some mixtures are more complex and some are simple is important. Pure water is easily obtained by distillation. Table salt is obtained from the solid deposits left by evaporation of ocean water. It requires greater effort and special techniques to separate the various mineral substances found in rocks and ores. Living systems are extremely complex mixtures containing thousands of chemicals. Analysis of even a relatively simple biological mixture is very complicated, time consuming, and expensive.

3-6 ELEMENTS AND COMPOUNDS

Pure chemicals, like water and salt, are isolated by separating them from mixtures. See Figure 3-1. Using special techniques, pure substances are made to decompose into even simpler substances. Water, for instance, with some effort and energy, decomposes to form hydrogen gas and oxygen gas. Similarly, salt, with some effort and energy, decomposes into sodium metal and chlorine gas.

Throughout history, scientists have investigated matter and over the years have isolated millions of pure substances. Some important discoveries were made by accident. An example is illustrated by a quote from Laura Esquirel's book *Like Water for Chocolate*: "Phosphorus was discovered in 1669, by Brant, a Hamburg chemist who was looking for the philosopher's stone. He believed that metal could be transmuted into gold by mixing it with extract of urine. Using this method, he obtained a luminous substance that burned with an intensity that had never been seen before. For a long time phosphorus was obtained by vigorously heating the residue from evaporating urine in an earth retort the neck of which was submerged in water."

Close study and testing of substances reveal there are just two kinds of pure chemicals: compounds and elements. **Compounds** are pure chemicals composed of two or more even simpler chemicals. Water and salt are common examples. **Elements** are pure chemicals that cannot be further separated into simpler chemicals. Examples are hydrogen, oxygen, sodium, and chlorine. In general, any chemical that cannot be broken

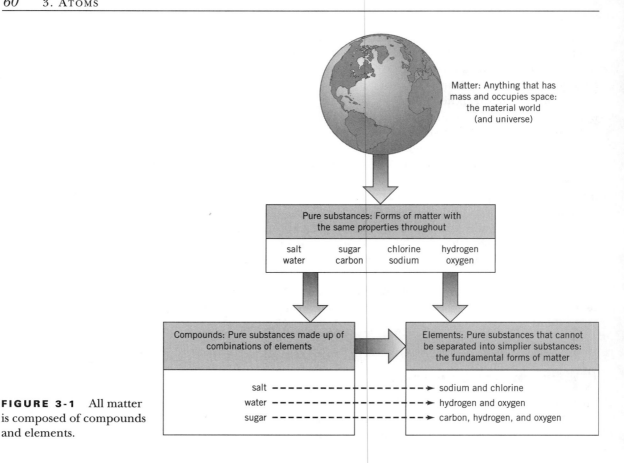

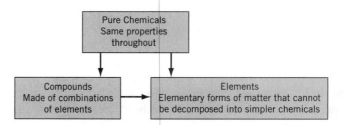

FIGURE 3-1 All matter is composed of compounds and elements.

down by chemical processes into simpler substances is an element, and any pure chemical that can be broken down into simpler substances is a compound.

Give the name of one common element and one common chemical compound.

The word *compound* literally means made up of parts. A **chemical compound** has definite properties and composition, but it is possible, with some effort, to separate it into two or more chemical elements. The word *element* comes from the Latin *elementum* (fundamental or basic principle). Elements are the fundamental or simplest forms of matter. Water is a compound, and it is possible to decompose it into the elements hydrogen and oxygen that are fundamental forms of matter. They cannot be further separated into simpler chemicals. They cannot be decomposed into simpler chemical forms.

A typical chemical compound is a combination of two or more elements, and it always contains these elements in definite proportions by mass. Water always contains the elements hydrogen and oxygen, and the elements are always present in a proportion of 1.0 g of hydrogen for every 8.0 g of oxygen. Another way of describing the composition of water is to say that it is 11% hydrogen and 89% oxygen by mass. As we will see, compounds are not just simple mixtures of elements. They are special chemical combinations of elements. Liquid water is obviously not a simple mixture of hydrogen gas and oxygen gas. How elements chemically combine is discussed in Chapter 9.

Over several centuries, scientists have identified 112 elements. Each element has unique properties that distinguish it from the other elements. For example, at room temperature the element oxygen is a colorless gas and the element gold is a distinctly colored solid. Oxygen readily combines with other elements. Oxygen is quite chemically active and forms

Percent

The term *cent* represents 100, as in the word *century*. Percent corresponds to parts per hundred. It is a common way of expressing what part one quantity is of a whole. The word *percent* means parts per hundred as represented by the symbol %. Thus, 20% means 20 parts per 100 parts or 20/100. A percent is expressed as a decimal fraction by dividing by 100. So 20% as a decimal fraction is 0.20.

If 59 g of water are found to contain 6.6 g of the element hydrogen, the percent by mass of hydrogen in water is found by dividing the mass of hydrogen by the mass of water, then multiplying by 100.

$$\% \text{ H} = \left(\frac{6.6 \text{ g}}{59 \text{ g}}\right) \times 100 = 11\%$$

Generally, to calculate what percent one quantity a is of the whole b, a is divided by b and then multiplied by 100.

$$\% \text{ a} = \frac{a}{b} \times 100$$

Percents are expressed using various bases of comparisons. For example, if it is found that in a large group of people 3 out of every 100 are left-handed, we say that 3% are left-handed. The basis of comparison is numbers of people. We most often use percents expressed on a basis of mass, called percent by mass. For a given part of a whole, its percent by mass is expressed as a factor showing the number of grams of the part per 100 g of the whole. For instance, the percent by mass of hydrogen in water is 11%. This percent expressed as a factor relates the mass of H in 100 g of water.

$$\left(\frac{11 \text{ g H}}{100 \text{ g water}}\right)$$

one or more compounds with most elements. Gold is not as active and forms a limited variety of compounds.

EXAMPLE 3-1

If a 65.3 g sample of water is found to contain 58.0 g of oxygen, what is the percent by mass of oxygen in water?

The percent is found by dividing the oxygen part by the whole water sample and multiplying by 100.

$$\% \, O = \left(\frac{58.0 \, g}{65.3 \, g} \right) 100 = 88.8\%$$

EXAMPLE 3-2

If water is 11% by mass hydrogen, how many grams of hydrogen are in 25 g of water?

First, we express the percent as a factor.

$$\left(\frac{11 \, g \, H}{100 \, g \, water} \right)$$

Multiplying the mass of water by this factor gives the number of grams of hydrogen in the sample.

$$X \, g \, H = 25 \, g \, water \left(\frac{11 \, g \, H}{100 \, g \, water} \right) = 2.8 \, g \, H$$

3-7 THE ELEMENTS

Since the elements are fundamental materials, which make up all matter, understanding their similarities and differences is important. Each element is given a unique **name** and **symbol.** The names and symbols of the elements have various historical origins reflecting their discoveries over the centuries. Inside the back cover of this book is an alphabetical list with each element's name and corresponding symbol. As you look over the list, note that most of the symbols come directly from the names of the elements. For example, the symbol for platinum is Pt. The name for this element comes from the Spanish word for silver, *platina*. When first discovered, this silvery-colored metal was mistaken for silver. Sometimes the symbol for an element comes from its Latin or Greek name. For instance, consider potassium, symbol K; its name comes from the English word *potash* and its symbol comes from *kalium*, the Latin name for potash. Some of the first-known potassium compounds came from material obtained by extracting wood ashes with water, then boiling off the water in a large pot. The material in the pot was potash.

ACTIVITY
3-2

Choose one of the elements and use a reference book or Web sources to find some facts about it.

The symbol for the element silver is Ag, from the Latin *argentum*. The symbol for the element mercury is Hg, from the Greek *hydrargyros* (liquid silver), an apt name for this shiny liquid metal. The symbol for the element tungsten is W, from the German *Wolfram*. The name of the element iron has Anglo-Saxon origins but its symbol, Fe, comes from the Latin *ferro*.

The names of the elements can differ in various languages. The name for iron is *fer* in French, *hierro* in Spanish, and *Eisen* in German, but its international chemical symbol is Fe. The **symbol for an element,** however, is a specific letter or set of two letters used internationally to represent the element. Each symbol always has a capital for its first letter; if there is a second letter, it is always lowercase. Knowledge of the names of the elements and the symbols that represent them is fundamental to any discussion of chemistry. To aid your study and learning of chemistry, you need to memorize some element names and symbols. Table 3-1 is a list of some common elements.

Normally, elements occur in the environment as components of compounds. Of the 112 known elements, only 89 are found on Earth in sufficient amounts to isolate samples of them. The remaining elements are rare; some exist for only a short time because they are radioactive. See Chapter 16 for a discussion of radioactivity. A few elements do not exist in

Table 3-1 *Some Important Elements*

ELEMENT	SYMBOL
Aluminum	Al
Arsenic	As
Barium	Ba
Boron	B
Bromine	Br
Cadmium	Cd
Calcium	Ca
Carbon	C
Chlorine	Cl
Chromium	Cr
Cobalt	Co
Copper	Cu
Fluorine	F
Gold	Au
Helium	He
Hydrogen	H
Iodine	I
Iron	Fe
Lead	Pb
Lithium	Li
Magnesium	Mg
Manganese	Mn
Mercury	Hg
Nickel	Ni
Nitrogen	N
Oxygen	O
Phosphorus	P
Potassium	K
Silicon	Si
Silver	Ag
Sodium	Na
Sulfur	S
Tin	Sn
Uranium	U
Zinc	Zn

Table 3-2 *Percent by Mass of Elements in Earth's Surface*

ELEMENT	PERCENT	ELEMENT	PERCENT
Oxygen	49.20	Chlorine	0.19
Silicon	25.67	Phosphorus	0.11
Aluminum	7.56	Manganese	0.09
Iron	4.71	Carbon	0.09
Calcium	3.39	Sulfur	0.06
Sodium	2.63	Barium	0.04
Potassium	2.40	Fluorine	0.03
Magnesium	1.93	Nitrogen	0.03
Hydrogen	0.87		
Titanium	0.58	Others	0.47

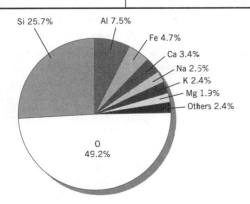

Table 3-3   *Percent by Mass of Elements in the Human Body*

ELEMENT	PERCENT	ELEMENT	PERCENT
Oxygen	64.6	Sodium	0.15
Carbon	18.0	Magnesium	0.03
Hydrogen	10.0	Iron	0.004
Nitrogen	3.1	Zinc	0.002
Calcium	1.9	Copper	0.0004
Phosphorus	1.1	Iodine	0.0004
Potassium	0.36	Tin	0.0001
Sulfur	0.25	Manganese	0.0001
Chlorine	0.15	Others	0.36

nature and have been synthesized using special methods. About 40 of the elements have geological or biological importance. Table 3-2 gives the percents by mass of the 18 elements that make up 99.6% of Earth's environment. For comparison, Tables 3-3 and 3-4 list the percents by mass and percent by atom count of the most common elements found in the human body. Some elements are fundamental in living systems. Carbon, hydro-

Table 3-5   *Some Important Elements in the Human Body*

ELEMENT	PERCENT OF BODY WEIGHT	ELEMENT	PERCENT OF BODY WEIGHT
Calcium	1.5–2.2	Chlorine	0.15
Phosphorus	0.8–1.2	Sodium	0.15
Potassium	0.35	Magnesium	0.03
Sulfur	0.25	Iron	0.004

Table 3-4   *Percent by Number of Atoms of Elements in the Human Body*

ELEMENT	PERCENT BY ATOM COUNT
Hydrogen	62.88
Oxygen	25.59
Carbon	9.50
Nitrogen	1.40
Calcium	0.30
Phosphorus	0.22
Potassium	0.06
Sulfur	0.05
Sodium	0.01
Chlorine	0.01
Magnesium	0.01
Others	0.03

Calcium: Found in compounds in bones, teeth, and body fluids.

Phosphorus: About 85% found in combination with calcium in bones and teeth. The rest incorporated in compounds in body fluids and DNA and RNA in cells.

Magnesium: Found in compounds contained in bone and body fluids.

Sodium: Mainly found as dissolved salt contained in extracellular fluids. Also found in cellular fluids and involved in transmission of nerve impulses.

Chlorine: Mainly found as dissolved salt contained in extracellular fluids. Also found in gastric juices in the stomach.

Potassium: A major element found in compounds contained in cellular fluids. Also involved in transmission of nerve impulses.

Sulfur: Found in amino acids and proteins of the body.

Iron: An important component of blood hemoglobin and muscle myoglobin. Stored in compounds found in liver, spleen, and bone.

Table 3-6   *Important Trace Elements*

ELEMENT	APPROXIMATE AMOUNT IN BODY (MILLIGRAMS PER KILOGRAM)	LOCATION OR FUNCTION IN THE BODY
Chromium	0.08	Related to function of insulin in glucose metabolism
Cobalt	0.04	Required for function of several enzymes, part of vitamin B_{12}
Copper	4	Required for function of respiratory enzymes and some other enzymes
Iodine	4	Located in thyroid gland, needed for hormone thyroxine
Manganese	1	Required for function of several digestive enzymes
Molybdenum	0.07	Required for function of several enzymes
Zinc	23	Required for function of many enzymes
Fluorine	Trace	Found in bones and teeth; thought to be essential, but function unknown
Selenium	Trace	Essential for liver function
Silicon	Trace	May be essential in humans
Tin	Trace	May be essential in humans

gen, oxygen, and nitrogen are the major elements included in compounds making up water, proteins, fats, and carbohydrates in the body. Other important elements in the body are described in Table 3-5. Trace elements are those elements needed in the diet in very small amounts. Table 3-6 describes important trace elements.

3-8 CHEMICAL EVIDENCE FOR ATOMS

Chemists developed precise methods to measure the relative masses of the various elements combined in compounds. Elements combine to form compounds, and it is possible to decompose compounds to form elements. When a sample of water is decomposed, the liquid is transformed into gases. Extensive observations of chemical processes in which elements form compounds, or compounds decompose, reveal that no measurable mass is lost or gained. For example, the decomposition of 100 g of water yields 89 g of oxygen and 11 g of hydrogen, and 89 g of oxygen combine with 11 g of hydrogen to give 100 g of water. That no mass change is observed in chemical processes is important. Experimental measurements show that in chemical processes or a chemical change, mass is conserved. Conservation of mass means that mass is not lost or gained. This idea became known as the law of conservation of matter as stated in the margin. The law of conservation of matter is another example of a physical law, an experimentally observed consistent pattern in nature. This law is fundamental in chemistry.

Another significant observation about the nature of matter is that a specific compound always contains the same elements. Let's consider this

ACTIVITY 3-3

How many grams of chemically combined hydrogen are contained in a 500 g sample of water? (Hint: Use the percent as a multiplying factor to relate grams of H to grams of water.)

Law of Conservation of Matter:
Matter is not created or destroyed in a chemical process.

idea in more detail. As an example, consider that sodium chloride, produced in the laboratory or obtained from a natural source, is always a compound of the elements sodium and chlorine. Furthermore, pure samples of salt always have the same relative amounts of the two elements. Any salt sample contains 39.4 g of sodium and 60.6 g of chlorine for every 100 g of salt. In percent, salt is always 39.4% sodium and 60.6% chlorine by mass. This example shows a consistent fixed pattern in compounds. The pattern is generally true for all compounds and is expressed as the law of constant composition, as given in the margin.

The laws of conservation of matter and constant composition provide evidence that elements are composed of some kind of characteristic building units or particles. These particles can join in definite patterns to form compounds. Thus, the composition of a compound has fixed proportions of elements. In other words, these laws make sense if we assume that elements are composed of characteristic particles: atoms.

3-9 ATOMIC THEORY

In 1803, John Dalton, drawing from the work of many early scientists, proposed a theory or model of the particle nature of matter. It is called the atomic theory and is stated as follows:

1. Matter is composed of tiny, fundamental particles called atoms. (Dalton used the term *atom* in recognition of the ideas of Democritus 2200 years earlier.)

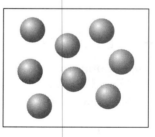

2. Atoms of an element are the same but differ from the atoms of all other elements. Each element has unique atoms.

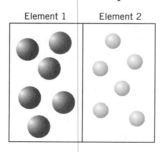

Element 1   Element 2

3. Atoms can combine to form compounds. Compounds contain atoms combined in definite whole-number ratios.

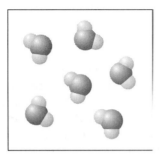

The atomic theory is very important in chemistry; besides explaining many properties of matter, it provides a mental image of matter. An atom is pictured as a tiny spherical particle, and matter is viewed as collections of atoms. An **atom** is the smallest representative particle of an element. Each kind of element has a unique kind of atom, and each atom has mass. A sample of an element contains only atoms of that element. The most obvious way in which the atoms of different elements differ is atomic mass. An aluminum atom has a different mass than an iron atom, an oxygen atom, or any other kind of atom. Look at a piece of aluminum. Imagine that it consists of a very large number of tiny aluminum atoms. If the sample is somehow subdivided into smaller and smaller pieces, the smallest possible piece is an aluminum atom. If the atom is subdivided, the pieces no longer represent the element. See Figure 3-2.

3-10 USING THE ATOMIC THEORY

The atomic theory served as a foundation for the development of chemistry and is one of the most fundamental theories of chemistry. A good theory is a simple idea that explains natural processes, introduces new ideas and concepts, and suggests some important questions.

FIGURE 3-2 Idealized subdivision of a piece of aluminum.

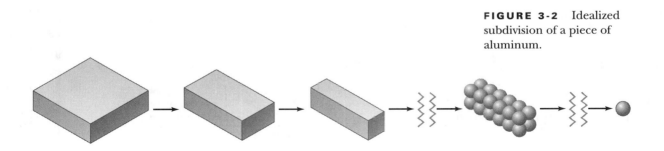

Explanations

Let's see how the atomic theory explains some observations mentioned in Section 3-8. We expect that elements are distinct and unique since each element is composed of atoms that are unlike the atoms of any other element. Aluminum atoms differ from iron atoms, and iron atoms differ from gold atoms. The law of conservation of matter makes sense if we view compounds as collections of combined atoms. Each atom has mass. Thus, the mass of a sample of a compound is the sum of the masses of all the atoms that compose the compound. Atoms are not destroyed when a compound is decomposed into its elements. Therefore, it is expected that the total mass of the elements is the same as the mass of the compound. In other words, the law of conservation of matter reflects that atoms are not created or destroyed in chemical processes. When elements combine to form compounds or compounds decompose to form elements, the atoms change their states of combination. They change combination but are not created or destroyed so the mass does not change. See Figure 3-3.

An explanation of the law of constant composition is that compounds involve definite and fixed combinations of atoms. That is, a given compound always includes the same kinds of atoms in the same relative numbers. The compound always contains the same proportions of the elements that make it up. See Figure 3-4. For instance, water always contains two combined hydrogen atoms for every combined atom of oxygen. Thus, it is represented by the formula H_2O. Since the atoms of elements differ from one another, they form different kinds of compounds with other elements. Occasionally, the atoms of any two elements can combine in distinct ways to form two or more different compounds. For example, hydrogen and oxygen combine to form the common compound water, but under certain conditions they also form the less common compound hydrogen peroxide, H_2O_2, a chemical used in medicinal preparations and hair bleaches. See Figure 3-4.

FIGURE 3-3 The synthesis and decomposition of a compound.

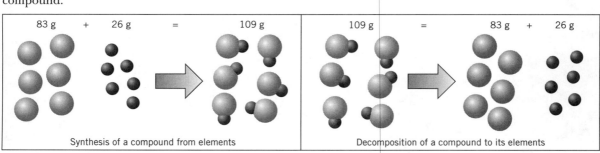

83 g + 26 g = 109 g	109 g = 83 g + 26 g
Synthesis of a compound from elements	Decomposition of a compound to its elements

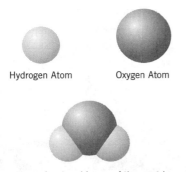

Hydrogen Atom   Oxygen Atom

Water is a very common solvent and is one of the most important compounds in the environment. It contains two combined hydrogen atoms for every one combined oxygen atom.

Hydrogen peroxide is a chemically active compound used as a paper pulp bleach, a hair bleach and a mild bactericide. It contains two combined atoms of combined hydrogen for every two combined atoms of oxygen.

FIGURE 3-4   The hydrogen and oxygen compounds, water and hydrogen peroxide.

Ideas

The atomic theory established the idea that elements were unique and had characteristic atoms. Based on this idea, chemists carefully searched nature for more elements. Thirty-one elements were known when Dalton proposed the atomic theory. Since then, as shown in Figure 3-5, 81 new elements have been added to the list.

The atomic theory states that atoms can combine to form compounds. Atoms are characteristic particles of elements, but compounds are composed of chemically combined atoms. In many kinds of compounds, atoms combine to make molecules. The formula of a compound made of molecules shows the kind and number of atoms that compose the molecules. Carbon dioxide, CO_2, is made of molecules in which one carbon atom is combined with two oxygen atoms. A sample of carbon dioxide is pictured as a collection of a vast number of CO_2 molecules. (As discussed in Chapter 6, not all compounds contain molecules.)

The atomic theory stimulated the imaginations of chemists and encouraged experimentation. That compounds are combinations of elements suggested that many different compounds could exist. Chemists isolated and purified many naturally occurring compounds. Today, many indus-

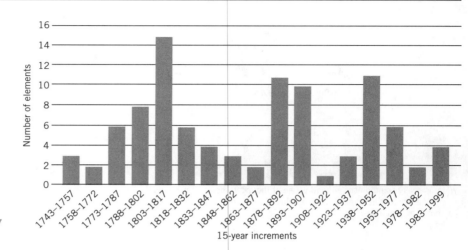

FIGURE 3-5 Discovery of elements.

trial processes involve the isolation of naturally occurring compounds from plant and mineral sources. These compounds are used in a variety of ways, ranging from raw materials for industry to food and medicines. As knowledge of chemistry developed, chemists began to explore possible combinations of elements that do not occur in nature. These combinations are made by a process called chemical synthesis. **Synthesis** means to make or put together. Today, many unique methods of synthesis are used to produce a large variety of "synthetic" chemicals. Hundreds of different kinds of synthetic chemicals are used as medicines, pesticides, and plastics. Synthesis reactions are also used to make chemicals that are identical to naturally occurring compounds. They are identical because they contain the same elements combined in the same way. There is no difference, for instance, between naturally occurring ammonia and ammonia synthesized by industrial processes, and there is no difference between pure synthetic vitamin C and pure natural vitamin C.

Questions

The atomic theory suggests some important questions. A few are given here. Can you think of additional questions?

How are the atoms of a given element the same, and how do atoms of the various elements differ from one another?

If atoms are characteristic chemical particles, are they composed of even more fundamental particles? If so, what kinds of particles structure atoms?

How are the formulas of compounds known?

How do atoms combine with one another?

Some friends tell you they only take natural vitamin C supplements. Explain to them why there is no difference between pure synthetic vitamin C and pure natural vitamin C.

Are there any patterns in the way in which elements combine to form compounds?

Answers to these questions are part of your continuing study of chemistry and are explored in future chapters.

3-11 CHEMICAL REACTIONS

Decomposition of compounds into elements and the combination of elements to make compounds are examples of chemical processes. Using the atomic theory, such processes are envisioned as changes in the combinations of atoms. Such transformations are called chemical reactions. The term *reaction* means to produce a result. A **chemical reaction** is a process in which one set of chemicals is mixed and transformed into a new set of chemicals. When charcoal, which is mostly elemental carbon, burns in air, the process is envisioned as carbon atoms combining with diatomic oxygen molecules to produce molecules of carbon dioxide gas. See Figure 3-6 for an idealized representation of this chemical reaction at the atomic level. Reactions are fundamental in chemistry and are discussed in more detail in Chapter 5.

As mentioned above, the law of conservation of matter expresses that atoms are not created or destroyed in chemical processes. In other words, atoms are conserved in chemical reactions. A reaction is a special process in which certain chemical combinations of atoms transform into new combinations of atoms. In a reaction, the total atom count does not change. Thus, destroying matter by chemical reactions is not possible. Burning a candle produces gases that become diluted and dispersed in the surrounding air, but the atoms are not destroyed. Solid chemicals in the candle are changed to gaseous forms of matter.

3-12 ELECTRICITY AND THE ATOM

In the early 1800s, atoms, as proposed by Dalton, were viewed as fundamental and indivisible particles. Experiments soon began to show that

FIGURE 3-6 Idealized view of a simple chemical reaction.

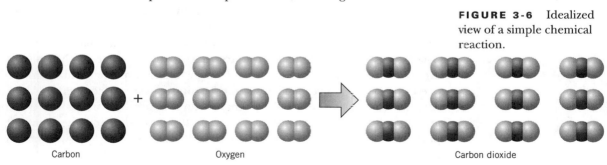

Carbon   Oxygen   Carbon dioxide

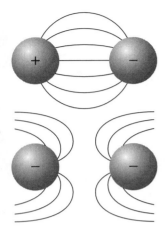

matter also had an electrical nature. A bolt of lightning is a dramatic example of the association of electricity and nature. You know the electrical effect of passing a comb through your hair, or you may have stuck a balloon to the wall after rubbing it on your hair. You may have experienced static cling. This behavior comes from electrical charges. There are two types of electrical charge: positive and negative. The terms *positive* and *negative* refer to different types of charge and are not the same as positive and negative as used in arithmetic. Yet, positive charge is represented by a + and negative charge by −. If an object has equal amounts of positive and negative charge, the charges balance and the object is electrically neutral. Recall that Coulomb's law, mentioned in Chapter 1, states that opposite electrical charges attract and like charges repel. Two pieces of clothing attract because of opposite electrical charges, whereas your hairs can repel when they carry like electrical charges.

What is electricity? Scientists, including the American Benjamin Franklin, developed ideas about the nature of electricity in the late 1700s and early 1800s. It was theorized that electrical current was carried by tiny, discrete, negatively charged particles. The ancient Greeks had noted that when amber (fossilized tree resin) was rubbed with cloth, it attracted small bits of other materials. The Greek name for amber was *elektron.* When the characteristic particles of electricity were discovered in the late 1800s, they were called electrons. Electrons are far too small to see, but electrical current flowing through a metal is pictured as the flow or movement of electrons.

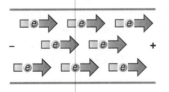

The particles found in atoms are simply called subatomic particles. Experiments with electricity and matter suggest that electrons are one type of subatomic particle. Electrons in atoms, however, are bound to the atoms and are not free-moving as are the electrons involved in electrical current. It appears that, under certain conditions, some electrons bound in atoms can leave the atoms and become free-moving electrons in electrical current. Normally, atoms are electrically neutral. Neutral atoms must have a balance of positive and negative charges. If electrons with negative charges are found in atoms, some kind of positively charged particle must also be present. Such positive particles, called **protons,** were discovered in 1914. An electron carries a negative charge and a proton carries an equal amount of positive charge. Years later, in 1932, another major subatomic particle was discovered. This particle is electrically neutral, so it was named the **neutron.**

3-13 SUBATOMIC PARTICLES

Atoms are now viewed as collections of subatomic particles. How do subatomic particles structure atoms, and how do atoms of the various elements differ from one another? Since the times of Dalton, scientists carried out extensive experiments and observations related to questions about atomic structure and the nature of subatomic particles. Experiments suggest that atoms are composed of electrons, protons, and neutrons.

Electrons:
Subatomic particles that carry negative electrical charge

Protons:
Subatomic particles that carry positive electrical charge

Neutrons:
Subatomic particles that are electrically neutral

PARTICLE	SYMBOL	CHARGE	MASS (g)	MASS RELATIVE TO PROTON
Electron	e^-	1−	9.110×10^{-28}	1/1835
Proton	p	1+	1.672×10^{-24}	1
Neutron	n	0	1.675×10^{-24}	1

Note that an electron has a relatively small mass and carries a fixed amount of negative charge. The electron carries the smallest possible unit of negative electrical charge, and we will refer to its charge as 1− or −. It is interesting that each proton carries an amount of positive charge that is equal to the negative charge on an electron. That is, electrons and protons have an equal but opposite electrical charge. The charge of a proton is represented as a 1+ or +. Since an atom is electrically neutral, it must contain equal numbers of protons and electrons. Atoms of the various elements differ in the number of protons and electrons they contain. Although electrons and protons are exact opposites with respect to electrical charge, they greatly differ in mass. An electron has a mass of 9.110×10^{-28} g, and a proton has a mass of 1.672×10^{-24} g. Thus, a proton has a mass that is 1835 times greater than the mass of an electron. Neutrons are also found in atoms, but since they are electrically neutral, they do not contribute any electrical charge. A neutron has a mass of 1.675×10^{-24} g, just slightly larger than a proton.

3-14 THE NUCLEAR ATOM

If atoms contain electrons, protons, and neutrons, the obvious question is, How do these subatomic particles give structure to an atom? A partial answer to this question was given by an English scientist, Lord Rutherford, in the early 1900s. Experimental observations done in his research laboratory led him to theorize that an atom consists of a small, extremely dense **nucleus** of positive charge surrounded by a swarm of negatively charged electrons of relatively low mass. The term *nucleus* means the center. This idea became known as the **nuclear model of the atom.** The vision of the atom became

FIGURE 3-7 An atomic nucleus.

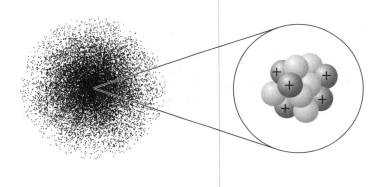

The ratio of the diameter of the atom to the diameter of the nucleus reveals how much larger the atom is compared with the nucleus. Calculate this ratio using the numbers given in Section 3-14.

one in which very tiny electrons are in motion about a nucleus. The vision of the nucleus became a cluster of protons and neutrons (see Figure 3-7).

Using unique experimental methods, Rutherford calculated the approximate sizes of atoms and nuclei. His calculations gave 2×10^{-8} cm for the diameter of a typical atom and 1×10^{-13} cm for the diameter of a typical nucleus. These numbers reveal that atoms are very tiny particles and that atomic nuclei are extremely small parts of atoms. Thus, the nucleus of an atom makes up an extremely small portion of the entire atom and the electrons populate a very large portion.

To get a feeling for or picture of the size of an atom, consider that a speck of carbon barely visible to the unaided eye contains 6×10^{16} carbon atoms (60 million billion atoms). In comparison, the population of Earth is about 6 billion (6×10^{9}) humans. It takes 10 million such populations to equal the number of atoms in this single speck of carbon.

Use a pencil or a pen to make a dot on a piece of paper. If a single atom is somehow expanded so that its nucleus had a diameter equal to this dot, the outer regions of the atom extend to about 100 m from the dot. In contrast to the differences in atomic and nuclear sizes, most of the mass of an atom is concentrated in the nucleus. The nucleus makes up more than 99.9% of the mass of an atom. Nuclei are very dense concentrations of mass with positive electrical charge. They are dense clusters of protons and neutrons. The term *nucleon* is used to refer to a proton or neutron found in the nucleus of an atom.

3-15 ATOMIC STRUCTURE

You may wonder why an atom is viewed as a fundamental particle of matter. It is, after all, composed of subatomic particles. From a chemical point of view, atoms are fundamental. An **atom of an element** is the characteristic particle of the element. Since the behavior of atoms explains chemical processes, **chemistry** is the science of the nature and behavior of atoms. (Recall the definition of chemistry given in Chapter 1. Why is that definition essentially the same as the definition given here?)

An atom of a particular element differs from the atoms of all other elements in the numbers of subatomic particles it contains. Since these particles have mass, the atoms of various elements also differ in mass. How do the structures of the atoms of the various elements differ? An atom of a specific element has a characteristic number of protons in the nucleus. For example, nuclei of oxygen atoms contain eight protons. This number is characteristic of oxygen, so any atom that has a nucleus with eight protons is an oxygen atom. Thus, atoms with the same number of protons in the nuclei are atoms of the same element. Atoms with different numbers of protons in the nuclei are atoms of different elements. All hydrogen atoms have nuclei with one proton compared with oxygen atoms that have eight protons. A neutral atom has the same number of electrons as protons. Hydrogen atoms have one electron and one proton, and oxygen atoms have eight electrons and eight protons. For a neutral atom, it is always true that the number of protons equals the number of electrons.

The number of protons in a nucleus of an atom is called the **atomic number.** Each element has a unique atomic number. Look at the table inside the back cover of this book. The elements are listed alphabetically. Note that the atomic number is listed for each element. For instance, hydrogen has an atomic number of 1, oxygen 8, sulfur 16, and uranium 92. The atomic number of an element reveals the number of protons and the number of electrons in an atom of the element.

3-16 ISOTOPES

That atoms of different elements have different numbers of protons and electrons partially account for the differences in masses of the atoms. The neutrons also contribute to the masses of atoms. The atoms of most elements have nuclei consisting of more than one combination of neutrons and protons. For example, study of elemental chlorine shows that it is made up of two kinds of atoms. About 76% of chlorine atoms have nuclei with 18 neutrons and 17 protons, and 24% have nuclei with 20 neutrons and 17 protons. Since the number of protons is 17 in each case, both types of atoms are chlorine atoms. Thus, two kinds of chlorine atoms are found in nature, and they have slightly different masses since they contain different numbers of neutrons. Atoms of the same element with differing masses are called isotopes. **Isotopes** are atoms of the same element having the same number of protons but different numbers of neutrons.

The element oxygen has three natural isotopes. Let's see how they differ. The atomic number of oxygen is 8, so all oxygen atoms have eight protons and eight electrons. One isotope of oxygen has atoms with eight neutrons, the second isotope has nine neutrons, and the third has 10 neutrons. For any isotope, the sum of the number of protons and neutrons is called the **nucleon number** or **mass number.** The nucleon numbers of the oxygen isotopes are 16, 17, and 18, respectively. Nucleon numbers distinguish one iso-

PRO-TONS	NEU-TRONS	ELEC-TRONS
8	8	8
8	9	8
8	10	8

tope from another. Thus, the isotopes of oxygen are oxygen-16, oxygen-17, and oxygen-18. Isotopes of elements are sometimes shown by the symbol of the element with a superscript nucleon number on the left. Using this notation, the natural isotopes of oxygen are ^{16}O, ^{17}O and ^{18}O. A summary of the composition of the oxygen isotopes is given in the margin on page 75. Incidentally, to find the number of neutrons in an isotope, subtract the atomic number from the nucleon number. For example, since oxygen has an atomic number of 8, its isotopes have 8, 9, and 10 neutrons, respectively.

FIGURE 3-8 The naturally occurring isotopes of some elements.

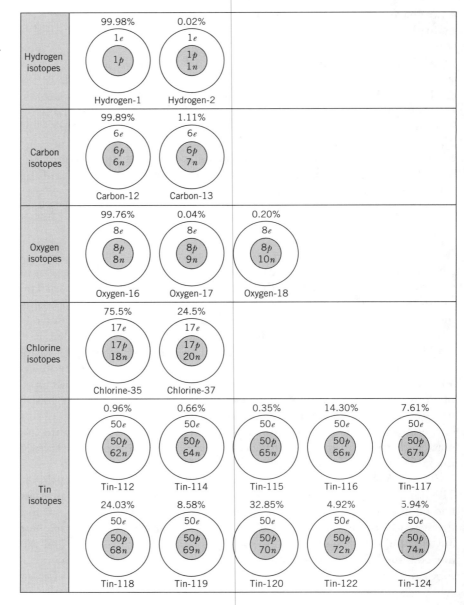

Some elements found in nature exist as only one type of atom. These elements have no natural isotopes. For instance, only one kind of aluminum atom occurs in nature. Other elements have two or more naturally occurring isotopes. Examples, shown in Figure 3-8, are hydrogen with two natural isotopes, carbon with two, oxygen with three, chlorine with two, and tin with 10. Notice that the figure shows the composition of each isotope and the percent each isotope contributes to the naturally occurring element. The isotopes of an element occur in varying but consistent percents.

The nucleon number of an isotope is a number corresponding to the sum of the number of protons and the number of neutrons in the nucleus of an atom. Nucleon numbers are used to refer to a specific isotope of an element to distinguish it from other isotopes of the element. Some common examples are plutonium-239, ^{239}Pu, used in nuclear weapons; uranium-235, ^{235}U, used in nuclear reactors; americium-241, ^{241}Am, used in smoke detectors, cobalt-60; ^{60}Co, used in cancer therapy; and carbon-14, ^{14}C, used in archaeological age dating. When we use the name or symbol of a specific element, however, we are referring to the naturally occurring element that is a collection of natural isotopes. That is, when we refer to oxygen, O, we mean that collection of atoms (^{16}O, ^{17}O, and ^{18}O) that includes all the natural isotopes. When you breathe oxygen, you take in all the natural isotopes of oxygen. Oxygen-16 is the most abundant of these isotopes, but the others are always present in natural samples of oxygen. The chemical behaviors and chemical properties of these three isotopes are the same. Generally, all isotopes of a given element have the same chemical behavior.

ACTIVITY 3-6

The nucleus of an atom is envisioned as a cluster of protons and neutrons. Hydrogen has the isotopes hydrogen-1, hydrogen-2, and hydrogen-3. Draw representations of the nuclei for each of these isotopes and show the number of protons and neutrons.

3-17 ATOMIC MASSES

Atoms are very tiny, and we cannot make the same kind of measurement on atoms that we can make on large samples that are collections of atoms. We cannot weigh individual atoms. Yet measuring the relative masses of the atoms of elements is possible. By relative mass, we mean the mass of an atom of one element compared with the mass of an atom of another element. For example, an atom of carbon-12 (the most common isotope of carbon) is about three times the mass of an atom of the helium-4 isotope. That is, the ratio of the mass of such a helium atom to a carbon-12 atom is 1 to 3.

The most accurate method of measuring the relative masses of atoms is mass spectroscopy. This very sensitive method uses an instrument called a mass spectrometer in which electrically charged atoms are forced to move through a magnetic field. As shown in Figure 3-9, the moving atoms take on curved paths directly related to their masses. As an analogy (see Figure 3-10), imagine standing on a high bridge with a brisk wind blowing below. If you drop balls of different masses—say a tennis ball, a handball, and a baseball—off the bridge, they are deflected by the wind differently and take on different curved paths. By carefully measuring the curvature of the paths of atoms in a mass spectrometer, it is possible to measure the relative masses of the atom. When a sample of a specific element is placed in a

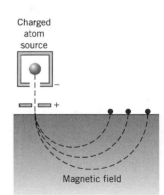

Charged atom source

Magnetic field

FIGURE 3-9 A type of mass spectrometer: charged atoms take on specific circular paths depending on their masses.

FIGURE 3-10 A mass spectrometer analogy: balls of different mass take on different paths as they are dropped into the wind.

Wind

Refer to Figure 3-8. Imagine a sample containing 10,000 hydrogen atoms. How many of these atoms are hydrogen-1 and how many are hydrogen-2? (Actually, 10,000 atoms is an extremely small sample, much too small to see or weigh.)

mass spectrometer, it is possible to measure the relative masses of its isotopes. Furthermore, mass spectrographic analysis reveals the percent of each of these isotopes in the sample. For instance, when chlorine is analyzed, it is found to contain two isotopes having slightly different relative masses. The isotopes have nucleon numbers of 35 and 37, so their relative masses are close in value but not equal. The analysis also reveals that chlorine is 75.77% ^{35}Cl and 24.23% ^{37}Cl. These percents mean that for every 10,000 chlorine atoms, 7577 are ^{35}Cl and 2423 are ^{37}Cl. These isotopes of chlorine have different masses but both have the same chemical behavior. Each isotope has 17 protons and 17 electrons.

Although an element has isotopes, all the atoms of that element have the same chemical behavior. Thus, from a purely chemical point of view, knowing how many isotopes an element has or even whether it has isotopes at all is not important. It is true, however, that the existence of isotopes is reflected in the masses of the atoms of elements. Consequently, when the relative masses of the atoms of various elements are needed, the existence of isotopes is relevant.

3-18 ATOMIC WEIGHTS

Because atoms are very tiny particles, expressing masses of individual atoms in grams is not convenient or useful. A special unit of mass, called the **atomic mass unit (u)**, is used as a unit to express the masses of atoms. To establish the atomic mass unit, an atom of the common isotope carbon-12 is used as a reference. By definition, the mass of a carbon-12 atom is exactly 12 u. Thus, one atomic mass unit is one twelfth the mass of a carbon-12 atom. The atomic mass unit is defined using carbon-12 so that no atom has a mass less than 1 atomic mass unit. An atom of the lightest isotope of hydrogen, which is the least massive of all isotopes, has a mass around 1 u relative to carbon-12. Measuring the masses of isotopes of any element relative to carbon-12 is possible, and these masses are expressed in atomic mass units. Recall that the relative mass of a helium-4 atom to a carbon atom is 1 to 3. Thus, a helium-4 atom has a mass in atomic mass units that is one third the mass of a carbon-12 atom, or 4 u. The masses of atoms are usually measured to many more significant digits than given in this example.

In chemistry, having a numerical value for the average mass of the atoms of an element is very useful. Although this may seem difficult in view of the many isotopes of the elements, these average masses are easily calculated from the mass and the percent abundance of each natural isotope of an element. These data are obtained from mass spectroscopy experiments. For instance, the average mass of an atom of chlorine is calculated using the masses of chlorine-35 and chlorine-37 and the percent distribution of each isotope. Each isotope contributes a share to the average mass of a chlorine atom. As an analogy, imagine four students take a quiz and get scores of 35, 35, 35, and 37. The average score of 35.5 is found by adding the scores and dividing by 4. Another way to express the quiz results is to use the percent of students having a given score.

$$
\begin{array}{ll}
35 & 75\% \\
37 & 25\%
\end{array}
$$

Another way to find the average score is to multiply each score by the corresponding percent divided by 100 and add these products.

$$
\begin{array}{rl}
35(0.75) = & 26.25 \\
37(0.25) = & \underline{9.25} \\
& 35.5
\end{array}
$$

Notice that this is a weighted average score and that the value of 35.5 does not represent any actual score, just the average.

The **atomic weight** of an element is found by multiplying the mass of each natural isotope by its fractional abundance and adding these products. The fractional abundance is the percent divided by 100. The weighted average mass calculated in this way is the mass of a hypothetical average atom. As an example, consider the atomic weight of chlorine found from the masses and percent abundances of its two natural isotopes.

$$
\begin{array}{lll}
\text{chlorine-35} & 34.9689 \text{ u} & 75.770\% \\
\text{chlorine-37} & 36.9659 \text{ u} & 24.230\%
\end{array}
$$

The mass of each isotope is multiplied by its fractional abundance, and the sum of these products gives the weighted average mass, which is the atomic weight.

$$
\begin{array}{lll}
\text{chlorine-35} & (34.9689 \text{ u})(0.75770) = & 26.496 \text{ u} \\
\text{chlorine-37} & (36.9659 \text{ u})(0.24230) = & \underline{8.9568 \text{ u}} \\
& & 35.453 \text{ u atomic weight of chlorine}
\end{array}
$$

Use the following information to calculate the atomic weight of magnesium.

ISOTOPE	PERCENT	MASS
^{24}Mg	78.99%	23.985042 u
^{25}Mg	10.00%	24.985838 u
^{26}Mg	11.01%	25.982594 u

Note that no actual chlorine atom has this mass, but the value represents an average mass.

By definition, **atomic weight** is the average mass of an atom of an element determined by considering the contribution of its natural isotopes. Each element has a unique and characteristic atomic weight. The alphabetical list of elements inside the back cover of this book also gives the atomic weights. Precise atomic weights for the elements are the result of many years of tedious experimental work. You'll find a list of atomic weights very useful for some interesting kinds of chemical computations discussed in future chapters. When you need an atomic weight, just look it up.

You often see tables of elements posted on the walls of chemistry laboratories and lecture rooms. Such a table is called the **periodic table of the elements**. The periodic table is discussed in more detail in Chapter 5. Elements in the table are listed from left to right by increasing atomic number. Typically, the table gives the symbol for each element, its atomic number, and its atomic weight. A periodic table is given inside the front cover of this book. You can use a periodic table as a convenient and quick source of atomic weights.

A close look at a table of atomic weights shows that atomic weights vary in the number of significant digits. Atomic weights are determined using experimental measurements of isotopic masses and percent abundances of the isotopes. Variations in these measurements for a given element dictate the number of significant digits in the atomic weight. Typically, atomic weights have many significant digits. When we use atomic weights in calculations, we often round them to a useful number of digits. Four or five significant digits are sufficient since most calculations involving atomic weights include experimental data that are in this range of significant digits. A few atomic weight values are listed below. Note that atomic weights are not whole numbers, and elements of higher atomic number typically have higher atomic weights, than those of lower atomic numbers.

Express the atomic weights of the following elements to four significant digits: carbon, nitrogen, sodium, aluminum, copper, zinc, and mercury.

ATOMIC NUMBER	ELEMENT	ATOMIC WEIGHT	ROUNDED TO FOUR DIGITS
1	Hydrogen	1.00794	1.008
8	Oxygen	15.9994	16.00
9	Fluorine	18.9984032	19.00
16	Sulfur	32.066	32.07
17	Chlorine	35.4527	35.45
26	Iron	55.845	55.85
92	Uranium	238.0289	238.0

FOCUS
3

Some Facts about Atoms

1. The atoms of a given element always have the same number of protons. A neutral atom contains equal numbers of protons and electrons. Example: Atoms of hydrogen have one proton and one electron.

2. The atomic number of an element is the number of protons in the nuclei of the atoms of that element. If we know the atomic number of an element, we know how many protons and electrons are in atoms of that element. Example: Iron has atomic number 26, so iron atoms must contain 26 protons and 26 electrons.

3. The nucleus of an atom is a cluster of protons and neutrons. Isotopes are atoms of the same element that have different numbers of neutrons. The nucleon number of an isotope of an element is the number of protons plus the number of neutrons. To find the number of neutrons in an isotope, subtract the atomic number from the nucleon number. Example: The isotope plutonium-239 has nucleon number 239. The atomic number of plutonium is 94, so there are 94 protons and 145 neutrons in an atom of the isotope ($239 - 94 = 145$).

4. The atomic weight of an element is the average mass of an atom of an element determined by considering the contribution of each natural isotope. The atomic weights of various elements are determined from experimental data. The table inside the back cover lists values of the atomic weights of the elements, and atomic weights are given in the periodic table shown in the front cover. Atomic weights are rounded to a desired number of digits. Example: The listed atomic weight of silver, Ag, is 107.8682. Rounding to four digits gives 107.9.

Seeing Atoms

How small are atoms? One nanometer is the name for one billionth of a meter (10^{-9} m). Six carbon atoms could fit in a space about one nanometer wide. If six million carbon atoms could somehow be lined up, the line would be about one millimeter long. It would take about 150 million carbon atoms to make a line one inch long. Imagine drawing a line one inch long and dividing the line into 150 million parts. You see atoms are quite small. A sample of carbon, slightly larger than a sugar cube, with a mass of 12 grams, contains about 600 million-million-billion atoms (6.0×10^{23} atoms).

Is it possible to see atoms? Atoms are submicroscopic in size and cannot be seen, even by the most powerful conventional microscope. In one sense, atoms are conceptual particles imagined in our minds. How do we know that they exist if we can't see them?

When a small piece of dust or pollen is suspended in a liquid and observed using a powerful microscope, it follows a random, erratic path as though it is being pushed about. This phenomenon is called Brownian motion after Robert Brown, who first observed it in the early 1800s. In 1905, Albert Einstein explained this behavior by assuming the piece of dust is continuously bombarded by the particles (molecules) of the liquid. Consequently, the random motion of the

FIGURE 3-11 The diffraction of X rays is evidence for the location of atoms.

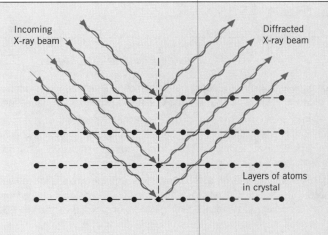

IBM STM Image Gallery

FIGURE 3-12 An image produced by a scanning tunneling microscope (STM). (Courtesy of IBM)

small piece of dust provided a way to "see" atoms. Incidently, Einstein's explanation also serves as evidence for the view of the incessant motion of particles of matter as suggested poetically by Lucretius.

Another method used to see atoms is X-ray diffraction. X rays are a form of high-energy light that move in a wavelike manner. When materials are exposed to X rays, the X rays reflect or diffract when they encounter highly dense masses like nuclei. See Figure 3-11. As an analogy, note that water waves diffract as they pass through the pilings on a pier. If the pilings are equally spaced, a diffraction pattern is observed. The reflection pattern of the waves depends on the spacing of the pilings. Similarly, when X rays impinge upon materials, they diffract upon encountering nuclei. The resulting diffraction pattern reveals information about the spacing of the atoms containing the nuclei. The existence of a diffraction pattern reflects the existence of atoms.

The most modern way to see atoms is by use of special microscopes called scanning tunneling microscopes. These devices have precise probes so fine that the tip of the probe is a single atom. The probe is moved over a surface. As it moves, a measure is made of the electrical current between the probe and atoms on the surface of a sample of material. Such microscopes allow us to see computerized images of atoms by visualizing the electrical force fields around them. The currents are analyzed by computer and atomic images are displayed on computer monitors. Specially designed scanning tunneling microscopes are used to pick up atoms and move them about. Figure 3-12 shows atomic images obtained from a scanning tunneling microscope.

3-19 SOME COMMON ELEMENTS

Of the 112 different elements, some are much more common and useful than others. Several of these elements are described below.

Carbon, C

In the pure form, carbon occasionally occurs as diamond, but more commonly as graphite or in the amorphous forms of charcoal, coke (coal heated in the absence of air), and carbon black (a finely divided black sootlike powder). Carbon is also present in nature as the major component of compounds found in plants and animals. There are millions of chemical compounds that contain carbon in combination with hydrogen and oxygen, nitrogen, phosphorus, or sulfur. Such compounds are organic compounds, and the study of the compounds of carbon is called **organic chemistry.** Carbon is an element essential to life, and there is an ample supply in the food that we eat since carbohydrates, proteins, and fats all contain carbon compounds.

Diamonds have many industrial uses because this form of carbon is the hardest substance known. Graphite is used as a lubricant and as the major component of pencil leads. The word *graphite* comes from the Greek *graphein* (to write). Charcoal and coke are used as fuels, and carbon black is added to rubber to give it strength for use in tires (this is why tires are black).

Chlorine, Cl

Chlorine is a pale, greenish yellow gas that occurs as diatomic molecules, Cl_2. It rarely occurs as the free element in nature, but it is obtained from the many compounds in which it is found. Combined chlorine is found in various minerals called chlorides, the most notable of which is common table salt or sodium chloride. Vast salt deposits are found in nature, and dissolved salt is found in abundance in seawater. Elemental chlorine can kill bacteria, so it is used to treat drinking water and swimming pool water. It is also used to make household chlorine bleach. For safety, never mix chlorine bleach with acid cleaners or ammonia since such mixtures produce chlorine gas and other poison gases. Other uses of chlorine are in the manufacture of the plastic polyvinyl chloride (PVC) and the plastic polyvinyldiene chloride, used in Saran wrap.

Copper, Cu

Copper, a distinctive, reddish-colored metal, was known to ancient civilizations because it sometimes occurs free in nature and is rather easily extracted from minerals that contain it. The Stone Age man of Chapter 1 had an ax that was nearly pure copper. The ancients used copper for tools, weapons, and coins. They even discovered that when copper is mixed with other metals, it becomes more durable and useful. Mixtures of metals that have properties that are different from the individual metals that make them up are called **alloys.** Bronze is an alloy of copper and tin (around 80% copper by mass), and brass is an alloy of copper and zinc (ranging from 60% to 85% copper by mass). Copper is a very good electrical conductor, so it is often used in electrical wires.

Gold, Au

The symbol for the enchanting metal gold comes from the Latin *aurum* (shining dawn), which also gives the English word aura. Relatively rare in nature, gold has been a precious and valuable element for millennia. It is used in jewelry and coins and is hoarded as a symbol of wealth. This bright yellow metal can occur in nature as the free metal and is noted for its ability to keep its luster since it does not easily combine with oxygen in the air. It is also found in gold ore deposits. Gold is used in the manufacture of electronic devices, dental crowns, and objects of art and jewelry.

Hydrogen, H

Hydrogen is a low-density, colorless, odorless gas that occurs as diatomic molecules, H_2. It is apparently the most abundant element in the universe. It is a component of stars and is abundant in regions of interstellar space. On Earth, very little hydrogen is found in nature as the free element, but it is an important component of most of the chemicals found in plants and animals. It is also found, combined with carbon, in the fossil fuels, coal, petroleum, and natural gas. Petroleum and natural gas are the major industrial sources of elemental hydrogen. The name hydrogen literally means water former, and, as you know, it is a component of water.

Iron, Fe

Iron is a silvery metal found compounded in a variety of minerals and, more important, in iron ores. Processing of the ores yields iron with its many uses, ranging from tools to cars to skyscrapers to bridges. Since pure iron is somewhat soft and easily corroded, it is made into various alloys that are strong, hard, and malleable. **Malleable** means easily formed by bending or pounding into various shapes. A variety of iron alloys are made by mixing various metals and carbon with iron. The most common iron alloys are **carbon steel,** containing a small percent of carbon, and stainless steel. One of the most common types of **stainless steel** contains iron with around 18% chromium and 8% nickel by mass and some carbon.

Nitrogen, N

Nitrogen, a colorless, odorless gas, is found as the free element in air. As you know, nitrogen, in the form of diatomic molecules, N_2, is the major component of air (78.08% by volume). Nitrogen compounds are found as minerals and are important constituents of plants and animals since they occur as amino acids that make up proteins. Nitrogen, along with phosphorus, P, and potassium, K, is an important plant nutrient. Some nitrogen-containing compounds, such as ammonia, ammonium sulfate, and urea, are used as plant fertilizers.

Oxygen, O

Oxygen is the most abundant element on the surface of Earth. A colorless, odorless gas, it is an important component of air (20.95% by volume). This gas in the form of diatomic molecules, O_2, is essential to oxygen-breathing animal life. Many substances can react chemically with the oxygen of the air. When the reaction is rapid and vigorous, it is called burning or combustion. In any case, reaction with oxygen is called oxidation and the compounds formed are called oxides. Oxygen is also found as a component of water and in compounds that compose rocks, minerals, plants, and animals.

8
O
15.9994

Silicon, Si

The element silicon is a shiny, bluish gray, brittle solid. Silicon has unique electrical properties. Certain elements mixed with silicon, at the parts per million level, increase its effectiveness as a semiconductor of electricity. Semiconducting silicon is used in transistors and other electronic components. Large numbers of these electronic components are fabricated, at the microscopic level, on the surface of tiny chips of silicon. These chips contain integrated circuits having thousands of electronic components in an area measuring only a few millimeters on an edge. As you know, these chips are used in computers, calculators, and many other consumer products. Some experts claim that the development of the integrated circuit on a chip will affect society on a level rivaling that of the Industrial Revolution. Clays, sand, and other compounds containing silicon are widely used in making ceramics, ceramic glazes, and various types of glass.

14
Si
28.0855

Silicones are a group of compounds containing silicon, oxygen, carbon, and hydrogen that range from oily liquids to rubbery plastics. Silicones are used for medical implants, as the centers of golf balls, and in various industrial products such as silicone caulking compounds. Some swimmers use soft pliable silicone rubber earplugs.

Sodium, Na

Sodium is a soft, silvery metal that does not occur naturally as the free element. Its compounds are found in a variety of minerals, especially in vast deposits of salt formed by the evaporation of ancient seas. Salt is an essential compound in our diets. Normally, we get sufficient salt from the foods that we eat. It is often added to foods to enhance flavors. The salt content of prepared foods is listed on nutritional labels as milligrams of sodium.

11
Na
22.98977

The recommended daily intake (called the daily value) of sodium is less than 2400 mg. Nutritional labels on prepared foods list the sodium content and expresses the percent contribution to the daily value of sodium.

Sulfur, S

Sulfur is a brittle, yellow solid sometimes called brimstone. It has been known for centuries because it sometimes occurs as the free element. Occasionally, it accumulates on the brim of water pools in volcanic regions. In such regions, the "sulfur smell" is actually the odor of sulfur compounds, such as hydrogen sulfide (a rotten egg odor) or sulfur dioxide (sharp, choking odor sometimes produced upon striking a match). Sulfur is also found combined in minerals such as gypsum, used in wallboard, and iron pyrites, sometimes called fool's gold. Sulfur compounds are essential to plant and animal life since sulfur is a component of some amino acids found in proteins. Sulfur has many industrial uses, including the manufacture of matches. That's why you sometimes smell sulfur dioxide when you strike a match. The major use of sulfur is in the manufacture of sulfuric acid, one of the most important industrial chemicals.

QUESTIONS

1. Match the following terms with the appropriate lettered definitions.

atom

a. the sum of the number of protons and neutrons in the nucleus of an atom

solid

b. the average mass of an atom of an element

liquid

c. the number of protons in an atom of an element

gas

d. atoms of the same element with different numbers of neutrons

heterogeneous

e. the center of atom composed of a cluster of nucleons

homogeneous

f. neutral subatomic particle

solution

g. subatomic particle carrying a positive charge

pure chemical

h. subatomic particle carrying a negative charge

element

i. the study of compounds of carbon

compound

j. mixtures of metals and other elements that have unique properties

chemical reaction

k. a process in which one set of chemicals is transformed into another set

electron

l. pure chemical composed of two or more elements

proton

m. pure chemical that cannot be further broken down into simpler chemicals

neutron

n. chemical that has the same properties throughout

nucleus

o. a homogeneous mixture containing solutes dissolved in a solvent

isotopes

p. made up of the same parts throughout

atomic number

q. made up of a variety of different parts

nucleon number

r. the state of matter with no definite shape or volume

atomic number   s. the state of matter with a fixed volume and a flexible shape

organic chemistry   t. the state of matter with a fixed volume and a fixed shape

alloys   u. the smallest characteristic particle of an element

Section 3-1

2. Make a time line from 500 B.C.E. to the present and show the dates given in Section 3-1 that represent the passage of the idea of an atom through time.

Section 3-2

3. What are the different states of matter? Give an example of each type.

4. Draw sketches to illustrate the differences between solids, liquids, and gases according to the particle view.

Sections 3-4 to 3-6

5. What does homogeneous mean? What does heterogeneous mean?

6. What is a solution?

7. What is a solvent? What is a solute?

8. What is a pure substance or chemical?

9. What is an element? What distinguishing characteristic applies to an element?

10. What is a compound? What distinguishing characteristic applies to a compound?

11. What is the difference between an element and a compound?

12. What is a percent?

13. *Express the following as percent by mass.

    (a) 56.0 g of baking soda contains 8.00 g of carbon

    (b) 235 g of water contains 26.3 g of hydrogen

    (c) 28.25 g of table sugar contains 11.9 g of carbon

14. Express the following as percent by mass.

    (a) 43.6 g of salt contains 26.41 g of Cl

    (b) 79.81 g of ammonia contains 65.66 g of N

    (c) 17.75 g of fructose contains 7.09 g of C

15. *Express the following percents as factors. For example, if water is 11.2% H, then there are (11.2 g H/100 g water).

    (a) acetic acid (vinegar) contains 40.0% C

    (b) table sugar contains 42.1% C

    (c) baking soda contains 14.3% C

16. Express the following percents as factors. For example, if water is 11.2% H, then there are (11.2 g H/100 g water).

    (a) glucose contains 53.3% O

    (b) ammonia contains 82.3% N

    (c) table sugar contains 6.48% H

17. How many (see Questions 15 and 16)

    (a) grams of C are in 50.0 g of acetic acid?

    (b) grams of C are in 50.0 g of table sugar?

    (c) grams of O are in 25.0 g of glucose?

    (d) grams of N are in 200 g of ammonia?

Section 3-7

18. The most common elements found in Earth's crust are oxygen, silicon, aluminum, iron, calcium, sodium, potassium, and magnesium. Give the symbol for each of these elements.

19. The elements that make up your body include C, H, O, N, P, S, Ca, K, Cl, Na, Mg, and Fe. Name these elements.

20. Trace elements found in the body include iodine, copper, zinc, manganese, cobalt, chromium, selenium, molybdenum, fluorine, tin, silicon, and vanadium. Write the symbols for these elements.

21. As, Pb, Hg, Cd, Se, and Pu are some highly toxic elements. Give the names of these elements.

22. The following metals are used in jewelry and coins: Au, Ag, Cu, Zn, Al, Sn, Ni, and Pt. Give the names of these elements.

23. The elements nitrogen, phosphorus, potassium, and sulfur are important plant nutrients. Write the symbols for these elements.

24. Silicon, germanium, tellurium, gallium, indium, arsenic, antimony, and terbium are used in the semiconductor industry. Write the symbols for these elements.

25. List the elements that have symbols consisting of one letter only.

26. List the elements that have names ending in -ine.

27. List the elements that have symbols that are not derived from the English name of the element.

Section 3-8

28. Give statements of the following:

    (a) law of conservation of matter

    (b) law of constant composition

29. *A widely traveled scientist collected water samples from many parts of the world. After returning to the lab, the scientist purified the samples, subjected them to elemental analysis, and recorded the data. What laws do these data support? Why?

SOURCE	% H BY MASS	% O BY MASS
Water from the Antarctic ice cap	11.2	88.8
Water from the Stensvad water well in Montana	11.2	88.8
Water from the South China Sea	11.2	88.8
Water freshly synthesized from hydrogen and oxygen	11.2	88.8

30. While the scientist of Question 29 was away, an assistant stayed in the laboratory and combined lithium metal and fluorine gas to make a compound. What laws do the assistant's data support? Why?

MASS Li (g)	MASS F (g)	MASS LITHIUM FLUORIDE COMPOUND (g)
0.234	0.642	0.876
0.489	1.339	1.828
0.360	0.986	1.346

31. It is sometimes said that you can never get rid of wastes. What scientific law is the basis of this statement?

Section 3-9

32. What is a scientific theory or model?

33. Describe or state the atomic theory. Use pictures to illustrate.

34. Why is the atomic theory fundamental to chemistry?

35. Describe an atom.

36. Who is noted for first suggesting the existence of atoms?

37. Look up John Dalton's atomic theory on the Web or in an encyclopedia or a history of science text and describe the theory as Dalton stated it. Describe the historical background of the theory.

Section 3-10

38. Explain the law of conservation of matter in terms of the atomic theory.

39. Explain the law of constant composition in terms of the atomic theory.

Sections 3-11 and 3-12

40. What is a chemical reaction?

41. What characteristics make one element different from other elements?

42. What is electricity?

43. What are electrons?

44. Sketch the flow of electrical current in a metal wire.

Section 3-13

45. List the three basic kinds of subatomic particles. How do they differ?

46. An electron was once described as "a negative twist of nothingness." Describe some properties of an electron.

Section 3-14

47. Describe the nuclear model of the atom.

48. What particles are present in atomic nuclei?

49. Imagine a typical atom enlarged is the size of a beach ball having a radius of 8 in. What radius in inches does the nucleus have in such an atom?

Sections 3-15 to 3-18

50. What is an atom in chemical terms?

51. Define or describe the following:

(a) atomic number

(b) nucleon number

(c) isotope

52. *Describe the following isotopes in terms of the numbers of protons, neutrons, and electrons they contain (atomic numbers are found on the table inside the back cover of this book).

(a) carbon-14   (b) chlorine-37

(c) ^{234}U (d) ^{18}O

(e) phosphorus-32   (f) calcium-45

(g) ^{90}Sr (h) iron-56

(i) ^{125}I (j) ^{137}Ba

53. Describe the following isotopes in terms of the numbers of protons, neutrons, and electrons they con-

tain (atomic numbers are found inside the back cover of this book).

(a) hydrogen-3   (b) potassium-40
(c) ^{235}U   (d) ^{16}O
(e) cesium-137   (f) cobalt-60
(g) I-133   (h) helium-2
(i) ^{239}Pu   (j) ^{226}Ra

54. Hydrogen is the only element for which the isotopes have their own names. Hydrogen-1 is protium, hydrogen-2 is deuterium, and hydrogen-3 is tritium. Tritium is a radioactive isotope of hydrogen that does not occur naturally. Describe the difference between the three isotopes of hydrogen.

55. Three neutral atoms differ only in that one contains 23 electrons, one contains 24 electrons, and one contains 26 electrons. Are these atoms of the same or different elements? Why?

56. What is a mass spectrometer?

57. What is an atomic weight?

58. What information is needed to calculate the atomic weight of an element?

59. What is an atomic mass unit?

60. Why is the atomic mass unit and not the gram used for atomic masses?

61. The ratio of the mass of a fluorine atom to a carbon-12 atom is 1.58320. Calculate the atomic weight of fluorine.

62. Use the following data to calculate the atomic weight of Li.

ISOTOPE	ISOTOPE MASS	% ABUNDANCE
Lithium-6	6.01513 amu	7.42
Lithium-7	7.01601 amu	92.58

63. *Use the following data to calculate the atomic weight of silicon.

ISOTOPE	ISOTOPE MASS	% ABUNDANCE
Silicon-28	27.98	92.21
Silicon-29	28.98	4.70
Silicon-30	29.97	3.09

Section 3-19

64. Give brief descriptions of each of the following elements.

(a) carbon   (b) chlorine   (c) copper
(d) gold   (e) hydrogen   (f) iron
(g) nitrogen   (h) oxygen   (i) silicon
(j) sodium   (k) sulfur

65. What are alloys? Give an example.

66. What is carbon steel? What is stainless steel?

67. What is burning or combustion?

68. Describe a major use of the element silicon.

Questions to Ponder

69. Are atoms real? Explain your answer.

70. How would atoms and matter be different if electrons had positive electrical charge and protons had negative electrical charge?

71. Using your knowledge of the atomic nature of matter and the motion of particles, explain why materials differ in density. Would you expect lead to be more dense than tin? Why?

72. Why does hydrogen rank third in Table 3-3 and first in Table 3-4?

FORMULAS

4-1 CHEMICAL FORMULAS

Table salt or sodium chloride has the appearance of a clear crystalline solid. Elemental sodium is a silvery metal, and chlorine is a greenish yellow gas. Both elements are quite toxic and dangerous, yet they combine chemically to form sodium chloride, the same salt that we need in our diets to sustain life. When a sodium atom changes from an atom of sodium metal to a combined atom in salt, its properties change dramatically. Similarly, chlorine atoms change their properties as they go from the elemental form of diatomic molecules to the combined form in salt. Atoms are the building blocks of nature, and all compounds are composed of chemically combined atoms. The properties of a compound are quite different from the properties of the elements that it contains. In the elemental form, hydrogen and oxygen are both gases. We know that water, the compound formed by the chemical combination of hydrogen and oxygen, is typically a liquid with a specific density, freezing point, boiling point, and familiar appearance. Elemental hydrogen and oxygen are both gases composed of diatomic molecules and have different properties compared with water.

A given compound always contains the same elements present in the same percentages because compounds contain atoms combined in definite proportions. Sodium chloride contains one combined atom of sodium for every one combined atom of chlorine and has the formula NaCl. This formula concisely states that there is a one-to-one atomic proportion in the compound. Any compound can be represented by a **formula** that shows which elements are present using the symbols and the relative number of combined atoms of each element using numerical subscripts. For any subscript that is 1, as in NaCl, the number 1 is not written but implied. The formula of water H_2O shows two atomic parts of hydrogen to one of oxygen. The formulas of elemental oxygen, O_2, and elemental hydrogen, H_2, show that these elements occur as diatomic molecules.

The word *formula* may conjure up images of a secret formula in the possession of a mad scientist. Mad scientists are quite rare, and a formula is

H_2O
water

NaCl
sodium chloride
table salt

$C_{12}H_{22}O_{11}$
sucrose
table sugar

$NaHCO_3$
sodium
bicarbonate

NH_3
ammonia

CO_2
carbon dioxide

not some obscure recipe but instead simply serves as a concise symbolic description of a compound. It tells which elements are present and the relative amounts of each element. Consider the formulas of some common compounds. (See Figure 4-1.) You can read a formula by just pronouncing the letters and any subscript numbers. NaCl reads N-A-C-L, and H_2O reads H-2-O. Common table sugar is the chemical compound sucrose, $C_{12}H_{22}O_{11}$ (read C-12-H-22-O-11). Baking soda is the compound sodium hydrogen carbonate, $NaHCO_3$ (read N-A-H-C-O-3). It is sometimes called bicarbonate of soda or sodium bicarbonate. Clear household ammonia is a water solution of the gaseous compound ammonia, NH_3 (read N-H-3). Seltzer water is a water solution of carbon dioxide gas, CO_2 (read C-O-2).

How do we go about finding the formula of a compound? How is it known that the formula of water is H_2O? In this chapter, you'll learn about experimental methods used by chemists to find formulas of compounds. These methods use the quantitative analysis of a compound in terms of the percent by mass of its elements. For example, we can use the experimentally determined fact that water is 11.2% hydrogen and 88.8% oxygen by mass to find its formula.

FIGURE 4-1   Formulas of some compounds.

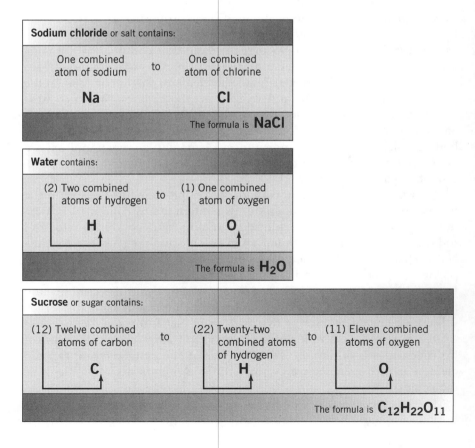

4-2 THE MOLE

A 1-carat diamond weighs 0.200 gram. A diamond is composed of a three-dimensional array of chemically bonded carbon atoms. How many carbon atoms are in a 1-carat diamond? An answer to this question may seem impossible. You cannot count atoms one at a time. A method is available to find the number of atoms in a measured mass of an element, however. Balances are used to measure precise masses of samples. Atoms are very tiny particles. Even a sample of an element weighing less than a gram contains a huge number of atoms. How many? Each atom has mass, and, as we know, the atomic weight serves as a measure of this mass. If we have a sample of an element, its mass must be the sum of the masses of the atoms in the sample. This is the key to counting the number of atoms in a sample of known mass.

To see how counting atoms is possible, we can use an analogy involving nuts and bolts. Suppose you are working in a nut-and-bolt factory that makes only one kind of nut and one kind of bolt. By experiment, you find that the weight of a nut is 12.0 grams and the weight of a bolt is 16.0 grams. These "hardware weights" expressed as factors are

$$\left(\frac{12.0 \text{ g}}{1 \text{ nut}}\right) \quad \text{and} \quad \left(\frac{16.0 \text{ g}}{1 \text{ bolt}}\right)$$

We use such factors to count nuts and bolts by weighing samples instead of counting them one at a time. The important fact is that samples of nuts and bolts with the same numerical values as the relative weights given in the factors have equal numbers of nuts and bolts. For instance, if we weigh 12.0 pounds of nuts and 16.0 pounds of bolts each contains the same number of items.

$$12.0 \text{ lb} \left(\frac{454 \text{ g}}{1 \text{ lb}}\right) \left(\frac{1 \text{ nut}}{12.0 \text{ g}}\right) = 454 \text{ nuts}$$

$$16.0 \text{ lb} \left(\frac{454 \text{ g}}{1 \text{ lb}}\right) \left(\frac{1 \text{ bolt}}{16.0 \text{ g}}\right) = 454 \text{ bolts}$$

If we need a washer for each set of nuts and bolts, we just need to know the weight per washer. For example, if the weight per washer is 1.00 gram (1.00 g/washer), we know that we need 1.00 pound of washers to give 454 washers. For practice, you can do the calculation that proves this conclusion.

Although atoms are very tiny, we can count them using atomic weights in a way that is similar to the nut-and-bolt analogy. We typically express masses of chemicals in units of grams. Imagine that we had a sample of 12.0 grams of carbon and a sample of 16.0 grams of oxygen. Remember that the atomic weights of elements differ, but they do reflect the relative masses of atoms. If we have 12.0 g of carbon and 16.0 g of oxygen, the number of atoms of each element must be the same since the sample masses are numerically the same as their atomic weights. The atomic weights of these elements expressed as factors are

$$\left(\frac{12.0 \text{ u}}{1 \text{ C atom}} \right) \text{ and } \left(\frac{16.0 \text{ u}}{1 \text{ O atom}} \right)$$

Recall that u is the atomic mass unit. Calculating the number of atoms in each sample using the atomic weights as factors can be done in a similar way to the nut-and-bolt calculations shown above.

$$12.0 \text{ g} \left(\frac{X \text{ u}}{1 \text{ g}} \right)\left(\frac{1 \text{ C atom}}{12.0 \text{ u}} \right) = X \text{ C atoms}$$

$$16.0 \text{ g} \left(\frac{X \text{ u}}{1 \text{ g}} \right)\left(\frac{1 \text{ O atom}}{16.0 \text{ u}} \right) = X \text{ O atoms}$$

The factor $(X \text{ u}/1 \text{ g})$ is used here because we do not know the actual numerical value of this factor. We use X to represent the number of atomic mass units per gram. The point is that the same value of X is the result in both calculations, which means that 16.0 g of oxygen has the same number of atoms as found in 12.0 g of carbon. Each element has a unique atomic weight. Thus, as a general principle we can say that a sample of any element having a mass in grams the same as its atomic weight contains the same number of atoms as the above samples: X atoms.

We commonly use special names for counting units. For instance, a dozen is the name for 12 and gross is the name for 144. In chemistry, the atomic counting unit is the mole. The term *mole* comes from the Latin word for heap or pile. A mole contains a specific number of atoms. Counting atoms by weighing samples of elements is based upon the definition of the mole as given in the margin.

The isotope carbon-12 is in the definition because it is also the reference used in the atomic weight scale. Twelve grams of carbon-12 is a mole of carbon atoms. The atomic weight of oxygen, to three digits, is 16.0. So we know that 16.0 g of oxygen contains as many atoms as 12 g of carbon-12. This number of atoms is a mole of oxygen. From the definition of the mole, the mass in grams of an element equal to the numerical value of its atomic weight contains a mole of atoms. Furthermore, a mole of any element contains the same specific number of atoms called X in the above discussion. Soon we'll learn the actual numerical value of X.

If you look up the word *mole* in the dictionary, you'll find that it has four meanings. It is interesting that these words are all related. A mole is a stone structure used as breakwater, a pile of stones. A mole is an elevated colored spot on the skin, a pile of skin. A mole is a small, burrowing animal that makes piles of dirt. Finally, in chemistry, a mole is a special kind of pile of atoms or chemical particles, such as molecules. The word *molecule* comes from a small part of a mass. In other words, a molecule is a small but characteristic part of a mole of a compound.

Mole of an Element:

The amount of an element that contains as many atoms as there are carbon atoms in exactly 12 grams of the isotope carbon-12.

4-3 AVOGADRO'S NUMBER

How many atoms are in a mole of an element? In other words, what is the value of X referred to above? The definition of the mole does not reveal this number. It is, however, determined using special experimental data. A precise value comes from X-ray diffraction studies of samples of ultrapure silicon. See Section 3-18 for a discussion of X-ray diffraction. Ultrapure silicon is material that has very low levels of impurities so that samples are assumed to contain only silicon atoms. Using X-ray diffraction, counting atoms in a sample of carefully measured mass is possible. This value is used, in a proportion, to calculate the number of atoms that are in a sample of silicon having a mass equal to its atomic weight. This experimentally determined number, as we might expect, is very large. The number of atoms in a mole of an element is 6.0221367×10^{23}. This number is called **Avogadro's number** in honor of the Italian scientist, Amedeo Avogadro, who first suggested, in 1811, that counting atoms might be possible. In normal form, Avogadro's number is 602,213,670,000,000,000,000,000. This very large number reflects that atoms are very small. For convenience in calculations, Avogadro's number rounded to four digits is expressed as a factor.

$$\left(\frac{6.022 \times 10^{23} \text{ atoms}}{1 \text{ mol atoms}} \right)$$

$$\left(\frac{6.022 \times 10^{23} \text{ atoms}}{1 \text{ mol atoms}} \right)$$

To get an idea of how very large Avogadro's number is, imagine measuring the oceans of the world with a teaspoon. After measuring 6.022×10^{22} teaspoons you would be half done. In other words, there are about twice Avogardro's number of teaspoons of water in the oceans of the world. Don't try this experiment at home because, even if you measured one teaspoon each second, it would take you about 10^{16} years. Because atoms are very small particles, there are tremendous numbers of them in typical samples of matter. There are more atoms in a grain of sand than the estimated number of stars in the universe.

Referring to the discussion in Section 4-2, Avogadro's number corresponds to X, the number of atomic mass units per gram.

$$12.0 \text{ g} \left(\frac{X \text{ u}}{1 \text{ g}} \right) \left(\frac{1 \text{ carbon-12 atom}}{12.0 \text{ u}} \right) = X \text{ carbon-12 atoms}$$
$$= 6.022 \times 10^{23} \text{ carbon-12 atoms}$$

Twelve grams of carbon-12 contain Avogadro's number of atoms. A mole of any element is the mass that contains Avogadro's number of atoms. This mass for a specific element is the atomic weight of the element in grams. In other words, an amount of an element corresponding to its atomic weight in grams contains 6.022×10^{23} atoms.

4-4 MOLAR MASS

The amount of an element in grams that contains as many atoms as there are in 12.0 g of carbon-12 is numerically equal to the atomic weight of the element. For any element, this mass in grams is the mass of a mole of atoms. Thus, for any element, we can express the **number of grams per mole** or the **molar mass** using the value of its atomic weight. Molar amounts of elements range from a few grams to hundreds of grams. The molar mass of an element is useful as a conversion factor. Four, or some-times three, significant digits are usually sufficient for molar masses used in calculations. Some examples of molar masses as factors are given below. Note that the abbreviation for mole is mol and that these factors are ex-pressed to four significant digits.

$$\left(\frac{16.00 \text{ g}}{1 \text{ mol O}}\right) \quad \left(\frac{1.008 \text{ g}}{1 \text{ mol H}}\right) \quad \left(\frac{55.85 \text{ g}}{1 \text{ mol Fe}}\right) \quad \left(\frac{238.0 \text{ g}}{1 \text{ mol U}}\right)$$

To write the molar mass as a factor, just look up the appropriate atomic weight and round it to the desired number of digits. A molar mass, as a fac-tor, has a numerator that is the atomic weight expressed in grams and a denominator of 1 mole. Molar masses are very similar to densities as fac-tors. Both relate two properties. The density of a substance expresses the number of grams per milliliter and relates mass and volume. The molar mass expresses the number of grams per mole and relates mass to moles. The term *molar mass* is used because it refers specifically to the mass of a mole of an element.

4-5 THE MOLAR MASS AS A CONVERSION FACTOR

The amazing thing about the mole concept is that it allows easy counting of atoms. The molar mass of an element is used as a conversion factor to accomplish this feat. For instance, if we have 24 g of carbon, we know that there are 2 moles of carbon atoms. If we have 6.0 g, we know that there is $\frac{1}{2}$ mole of carbon atoms. The number of atoms of an element is usually ex-pressed in terms of moles instead of the absolute number of atoms, like the use of a dozen for the number 12 or a gross for the number 144. We commonly use the term *dozen* when referring to tortillas or eggs rather than saying 12. In a sense, the mole is the chemists "dozen" since it is a name for the specific numerical value of 6.022×10^{23}.

Molar masses as factors are used in the same way unit prices or densi-ties are used as conversion factors. To find the number of pounds of ap-ples you can buy for a given amount of money, multiply the amount of money by the inverse of the price expressed as a factor:

$$X \text{ lb} = Y\cancel{\text{¢}}\left(\frac{1 \text{ lb}}{\#\cancel{\text{¢}}}\right)$$

To find the number of milliliters of a given mass of a substance, multiply by the inverse of the density:

$$X \text{ mL} = Y\cancel{g}\left(\frac{1 \text{ mL}}{\#\cancel{g}}\right)$$

A molar mass is used as a factor to find the number of moles of an element in any sample of known mass or the mass of any given number of moles. To find the number of moles in a given mass of a substance, multiply by the inverse of the molar mass:

$$X \text{ mol} = Y\cancel{g}\left(\frac{1 \text{ mol}}{\#\cancel{g}}\right)$$

To find the cost of a given number of pounds of apples, multiply by cost expressed as a factor:

$$Y\text{¢} = X\cancel{lb}\left(\frac{\#\text{¢}}{1\cancel{lb}}\right)$$

To find the mass of a given number of milliliters of a substance, multiply by the density:

$$Y \text{ g} = X\cancel{mL}\left(\frac{\#\text{g}}{1\cancel{mL}}\right)$$

To find the mass of a given number of moles of a substance, multiply by the molar mass:

$$Y\text{g} = X\cancel{mol}\left(\frac{\#\text{g}}{1\cancel{mol}}\right)$$

Suppose we have a chunk of iron and want to know how many iron atoms are in the sample. First, we weigh the sample to find its mass in grams. Then, we look up the atomic weight of iron so that we have a value for the molar mass. The molar mass is used as a factor to find the number of moles of iron. As an example, let's find the number of moles of iron atoms in an 85.3 g sample of iron. Using its atomic weight, the molar mass of iron is

$$\left(\frac{55.85 \text{ g}}{1 \text{ mol Fe}}\right)$$

Usually, we use at least one extra digit in the molar mass than is needed in the answer to avoid potential calculation error resulting from rounding too soon. The molar mass of iron is inverted or "turned over" to give the factor used to find the number of moles from the grams.

$$X\text{ mol Fe} = 85.3\cancel{g}\left(\frac{1 \text{ mol Fe}}{55.85\cancel{g}}\right) = 1.53 \text{ mol Fe}$$

Note how the gram units cancel and the moles remain. Keep in mind that if we have a mass in grams and we want the number of moles, we have to multiply by the factor having moles in the numerator and grams in the denominator. In terms of numbers, this amount of iron is 1.53 times Avogadro's number of iron atoms. In chemistry, we typically use moles instead of the actual number of atoms.

EXAMPLE 4-1

How many moles of copper atoms are in an old copper penny which contains 3.2 g of copper? (Pennies minted after 1982 are made of zinc with copper plate.) We look up the atomic weight of copper and express the molar mass to three digits as (63.5 g/mol Cu). Using this as a conversion factor gives

$$X \text{ mol Cu} = 3.2 \, \cancel{g} \left(\frac{1 \text{ mol Cu}}{63.5 \, \cancel{g}} \right) = 0.050 \text{ mol Cu} \quad \text{or} \quad 5.0 \times 10^{-2} \text{ mol Cu}$$

Note that it is good practice to include the units in the factor to ensure that the gram units cancel and we get moles in the answer. Remember, always include a unit with the numerical part of an answer.

The molar mass is also used to find the mass of an element corresponding to a specific number of moles. That is, if we know the number of moles in a sample of an element and want to know the corresponding mass, we multiply by the molar mass to convert moles to grams. Remember to use at least one extra digit in the molar mass than is needed in the final answer.

ACTIVITY 4-1

A 1-carat diamond has a mass of 0.200 g. How many moles of carbon are in such a diamond? What is the mass of 1.00×10^3 moles of carbon?

EXAMPLE 4-2

What is the mass of a 0.525 mole sample of sulfur?

First, look up the molar mass of sulfur. Multiplying the number of moles by the molar mass gives the number of grams.

$$X \text{ g} = 0.525 \, \cancel{\text{mol S}} \left(\frac{32.06 \text{ g}}{1 \, \cancel{\text{mol S}}} \right) = 16.8 \text{ g}$$

4-6 NUMBER OF ATOMS IN A SAMPLE

The molar mass of an element serves as a factor used to calculate the number of moles of atoms in any sample of an element. We know that the number of atoms in a mole is Avogadro's number as represented by the factor

$$\left(\frac{6.022 \times 10^{23} \text{ atoms}}{1 \text{ mol}} \right)$$

This factor is used to find the actual number of atoms in a sample of an element. It can also be used to find the number of moles corresponding to a given number of atoms. As a factor, it is directly used to find the actual number of atoms in a sample if the number of moles is known.

EXAMPLE 4-3

Graphite, one form of carbon, is a powdery black solid used to make pencil lead. A speck of graphite powder just visible to the unaided eye has a mass of about 1×10^{-6} gram. How many carbon atoms are there in such a speck of graphite?

First, we find the number of moles of carbon in the sample using the molar mass of carbon. The mass is multiplied by the inverted molar mass so that the grams cancel to give moles.

$$1 \times 10^{-6} \, \cancel{g} \left(\frac{1 \, \text{mol C}}{12.01 \, \cancel{g}} \right)$$

Next, we multiply this product by Avogadro's number to give the actual number of carbon atoms.

$$X \text{ C atoms} = 1 \times 10^{-6} \, \cancel{g} \left(\frac{1 \, \cancel{\text{mol C}}}{12.01 \, \cancel{g}} \right) \left(\frac{6.022 \times 10^{23} \, \text{C atoms}}{1 \, \cancel{\text{mol C}}} \right)$$

$$= 5 \times 10^{16} \text{ C atoms}$$

It is interesting that even a minute speck of an element contains millions upon billions of atoms. The product of 5×10^{16} is 50 million billion.

ACTIVITY 4-2

A troy ounce of gold has a mass of 31.1 g. How many gold atoms are there in such a sample?

The absolute number of atoms in a sample is found using the method given in the example. As we will see, however, expressing the number of atoms as the number of moles rather than the actual number of atoms is convenient and useful. We seldom deal with actual numbers of atoms but instead use the mole as a counting unit for atoms.

4-7 EMPIRICAL FORMULAS

Formulas are important in chemistry because they provide a concise way to describe the chemical composition of a compound. Formulas represent the symbolic language of chemistry. We use a formula to refer to a compound. How do we know the formula of a compound? The formula is not apparent from the appearance and properties of the compound. We can't look at a compound under a microscope and see little formulas. To find the formula, a compound is first analyzed chemically to discover which elements it contains. Following this analysis, the mass or percent of each element is determined. This information is found by decomposing a carefully weighed sample of a compound into its constituent elements and then weighing them or by synthesizing the compound using weighed amounts of each element. As we know, the mass of an element is used to find the number of moles of that element. The mole concept is the key to formula determination.

To illustrate how to find a formula, imagine that we do not know the "formula" of a bicycle. A shipment of 1.00 ton of parts contains 0.70 ton of frames and 0.30 ton of wheels. Thus, the mass composition is 30% wheels and 70% frames. We find that one wheel weighs 3.0 pounds and one frame weighs 14 pounds. The formula of a bicycle is found by first changing each percent to pounds. Next, the part weights are used to find the number of parts in the number of pounds. Finally, the ratio of the number of wheels to frames reveals the subscript in the "formula." First, 30% wheels corresponds to 30 lb. Multiplying by the reciprocal of the wheel weight gives the number of wheels.

$$30 \text{ lb} \left(\frac{1 \text{ wheel}}{3.0 \text{ lb}} \right) = 10 \text{ wheels}$$

Next, 70% frames corresponds to 70 lb. Multiplying by the reciprocal of the frame weight gives the number of frames.

$$70 \text{ lb} \left(\frac{1 \text{ frame}}{14 \text{ lb}} \right) = 5 \text{ frames}$$

The formula comes from the ratio of wheels to frames.

$$\left(\frac{10 \text{ wheels}}{5 \text{ frames}} \right) = \left(\frac{2 \text{ wheels}}{1 \text{ frame}} \right) \quad \text{or} \quad \text{wheel}_2\text{frame}$$

The formula of a chemical compound gives the relative number of combined atoms of each element. Since moles of elements correspond to exact numbers of atoms, we can say that the same relative ratios exist for moles as exist for atoms. From an atomic view, water, for instance, has two parts hydrogen to one part oxygen. The subscripts in a formula reveal the ratio of combined atoms in the compound. Alternatively, the subscripts reveal the ratio of combined moles of each element in the compound. Thus, a formula is interpreted as an expression of the molar relation between the constituent elements of the compound. The formula of water, H_2O, shows two hydrogen atoms for one oxygen atom. Another interpretation is that water has two moles of combined hydrogen for every one mole of combined oxygen. In other words, if we know the formula for water, then we know the **molar ratio** is

$$\left(\frac{2 \text{ mol H}}{1 \text{ mol O}} \right)$$

If the percent-by-mass composition or the mass of each element in a sample of a compound is known, calculating the number of moles of each element in a given mass of the compound is possible. Dividing the number of moles of one element by the number of moles of another element gives the molar ratio. Molar ratios show the subscripts used in the formula of the compound. A formula found in this way is called an **empirical formula.** The term *empirical* means derived only from experimental data.

4-8 EMPIRICAL FORMULA CALCULATIONS

The elemental composition of a compound is needed to find the empirical formula of the compound. The elemental composition is measured by experiment. Composition is expressed as the percent of each element or the mass of each element in a sample of the compound An example calculation illustrates the reasoning involved in the determination of a formula. The mineral galena contains a compound of lead and sulfur. By experiment, it is found that this lead-sulfur compound contains 86.6% lead and 13.4% sulfur. To find the formula, first change the percents to masses: 86.6% is 86.6 g, and 13.4% is 13.4 g. Next, find the number of moles of each element in these masses using the appropriate molar masses as conversion factors. You need to consult a table of atomic weights for the molar masses. To change from grams to moles, use the inverted molar mass as a factor. In this way, the grams cancel and give units of moles.

$$\text{lead} \quad 86.6 \text{ g} \left(\frac{1 \text{ mol Pb}}{207.2 \text{ g}} \right) = 0.418 \text{ mol Pb}$$

$$\text{sulfur} \quad 13.4 \text{ g} \left(\frac{1 \text{ mol S}}{32.06 \text{ g}} \right) = 0.418 \text{ mol S}$$

Dividing the number of moles of lead by the number of moles of sulfur gives the molar ratio.

$$\left(\frac{0.418 \text{ mol Pb}}{0.418 \text{ mol S}} \right) = \left(\frac{1.00 \text{ mol Pb}}{1 \text{ mol S}} \right)$$

The molar ratio is simply 1:1. Thus, the compound contains one combined lead atom for one combined sulfur atom. What is true for moles of atoms is true for individual atoms. Consequently, the empirical formula of the compound is PbS.

EXAMPLE 4-4

What is the empirical formula of a compound that contains 82.2% by mass nitrogen and 17.8% by mass hydrogen?

The pattern includes changing the percents to masses, finding the moles of each element and then finding the molar ratios. Since we want to deal with masses, we can assume that a 100 g sample of the compound contains 82.2 g of combined nitrogen and 17.8 g of combined hydrogen. We could use any mass, but 100 g is the most convenient. Find the number of moles of each element from its mass using the molar mass from a table of atomic weights. The number of moles of the elements is then used to find the molar ratio. The calculation sequence of percent to mass, mass to mole, and molar ratios to formula is summarized as:

	NITROGEN, N	HYDROGEN, H
Percent	82.2%	17.8%
Mass	82.2 g	17.8 g
Molar mass	14.01 g/mol N	1.008 g/mol H

Convert each mass to number of moles using the molar masses.

	NITROGEN, N	HYDROGEN, H
Moles	5.867 mol N	17.66 mol H

Divide by smallest moles to give the molar ratio.

	NITROGEN, N	HYDROGEN, H
Molar ratio	1	3.01
Formula	NH_3 Use the molar ratio for the formula.	

In the above example, the molar ratio reveals that the formula for the compound is NH_3. The molar ratio means that there are three moles of combined hydrogen per mole of combined nitrogen or three combined hydrogen atoms for each nitrogen atom. When finding molar ratios, always divide by the smallest number of moles so that ratios are never less than 1. Notice that the molar ratio calculated above had a value of 3.01 and we assumed that it was 3. Ignoring such extra digits is reasonable since they come from experimental error in the data or from rounding errors in the calculations. Usually, subscripts in formulas are small whole numbers, so neglect any extraneous digits.

Typically, the molar ratios in empirical formula calculations are simple whole numbers. Occasionally, some calculated molar ratios are fractions rather than whole numbers and must be converted to the appropriate whole number. Let's consider an example of this situation. What is the formula for a compound that has the following percent-by-mass composition: 26.5% combined potassium, K; 35.4% combined chromium, Cr; and 38.1% combined oxygen, O?

Express the number of grams of each element present in 100 g of the compound by simply using the value of each percent. Next, the number of moles of each element is determined using their molar masses as factors. Then, the molar ratios are found from the numbers of moles.

"Percent to mass
Mass to mole
Divide by small
Multiply 'til whole."
Anonymous

	POTASSIUM, K	CHROMIUM, Cr	OXYGEN, O
Percent	26.5%	35.4%	38.1%
Mass	26.5 g	35.4 g	38.1 g
Molar mass	39.10 g/mol K	52.00 g/mol Cr	16.00 g/mol O
Moles	0.678 mol K	0.681 mol Cr	2.38 mol O
Divide by 0.678 for the molar ratios.	1	1	3.51

We write the formula from the molar ratios, so the formula is $KCrO_{3.5}$. Note that 3.5 oxygens does not make sense from an atomic point of view. Having 3.5 oxygen atoms is not possible. The value 3.5 is obtained because the calculations give the simplest ratio of the elements. By convention, only whole numbers are used in formulas, so any fractional ratios are changed to whole numbers. Here, to obtain whole-number subscripts, double each subscript (multiply each by 2). Thus, the best empirical formula is $K_2Cr_2O_7$.

The reason fractions sometimes occur is that the method of calculation gives the simplest or most reduced molar ratios. These reduced ratios might be decimal fractions, such as 1.33, 1.25, 1.5, or 1.67. These numbers expressed as fractions are $\frac{4}{3}$, $\frac{5}{4}$, $\frac{3}{2}$, or $\frac{5}{3}$. To obtain whole-number subscripts, all subscripts are multiplied by the appropriate numerical value to clear the fractions (e.g., 2, 3, 4, . . .).

EXAMPLE 4-5

A sample of a compound of iron and oxygen is found to contain 145 g of iron, Fe, and 55.3 g of oxygen, O. What is the empirical formula of the compound?

Here, use the experimental data as mass composition and convert directly to number of moles using the molar masses as factors. That is, use the given masses instead of percents.

	IRON, Fe	OXYGEN, O
Mass	145 g	55.3 g
Molar mass	55.85 g/mol Fe	16.00 g/mol O
Moles	2.60 mol Fe	3.46 mol O
Divide by 2.60 for molar ratios.	1	1.33

Thus, the initial formula is $FeO_{1.33}$. To make whole-number subscripts, dealing in fractions is best. The subscript 1.33 is $1\frac{1}{3}$ or $\frac{4}{3}$. Thus, using a fraction, the formula could be written as $FeO_{4/3}$. Multiplying each subscript by 3 gives whole-number subscripts and a formula of Fe_3O_4.

ACTIVITY 4-3

(a) What is the empirical formula of a compound that contains 69.94% Fe and 30.06% O? (b) What is the empirical formula of cholesterol? It contains 83.87% C, 11.99% H, and 4.14% O.

The empirical formula determined by the methods given here is always the simplest formula, and in some cases it is not the most useful formula of a compound. The empirical formula always reflects the simplest whole-number ratio of elements. The formula of a compound depends on the way in which the atoms combine to form particles of the compound. Consequently, the real formula may be some multiple of the simplest formula. The empirical formula, however, is very useful and is often the appropriate formula for a compound. See Section 4-10 for further discussion of compound formulas.

4-9 FORMULA UNITS AND MOLES OF COMPOUNDS

What does the formula of a compound represent other than the elemental composition? We can interpret the formula in terms of the number of atoms or the number of moles of each element. That is, the formula for sodium chloride, NaCl, shows that this compound contains one atom of combined sodium to one atom of combined chlorine. The formula for sucrose, $C_{12}H_{22}O_{11}$, shows 12 atoms of combined carbon to 22 atoms of combined hydrogen to 11 atoms of combined oxygen. Combined atoms make up compounds, but what about particles or units of compounds? How atoms combine in compounds is the subject of Chapter 9. For now, we can view a compound as a collection of units corresponding to the formula.

A **formula unit** is the amount of a compound that contains the number of atoms of each element given in the formula. As an analogy, imagine you had a pile of nut-and-bolt combinations having two nuts on each bolt. A "formula unit" is nut_2bolt. In other words, one of these combinations is representative of the entire collection. A formula unit of a compound is the smallest representative unit or portion of the compound. For compounds that are made of molecules, the formula unit is the formula of the molecule. Just as working with individual atoms in the laboratory is not possible, working with a single formula unit is not possible. Samples of a compound measured in grams contain huge numbers of formula units. The mole relates the mass of a sample of a compound to the number of formula units. It also relates the mass of a sample of a compound to the formula. See the alternative definitions for the mole given in the margin.

A compound contains a specific number of combined moles of each element. Thus, a mole of a compound is viewed as the amount that contains the number of moles of each element given by the subscripts in the formula. Each compound has a unique number of grams per mole. A mole of sodium chloride, NaCl, is the amount that contains 1 mole of combined sodium and 1 mole of combined chlorine. A mole of sucrose, $C_{12}H_{22}O_{11}$, is the amount that contains 12 moles of combined carbon, 22 moles of combined hydrogen, and 11 moles of combined oxygen.

Mole of a Compound:
The amount of a compound that contains Avogadro's number of formula units.

Mole of a Compound:
The mass in grams of a compound that contains the number of moles of each combined element as given in the formula.

Let us return to the nut-and-bolt analogy mentioned in Section 4-2. For a compound made of two nuts and one bolt, we could express the mass of the compound as a summation of the masses of the components that it contains. Since one nut is 12.0 grams and one bolt is 16.0 grams, then one nut$_2$bolt unit is 40.0 grams. Furthermore, we know that a sample of nut$_2$bolt combinations weighing 40.0 pounds contains 454 nut$_2$bolt units. See Section 4-2. In a similar way, we can refer to the molar mass of a compound. The **molar mass of a compound** expresses the number of grams per mole of the compound. Sometimes, chemists call molar masses of compounds molecular weights or formula weights. The term *molar mass* is used in this book. The molar mass of a compound is found from the formula of the compound. Each element in the compound contributes to the molar mass of the compound according to its subscript in the formula. To calculate the molar mass of a compound given the formula:

1. Multiply the molar mass of each element by its subscript in the formula.
2. Add the contributions of each element and express the result as the number of grams per mole of the compound.

EXAMPLE 4-6

Determine the number of grams per mole, or molar mass, of water, H_2O (to four digits).

To find the molar mass of the compound, multiply the molar mass of each element by its subscript and add the products. Note that since we want a four-digit answer, we use four or five digits (or more) in the various molar masses and then round after the calculation.

$$
\begin{array}{lll}
\text{hydrogen} & 2(1.008) = & 2.016 \\
\text{oxygen} & 1(16.00) = & \underline{16.00} \\
& & 18.016 \quad \text{or} \quad 18.02
\end{array}
$$

The molar mass is expressed as the number of grams per mole of the compound. When writing the molar mass as a factor, always show the formula of the compound following the term *mol*.

$$
\text{Molar mass of water} = \left(\frac{18.02 \text{ g}}{1 \text{ mol } H_2O} \right)
$$

Note that when calculating a molar mass, it is not necessary to include the units on each factor. We can write the units after we add up the numbers and round. The units are grams in the numerator and 1 mole in the denominator. Remember to include the formula of the compound in the denominator.

A C T I V I T Y
4-4

Calculate the molar mass of sodium hydrogen carbonate (baking soda), $NaHCO_3$ (to four digits).

EXAMPLE 4-7

Determine the molar mass of sucrose, $C_{12}H_{22}O_{11}$ (to four digits).

$$
\begin{array}{lll}
C & 12(12.01) = & 144.12 \\
H & 22(1.008) = & 22.176 \\
O & 11(16.00) = & \underline{176.0} \\
& & 342.296
\end{array}
$$

or $\left(\dfrac{342.3 \text{ g}}{1 \text{ mol } C_{12}H_{22}O_{11}} \right)$

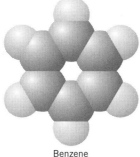

Benzene

U.S. Occupational Health and Safety Administration

4-10 FORMULAS FOR COMPOUNDS

The empirical formula of a compound expresses the simplest whole-number ratio of the atoms of the elements in the compound. As mentioned in Section 4-8, the empirical formula of a compound is not always the best representative formula. The real or representative formula of a compound depends on the way in which the atoms combine to make molecules or ions in the compound. The formula may be the empirical formula or some simple whole-number multiple of the empirical formula. Benzene is a carbon-hydrogen compound used industrially as a solvent and to make certain plastics and detergents. It is interesting that as a compound benzene is carcinogenic but that most more complex compounds that have molecules derived from benzene are not carcinogenic. Chemical analysis of benzene reveals that it has the empirical formula of CH. Based on experimental measurements, benzene is known to have molecules containing chemical combinations of six atoms of carbon and six atoms of hydrogen. The molecular formula, C_6H_6, reflects this composition. The simplest ratio of carbon to hydrogen is 1 to 1 as the empirical formula shows, but the molecular formula is six times the empirical formula.

Carcinogenic Chemicals and Material Safety Data Sheets (MSDS)

Carcinogens are chemicals that cause abnormal growth of living cells and produce tumors and cancers. Malignant cancers undergo uncontrolled growth and may invade and destroy neighboring tissues. Not all cancers are caused by contact with chemicals, but it is estimated that about 85% of cancers result from environmental sources. The other cancers relate to genetic factors and some possibly to viruses. Cigarette smoking accounts for about 40% of the environmental causes, dietary sources for about 28%, occupational sources for about 10%, and the remainder most likely from natural and synthetic environmental chemicals.

Obviously, testing the carcinogenic properties of chemicals in humans using direct tests is not possible. Testing for carcinogenic properties is done by bacterial screening in which a strain of bacteria is dosed with a chemical and any genetic mutation is observed. This test is called the Ames test since it was developed by Bruce Ames, an American biologist. The test in not always accurate and often additional tests are needed. Animal tests (normally using laboratory-bred rats) are used for additional screening of suspected carcinogens. Again, such tests are not always reliable, but there is a reasonably good correlation between animal tests and carcinogenic behavior in humans.

The use of epidemiological data is one of the best methods to identify carcinogenic chemicals. Population groups with higher than normal incidences of certain cancers are studied to try to identify a causative factor. These studies often give results in which there is a significant correlation between the incidence of cancer and the presence of a specific chemical or group of chemicals. Extensive epidemiological studies carried out over several years established that cigarette smoking is a direct cause of lung cancer. The only precaution in epidemiological studies is that some factor or chemical that was not considered might be the carcinogenic agent.

More than 300 chemicals are known carcinogens from animal studies, and only about 30 of those chemicals are classified as known human carcinogens. Some carcinogens are naturally occurring chemicals, some are synthetic chemicals. Normally, the control of amounts of naturally occurring chemicals is not easy, but the use of undesirable synthetic chemicals can be monitored and controlled. Occupational exposure to carcinogenic chemicals is a problem for some workers. The U.S. Occupational Safety and Health Administration (OSHA) requires that industries, businesses, and institutions inform workers about the dangers and risks of chemicals that they use. Material data safety sheet (MSDS) or toxic substance data sheets must be available for each of these chemicals. Part of a toxic substance data sheet for benzene is shown in Figure 4-2.

Common Name: Benzene
CAS Number: 71-43-2
DOT Number: UN 1114
Date: January 1988

HAZARD SUMMARY
  Benzene can affect you when breathed in and by passing through your skin.
  Benzene is a CARCINOGEN. HANDLE WITH EXTREME CAUTION.
  Exposure can cause you to become dizzy and light-headed. Higher levels can cause convulsions and death.
  Exposure can irritate the nose and throat and may cause an upset stomach and vomiting.
  Benzene can cause an irregular heartbeat that can lead to death.
  Prolonged exposure can cause fatal damage to the blood (aplastic anemia).
  Benzene is a FLAMMABLE LIQUID and a FIRE HAZARD.

IDENTIFICATION
Benzene is a colorless liquid with a pleasant odor. It is used mainly in making other chemicals, is used as a solvent, and is found in gasoline.

REASON FOR CITATION
  Benzene is on the Hazardous Substance List because it is regulated by OSHA and cited by ACGIH, DOT, NIOSH, IARC, NTP, CAG, DEP, NFPA, and EPA.
  It is on the Special Health Hazard Substance List because it is a CARCINOGEN, is a MUTAGEN, and is FLAMMABLE.

FIGURE 4-2   Part of a toxic substance data sheet for benzene.

You know how to calculate the empirical formula from experimental data. Now, let us see how the real formula is found from the empirical formula. To do this, we need an experimental determination of the molar mass of the compound. That is, we must have a measure of the molar mass of the compound that is found by experiment rather than by knowing the formula beforehand. The idea is that if we have a molar mass, we can use it to find the actual formula. Several experimental methods are available by which the molar mass of a compound is found without first knowing its formula. The most accurate method is mass spectroscopy, discussed in Chapter 3. One identifying characteristic revealed by the mass spectroscopic analysis of a compound is its molar mass.

Suppose we do not know the molecular formula of benzene but only the empirical formula of CH. Mass spectroscopic analysis reveals the molar mass of benzene is 78 g/mol benzene. Notice that the empirical formula is CH, with 1 mole of C and 1 mole of H, giving a molar mass for the empirical formula of 13 g/mol CH. To find the molecular formula of the compound, divide the molar mass of the compound by the molar mass of the empirical formula. This division shows how many empirical formulas there are in the molecular formula of the compound.

$$\left(\frac{78 \, \cancel{g}/\text{mol benzene}}{13 \, \cancel{g}/\text{mol CH}}\right) = \left(\frac{6 \, \text{mol CH}}{1 \, \text{mol benzene}}\right)$$

The result suggests that the formula of the compound contains six CH. Thus, the molecular formula is six times the empirical formula. The representative formula of benzene is C_6H_6, not 6CH. Notice that, as expected, the molar mass of benzene calculated from its formula is

$$78 \, \text{g/mol} \, (6 \times 12 + 6 \times 1 = 78).$$

The formula of a compound is either the same as the empirical formula or some simple whole-number multiple of the empirical formula. Whenever you see a chemical formula, you should appreciate that someone had to find the formula using experimental data to find both the empirical formula and the representative or molecular formula.

EXAMPLE 4-8

Octane, a compound found in gasoline, has the empirical formula of C_4H_9. Experimental measurements give a molar mass value of about 120 g/mol octane. What is the molecular formula?

By dividing the molar mass of the compound by the molar mass corresponding to the empirical formula, C_4H_9, we can deduce the formula of the compound. The first step is to find the molar mass corresponding to the formula C_4H_9.

C	4(12.01) =	48.04
H	9(1.008) =	9.072
		57.112 or 57.11

Then, this molar mass is divided into the molar mass of the compound.

$$\left(\frac{120\text{ g/mol octane}}{57.11\text{ g/mol } C_4H_9}\right) = \left(\frac{2.10\text{ mol } C_4H_9}{1\text{ mol octane}}\right)$$

We know that a formula is a simple, whole-number multiple of the empirical formula, so we round the answer to the nearest whole number. The extra fraction in the ratio is likely a result of experimental error. We conclude that the molecular formula is double the empirical formula or C_8H_{18}.

Glucose, a common sugar, has the empirical formula of CH_2O. What is the molecular formula if the experimental molar mass is 180 g/mole glucose?

4-11 WRITING FORMULAS

A common question asked about the formula of a compound is, Which element symbol is written first in the formula? Look at a periodic table of the elements. By convention, the elements in the formula of a compound are typically listed from bottom to top in a vertical column of the periodic table (e.g., SO_2 rather than O_2S) and from left to right in a horizontal row of the table (e.g., SCl_2 rather than Cl_2S). One exception to this pattern is that hydrogen follows the vertical column starting with nitrogen and comes before the vertical column starting with oxygen (e.g., NH_3 and H_2O). Another exception is that oxygen follows all elements except fluorine (e.g., Cl_2O and OF_2).

Sometimes formulas of compounds are written in alternative ways. For example, the formula of acetic acid is written as $HC_2H_3O_2$, CH_3COOH, or CH_3CO_2H. Note that each formula has 2 C, 4 H, and 2 O. Another example is ethyl alcohol, which is written as C_2H_6O, C_2H_5OH, or CH_3CH_2OH. Sometimes formulas are written using parentheses to reflect multiples of specific groups of atoms. Two examples are ammonium sulfate, written as $(NH_4)_2SO_4$, and calcium acetate, written as $Ca(C_2H_3O_2)_2$. To count the number of atoms of a given element within parentheses, multiply its subscript by the subscript outside the parentheses. For instance, calcium acetate contains 1 Ca, 4 C, 6 H, and 4 O. Parentheses are used to show multiples of certain atomic sequences in some compounds. Further examples of the use of parentheses in formulas are given in Chapter 6.

4-12 MASS AND MOLE RELATIONS FROM FORMULAS

Suppose we have 398 grams of Fe_2O_3, a compound found in iron ore, and we want to calculate the number of grams of iron in this sample. Questions of this kind are answered by using molar interpretations of formulas and the idea that moles and mass are easily related.

A formula reveals the number of moles of each element per mole of the compound. For instance, Fe_2O_3 has 2 moles of Fe per mole of Fe_2O_3

and 3 moles of oxygen per mole of Fe_2O_3. These relations are expressed as molar ratios.

$$\left(\frac{2 \text{ mol Fe}}{1 \text{ mol Fe}_2O_3} \right) \text{ and } \left(\frac{3 \text{ mol O}}{1 \text{ mol Fe}_2O_3} \right)$$

Molar ratios are written simply by noting the subscripts for the various elements in a compound. The three common ways in which molar ratios of this type are useful are described below.

1. Mole to mole: A molar ratio is used to find the number of moles of an element in a given number of moles of a compound. For example, how many moles of iron, Fe, are in 2.49 moles of Fe_2O_3? To find the number of moles of Fe, just multiply the number of moles of compound by the molar ratio relating Fe and Fe_2O_3.

$$2.49 \text{ mol Fe}_2O_3 \left(\frac{2 \text{ mol Fe}}{1 \text{ mol Fe}_2O_3} \right) = 4.98 \text{ mol Fe}$$

EXAMPLE 4-9

How many moles of combined hydrogen are in 5.73 moles of methane, CH_4?

First, note that the molar relation between hydrogen and methane is

$$\left(\frac{4 \text{ mol H}}{1 \text{ mol CH}_4} \right)$$

The number of moles of methane is multiplied by this ratio to give the number of moles of hydrogen.

$$5.73 \text{ mol CH}_4 \left(\frac{4 \text{ mol H}}{1 \text{ mol CH}_4} \right) = 22.9 \text{ mol H}$$

---

2. Mass to mole: Suppose we want to find the number of moles of an element in a given mass of a compound. First, the mass of the compound is changed to the number of moles using its molar mass. Then, the number of moles of a specific element is found from the number of moles of compound using the molar ratios. For example, how many moles of iron are in a 398 g sample of Fe_2O_3? To begin, we need the molar mass of the compound as found from its formula. The molar mass is $[2(55.85) + 3(16.00) = 159.70]$.

$$\left(\frac{159.7 \text{ g}}{1 \text{ mol Fe}_2O_3} \right) \quad \text{or in the inverted form} \quad \left(\frac{1 \text{ mol Fe}_2O_3}{159.7 \text{ g}} \right)$$

We want to change mass to the number of moles, so the given mass of the compound is multiplied by the inverted form of the molar mass.

$$398 \, g \left(\frac{1 \text{ mol Fe}_2\text{O}_3}{159.7 \, g} \right)$$

Since we are not finished, calculating an answer at this point is not necessary. If you want, however, you can calculate the number of moles of Fe_2O_3 and use the answer in the next step. Remember to carry more digits than you need, and when you get the final answer, round to the correct number of significant digits. To find the number of moles of Fe from the number of moles of Fe_2O_3, multiply the above product by the appropriate molar ratio as revealed by the formula.

$$398 \, g \left(\frac{1 \text{ mol Fe}_2\text{O}_3}{159.7 \, g} \right) \left(\frac{2 \text{ mol Fe}}{1 \text{ mol Fe}_2\text{O}_3} \right) = 4.98 \text{ mol Fe}$$

EXAMPLE 4-10

Argentite, a common silver ore, contains the compound Ag_2S. How many moles of silver are in a 25.4 g sample of this compound?

Given: 25.4 g of Ag_2S Goal: ? moles of Ag   Need: molar mass of Ag_2S

The molar mass of Ag_2S is [2(107.9) + 32.06 = 247.86] or 247.9 g/1 mol Ag_2S. The given mass of the compound is multiplied by the inverted form of the molar mass and the appropriate molar ratio to give the moles of Ag.

$$25.4 \, g \left(\frac{1 \text{ mol Ag}_2\text{S}}{247.9 \, g} \right) \left(\frac{2 \text{ mol Ag}}{1 \text{ mol Ag}_2\text{S}} \right) = 0.205 \text{ mol Ag}$$

---

3. Mass to mass: The two methods illustrated above serve as a prelude to a common question related to a formula. The question is, How many grams of an element are in a given number of grams of a compound? Let us return to the question asked at the beginning of this section. How many grams of iron are in 398 g of Fe_2O_3? To find the answer, first change the number of grams of the compound to the number moles of compound. The molar ratio is used to find the number of moles of Fe; then, the molar mass of iron is used to change the moles of Fe to its mass. Note that the first part is just the mass-to-mole problem illustrated above.

$$398 \, g \left(\frac{1 \text{ mol Fe}_2\text{O}_3}{159.7 \, g} \right) \left(\frac{2 \text{ mol Fe}}{1 \text{ mol Fe}_2\text{O}_3} \right)$$

This calculation gives the number of moles of Fe. Again, we do not have to calculate at this point. This setup is multiplied by the molar mass of iron that we get from its atomic weight.

$$398 \text{ g} \left(\frac{1 \text{ mol Fe}_2\text{O}_3}{159.7 \text{ g}} \right) \left(\frac{2 \text{ mol Fe}}{1 \text{ mol Fe}_2\text{O}_3} \right) \left(\frac{55.85 \text{ g}}{1 \text{ mol Fe}} \right) = 278 \text{ g}$$

A C T I V I T Y
4-6

A compound of iron and oxygen has the formula Fe_3O_4. (a) How many moles of iron are in 15 moles of Fe_3O_4? (b) How many moles of iron are in a 875 g sample of Fe_3O_4? (c) How many grams of iron are in a 875 g sample of Fe_3O_4?

EXAMPLE 4-11

How many grams of silver are in 25.4 g of Ag_2S?

Given: 25.4 g of Ag_2S Goal: ? g of Ag   Need: molar masses of Ag_2S and Ag

The calculation sequence is

$$\text{g Ag}_2\text{S} \xrightarrow{\text{molar mass}} \text{mol Ag}_2\text{S} \xrightarrow{\text{molar ratio}} \text{mol Ag} \xrightarrow{\text{molar mass}} \text{g Ag}$$

$$25.4 \text{ g} \left(\frac{1 \text{ mol Ag}_2\text{S}}{247.9 \text{ g}} \right) \left(\frac{2 \text{ mol Ag}}{1 \text{ mol Ag}_2\text{S}} \right) \left(\frac{107.9 \text{ g}}{1 \text{ mol Ag}} \right) = 22.1 \text{ g}$$

4-13 PERCENT-BY-MASS COMPOSITION

It is sometimes convenient to express the composition of a compound as the percent of one or more elements in the compound. For example, to four significant digits Fe_2O_3 contains 69.94% iron by mass. We have seen that finding the formula of a compound by using the experimentally determined percent-by-mass composition is possible. Now let us consider how to calculate the percent of an element in a compound if we start with the formula of the compound.

The mass and the number of moles of a compound are related by the molar mass. Thus, the formula of the compound, which gives the number of moles of each constituent element, is used to find the percent by mass of each element. It makes sense that if formulas are originally determined from the percent by mass of each element, then finding each percent from a formula is possible.

The compound Fe_2O_3 contains 2 moles of combined iron per mole of the compound, which, expressed as a molar ratio, is

$$\left(\frac{2 \text{ mol Fe}}{1 \text{ mol Fe}_2\text{O}_3} \right)$$

The percent of iron is found by converting the 2 moles of iron to mass and 1 mole of Fe_2O_3 to mass. The molar masses of each are used for this purpose. To begin, we need to calculate the molar mass of Fe_2O_3. Look up the molar mass of iron, multiply it by 2, and then add this to three times the molar mass of oxygen. The resulting value is (159.7 g/1 mol Fe_2O_3). To calculate the percent of iron, first multiply the molar ratio (given above) by the molar mass of iron, which converts the 2 moles of iron to grams.

$$\left(\frac{2 \text{ mol Fe}}{1 \text{ mol Fe}_2O_3} \right) \left(\frac{55.85 \text{ g}}{1 \text{ mol Fe}} \right)$$

Next, convert the 1 mole of Fe_2O_3 to mass using its molar mass in the inverted form (note the units of the factor).

$$\left(\frac{2 \text{ mol Fe}}{1 \text{ mol Fe}_2O_3} \right) \left(\frac{55.85 \text{ g}}{1 \text{ mol Fe}} \right) \left(\frac{1 \text{ mol Fe}_2O_3}{159.7 \text{ g}} \right)$$

Canceling the units and multiplying by 100 gives the percent of iron in the compound. The first part of the calculation expresses the amount of the element as a fraction of a mole of the compound. To express as parts per hundred, it is necessary to multiply by 100. Canceling the units gives

$$\frac{2 \times 55.85}{159.7} \times 100 = 69.94\% \text{ Fe}$$

Note the general pattern revealed by this example. To find the percent by mass of an element, multiply the molar mass of the element by its subscript in the formula. Then, divide by the molar mass of the compound. Finally, multiply by 100 to express the percent.

$$\text{percent element} = \frac{s \text{ (molar mass of element)}}{\text{molar mass of compound}} \times 100$$

In this expression, s is the subscript of the element in the formula. Units are not included in a percent, so you don't have to include units in the calculation.

EXAMPLE 4-12

What is the percent-by-mass composition of water, H_2O (to four digits)?

We want to find the percent of each element in the compound. First, find the molar mass of water to five digits: $2(1.0079) + 15.9994 = 18.015$

Use at least one more digit than the desired number of digits in the percent. Multiply the molar mass of hydrogen by its subscript in the formula and divide by the molar mass of the compound. Finally, multiply by 100 to give the percent of hydrogen.

$$\% \text{ H} = \frac{2(1.0079)}{18.015} \times 100 = 11.19\%$$

Since there are only two elements in the compound, the percent by mass of oxygen is found by subtracting the percent by mass of hydrogen from 100.00.

$$\% \text{ O} = 100.00\% - 11.19\% = 88.81\%$$

The formula of urea, a compound used as fertilizer, is N_2H_4CO. What is the percent of nitrogen in urea to four digits?

(4-14) THE USES OF FORMULAS

A formula of a compound provides a wealth of useful information. The following items summarize the usefulness of a formula.

1. A formula tells which elements are in a compound, and the subscripts show the relative number of atoms of each element in the compound.
2. A formula shows the relative number of each element contained in a representative unit or formula unit of the compound.
3. A mole of a compound is that amount that contains Avogadro's number of formula units.
4. The molar mass or number of grams per mole of a compound is found by adding the products of the molar mass of each element and its corresponding subscript in the formula.
5. The molar mass of a compound is used as a factor to change a given mass of a compound to the number of moles of the compound or vice versa.
6. The subscripts in a formula are used to express molar ratios relating the number of moles of an element per mole of the compound or the number of moles of an element relative to another element.
7. Molar ratios and molar masses are used to find the number of moles of an element or the mass of an element in a known number of moles or a known mass of a compound.
8. The percent by mass of an element in a compound is found by multiplying the molar mass of the element by its subscript in the formula, dividing this number by the molar mass of the compound, and multiplying the result by 100.

FOCUS
4

Moles and Formulas

Mole of an element: The amount of an element that contains as many atoms as there are carbon atoms in exactly 12 g of carbon-12.

Avogadro's number: The number atoms in a mole of an element or the number of formula units in a mole of a compound: 6.022×10^{23}.

Mole of a compound: The amount of a compound that contains Avogadro's number of formula units.

Molar mass of an element: The number of grams per mole of an element as given by the numerical value of its atomic weight.

Molar mass of a compound: The number of grams per mole as found by multiplying the molar masses of elements in the compound by their subscripts in the formula and summing the products. Molar mass is used to find the number of moles in a given mass or the mass of a given number of moles.

Given the mass of an element or a compound, find the number of moles by multiplying the mass by the inverse of the molar mass

$$X \, g \left(\frac{1 \text{ mol}}{\# \, g} \right) = ? \text{ mol}$$

To find the empirical formula of a compound:

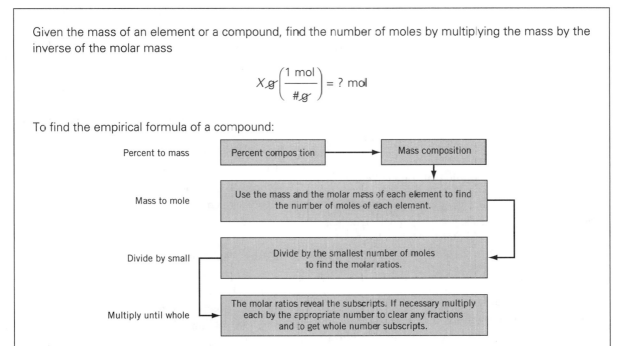

To find the molecular formula of a compound, first divide the molar mass of the compound by the molar mass corresponding to the empirical formula for the compound. Then, multiply the subscripts in the empirical formula by the number obtained in the division.

The molar ratio relating the number of moles of an element in a mole of a compound is used to find:

1. The number of moles of an element in a given number of moles of a compound.

    (moles of compound)(molar ratio) = moles of element

2. The number of moles of an element in a given mass of a compound.

    (grams compound)(inverse of molar mass of compound)(molar ratio) = moles of element

3. The mass of an element contained in a given mass of a compound.

    (grams compound)(inverse of molar mass of compound)(molar ratio) (molar mass element) = grams of element

QUESTIONS

1. Match the following terms with the appropriate lettered definitions.

formula

a. a formula that reflects the number of atoms in a molecule

mole

b. a formula found from the percent of each element in a compound

Avogadro's number

c. the number of grams per mole of an element or a compound

molar mass   d. 6.022×10^{23}

empirical formula   e. symbolic representation
                                       of a compound showing
                                       its elements

molecular or real   f. the amount of a
formula   chemical that contains
                                       as many chemical
                                       particles
                                       as there are atoms in 12
                                       grams of carbon-12

You will need to use a periodic table or a table of atomic weights for some of the following questions. In some of the questions, the names of compounds are given for reference. You are not expected to name compounds at this time. Chemical nomenclature is discussed in Chapter 6.

Section 4-1

2. What is a chemical formula? Give an example.

3. Describe the information that each of the following formulas contains in terms of atomic proportions of elements.

   (a) ZnO (zinc oxide)

   (b) $HC_2H_3O_2$ (acetic acid)

   (c) $C_9H_8O_4$ (aspirin)

   (d) SO_2 (sulfur dioxide)

   (e) $MgCO_3$ (magnesium carbonate)

   (f) CCl_2F_2 (Freon, a chlorofluorocarbon or CFC)

Sections 4-2 to 4-6

4. Define or explain these terms.

   (a) mole of an element

   (b) molar mass

5. What is Avogadro's number, and what units are used with it?

6. Explain why the molar mass of an element is numerically equal to the atomic weight of that element.

7. *Calculate the number of moles of atoms contained in each sample given below.

   (a) a 24.8 g silver bracelet

   (b) a 53.5 kg iron magnet

   (c) 6.71 g of powdered sulfur

   (d) a 0.0699 g tungsten filament

   (e) 235 metric tons of recycled aluminum (1 metric ton = 1000 kg)

   (f) a 4.81 g sample of helium gas

   (g) a 131 mg silicon chip

   (h) a 454 g spool of magnesium ribbon

   (i) a 2.76 kg copper kettle

   (j) an 875 g lead pipe

   (k) a 0.30 g sample of beryllium

8. Calculate the number of moles of atoms contained in each sample given below.

   (a) a 0.976 g sample of neon gas

   (b) a 14.7 g sample of mercury

   (c) a 0.354 g zinc wire

   (d) a 2.86 g chunk of carbon

   (e) a 32.01 g piece of gold

   (f) a 50.0 kg lead ball

   (g) an 8.92 g sample of calcium

   (h) a 237 g sample of liquid bromine

   (i) a 10.0 g sample of boron

   (j) a 69.3 g sample of nickel

   (k) a 1.00 mL sample of mercury

9. *Calculate the mass of the following samples of elements.

   (a) 124 mol Pt

   (b) 3.18×10^{-6} mol Mo

   (c) 6.04×10^{-7} mol Si

   (d) 4.24×10^{5} mol Na

   (e) 11.9 mol Ar

   (f) 7.58×10^{-10} mol Ga

10. Calculate the mass of the following samples of elements.

    (a) 84 mol K

    (b) 7.60×10^{-3} mol U

    (c) 0.0395 mol Li

    (d) 2.25×10^{6} mol Ba

    (e) 4.69 mol B

    (f) 5.78×10^{-6} mol Y

11. *Calculate the number of atoms in each sample given below.

    (a) an 84.8 g sample of Hg in a barometer

    (b) a 0.249 g Ti metal artificial heart valve

    (c) a 5.25 g sample of Kr gas

    (d) a 3709 g Ni meteorite

    (e) an 8.23 mg sample of Ag

    (f) 1.00 troy ounce of Au (1 troy ounce = 31.1 g)

12. Calculate the number of atoms in each sample given below.

(a) 3.18 g of Cu in a copper penny

(b) a 0.171 g W filament

(c) a 12.5 g sample of He

(d) a 453.6 g chunk of iron

(e) a 0.200 g or 1-carat diamond (diamond is pure carbon)

(f) a 1.31 g aluminum nail

13. *Use the molar mass of hydrogen and Avogadro's number to calculate the mass in grams of a typical H atom in units of grams per H atom. Use this result to calculate the mass in grams of a typical O atom. (Remember that atomic weights express the relative masses of atoms.)

Sections 4-7 and 4-8

14. What is an empirical formula, and what experimental data are needed to calculate the empirical formula of a compound?

15. *Calculate the empirical formulas for the following compounds for which the percent by mass or mass composition is given.

(a) lysine: 19.1% N, 9.6% H, 22.0% O, 49.3% C

(b) potassium tartrate found in baking powder: 20.8% K, 2.7% H, 25.5% C, 51.0% O

(c) calcium fluoride found in the mineral fluorspar: 119.4 g Ca, 114.7 g F

(d) hydroquinone used in photography: 72.06 g C, 6.05 g H, 32.0 g O

(e) borax: 12.1% Na, 11.3% B, 71.3% O, 5.29% H

(f) saccharin, an artificial sweetener: 45.9% C, 2.76% H, 26.2% O, 7.65% N, 17.5% S

(g) barium phosphate: 68.4% Ba, 10.3% P, 21.3% O

(h) vitamin C: 54.5% O, 40.9% C, 4.6% H

16. Calculate the empirical formulas for the following compounds for which the percent by mass or mass composition is given.

(a) butane: 82.7% C, 17.3% H

(b) sodium nitrate: 27.0% Na, 16.5% N, 56.5% O

(c) adenosine triphosphate (ATP): 71.04 g C, 9.54 g H, 41.43 g N, 54.96 g P, 123 g O

(d) citric acid: 56.3 g C, 6.3 g H, 87.4 g O

(e) urea: 20.0% C, 46.8% N, 6.7% H, 26.5% O

(f) potassium chromate: 110.8 g K, 73.7 g Cr, 90.75 g O

(g) ethyl alcohol or ethanol: 52.2% C, 13.1% H, 34.7% O

(h) propane: 4.08 g C, 0.915 g H

17. *Determine the empirical formula of aspartame (Nutrasweet), an artificial sweetener made from amino acids, if a 250.0 g sample of the compound contains 142.85 g C, 15.40 g H, and 23.80 g N, and the rest is O.

18. Cisplatin is a drug used in chemotherapy for some cancer patients. A 112.5 g sample contains 73.2 g Pt, 26.6 g Cl, and 2.27 g H, and the rest is N. What is the empirical formula of cisplatin? If the molar mass cisplatin is 300 g/mol, what is the molecular formula of cisplatin?

19. *A 32.38 g sample of aluminum reacts with oxygen to form 61.18 g of an aluminum–oxygen compound. Calculate the empirical formula of the compound.

20. Nitroglycerine is used as an explosive and as a heart medicine. Determine the empirical formula if a sample is found to contain 19.88 g C, 2.75 g H, 23.13 g N, and 79.25 g O.

Section 4-9

21. How is a mole of a compound defined?

22. What is a formula unit?

23. If you have 1 mole of water, what does this mean in terms of Avogadro's number?

24. If you have a sample of 2.00 moles of NaCl, what does this mean in terms of Avogadro's number?

25. *Determine the molar mass of the following compounds to four significant digits.

(a) potassium oxalate, $K_2C_2O_4$

(b) pentane, C_5H_{12}

(c) methyl bromide, CH_3Br (used as a soil fumigant, but may be banned by the EPA)

(d) butane, C_4H_{10} (used as a fuel)

(e) calcium fluoride, CaF_2

(f) acetylcholine (a neurotransmitter), $C_7H_{16}NO_2$

(g) trinitrotoluene, $C_7H_5N_3O_6$ (TNT, an explosive)

(h) magnesium carbonate, $MgCO_3$

26. Determine the molar mass of the following compounds to four significant digits.

(a) isopropyl alcohol, C_3H_8O (rubbing alcohol)

(b) barium sulfate, $BaSO_4$ (used as X-ray absorber during intestinal tract X rays)

(c) sodium nitrate, $NaNO_3$ (found in the mineral Chile saltpeter)

(d) potassium dichromate, $K_2Cr_2O_7$

(e) titanium tetrachloride (used in smoke screens), $TiCl_4$

(f) sodium cyanide, NaCN (a deadly poison)

(g) sodium bisulfate, $NaHSO_4$ (used in some toilet bowl cleaners)

(h) ethane, C_2H_6

(i) sodium hydrogen carbonate or sodium bicarbonate, $NaHCO_3$

27. *Determine the number of moles of formula units in the following samples of compounds.

(a) 819 g NH_3

(b) 12.3 g $K_2Cr_2O_7$

(c) 0.1086 g NaCN

(d) 28.3 g $C_{12}H_{22}O_{11}$

(e) 2.19×10^6 g NaCl

(f) 6.37 g acetylene, C_2H_2

(g) 56.8 g phenylalanine, $C_9H_{11}NO_2$, an amino acid

28. Determine the number of moles of formula units in the following samples of compounds.

(a) 311 g $NaNO_2$ (sodium nitrite)

(b) 75.0 g of muskone, $C_{16}H_{30}O$ (a pheromone that acts as a sex attractant)

(c) 2.88×10^5 g of SO_3

(d) 2.24×10^{-7} g Au_2S_3

(e) 9.10 g H_2O_2 (hydrogen peroxide)

(f) 250 g $C_6H_{12}O_6$ (glucose)

(g) 675 g $NaHCO_3$

29. *Determine the mass of each of the following samples of compounds. (When you have completed Question 25, you will have the molar masses needed for this question.)

(a) 3.7 mol $K_2C_2O_4$

(b) 262 mol C_5H_{12}

(c) 5.87 mol CH_3Br

(d) 0.946 mol C_4H_{10}

(e) 7.46×10^{-3} mol CaF_2

(f) 0.215 mol acetylcholine, $C_7H_{16}NO_2$

(g) 4.35×10^5 mol trinitrotoluene, $C_7H_5N_3O_6$

(h) 6.52 mol $MgCO_3$

30. Determine the mass of each of the following samples of compounds. (When you have completed Question 26, you will have the molar masses needed for this question.)

(a) 2.13×10^{-7} mol C_3H_8O

(b) 4.72 mol $BaSO_4$

(c) 3.41×10^{-2} mol $NaNO_3$

(d) 58 mol $K_2Cr_2O_7$

(e) 6.9×10^5 mol $TiCl_4$

(f) 1.22 mol NaCN

(g) 8.47×10^4 mol $NaHSO_4$

(h) 7.29×10^{-6} mol ethane, C_2H_6

(i) 9.05 mol $NaHCO_3$

31. *Epsom salt can be purchased in 5.00 lb boxes. How many moles of $MgSO_4$ are in such a box if Epsom salt is 48.8% $Mg SO_4$?

32. How many moles of iron are there in 2.00 pounds of iron nails? How many iron atoms are there in 2.00 pounds of nails?

33. *A box of baking soda contains 16 oz of sodium hydrogen carbonate, $NaHCO_3$. How many moles of $NaHCO_3$ are there in the box?

34. A pain reliever tablet contains 5.0 grains of phenacetin, $C_{10}H_{13}N_2O$. How many moles of phenacetin are there in a tablet? (0.0648 g = 1 grain)

35. *A tablespoon of sugar has a volume of 15.0 mL. If the density of sugar is 1.59 g/cm^3, how many moles of sucrose, $C_{12}H_{22}O_{11}$, are in the sugar sample?

36. Ethanol, C_2H_6O, has a density of 0.79 g/mL. How many moles of ethanol are in a 4.00 oz sample? (29.57 mL = 1 fluid oz)

Section 4-10

37. *Calculate the empirical formulas of the compounds given below using the mass composition. Then, use the molar mass given to calculate the representative or molecular formula of each compound.

(a) tryptophan, an amino acid: 32.35 g C, 2.96 g H, 6.86 g N, 7.84 g O; molar mass 204 g/mol

(b) Freon 12, a chlorofluorocarbon (CFC): 5.41 g C, 31.90 g Cl, 17.10 g F; molar mass 121 g/mol

(c) vanillin, a flavoring agent: 50.56 g C, 4.21 g H, 25.28 g O; molar mass 152 g/mol

(d) calomel: 12.75 g Hg, 2.25 g Cl; molar mass 472 g/mol

(e) ribose sugar: 14.8 g C, 2.49 g H, 19.8 g O; molar mass 150 g/mol

(f) octane, found in gasoline: 84.1% C, 15.9% H; molar mass 114.2 g/mol

38. Calculate the empirical formulas of the compounds given below using the mass composition. Then, use the

molar mass given to calculate the representative or molecular formula of each compound.

(a) cholesterol: 36.03 g C, 5.05 g H, 1.78 g O; molar mass 386 g/mol

(b) hydrogen peroxide: 5.93% H, 94.1% O; molar mass 34 g/mol

(c) sodium phosphate, trisodium phosphate TSP, used to clean walls: 28.14 g Na, 12.66 g P, 26.13 g O; molar mass 164 g/mol

(d) aluminum carbonate (antacid): 13.49 g Al, 9.01 g C, 36.00 g O; molar mass 234 g/mol

(e) phenolphthalein, an acid–base indicator: 75.46% C, 4.43% H, 20.10% O; molar mass 318 g/mol

(f) gold (III) sulfide: 40.20 g Au, 9.81 g S; molar mass 490 g/mol

Sections 4-11 and 4-12

39. *Give the molar ratio that expresses the number of moles of the specified element per mole of compound; for example: hydrogen in water, H_2O. The molar ratio is

$$\left(\frac{2 \text{ mol H}}{1 \text{ mol } H_2O} \right)$$

(a) molar ratio of hydrogen in methane, CH_4
(b) molar ratio of oxygen in sucrose, $C_{12}H_{22}O_{11}$
(c) molar ratio of carbon in aspirin, $C_9H_8O_4$
(d) molar ratio of hydrogen in propane, C_3H_8
(e) molar ratio of calcium in calcium carbonate, $CaCO_3$
(f) molar ratio of sodium in sodium sulfate, Na_2SO_4

40. Give the molar ratio that expresses the number of moles of the specified element per mole of compound; for example: hydrogen in water, H_2O. The molar ratio is

$$\left(\frac{2 \text{ mol H}}{1 \text{ mol } H_2O} \right)$$

(a) molar ratio of carbon in methane, CH_4
(b) molar ratio of hydrogen in sucrose, $C_{12}H_{22}O_{11}$
(c) molar ratio of oxygen in aspirin, $C_9H_8O_4$
(d) molar ratio of carbon in propane, C_3H_8
(e) molar ratio of oxygen in calcium carbonate, $CaCO_3$
(f) molar ratio of sulfur in sodium sulfate, Na_2SO_4

41. *Use the formulas and amounts of the following compounds to find the number of moles or grams of the element indicated.

(a) How many moles of hydrogen are in 2.00 moles of methane, CH_4?

(b) How many moles of hydrogen are in 64.0 g of methane, CH_4?

(c) How many grams of hydrogen are in 25.4 g of methane, CH_4?

(d) How many moles of oxygen are in 278 g of sucrose, $C_{12}H_{22}O_{11}$?

(e) How many grams of oxygen are in 454 g of sucrose, $C_{12}H_{22}O_{11}$?

(f) How many moles of calcium are in 0.725 g of calcium carbonate, $CaCO_3$?

(g) How many grams of calcium are in 48 g of calcium carbonate, $CaCO_3$?

(h) How many moles of hydrogen are in 0.256 g of glucose, $C_6H_{12}O_6$?

(i) How many grams of hydrogen are in 125 g of glucose, $C_6H_{12}O_6$?

(j) How many moles of sodium are in 12.0 moles of sodium sulfate, Na_2SO_4?

(k) How many moles of sodium are in 5.00×10^2 g of sodium sulfate, Na_2SO_4?

(l) How many grams of sodium are in 5.00×10^2 g of sodium sulfate, Na_2SO_4?

42. Use the formulas and amounts of the following compounds to find the number of moles or grams of the element indicated.

(a) How many moles of carbon are in 2.00 moles of methane, CH_4?

(b) How many moles of carbon are in 64.0 g of methane, CH_4?

(c) How many grams of carbon are in 25.4 g of methane, CH_4?

(d) How many moles of hydrogen are in 278 g of sucrose, $C_{12}H_{22}O_{11}$?

(e) How many grams of hydrogen are in 454 g of sucrose, $C_{12}H_{22}O_{11}$?

(f) How many moles of oxygen are in 0.725 g of calcium carbonate, $CaCO_3$?

(g) How many grams of oxygen are in 48 g of calcium carbonate, $CaCO_3$?

(h) How many moles of carbon are in 0.256 g of glucose, $C_6H_{12}O_6$?

(i) How many grams of carbon are in 125 g of glucose, $C_6H_{12}O_6$?

(j) How many moles of sulfur are in 12.0 moles of sodium sulfate, Na_2SO_4?

(k) How many moles of sulfur are in 5.00×10^2 g of sodium sulfate, Na_2SO_4?

(l) How many grams of sulfur are in 5.00×10^2 g of sodium sulfate, Na_2SO_4?

Section 4-13

43. *You are offered a choice between equal masses of two compounds of gold, Au_2S_3 and Au_2S. What is the percent by mass of gold in each compound? Which compound would you choose?

44. Two nitrogen-containing fertilizers are urea, N_2H_4CO, and ammonium nitrate, $N_2H_4O_3$. Which of these compounds has the greater percent by mass of nitrogen?

45. *Calculate the percent by mass of the following.

(a) nitrogen in saccharin, $C_7H_5SNO_3$

(b) iodine in thyroxine a thyroid hormone, $C_{15}H_{11}I_4NO_4$

(c) sulfur in sulfur trioxide, SO_3

(d) carbon in the amino acid arginine, $C_6H_{16}N_4$

(e) sodium in baking soda, $NaHCO_3$

46. Use your answers to Question 45 to calculate the number of grams of the following.

(a) nitrogen in 0.186 g of saccharin

(b) iodine in 425 g of thyroxine

(c) sulfur in 250 g of sulfur trioxide

(d) carbon in 75 mg of the amino acid arginine

(e) sodium in 500 mg of baking soda

47. Calculate the percent by mass of the following.

(a) oxygen in peroxyacetyl nitrate, found in smog, $C_2H_3NO_5$

(b) silver in argentite a silver ore, Ag_2S

(c) magnesium in chlorophyll, $C_{55}H_{72}MgN_4O_5$

(d) nitrogen in nicotine, $C_{10}H_{14}N_2$

(e) chromium in chromium (III) oxide, Cr_2O_3

48. *Determine the percent-by-mass composition of the following compounds (to four significant digits).

(a) nitrogen oxide, NO

(b) urea, N_2H_4CO

(c) potassium chlorate, $KClO_3$

(d) propane, C_3H_8

(e) magnesium carbonate, $MgCO_3$

(f) amphetamine, $C_9H_{13}N$

49. Determine the percent-by-mass composition of the following compounds (to four significant digits).

(a) sulfur dioxide, SO_2

(b) methyl chloride, CH_3Cl

(c) sodium acetate, $NaC_2H_3O_2$

(d) lead (II) oxide, PbO

(e) copper (I) nitrate, $CuNO_3$

(f) benzene, C_6H_6

50. *Which of the following compounds contains the highest percent by mass of fluorine?

(a) LiF (b) BaF_2 (c) CaF_2 (d) NaF

51. Which of the following compounds contains the lowest percent by mass of cobalt?

(a) CoO (b) Co_2O_3 (c) Co_3O_4

Questions to Ponder

52. If you had Avogadro's number of pennies and divided them among all the 6 billion people on Earth, how many dollars would each person have?

53. Suppose you had a square piece of aluminum foil having a mass of 2.7 g. How many times does the foil need to be cut in half so that one atom of aluminum remains? (Hint: First find the number of aluminum atoms in the sample.)

CHAPTER

5

CHEMICAL

REACTIONS AND

EQUATIONS

5-1 THE PERIODIC TABLE

Very often, curiously shaped periodic tables of the elements adorn the walls of chemistry laboratories and lecture rooms. In the 1800s, the atomic theory captured the imaginations of chemists. As shown in Figure 3-5, many new elements were discovered and added to the list of previously known elements. During this time, chemists successfully determined the relative atomic weights of the elements. As each new element was discovered, it was investigated and its properties noted. Information about elements began to accumulate. During these times, scientists became aware that the properties of some elements were very similar. Eventually, they noticed that some periodic or repeating pattern of properties existed among the elements. Periodic means repeating again and again at regular intervals. A periodical is a magazine published weekly or monthly. A menstrual period typically follows a 28-day cycle, and the moon follows a 27-day cycle.

Periodic behavior of the elements refers to a pattern in which elements of higher atomic number have chemical properties like those of elements of lower atomic number. Elements with similar properties undergo similar chemical combinations and form compounds having corresponding formulas. Elements that have similar properties and form compounds with similar formulas are classified as families or groups of elements. Examples of compounds with similar formulas formed by some families of elements with oxygen are given below. Compounds between various elements and oxygen are collectively called oxides. Atomic numbers are shown for reference, and some of the formulas are empirical formulas. In some families, elements of compounds with similar formulas have similar properties. In some cases, however, the properties of the compounds are quite different.

For example, CO_2 is a colorless gas, whereas SiO_2 is the empirical formula of a complex solid containing silicon and oxygen.

3	4	5	6	7	8	9
Li_2O	BeO	B_2O_3	CO_2	N_2O_3	O_3	OF_2
11	12	13	14	15	16	17
Na_2O	MgO	Al_2O_3	SiO_2	P_2O_3	SO_2	Cl_2O
19	20	31	32	33	34	35
K_2O	CaO	Ga_2O_3	GeO_2	As_2O_3	SeO_2	Br_2O
37	38	49	50	51	52	53
Rb_2O	SrO	In_2O_3	SnO_2	Sb_2O_3	TeO_2	I_2O
55	56	81	82	83	84	
Cs_2O	BaO	Tl_2O_3	PbO_2	Bi_2O_3	PoO_2	

Observations of the similarities and differences in the behavior of elements stimulated the curiosity of many chemists. Was there a grand pattern to such similarities? If such a pattern existed, what message did it convey about the nature of matter?

When the elements are listed according to increasing atomic number, the eleventh element, sodium, is similar to the third element, lithium; the twelfth element, magnesium, is similar to the fourth element, beryllium; the thirteenth element, aluminum, is similar to the fifth element, boron; the fourteenth element, silicon, is similar to the sixth element, carbon; and so on. The pattern is repeated with the nineteenth element, potassium, having properties similar to the eleventh element, sodium, and so on. The pattern is shown by arranging the elements in columns according to the periodic repetition in properties. If an element has no similarities with any of the preceding elements, it is given a new column. The result is an arrangement called the **periodic table of elements** (see the inside front cover for a typical periodic table). Elements in the same vertical **column** in the table have similar properties and belong to the same group or family of elements. We refer to vertical columns of elements as groups.

The periodic table, much as we know it today, was first proposed in 1869 by a Russian chemist, Dmitri Ivanovich Mendeleev. In an attempt to organize the known elements into a useful arrangement, Mendeleev used information about elements gathered by numerous scientists. His table was based on studies of the similarities in the properties of the elements and the compounds that they form. He based his table of his idea that "if all the elements are arranged in order of their atomic weights a periodic repetition of properties is obtained." Today, however, it is recognized that the periodic behavior is related to increasing atomic numbers rather than atomic weights. Mendeleev's table represented a revolutionary step in the development of chemical science. When he created the periodic table,

Periodic Tables
Mendeleev

Mendeleev ingeniously left open boxes in families that seemed to have missing members. In time, chemists discovered elements that fit into the open positions. In addition, chemists also discovered elements having higher atomic numbers than any of those known to Mendeleev. These new elements fit nicely into the evolving periodic table. Today, the periodic table is a fundamental icon of chemistry. The reason the periodic table has such an interesting and curious shape is discussed in Chapter 8.

Elements are listed in the periodic table from left to right by increasing atomic numbers. Each element has its own box in the table. A box usually contains the symbol of the element with the atomic number above the symbol and atomic weight below the symbol. You can use the table as a source of useful information. For instance, you know it as a convenient source of atomic weights.

Vertical columns of elements or groups are distinguished in the table by group numbers. Two numbering systems are used. The groups are simply numbered from 1 to 18 from left to right. Alternatively, some groups are designated with roman numerals. One set is numbered from IA to VIIA and another set is referenced as B groups. The block of elements including the B groups are called the transition elements or transition metals.

5-2 METALS, NONMETALS, AND METALLOIDS

Elements are members of specific groups or families in the periodic table. Elements are also separated into three important general classes or categories. Some elements have metallic properties. As **metals** they are:

good conductors of electricity

good conductors of heat

malleable, easily formed into a variety of shapes

ductile, easily drawn into wires

shiny, having metallic luster

Any element that has these properties is a metal. Figure 5-1 shows that over three-fourths of the elements are metals.

Most metals have names that end with *-ium*. Some obvious exceptions are iron, cobalt, nickel, copper, silver, gold, zinc, mercury, tin, and lead. Aluminum is a good example of a metal. It conducts electricity and is used in a variety of electrical applications, such as long-distance power transmission lines. The use of aluminum in cookware shows that this metal is a good conductor of heat. Aluminum is cast or formed into a variety of shapes and drawn into wires. It is formed into beams and tubes used in construction, and it is easily pressed into sheets and foil. The shiny, silvery appearance of aluminum is a metallic characteristic.

1	2	3	4	5	6	7	8	9	10	11	12	13	14	15	16	17	18 NOBLE GASES
H / H₂																	He
Li metal	Be metal											B	C	N / N₂	O / O₂	F / F₂	Ne
Na metal	Mg metal											Al metal	Si	P	S	Cl / Cl₂	Ar
K metal	Ca metal	Sc metal	Ti metal	V metal	Cr metal	Mn metal	Fe metal	Co metal	Ni metal	Cu metal	Zn metal	Ga metal	Ge	As	Se	Br / Br₂	Kr
Rb metal	Sr metal	Y metal	Zr metal	Nb metal	Mo metal	Tc metal	Ru metal	Rh metal	Pd metal	Ag metal	Cd metal	In metal	Sn metal	Sb	Te	I / I₂	Xe
Cs metal	Ba metal	La metal	Hf metal	Ta metal	W metal	Re metal	Os metal	Ir metal	Pt metal	Au metal	Hg metal	Tl metal	Pb metal	Bi metal	Po	At	Rn
Fr metal	Ra metal	Ac metal	Rf metal	Db metal	Sg metal	Bh metal	Hs metal	Mt metal									

Legend: Nonmetals, Metalloids, Metals

Ce metal	Pr metal	Nd metal	Pm metal	Sm metal	Eu metal	Gd metal	Tb metal	Dy metal	Ho metal	Er metal	Tm metal	Yb metal	Lu metal
Th metal	Pa metal	U metal	Np metal	Pu metal	Am metal	Cm metal	Bk metal	Cf metal	Es metal	Fm metal	Md metal	No metal	Lr metal

FIGURE 5-1   A periodic table showing metals, nonmetals, and metalloids.

ACTIVITY 5-1

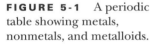

Identify each of the following elements as a metal, nonmetal, or metalloid.
(a) lithium
(b) arsenic
(c) bromine
(d) chromium
(e) lead
(f) sulfur

Other elements do not have metallic properties. They are poor conductors or nonconductors and occur as gases or as brittle, nonlustrous solids. These elements are the **nonmetals.** As shown in Figure 5-1, the nonmetals occupy the upper right corner of the periodic table. Examples are the gases nitrogen, oxygen, fluorine, and chlorine; the liquid bromine; and the solids carbon, sulfur, and iodine.

A few elements display properties that are both metallic and nonmetallic. A good example is silicon, which is a shiny, bluish gray metallic-appearing element, but it is quite brittle and cannot be formed into a variety of shapes like a metal. Silicon does not conduct electricity in the same way that metals do. It conducts in a special way; it is a semiconductor. Silicon, mixed with small amounts of selected impurities to enhance its conductivity, is used to fabricate transistors and other electronic components. These electronic components are made in miniature directly on the surfaces of small chips or thin squares of silicon to form integrated circuits used in solid-state electronics. See Section 11-12 for further discussion of silicon. Elements, like silicon, that have some metallic and some nonmetallic properties are called **metalloids** or **semimetals.** Figure 5-1 shows that the metalloids form a steplike demarcation between the metals and nonmetals.

5-3 THE PHYSICAL STATES AND CHEMICAL FORMULAS OF THE ELEMENTS

Samples of most elements have been isolated from nature, so we know whether they occur in the solid, liquid, or gaseous physical state. Most elements are solids. Two elements, mercury and bromine, occur as liquids, and gallium and cesium becomes liquids at temperatures slightly above typical room temperatures. Eleven of the elements are gases. Elements that occur as gases include hydrogen, nitrogen, oxygen, fluorine, chlorine, and the group 18 elements called the noble gases. The noble gases are helium, He; neon, Ne; argon, Ar; krypton, Kr; xenon, Xe; and radon, Rn. They occur in nature as simple monatomic gases.

As you know, copper and gold metals have notable colors. Other colored elements are fluorine, a pale yellow gas; chlorine, a greenish yellow gas; bromine, a dark-red liquid; and iodine, a deep-purple solid. Nitrogen, hydrogen, oxygen, and the noble gases are colorless.

Knowing the appropriate chemical formula for an element when it is not combined with other elements is useful. Most elements occur as collections of atoms. We can represent such elements simply by using the elemental symbols. For instance, a sample of copper metal is represented as Cu. Among the elements, seven occur in unique chemical forms that we need to consider. The elements hydrogen, nitrogen, oxygen, fluorine, chlorine, bromine, and iodine occur as collections of diatomic molecules. Formulas for the **diatomic molecules** of these elements are H_2, N_2, O_2, F_2, Cl_2, Br_2, and I_2.

Visual Periodic Table

DIATOMIC ELEMENTS

Hydrogen	H_2
Nitrogen	N_2
Oxygen	O_2
Fluorine	F_2
Chlorine	Cl_2
Bromine	Br_2
Iodine	I_2

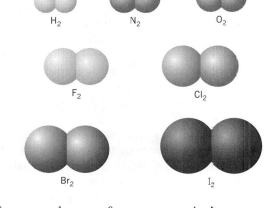

If we want, for example, to refer to oxygen in its natural elemental state (when it is not combined with other elements), we use the formula O_2. A sample of oxygen gas is pictured as a collection of diatomic molecules. This does not mean that oxygen is O_2 when it is combined with other elements in compounds. It does mean that when we refer to atmospheric oxygen we use the formula O_2. As you know, the two major components of air are ele-

Identify and sketch molecules of the two elements found in air as diatomic molecules.

mental nitrogen, N_2, and elemental oxygen, O_2. A cubic foot of dry air contains about 6×10^{23} nitrogen molecules and 2×10^{23} oxygen molecules along with a variety of molecules of minor components, such as approximately 3×10^{20} molecules of carbon dioxide.

All other elements in their elemental or uncombined state are normally represented only by their symbols. This point of symbolism is very important in chemistry. In general, an element in elemental uncombined state is represented by the element symbol unless the element is one of the seven diatomic elements. For example, we use the symbol Fe for a sample of iron metal and the symbol Cl_2 for a sample of chlorine. A sample of iron is pictured as a collection of iron atoms and a sample of chlorine as a collection of diatomic Cl_2 molecules. When elements occur in compounds, we use the normal element symbols along with the appropriate subscript in the formula. For example, the formulas of some compounds of chlorine are KCl, $MgCl_2$, $AlCl_3$, SCl_4, and PCl_5. The symbol Cl_2 by itself represents the uncombined element. Memorizing the names and formulas of all the seven diatomic elements is useful. Notice the locations of these elements in the periodic table. See Figure 5-1. Excluding hydrogen, their locations form a shape similar to the number seven.

5-4 CHEMICAL REACTIONS

When a silvery, solid sample of the element sodium is mixed with a sample of the greenish yellow gas chlorine, a vigorous chemical transformation takes place producing common table salt, NaCl. In the reaction, two chemicals having unique characteristics interact to form a new chemical having very different characteristics. When natural gas is mixed with air and ignited, vigorous combustion occurs with the evolution of heat and light. Methane, CH_4, the main component of natural gas, and oxygen in the air are transformed into carbon dioxide and water. One set of chemicals has changed into a new set of chemicals. The same elements are in each set of chemicals, but new compounds are formed from the starting chemicals as the atoms recombine. New combinations of atoms replace the original combinations.

Transformations in which one set of chemicals is mixed and a new set of chemicals is produced are fundamental in chemistry. Such transformations are chemical reactions. By definition, a **chemical reaction** is a process in which one set of chemicals is transformed into a new set of chemicals. A chemical reaction may involve elements combining to form a compound, a compound decomposing to form elements, or chemical compounds interacting to form new chemical compounds.

Chemical reactions are truly amazing processes. They represent dynamic chemical actions. These actions range from the rusting of iron to the many complex reactions involved in life processes. In any chemical reaction, the initial chemicals, called **reactants,** react to form the new chemicals, called **products.** New substances are formed. Chemical reactions are occurring in and around us all the time. The digestion and metabolism of food involve chemical reactions. The burning of fuels are chemical reactions. Experimental chemistry includes the observation and description of various chemical reactions. Reactions are also fundamental to industrial processes since they provide ways to extract elements from compounds and ways to synthesize new and useful compounds.

5-5 CHEMICAL REACTIONS AND CHEMICAL EQUATIONS

In a chemical reaction, the starting materials are those chemicals that react, the reactants. The reactants have specific properties, such as color, texture, density, and physical state. A reaction produces chemical products. Reactants and products have distinct and different properties.

A chemical reaction involves the breaking of one set of atomic combinations and the formation of a new set of combinations. That is, a reaction involves the change of an initial arrangement of chemically bound atoms into a new arrangement of atoms. Many chemical reactions are observed by noting that the starting materials change into new materials. When sodium and chlorine react, a silvery solid and a greenish yellow gas are transformed into the white crystals characteristic of salt. Reactions of colorless gases, such as the burning of methane in air, are not directly observed, but the release of heat and light in the reaction suggests that something is happening. Describing a chemical reaction by naming the reactants and products is possible. When we burn charcoal briquettes in a barbecue, the main chemical reactants are carbon in the charcoal and oxygen gas in the air. The product of this reaction is carbon dioxide gas. The reaction is described by stating that carbon reacts with oxygen to form carbon dioxide. Note also that heat is released in the burning of charcoal. Many reactions release energy as heat, so a heat exchange is used as evidence of a reaction.

A very useful way to describe a chemical reaction is to use a **chemical equation.** To write an equation, we use precise formulas for the reactants and products. In the reaction of carbon and oxygen to give carbon dioxide, the formulas of the reactants are C for carbon and O_2 for oxygen. (Remember that oxygen is a diatomic element.) The formula of the product, carbon dioxide, is CO_2. The equation for the reaction is

$$C + O_2 \longrightarrow CO_2$$

By convention, the formulas of the reactants are separated by plus signs. Such plus signs are read as "plus" or "and." Similarly, the formulas of the products, if more than one, are separated by plus signs. Typically, the reactants and products are separated by an arrow, which is read as "react to yield," "react to give," "react to produce," or "react to form." The equation above is read as "carbon and oxygen react to produce carbon dioxide."

A chemical equation is a symbolic description of a chemical reaction. The term *equation* is full of meaning. In mathematics, equation refers to setting equal or having an equality on both sides of an equal sign. Of course, a chemical equation is different from a mathematical equation, but some important equalities are associated with chemical equations. The numbers of atoms of a specific element are equal on both sides of any equation, which makes sense if we remember that atoms are not created or destroyed in chemical processes. Of course, a chemical reaction does not simply involve separate atoms, so in an equation all chemicals are shown by the formulas of the elements or compounds involved in the reaction. Realizing that a chemical equation is a before-and-after description of a reaction is important. The chemical equation shows the formulas of the reactants and the products. An equation does not tell how a reaction occurs.

5-6 COEFFICIENTS IN EQUATIONS

A chemical reaction is envisioned as one set of chemically combined atoms in the reactants changing to a new set of combinations in the products.

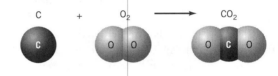

Atoms are conserved in a reaction, so in a chemical equation the number of atoms of each kind of element must be the same on the reactant and product side of the arrow. Atoms are not created or destroyed; they can, however, change their state of combination. The reaction that occurs when methane, CH_4, reacts with oxygen gas (burns in air) produces carbon dioxide, CO_2, and water, H_2O. The equation is written showing the formulas of the reactants and products separated by an arrow.

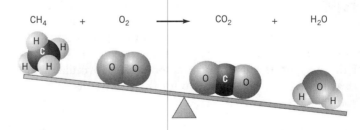

When you count the number of combined hydrogen atoms in the equation above you'll find four on the left and two on the right of the arrow. Similarly, the number of combined atoms of oxygen does not balance on both sides of the arrow. The equation is not correctly written. An equation must be written to reflect that atoms are conserved in chemical reactions.

To show the conservation of atoms in a reaction, an equation must be balanced. A **balanced equation** has the same number of combined atoms of each element on either side of the arrow. Let's consider how to balance an equation.

When **balancing an equation**, never change the formulas of compounds or any subscripts, since this would not represent the compounds correctly. We can, however, change the number of formula units of any compound that appears in the equation. We do this by placing a **numerical coefficient** in front of the formula of the compound. Refer to the reaction of methane and oxygen. To get four hydrogen atoms on each side of the equation, two H_2O molecules must be produced. To get four oxygen atoms on each side, two O_2 molecules must react. This change is represented by placing the necessary numbers, called coefficients, in front of the formula of the appropriate compounds or elements. Count the number of atoms on each side of the arrow to confirm that the equation is balanced.

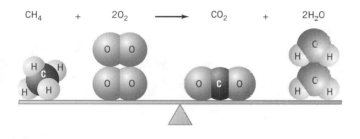

5-7 BALANCING CHEMICAL EQUATIONS

In any chemical equation, the coefficients must be adjusted so that the number of atoms of each element is the same on the left (reactant side) as on the right (product side). An equation in which the number of atoms of each element is the same on both sides of the arrow is called a balanced equation. The process of adjusting the coefficients is called balancing the equation. Often, balancing is done by trial and error, that is, by trying various coefficients and keeping a count of each kind of atom. You should never change the formulas or subscripts to balance an equation. The compounds are treated as units, since they involve chemically combined atoms. When balancing equations, you can, however, change the numerical coefficient in front of any formula. To balance a chemical equation, follow this pattern:

Balancing Equations

1. Write the correct formulas of the reactants and products. (For now, you'll be given this information. You are not expected to predict the products of a reaction at this point in your study of chemistry.)

2. Count the number of atoms of each element on the reactant side and on the product side.

3. If the number of atoms of any element differs from side to side, add necessary coefficients in front of the appropriate formulas.

   (a) Work with one element at a time and leave any diatomic elements until last.

   (b) Change any coefficient to maintain the balance. Change any coefficient as needed.

   (c) Never change a formula or a subscript in a formula.

   (d) Never add extra formulas to an equation.

Most equations are balanced using relatively small numbers for coefficients. Since balancing is a trial-and-error procedure, your first attempt to balance an equation may not be successful. If you find that the coefficients are getting too large or that the equation does not balance, just start over or consult someone else or a study group. With practice, you'll discover useful patterns for balancing equations.

EXAMPLE 5-1

During the rusting of iron, one reaction involves iron, Fe, combining with oxygen, O_2, to form solid Fe_2O_3. Write a balanced equation.

The unbalanced equation is

$$Fe + O_2 \longrightarrow Fe_2O_3$$

There are two O atoms and one Fe atom on the reactant side and three O atoms and two Fe atoms on the product side.

$$Fe + O_2 \longrightarrow Fe_2O_3$$
$$1\ Fe\ 2\ O \qquad 2\ Fe\ 3\ O$$

Balance the iron atoms by placing the coefficient of two in front of Fe on the reactant side.

$$2Fe + O_2 \longrightarrow Fe_2O_3$$
$$2\ Fe\ 2\ O \qquad 2\ Fe\ 3\ O$$

Note that there is an odd number of oxygen atoms on the product side and an even number on the reactant side. One way to balance the oxygen atoms is to use a coefficient of 3 for O_2 and a coefficient of 2 for Fe_2O_3, which gives an even number of oxygen atoms on both sides.

$$2Fe + 3O_2 \longrightarrow 2Fe_2O_3$$
$$2\ Fe\ 6\ O \qquad 4\ Fe\ 6\ O$$

The oxygen atoms are now balanced, with six on each side. Now, however, there are four Fe atoms on the product side and only two on the reactant side. To balance completely, we just change the coefficient of Fe on the left to 4.

$$4Fe + 3O_2 \longrightarrow 2Fe_2O_3$$

4 Fe 6 O   4 Fe 6 O

When working with a diatomic element, sometimes there is an even number of atoms on one side and an odd number on the other side. One way to balance the equation is to make the odd side even by doubling the coefficient; the other side is adjusted as needed. Another approach is to balance the atoms of the diatomic element using a fraction as a coefficient and then multiply all coefficients in the equation by 2. See Example 5-2.

ACTIVITY 5-3

Give the balanced equation for the reaction of iron metal and chlorine to give $FeCl_3$. Draw a picture to show the reaction of atoms of iron with diatomic molecules of chlorine to give formula units of the product.

EXAMPLE 5-2

Octane, C_8H_{18}, a component of gasoline, reacts with oxygen, O_2, to give carbon dioxide, CO_2, and water, H_2O. Balance the following unbalanced equation.

$$C_8H_{18} + O_2 \longrightarrow CO_2 + H_2O$$

First, balance the carbons using a coefficient of 8 for CO_2, then balance the hydrogens using a coefficient of 9 for H_2O. Note that 9 times the subscript of 2 for hydrogen gives a total of 18 hydrogens.

$$C_8H_{18} + O_2 \longrightarrow 8CO_2 + 9H_2O$$

Counting the oxygens, we note there are 25 oxygens on the product side (8 × 2 = 16 and 16 + 9 = 25) and 2 on the reactant side. Since we have O_2 on the reactant side, a way to get 25 oxygens on the left is to use a coefficient of 25/2 or 12.5.

$$C_8H_{18} + 25/2 O_2 \longrightarrow 8CO_2 + 9H_2O \quad \text{or} \quad C_8H_{18} + 12.5O_2 \longrightarrow 8CO_2 + 9H_2O$$

This equation is balanced in terms of atoms. There is nothing wrong with using fractional coefficients in equations, but often whole-number coefficients are used. Here, we change the coefficients to whole numbers by doubling each coefficient, that is, multiplying each by 2. The result gives the balanced equation

$$2C_8H_{18} + 25O_2 \longrightarrow 16CO_2 + 18H_2O$$

ACTIVITY 5-4

Balance the following unbalanced equations using whole-number coefficients.
(a) The reaction for the decomposition of water to give oxygen gas and hydrogen gas
(b) $Ca + NH_3 \longrightarrow$
$$CaH_2 + Ca_3N_2$$
(c) $C_2H_5O + O_2 \longrightarrow$
$$CO_2 + H_2O.$$

The last part of Example 5-2 illustrates another even–odd case that sometimes arises when a diatomic element is involved in a reaction. What happens is that during the balancing process, an odd number of atoms of the diatomic element occurs on the product side with two atoms of this element on the reactant side. When this occurs, the equation is balanced by using a fractional coefficient for the diatomic element (the odd number divided by 2). Double all the coefficients in the equation as the last step of balancing to give whole-number coefficients. Occasionally, however, you may see such equations written with fractional coefficients.

5-8   CHEMICAL EQUATIONS WITH ADDED INFORMATION

When writing certain chemical equations, it is sometimes useful to show the physical states of the various chemicals. Letter codes enclosed in parentheses following the formulas are used; (g) denotes gas, a script el, (ℓ), denotes liquid, and (s) denotes solid. Thus, the equation

$$2HgO(s) \xrightarrow{\text{heat}} 2Hg(\ell) + O_2(g)$$

reveals that when solid HgO is heated, liquid mercury and oxygen gas form. Note that a special condition for a reaction is written above or below the arrow in the equation as is done for heat in the above equation. The symbol (aq) is used to denote that a chemical is dissolved as a solute in aqueous solution. Aqueous refers to water as a solvent. Many important reactions take place in water. For instance, the equation

$$2HCl(aq) + Mg(OH)_2(s) \longrightarrow MgCl_2(aq) + 2H_2O(\ell)$$

represents the reaction that takes place in the stomach when you swallow some milk of magnesia, a common over-the-counter antacid and laxative.

Some reactions do not occur readily unless a catalyst is present. By definition a **catalyst** is a substance that increases the speed or rate of a reaction without undergoing a permanent chemical change. Some reactions occur too slowly to be useful unless a catalyst is present to speed it up. Since a catalyst is neither a reactant or a product, its formula is sometimes written above the arrow in the equation. For instance, the catalyst V_2O_5 is required to make the compound SO_3 efficiently when SO_2 and O_2 react. This situation is represented by the equation

$$2SO_2(g) + O_2(g) \xrightarrow{V_2O_5} 2SO_3(g)$$

See Section 14-3 for a more detailed discussion of catalysts.

You'll encounter many equations in your study of chemistry. As you gain experience with elements and compounds, you can interpret some equations by naming the reactants and products involved. A convenient way to read an equation is to pronounce each number and formula in a literal manner. For example, the equation

$$Fe + S \longrightarrow FeS$$

is read as F-E and S react to give F-E-S. The equation

$$CH_4 + 2O_2 \longrightarrow CO_2 + 2H_2O$$

is read as C-H-4 plus two O-2 react to form C-O-2 plus two H-2-O. Alternatively, it reads methane plus oxygen react to give carbon dioxide and water. The equation

$$2SO_2(g) + O_2(g) \xrightarrow{V_2O_5} 2SO_3(g)$$

is read as two S-O-2 plus O-2 react, using the catalyst V-2-O-5, to produce two S-O-3. Using names, it reads sulfur dioxide and oxygen react using a catalyst to give sulfur trioxide.

Sometimes equations are used in a general way to give chemical information. For example, that the family of compounds called carbohydrates react with oxygen to give carbon dioxide and water is conveyed by a general equation:

$$carbohydrates + O_2 \longrightarrow CO_2 - H_2O$$

Such symbolic equations are not balanced equations but are used as abbreviations of general kinds of reactions. They serve as succint representations of important types of reactions.

ACTIVITY 5-5

The active ingredient in vinegar is acetic acid, $HC_2H_3O_2(aq)$. Write a balanced equation for the reaction of zinc metal, Zn, and acetic acid to give hydrogen gas, H_2, and $Zn(C_2H_3O_2)_2(aq)$.

5-9 EVIDENCE FOR REACTIONS

Chemical processes are dynamic and active processes in which chemicals transform. Some transformations are noticeably dramatic; for instance, the explosion of nitroglycerine in dynamite is a fast and furious reaction. An **explosion** is a very fast chemical reaction that releases energy, some of which takes the form of heat and light. In an explosion, chemical products expand rapidly at speeds of many thousands of meters per second. An explosion scatters bits of solid material and produces a shock wave in the surrounding air, which is why a loud sound accompanies an explosion. In contrast, some reactions are slow but relentless. Iron in the environment slowly rusts; it reacts with oxygen and is transformed into a compound, the compound from which humans extracted the iron in the first place. It can take many years for an iron sample to rust completely.

How can you tell that a chemical reaction has occurred? When charcoal burns in air we notice that it glows, releases heat, and seems ultimately to disappear. The white ash remaining contains noncombustible compounds formed from impurities in the charcoal. From a chemical view, a reaction involves atoms changing associations. Atoms cannot be seen, but they are symbolically represented in equations. We cannot see atoms react, but we can witness reactions by noting physical changes in the chemicals as they are transformed. Some possible changes are:

A gas forms or disappears.
A solid forms or disappears.
A liquid forms or disappears.
A color change occurs.
Energy is released as heat and sometimes as light.
Energy is absorbed as heat from the surroundings.

Sometimes these kinds of changes can occur in processes that are not chemical reactions. Can you give some examples of such changes in which no chemical reaction occurs?

Consider a few examples of reactions in which obvious changes do occur. When magnesium metal is heated in air, the metal disappears and a white, powdery solid is produced. The reaction is accompanied by the noticeable release of heat and light. When iron metal with a metallic sheen rusts, the solid metal is transformed into a reddish (rusty) solid. When we treat hard-water scale in a teakettle (solid calcium carbonate) with vinegar, the solid slowly disappears and bubbles of gas form. When nitroglycerine explodes, the liquid chemical rapidly disappears and gases form along with a very noticeable release of energy.

ACTIVITY 5-6

You need two paper matches. Light one match and use it to ignite the second match. Record all that you see, hear, and feel when you ignite the match. Propose a hypothesis to explain why you hear the match igniting and burning.

5-10 ENERGY AND CHEMICAL REACTIONS

Chemical reactions that occur in our bodies provide energy for work and to maintain internal body temperature. Energy exchanges accompany all chemical processes. When we do work, we exert energy. Energy, however, is more than just physical work. The word *energy* developed as a composite of the Greek prefix *en* (in) and the Greek word *ergon* (work). An object that possesses energy has within it the capacity to do work. **Energy** is simply the capacity to do work.

When an object has energy, there is the possibility of work as some dynamic change. When you pick up an object, you must exert a force to overcome gravity. Your muscles have done work or expended energy. The object you picked up has stored energy or the potential for doing work. If you release the object, it expends the stored energy as it falls. The term *work* refers to a dynamic change; it may be useful, for example, picking something up or running a motor, but work does not have to be useful.

We experience various kinds of energy in energy flows, motion, and dynamic changes. Observations of energy changes show that energy has various forms. Changes from one form of energy to another result in action, motion, chemical changes, or heat. Work is done when energy changes from one form to another. Some common types of energy are described below.

Radiant Energy

Radiant energy is associated with light. It is sometimes called electromagnetic radiation and occurs in such forms as radiowaves, microwaves, infrared light, visible light, ultraviolet light, and X rays. Solar energy is radiant energy produced by processes on the sun. We experience solar energy as sunlight, which is mainly visible light along with some infrared and ultraviolet light.

Heat

Heat is a mode of energy transfer between objects. It is also called thermal energy. An object becomes hotter when it is heated by an energy source. We notice heat when a heat exchange causes an object to change its temperature. When a colder object is in contact with a warmer object, the colder object is heated and the warmer object is cooled. As mentioned in Chapter 2, heat is directly related to the motion of chemical particles in a material.

Kinetic Energy

Kinetic energy is energy of motion or energy associated with any moving object. A moving car, a moving wheel, in fact any moving object has kinetic energy. Of course, moving atoms and molecules have kinetic energy.

Mechanical Energy

Mechanical energy is energy associated with a motor, turbine, or engine. Mechanical energy is essentially a kind of kinetic energy associated with the motion of a flywheel or driveshaft of a motor, turbine, or engine.

Electrical Energy

Electrical energy is energy that comes from the flow of electrical current. Electrical current passing through the filament of a light bulb changes the electrical energy to heat and light.

Potential Energy

Potential energy is stored energy or energy of position. Energy of this type has the potential to do work by undergoing a change or flow. Water stored behind a dam has potential energy.

Chemical Energy

Chemical energy is energy stored in chemicals. It is a type of potential energy. Chemical changes allow the release of this energy, and it converts to other forms of energy. When natural gas burns, chemical energy is released as heat and light. A battery converts chemical energy to electrical energy and heat.

5-11 EXOTHERMIC AND ENDOTHERMIC REACTIONS

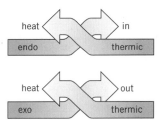

You might relish the warmth from the combustion of wood in a fire or use the heat from the combustion of natural gas to cook. Human societies use tremendous amounts of chemical energy for industrial purposes and everyday living. **Exothermic** reactions are chemical reactions that release energy as heat. The term *exothermic* means heat out. An exothermic reaction can heat its surroundings as it occurs. A simple way to show that a reaction is exothermic is to write the word *energy* on the product side of the equation. For example, the burning of charcoal is an exothermic reaction.

$$C + O_2 \longrightarrow CO_2 + energy$$

In contrast, some reactions require energy to occur. **Endothermic** reactions absorb heat from their surroundings. The term *endothermic* means heat in. To emphasize that a reaction is endothermic the word *energy* is written on the reactant side. For example, the reaction between nitrogen gas and oxygen gas to give nitrogen oxide gas is an endothermic reaction.

$$energy + N_2 + O_2 \longrightarrow 2NO$$

Potential energy is stored energy. Suppose we roll a boulder up a hill. At the top of the hill, the boulder has potential energy with respect to ground level. With a slight nudge, the boulder tumbles down the hill, converting its stored energy into energy of motion or kinetic energy. Energy is associated with the bonding of atoms in compounds; each chemical has a kind of potential energy called chemical energy. Simply stated, chemical energy is energy stored within chemical compounds.

Photosynthesis is the energy-capturing chemical process that occurs in all green plants. This endothermic reaction of carbon dioxide and water to make the carbohydrate glucose and oxygen gas is represented by the following equation.

$$energy + 6CO_2 + 6H_2O \longrightarrow C_6H_{12}O_6 + 6O_2$$

As shown in Figure 5-2, plants synthesize carbohydrates and oxygen from carbon dioxide and water. Energy for this endothermic process comes from sunlight. By this process solar energy is converted to chemical energy. The chemical energies of the carbohydrates and oxygen are higher than the chemical energies of the carbon dioxide and water. Under the correct conditions, however (somewhat like the boulder rolling down the hill), carbohydrates can react with oxygen to give carbon dioxide and water and release the stored energy in the process. The

ACTIVITY

5-7

Label each of the following processes as exothermic or endothermic.
(a) the burning of gasoline
(b) boiling water
(c) melting of ice

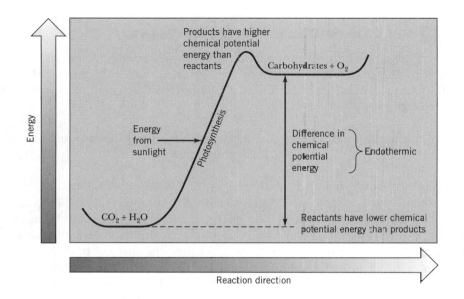

FIGURE 5-2 Change in chemical energy during photosynthesis, an endothermic reaction.

exothermic reaction between glucose and oxygen is represented by the following equation.

$$C_6H_{12}O_6 + 6O_2 \longrightarrow 6CO_2 + 6H_2O + energy$$

Figure 5-3 illustrates the energy change in such an exothermic reaction. Energy in chemical reactions is further discussed in Chapter 7.

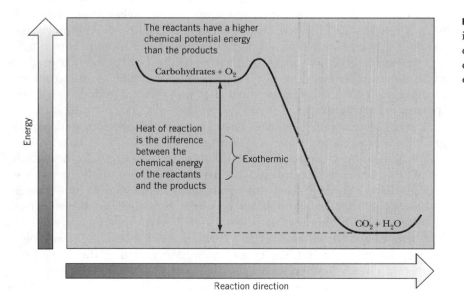

FIGURE 5-3 Change in chemical energy during combustion of a carbohydrate, an exothermic reaction.

5-12 THE LAW OF CONSERVATION OF ENERGY

Energy is a very important commodity in human societies. Scientists have studied energy and energy changes for centuries. Such study has resulted in a very important scientific law that is known as the law of conservation of energy:

> Energy is never created or destroyed but only changed from one form to another.

A burning candle does not create energy. As it burns, the chemical energy of the reactants is transformed to heat energy, light energy, and energy associated with gases produced. The energy is conserved since the chemical energy lost equals the total energy produced in the exothermic process. The law of conservation of energy is also called the first law of thermodynamics. In chemistry, it is sometimes stated in the form that energy is not created or destroyed in chemical processes. This law means that the difference in the chemical energies of the reactants and products manifests as an exothermic or endothermic process.

One practical implication of the conservation of energy is that because energy cannot be created, if we want useful energy, we need a convenient source of it. Conversion of energy from one form to another invariably involves losses. This statement does not mean that energy is destroyed; rather, when used for work, it does end up as the less useful form of heat. Heat is a low-grade form of energy because it is not efficiently converted to other energy forms. When you drive a car, the chemical energy of gasoline is converted to mechanical energy that provides the kinetic energy to move the car. During the drive, the energy is ultimately lost as heat. Such heat is not readily changed back to chemical energy. Furthermore, the gasoline spontaneously burns and gives us the energy we need, but the direct reverse process in which the heat from the environment is used to make the gasoline just does not happen. All energy that we use from the burning of fuels is changed to heat that ultimately is radiated to outer space as infrared energy.

Energy conversion typically results in more concentrated (high-grade) energy changing to a less concentrated (low-grade) energy. The practical implication is that significant amounts of energy can most easily be obtained from concentrated forms such as fossil fuels or nuclear sources. This statement does not mean that energy cannot be obtained from less concentrated sources, such as solar radiation, tidal flows, wind, or temperature differences in ocean water. It requires much greater effort and expenditure of energy to obtain useful energy from these sources, however. In fact, the cost of energy is related to the cost of the effort and energy needed to obtain the energy. Furthermore, it is obvious that the costs needed to obtain energy should not exceed the value of the energy.

Calories and Food

Food and Calories

Use a match to light a piece of popcorn, a potato chip, a corn chip, any kind of nut, or a cracker and then watch it burn. Foods are sources of energy. The three major kinds of chemicals found in foods are fats, carbohydrates, and proteins. They occur in food as a variety of compounds. Carbohydrates, for instance, range from simple sugars, such as glucose, to complex carbohydrates, such as starch. Analyzing a specific type of food to find the percent by mass of each of the three kinds of food chemicals is possible. Table 5-1 shows the composition of a few foods as examples. Foods also contain varying amounts of water and indigestible fiber as well as small amounts of vitamins and minerals.

Table 5-1   *Composition of Some Foods*

	PERCENT BY MASS		
Food	*Protein*	*Fat*	*Carbohydrate*
Meat (round steak)	20.5	16.5	0
Bacon	9.9	65	0
Milk (whole)	3.5	3.7	4.8
Eggs	13.4	10.5	0
Cheese (cheddar)	27.7	36.3	4.1
Butter	1.0	85	0
Peanuts	25.8	38.5	21.9
Beans (navy)	22.5	1.3	55.2

The three major food components are metabolized and serve as energy sources in our bodies. Metabolism refers to the numerous chemical reactions that occur in the body. The metabolic use of foods for energy is represented by the following general equations. These equations are symbolic, not balanced, representing the metabolic oxidation of food chemicals. Since proteins contain nitrogen, urea, $(NH_2)_2CO$, is produced from the metabolism of protein.

$$fats + O_2 \longrightarrow CO_2 + H_2O + energy$$

$$carbohydrates + O_2 \longrightarrow CO_2 + H_2O + energy$$

$$proteins + O_2 \longrightarrow CO_2 + H_2O + urea + energy$$

The amounts of energy released by reactions like these are measured by experiments using samples of typical fats, carbohydrates, and proteins. Food energy is often expressed in units of calories. One calorie corresponds to the heat needed to change 1 g of water by 1°C. Typically, fats provide 9 kcal/g. One dietetic calorie is defined as 1000 cal or 1 kcal. The symbol Cal (with a capital C) is used for the dietetic calorie. In terms of the dietetic calorie, fats provide 9 Cal/g. Carbohydrates provide about 4 Cal/g. Coincidentally, proteins also provide about 4 Cal/g.

All our body energy comes from the metabolism of fats, carbohydrates, and proteins. Much of the protein that we eat, however, is used as a source for our body to make its own protein, and some of the fats we eat are used to make body fat, which becomes fatty tissue. Nevertheless, in the determination of the caloric content of a food it is assumed that all the components of the food are used entirely for energy.

The body uses the energy from food for any endothermic metabolic reactions, physical work, and to maintain body temperature. The energy needs of a person depend on many factors, including basic metabolic processes to maintain life and energy for activity or physical work. As expected, the typical number of calories needed by a person depends on that person's age, weight, gender, and kinds of work done. Caloric requirements vary depending on the kinds of physical activity. A typical person requires about 2000 dietetic calories per day.

The percentages of a food that is carbohydrate, protein, and fat are used to find the caloric content of a food sample. You can use the following steps to find a sample's caloric content from the percent of each food component.

1. Express the number of grams of carbohydrate, protein, and fat in 100 grams of the food sample. Use percents as factors expressing the number of grams of a food component per 100 grams of food. For instance, if a food is 27.7% protein, the factor is 27.7 g protein/100 g food.

2. Multiply the total mass of the food sample by each of the factors found from the percents. These values give the mass of each food component.

3. Multiply the mass of each food component by the factor expressing the number of calories per gram.

4. Add the calories from each component to give the total number of calories.

As an example, let's calculate the caloric content of 1.00 oz of cheddar cheese (1.00 oz is 28.3 g). See Table 5-1 for the composition of cheddar cheese. Typical cheddar cheese is 27.7% protein, 4.1% carbohydrate, and 36.8% fat. Using the method described above for a 28.3 g sample of cheese gives

$$\text{protein} \quad 28.3 \text{ g cheese} \left(\frac{27.7 \text{ g protein}}{100 \text{ g cheese}} \right) \left(\frac{4 \text{ Cal}}{1 \text{ g protein}} \right) = 31.4 \text{ Cal}$$

$$\text{carbohydrate} \quad 28.3 \text{ g cheese} \left(\frac{4.1 \text{ g carbo}}{100 \text{ g cheese}} \right) \left(\frac{4 \text{ Cal}}{1 \text{ g carbo}} \right) = 4.6 \text{ Cal}$$

$$\text{fat} \quad 28.3 \text{ g cheese} \left(\frac{36.8 \text{ g fat}}{100 \text{ g cheese}} \right) \left(\frac{9 \text{ Cal}}{1 \text{ g fat}} \right) = 93.7 \text{ Cal}$$

Total Calories = 31.4 Cal + 4.6 Cal + 93.7 Cal = 129.7 or 130 Cal

The total caloric content of 1.00 oz of cheddar cheese is 130 Cal. Notice that we can also calculate the percent of calories provided by fat in the food. Do this by dividing the number of calories from fat by the total calories and multiplying by 100. In cheddar cheese, fat contributes 72% of the calories [(93.7/130) × 100]. The same methods are used to estimate the caloric content and percent of calories from fat for any food given its composition.

Many prepared foods have the composition listed on the label expressed in terms of the stated serving size. The label lists the mass of each food component per serving size instead of its percent. Thus, to find the caloric content, simply multiply the mass of each component by the appropriate calorie per gram factor and add the results. For example an 8-oz cup of typical low-fat milk contains 10 g protein, 12 g carbohydrates, and 2.5 g fat. The caloric content is

$$\text{protein} \qquad 10 \cancel{g}\,(4 \text{ Cal}/\cancel{g}) = 40 \text{ Cal}$$

$$\text{carbohydrate} \qquad 12 \cancel{g}\,(4 \text{ Cal}/\cancel{g}) = 48 \text{ Cal}$$

$$\text{fat} \qquad 2.5 \cancel{g}\,(9 \text{ Cal}/\cancel{g}) = 22.5 \text{ Cal}$$

Total Calories = 40 Cal + 48 Cal + 22.5 Cal = 110 Cal

The percent of calories from fat is found by dividing the fat calories by the total calories and multiplying by 100.

$$\% \text{ fat calories} = (22.5 \text{ Cal}/110 \text{ Cal}) \times 100 = 20\%$$

It is interesting that the low-fat milk contains 1% milkfat by mass but 20% of the calories come from the fat content.

Figure 5-4 shows an example food label. Check the caloric content using the information on the label and calculate the percent of calories that come from fat.

Nutrition Facts

Serving Size 5 Crackers (14g)
Servings Per Container About 16

Amount Per Serving

Calories 60	Calories from Fat 15

	% Daily Value*
Total Fat 1.5g	**3%**
Saturated Fat 0g	**0%**
Polyunsaturated Fat 0g	
Monounsaturated Fat 0.5g	
Cholesterol 0 mg	**0%**
Sodium 180mg	**7%**
Total Carbohydrate 10g	**3%**
Dietary Fiber Less than 1g	**2%**
Sugars 0g	
Protein 1g	

Vitamin A 0%	•	Vitamin C 0%
Calcium 2%	•	Iron 4%

*Percent Daily Values are based on a 2,000 calorie diet. Your daily values may be higher or lower depending on your calorie needs:

	Calories:	2,000	2,500
Total Fat	Less than	65g	80g
Sat Fat	Less than	20g	25g
Cholesterol	Less than	300mg	300mg
Sodium	Less than	2400mg	2400mg
Total Carbohydrate		300g	375g
Dietary Fiber		25g	30g

FIGURE 5-4 A typical food label.

A nutrition fact label is required by the U.S. Food and Drug Administration to contain specific information. It must give the reasonable serving size and the number of servings in the container. The total calories per serving and the calories from fat must be listed. To easily find the percent of calories from fat, divide the fat calories by the total calories and multiply by 100.

The nutrition fact label must also show the percent of the daily value of various food components. The daily values refer to number of grams or milligrams of these components recommended for a typical 2000-Cal diet. The grams of saturated fat are given as a subdivision of the total fat content. Simple sugars and dietary fiber are given as subdivisions of total carbohydrates. These subtracted from the total carbohydrates give the grams of complex carbohydrates or starch. There is no daily value for protein, but it should be consumed in moderate amounts. The content of combined sodium is given in milligrams along with the percent of the daily value of sodium. This value is essentially the salt content of the food expressed in terms of sodium. Likewise, the cholesterol content is given in milligrams along with the percent of daily value. Only foods containing meat, fish, eggs, and poultry can have cholesterol in them. Cholesterol never occurs in foods made only from vegetables or vegetable oils. Finally the content of vitamin A, vitamin C, and the minerals calcium and iron supplied by the food are given in terms of the percent of the daily value.

Chemical Reactions and Equations

Periodic table: The table in which the elements are listed according to increasing atomic numbers and are grouped in columns according to similarities in properties.

Metals: Elements that are flexible, have luster, and are good conductors of electricity and heat.

Nonmetals: Elements that are brittle solids or gases and do not conduct electricity. The nonmetals are located in the far right-hand region of the periodic table.

Metalloids: Elements, sometimes called semimetals, that have some properties of metals and some properties of nonmetals. The metalloids separate the metals and nonmetals in the periodic table.

Diatomic elements: Elements that occur as diatomic molecules in their elemental form. The diatomic elements are hydrogen, nitrogen, oxygen, fluorine, chlorine, bromine, and iodine. Formulas for these diatomic molecules of these elements are H_2, N_2, O_2, F_2, Cl_2, Br_2, and I_2.

Chemical reaction: A process in which one set of chemicals is transformed into another set of chemicals. The starting or initial chemicals are the reactants, and the final chemicals formed in a reaction are the products.

Chemical equation: A symbolic representation of a chemical reaction showing the formulas of the reactants and products separated by an arrow signifying "react to make, form, produce, and so forth."

Exothermic reaction: A chemical reaction that releases energy as it occurs.

Endothermic reaction: A chemical reaction that absorbs energy as it occurs.

Law of conservation of energy: Energy is never created or destroyed but only changed from one form to another. In chemistry, this law is sometimes stated in the form that energy is not created or destroyed in chemical processes.

Steps Used to Balance Equations

Propane, C_3H_8, a compound used as cooking gas, reacts with oxygen, O_2, to give carbon dioxide and water.

1. Start with the formulas of the reactants and products.

$$C_3H_8 + O_2 \longrightarrow CO_2 + H_2O$$

2. Count the number of atoms of each element on both sides.

$$C_3H_8 + O_2 \longrightarrow O_2 + H_2O$$

3C 8H 2O   1C 2O 2H 1O

3O

3. Use coefficients in front of formulas to balance the atoms of each element except the diatomic element. Keep a tally of atoms.

   (a) Change coefficients as needed.

$$C_3H_8 + O_2 \longrightarrow 3CO_2 + H_2O$$

3C 8H 2O   3C 6O 2H 7O

7O

(b) Work with one element at a time and leave diatomic elements until last.

$$C_3H_8 + O_2 \longrightarrow 3CO_2 + 4H_2O$$

3C 8H 2O   3C 6O 8H 4C

100H

$$C_3H_8 + 5O_2 \longrightarrow 3CO_2 + 4H_2O$$

3C 8H 10 O   3C 6O 8H 4O

100H

(c) Never change formulas or add formulas.

QUESTIONS

1. Match the following terms with the appropriate lettered definitions.

periodic table

element family

metal

nonmetal

metalloid

diatomic elements

chemical reaction

chemical equation

catalyst

energy

exothermic

a. the absorption of heat

b. the release of heat

c. the capacity to do work

d. elements that occur naturally as collections of diatomic molecules

e. a chemical that increases the speed of a reaction but is not chemically changed

f. the symbolic representation of a chemical reaction

g. a process in which one set of chemicals is transformed into another set of chemicals

h. elements that have luster and are good conductors of electricity and heat

i. elements that are brittle solids or gases and are poor conductors

j. elements that have some metallic and some nonmetallic properties

k. the tabular arrangement of elements according to their properties

endothermic

l. a vertical column or group of elements in the periodic table

Sections 5-1 and 5-2

2. What is the periodic table of elements? What is a column in the table called?

3. What is a metal and where are the metals located in the periodic table?

4. What is a nonmetal and where are the nonmetals located in the periodic table?

5. What is a metalloid and where are the metalloids located in the periodic table?

6. Identify each of the following elements as a metal, metalloid, or nonmetal.

(a) Na   (b) Cl   (c) As   (d) Si

(e) Al   (f) Ca   (g) S   (h) Au

Section 5-3

7. Give a list of the elements that are typically found as gases.

8. Which elements are normally found as liquids?

9. Give the names and molecular formulas for seven elements that occur in the form of diatomic molecules.

Sections 5-4 and 5-5

10. Define the following terms.

(a) chemical reaction

(b) reactants

(c) products

11. What is a chemical equation and how are the symbols + and $\longrightarrow$ used in equations?

Sections 5-6 and 5-7

12. What is a balanced chemical equation?

13. How is the law of conservation of matter related to the balancing of an equation?

14. Why is it not proper to balance an equation by changing the subscripts in the formulas of reactants or products?

15. Explain why the following equations are not correct.

   (a) $Fe + 3O_2 \longrightarrow Fe_2O_3$

   (b) $Fe + B \longrightarrow F + Be$

   (c) $Li + O_2 \longrightarrow 2Li_2O$

   (d) $Cu + Al \longrightarrow Au + Cl$

16. *Balance the following unbalanced equations.

   (a) $Zn + O_2 \longrightarrow ZnO$

   (b) $Cr + O_2 \longrightarrow Cr_2O_3$

   (c) $KClO_3 \longrightarrow KCl + O_2$

   (d) $Li + O_2 \longrightarrow Li_2O$

   (e) $B + F_2 \longrightarrow BF_3$

   (f) $C_3H_6 + O_2 \longrightarrow CO_2 + H_2O$

   (g) $Hg + N_2 \longrightarrow Hg_3N_2$

   (h) $N_2 + H_2 \longrightarrow NH_3$

   (i) $WO_3 + H_2 \longrightarrow W + H_2O$

   (j) $C_4H_{10} + O_2 \longrightarrow CO_2 + H_2O$

   (k) $Ca + NH_3 \longrightarrow CaH_2 + Ca_3N_2$

   (l) $P + S \longrightarrow P_2S_5$

   (m) $KNO_3 + C \longrightarrow K_2CO_3 + CO + N_2$

17. Balance the following unbalanced equations.

   (a) $Ca + N_2 \longrightarrow Ca_3N_2$

   (b) $HgO \longrightarrow Hg + O_2$

   (c) $Fe + H_2O \longrightarrow Fe_3O_4 + H_2$

   (d) $N_2 + O_2 \longrightarrow N_2O_4$

   (e) $Pb_3O_4 \longrightarrow Pb + O_2$

   (f) $P + H_2 \longrightarrow PH_3$

   (g) $NH_3 + O_2 \longrightarrow NO + H_2O$

   (h) $Ba + O_2 \longrightarrow BaO$

   (i) $CuS + O_2 \longrightarrow CuO + SO_2$

   (j) $CaCO_3 + C \longrightarrow CaC_2 + CO_2$

   (k) $CaCN_2 + H_2O \longrightarrow CaCO_3 + NH_3$

   (l) $FeCrO_4 + C \longrightarrow Cr + Fe + CO$

   (m) $MgO + Al \longrightarrow Mg + Al_2O_3$

18. *Balance the following unbalanced equations.

   (a) $Mg + O_2 \longrightarrow MgO$

   (b) $Fe + O_2 \longrightarrow Fe_2O_3$

   (c) $NaClO_3 \longrightarrow NaCl + O_2$

   (d) $Na + O_2 \longrightarrow Na_2O$

   (e) $Sb + Cl_2 \longrightarrow SbCl_3$

   (f) $C_2H_4 + O_2 \longrightarrow CO_2 + H_2O$

   (g) $Al + O_2 \longrightarrow Al_2O_3$

   (h) $F_2 + H_2 \longrightarrow HF$

   (i) $C + CaO \longrightarrow CaC_2 + CO$

   (j) $C_3H_6 + O_2 \longrightarrow CO_2 + H_2O$

   (k) $P + O_2 \longrightarrow P_4O_{10}$

19. Balance the following unbalanced equations.

   (a) $Mg + N_2 \longrightarrow Mg_3N_2$

   (b) $C_5H_{10}O_5 + O_2 \longrightarrow CO_2 + H_2O$

   (c) $Fe_3O_4 + CO \longrightarrow Fe + CO_2$

   (d) $Fe_2O_3 + CO \longrightarrow Fe + CO_2$

   (e) $P + S \longrightarrow P_2S_5$

   (f) $C_8H_{18} + O_2 \longrightarrow CO + H_2O$

   (g) $PCl_5 + H_2O \longrightarrow HCl + H_3PO_4$

   (h) $Ca + O_2 \longrightarrow CaO$

   (i) $ZnS + O_2 \longrightarrow ZnO + SO_2$

   (j) $FeCr_2O_4 + C \longrightarrow Cr + Fe + CO$

   (k) $BaO + Al \longrightarrow Ba + Al_2O_3$

Section 5-8

20. Explain the meaning of the following when used in a chemical equation.

   (a) (s)

   (b) (ℓ)

   (c) (g)

   (d) +

   (e) (aq)

   (f) heat $\longrightarrow$

   (g) Fe $\longrightarrow$

   (h) pressure $\longrightarrow$

21. *For each of the following reactions, describe the physical forms of the reactants and products. Also describe any diatomic element. Some names are given for your reference. For example,

Na(s) + Cl$_2$(g) $\longrightarrow$ NaCl(s)   Solid elemental sodium and diatomic molecular chlorine gas react to form solid sodium chloride.

(a) CH$_4$(g) + 2O$_2$(g) $\longrightarrow$ CO$_2$(g) + 2H$_2$O(g)
(CH$_4$ is methane.)

(b) S(s) + O$_2$(g) $\longrightarrow$ SO$_2$(g)
(SO$_2$ is sulfur dioxide.)

(c) Ca(s) + F$_2$(g) $\longrightarrow$ CaF$_2$(s)
(CaF$_2$ is calcium fluoride.)

$\qquad\qquad$ heat

(d) 2HgO(s) $\longrightarrow$ 2Hg(ℓ) + O$_2$(g)
(HgO is mercury(II) oxide.)

(e) PCl$_3$(ℓ) + Cl$_2$(g) $\longrightarrow$ PCl$_5$(s) (PCl$_3$ is phosphorus trichloride and PCl$_5$ is phosphorus pentachloride.)

22. For each of the following reactions, describe the physical forms of the reactants and products. Also describe any diatomic element. Some names are given for your reference. For example,

Na(s) + Cl$_2$(g) $\longrightarrow$ NaCl(s)   Solid elemental sodium and diatomic molecular chlorine gas react to form solid sodium chloride.

(a) HF(aq) + OH$^-$(aq) $\longrightarrow$ H$_2$O + F$^-$(aq)
(HF is hydrogen fluoride, OH$^-$ is hydroxide ion, and F$^-$ is fluoride ion.)

(b) 2K(s) + Br$_2$(ℓ) $\longrightarrow$ 2KBr(s)
(KBr is potassium bromide.)

(c) Al(s) + 3O$_2$(g) $\longrightarrow$ 2Al$_2$O$_3$(s)
(Al$_2$O$_3$ is aluminum oxide.)

$\qquad\qquad$ heat

(d) NH$_4$NO$_3$(s) $\longrightarrow$ N$_2$O(g) + 2H$_2$O(g)
(NH$_4$NO$_3$ is ammonium nitrate and N$_2$O is dinitrogen oxide.)

(e) C$_2$H$_4$(g) + 3O$_2$(g) $\longrightarrow$ 2CO$_2$(g) + 2H$_2$O(g)
(C$_2$H$_4$ is ethylene.)

Section 5-9

23. Describe some observations that serve as evidence for chemical reactions.

24. Imagine a burning candle. Describe any evidence for chemical reactions that occur when a candle burns.

25. What evidence is there for the chemical reactions involved in the metabolism of foods in our bodies?

Sections 5-10 to 5-12

26. Define or describe each of the following.

(a) energy

(b) radiant energy

(c) heat

(d) potential energy

(e) kinetic energy

(f) mechanical energy

(g) electrical energy

(h) chemical energy

(i) exothermic reaction

(j) endothermic reaction

27. How is energy involved in chemical reactions?

28. What is the difference between an exothermic reaction and an endothermic reaction?

29. Sucrose, C$_{12}$H$_{22}$O$_{11}$, can be burned by igniting a mixture of sucrose and pure oxygen. The products of the reaction are carbon dioxide and water. The reaction is exothermic (releases energy). Draw a diagram for this reaction similar to Figure 5-3. Show the reactant formulas and product formulas on a relative energy scale.

30. State the law of conservation of energy.

31. What is a dietetic calorie?

32. How is the caloric content of foods determined?

33. The label on an 8 oz container of yogurt lists 10 g protein, 46 g carbohydrates, and 4 g fat. What is the caloric content of the yogurt in the container? What percent of the calories come from fat?

34. Typical peanuts contain 25.8% protein, 36.8% fat, and 21.9% carbohydrates. What is the caloric content of 10 g of peanuts? What percent of the calories come from fat?

35. One 8 oz cup of reduced-fat milk contains 5 g of fat, 12 g of carbohydrate, and 8 g of protein. Calculate the caloric content of the serving and the percent of calories that come from fat.

36. A serving size of a brand of packaged cookies contains 4 g fat, 23 g carbohydrates, and 2 g of protein. Calculate the caloric content of the serving and the percent of calories that come from fat.

37. One quarter of a pound is about 113 grams. Determine the number of calories in 113 g of each of the foods listed in Table 5-1. Cheddar cheese was given as example in the discussion of food calories. Also calculate the percent of calories that come from fat for each of the foods.

Question to Ponder

38. Using the patterns of compounds formed by families and oxygen in Section 5-1, predict the formulas for the chlorine compounds of all the elements shown in that section. (Some of them may not exist.)

CHAPTER
6

MOLECULES,

IONS, AND

CHEMICAL

NAMES

6-1 MOLECULES

Atoms are fundamental and basic chemical particles. They are the building blocks of compounds. Once combined, atoms become parts of the other kinds of chemical particles or chemical species. The three kinds of **chemical particles or chemical species are atoms, ions, and molecules**. Recall that a molecule is a particle made of two or more chemically combined atoms. Compounds composed of molecules are simply called molecular compounds. A formula of a molecular compound shows number of atoms of each element that compose the molecule. The atoms that make up a molecule are combined in a definite three-dimensional pattern that is visualized using a picture or model. Let's consider models of some simple molecules that comprise some common chemical compounds. A few compounds are made of simple diatomic molecules. Examples are hydrogen chloride, HCl, and carbon monoxide, CO.

Molecules

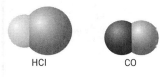

HCl   CO

Most molecular compounds have more complex molecules. Common examples are water, H_2O; ammonia, NH_3; methane, CH_4; formaldehyde, H_2CO; and carbon dioxide, CO_2.

147

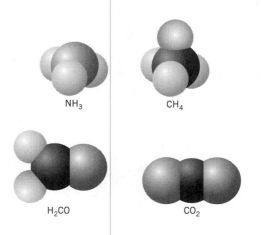

NH₃   CH₄

H₂CO   CO₂

The formulas of complex molecular compounds do not necessarily show how the atoms are bonded in a molecule of the compound. For instance, the formula for propane is often written as C_3H_8, and the molecule involves three carbon atoms bonded in sequence with the hydrogen atoms bonded to the carbons as shown in the model given here.

C_3H_8 or $CH_3CH_2CH_3$

To show the bonding sequence more clearly, some molecules are represented by framework models. The framework model of propane is shown below. Each center represents a carbon atom bonded to either other carbon atoms or to hydrogen atoms. As an alternative, ball-and-stick models show atoms as balls. The ball-and-stick model of propane is also shown here for comparison.

$CH_3CH_2CH_3$

The framework and ball-and-stick models of ethyl alcohol, C_2H_6O, show two bonded carbons with one bonded to three hydrogens and one bonded to two hydrogens and an oxygen that is also bonded to a hydrogen.

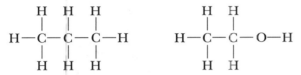

CH₃CH₂OH

Sometimes the formulas of molecular compounds are shown as two-dimensional (flat) projections that show which atoms are bonded by connecting the atomic symbols with lines. These are called structural formulas. The formulas for propane and ethyl alcohol are

$$
\begin{array}{ccc}
\text{H} & \text{H} & \text{H} \\
| & | & | \\
\text{H}-\text{C}-\text{C}-\text{C}-\text{H} \\
| & | & | \\
\text{H} & \text{H} & \text{H}
\end{array}
\qquad
\begin{array}{cc}
\text{H} & \text{H} \\
| & | \\
\text{H}-\text{C}-\text{C}-\text{O}-\text{H} \\
| & | \\
\text{H} & \text{H}
\end{array}
$$

Often the formulas for molecular compounds are written just to show the number of combined atoms of each element, and sometimes the formulas are written in a form that suggests the bonding sequence. The bonding sequence reveals which atoms are bonded. For example, the formula for propane is written as C_3H_8 or $CH_3CH_2CH_3$, and the formula of ethyl alcohol is written as C_2H_6O or CH_3CH_2OH. Note how the expanded formulas correspond to the bonding sequences in the compounds as shown above.

Which of the elements form molecular compounds? Generally, compounds containing nonmetals combined with other nonmetals are molecular. Also, compounds having metalloids combined with nonmetals are molecular. Thus, if the formula of a compound shows only nonmetals or metalloids and nonmetals, we can assume that the compound is composed of molecules corresponding to the formula. Use a periodic table to identify the kinds of elements in a compound. Recall that the nonmetals are found in the upper right corner of the periodic table. By identifying the kinds of elements in a formula you can decide whether it is a molecular compound. You are not, however, expected to deduce or figure out the bonding sequence in a molecule. Just remember that compounds containing only nonmetals or metalloids are molecular compounds

A molecule is a discrete particle, so the formula unit of a molecular compound is the formula of the molecule. The formula tells how many combined atoms of each type are in the molecule. The compound involving hydrogen and chlorine, called hydrogen chloride, is composed of HCl molecules. Molecules as formula units of compounds give us insight into why the empirical formulas of some compounds are different from the actual molecular formulas (see Section 4-10). The compound butane has molecules containing 4 carbon atoms and 10 hydrogen atoms. The molecular formula of butane is C_4H_{10}, which reflects the nature of butane molecules. The simplest ratio of carbon to hydrogen in this compound, however, is 2 to 5 (4/10 = 2/5). Thus, when the formula of butane is determined by experiment, it has an empirical formula of C_2H_5. Since butane contains molecules having 4 carbons and 10 hydrogens, it is represented by its actual molecular formula, C_4H_{10}. The structural formula of butane shows the bonding sequence in the molecule.

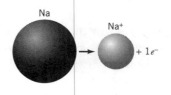

6-2 IONS

Some compounds contain combined ions rather than molecules. Simple ions are atoms that carry electrical charges. They form by the loss or gain of electrons from neutral atoms. Ionic compounds contain positive ions in combination with negative ions. Compounds containing ions are always electrically neutral because the charges of the positive ions balance the charges of the negative ions. For example, the ionic compound zinc oxide, used in medicinal and sun screen ointments, contains the zinc ion, Zn^{2+}, in combination with the oxide ion, O^{2-}. One zinc ion balances one oxide ion to give the formula ZnO.

Simple ions that form when atoms lose or gain electrons are sometimes called **monatomic ions.** Recall that an atom is neutral since it contains an equal number of electrons and nuclear protons. Ions formed by the loss of electrons from atoms carry positive charges since they have more nuclear protons than electrons. Positive ions are called cations. For example, a sodium atom loses one electron to form a sodium ion having a single positive charge.

Ions formed by the gain of electrons by atoms carry negative charges since they have more electrons than nuclear protons. Negative ions are called anions. For example, a chlorine atom gains one electron to form a chloride ion having a single negative charge.

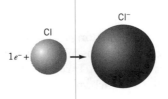

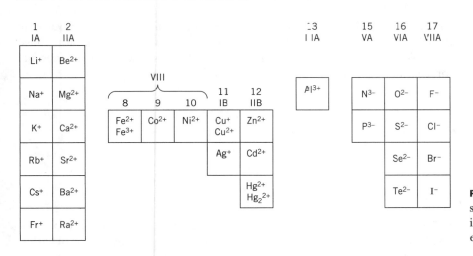

FIGURE 6-1 The simple ions or monatomic ions formed by common elements.

Various elements form ions with different charges. The formula of an ion shows the charge as a superscript following the symbol of the element that forms the ion. For example, the formula of sodium ion is Na^+ and the formula of chloride ion is Cl^-. The portion of the periodic table given in Figure 6-1 shows the formulas of some common simple ions. Note that metals typically form cations and nonmetals form anions. Ionic compounds are compounds that do not contain molecules. Instead, they contain distinct cations balanced by anions.

Some natural ionic compounds and some manufactured ionic compounds contain types of ions that are not simple ions. For example, a piece of limestone or marble is an ionic compound that contains the positive calcium ion, Ca^{2+}, in combination with a unique kind of negative ion. The negative ion is not a simple ion but a polyatomic ion containing more than one atom. A **polyatomic ion** is a group of two or more combined atoms that carries an electrical charge. The term *poly* means many, so a polyatomic ion contains many (two or more) atoms, whereas a simple ion is a single atom carrying a charge, or a monatomic ion. The compound found in limestone and marble is calcium carbonate. This compound contains calcium ions in combination with the polyatomic carbonate ion, CO_3^{2-} (read C–O–three–two–minus). The carbonate ion is an ion composed of three oxygen atoms combined with one carbon atom. This four-atom particle carries a double negative charge and can be visualized as shown here.

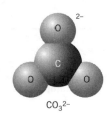

CO_3^{2-}

Table 6-1 includes the names and formulas of some common polyatomic ions (a similar table is inside the front cover). Most polyatomic ions have negative charges, but the ammonium ion, NH_4^+, is a common polyatomic ion that has a positive charge. A model is shown here.

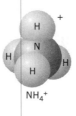

NH_4^+

Polyatomic ions are somewhat like molecules since they are composed of combined atoms. The important difference between ions and molecules, however, is that ions carry electrical charges. Furthermore, ions are always

Table 6-1   *Common Polyatomic Ions*

Cations

Ammonium ion	NH_4^+		
Hydronium ion[a]	H_3O^+		
Mercury(I) ion (mercurous ion)	Hg_2^{2+}		

Anions Single Charge

Acetate ion	$CH_3CO_2^-$ or $C_2H_3O_2^-$	Hydrogen sulfite ion	HSO_3^-
Chlorate ion	ClO_3^-	Hydroxide ion	OH^-
Chlorite ion	ClO_2^-	Hypochlorite ion	ClO^-
Cyanide ion	CN^-	Nitrate ion	NO_3^-
Formate ion	HCO_2^-	Nitrite ion	NO_2^-
Hydrogen carbonate ion (bicarbonate ion)	HCO_3^-	Perchlorate ion	ClO_4^-
Hydrogen sulfate ion	HSO_4^-	Permanganate ion	MnO_4^-

Anions Double Charge

Carbonate ion	CO_3^{2-}	Oxalate ion	$C_2O_4^{2-}$
Chromate ion	CrO_4^{2-}	Sulfate ion	SO_4^{2-}
Dichromate ion	$Cr_2O_7^{2-}$	Sulfite ion	SO_3^{2-}

Anion Triple Charge

Phosphate ion	PO_4^{3-}		

[a]The hydronium ion exists only in water solutions, not in compounds.

found in association with other ions of opposite charge. In other words, having a sample of a chemical containing molecules of one type is possible, but having a sample of a material containing only one type of positive ion or one type of negative ion is never possible. Positive and negative ions are always found in association with one another.

Since polyatomic ions are composed of combined atoms, showing their three-dimensional shapes by models is possible. Models of a few common polyatomic ions are shown here.

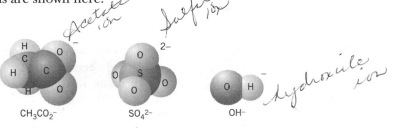

How do we know whether a compound contains molecules or ions? Generally, if the formula of a compound starts with a metal, it is assumed to be an ionic compound containing a metal cation and a nonmetal anion or a polyatomic ion. A common exception to this pattern involves those compounds that contain the ammonium ion, NH_4^+, in combination with a nonmetal anion or a polyatomic ion. When you see a formula that starts with the symbol of a metal or NH_4, you can assume that it is an ionic compound. Consult a periodic table to find metals, metalloids, and nonmetals. Recall that if the formula of a compound contains only nonmetals or a metalloid combined with a nonmetal, the compound is assumed to be molecular.

6-3 CHEMICAL NOMENCLATURE

What's in a chemical name? A name often contains some chemical information and sometimes a story. After Alexander the Great conquered parts of Egypt around 300 B.C.E., the Greeks built a temple dedicated to the god Ammon. Fires built in this temple used dried camel droppings as fuel. Over many years, a white crystalline material built up on the walls of the temple along with soot from the fires. This white, saltlike material was known as *sal ammoniac,* "salt of Ammon." Today, it is obtained from other sources and is usually named ammonium chloride, but sometimes the old name is used. In the early 1700s, sal ammoniac was used to produce a gaseous compound that was named ammonia (the spirit of Ammon) after its source. The compound is NH_3 and, centuries after it was named, ammonia is still its official name.

We can read a formula by pronouncing the element symbols as letters and the subscripts as numbers.

Na_2SO_3	N–A–two–S–O–three
$Ca(NO_3)_2$	C–A–paren–N–O–three–paren–two
K_2O	K–two–O
PBr_3	P–B–R–three
N_2O_5	N–two–O–five

Another approach is to translate the formula to the chemical name. Admittedly, this requires some knowledge of **chemical nomenclature.** Official systems of nomenclature are used to name various kinds of compounds. Names that convey information about the composition of compounds are **systematic names.** Names of many chemical compounds developed before any nomenclature rules were established. These names generally are not systematic and do not convey any information about the structure of the compound. They are **common** or **trivial names.** Some are so familiar that chemists normally use them, too. For example, the name water for H_2O is a common name that is invariably used. A few important compounds that have common names are

CH_4	methane	C_2H_6	ethane
C_3H_8	propane	C_4H_{10}	butane
NH_3	ammonia	H_2O_2	hydrogen peroxide
C_6H_6	benzene	$C_6H_{12}O_6$	glucose
$C_{12}H_{22}O_{11}$	sucrose	CH_3OH	methyl alcohol
CH_3CH_2OH	ethyl alcohol	C_2H_2	acetylene

Table 6-2 lists the trivial and chemical names for other familiar compounds. Occasionally, you may hear mineral names used for some chemicals. Examples are quartz (SiO_2), galena (a lead ore), bauxite (typical aluminum ore), and hematite and magnetite (iron ores). Mineral names developed historically and are not systematic chemical names. Most chemicals used as medicines and drugs are complex organic compounds. Normally, they are not referred to by their complex chemical names. Most have common names, trade names, or abbreviated names. For example, the pain reliever with the chemical name of acetylsalicylic acid is simply called aspirin. The painkiller found in the juice of opium poppies is called morphine, named for Morpheus, the Greek mythological god of dreams and sleep. The active ingredient in marijuana is called THC, an abbreviated form of the chemical name delta-9-tetrahydrocannabinol.

**Generic Drugs
Aspirin**

Table 6-2   *Trivial and Systematic Names for Some Common Compounds*

FORMULA	TRIVIAL NAME	SYSTEMATIC NAME
C_2H_2	Acetylene	Ethyne
$NaHCO_3$	Baking soda	Sodium hydrogen carbonate
$CaCO_3$	Calcite, marble, limestone, chalk	Calcium carbonate
$KHC_4H_4O_6$	Cream of tartar	Potassium hydrogen tartrate
C_2H_5OH	Grain alcohol, drinking alcohol	Ethyl alcohol or ethanol
$Na_2S_2O_3$	Hypo	Sodium thiosulfate
N_2O	Laughing gas, nitrous oxide	Dinitrogen oxide
PbO	Litharge	Lead(II) oxide
CaO	Quick lime	Calcium oxide
$NaOH$	Lye, caustic soda	Sodium hydroxide
K_2CO_3	Potash	Potassium carbonate
NH_4Cl	Sal ammoniac	Ammonium chloride
$NaNO_3$	Saltpeter (Chile)	Sodium nitrate
$Ca(OH)_2$	Slaked lime	Calcium hydroxide
$C_{12}H_{22}O_{11}$	Sugar	Sucrose or alpha-D-glucopyranosyl-beta-D-fructofuranoside
$NaCl$	Salt	Sodium chloride
$(CH_3)_2CHOH$	Rubbing alcohol	Isopropyl alcohol, 2-propanol
CH_3OH	Wood alcohol	Methyl alcohol or methanol

Generic Drug Names

Prescription and over-the-counter medicines contain chemicals that are complex organic compounds. These chemicals are sold under a variety of trade names or brand names. Many common drugs are marketed under specific brand names, but all brands contain the same drug. The generic name is the common chemical name that identifies the drug. Generic names cannot be used as exclusive brand names. Very often, a less advertised over-the-counter drug or a generic prescription drug has the same active ingredient but a lower price than a name brand. Ask a pharmacist for advice and you may save some money.

The most commonly used over-the-counter drugs are the pain relievers aspirin, ibuprofen, acetaminophen, and naproxen sodium. Aspirin relieves pain and reduces fever and inflammation. It is not useful for sharp, stabbing, intense pain caused by direct stimulation of nerves, but it is effective for dull, throbbing pain of inflammation or headache. Although aspirin was originally a brand name belonging to the Bayer corporation, it has, over time, become a generic name for the chemical acetylsalicylic acid.

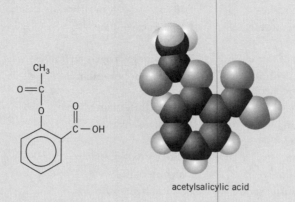

acetylsalicylic acid

Any brand of aspirin marked as USP (United States Pharmacopeia, a standard of purity) meets the same standard and has the same effect as any of the most heavily advertised brands.

Aspirin, like any drug, has side effects. Typically, some stomach bleeding occurs when you take an aspirin. Heavy drinkers and people with certain allergies, asthma, diabetes, and peptic ulcers should not take aspirin. Aspirin should not be used during the last three months of pregnancy since it "thins" the blood and can interfere with clotting. Children with flulike symptoms or with chicken pox should never be given aspirin since aspirin use is somehow related to Reye's syndrome, a serious but rare blood illness. Aspirin is poisonous in large doses, and thousands of cases of serious overdoses occur each year. Young children are particularly sensitive to aspirin overdose. Safely store aspirin and any over-the-counter or prescription drugs and keep them in childproof containers.

Acetaminophen is an aspirin substitute. It is in products like Tylenol, Anacin-3, and Midol. It can, however, be purchased under its generic name. Acetaminophen has fever- and pain-reducing effects but does not irritate the stomach as much as aspirin. Its minimal antiinflammatory effect recommends against its use for treating arthritis and rheumatism. Chronic use of acetaminophen or use by heavy drinkers can cause liver and kidney damage. Prolonged use or exceeding the recommended dose is inadvisable without consulting a physician, especially for children. Children should not be given adult doses.

Ibuprofen, another aspirin substitute, is a pain reliever and reduces fever and inflammation. You can purchase it under the generic name and under brand names such as Advil, Motrin IB, Midol 200, and Nuprin. Check the ingredients on the label when buying a pain reliever. Some people with arthritis use ibuprofen rather than aspirin. Chronic use, however, can cause serious kidney damage in heavy drinkers and persons having certain kidney diseases. Ibuprofen also "thins" the blood just like aspirin. Women should not use it during the last four months of pregnancy.

Naproxen sodium is an aspirin substitute sold under the brand name of Aleve. It is useful for pain from headache, menstrual cramps, or arthritis. One tablet provides relief for 8 to 12 hr. It can cause stomach irritation and allergic reactions. Furthermore, the recommended dose is not effective as an antiinflammatory. It should not be taken in doses above the recommended dose without consulting a pharmacist or medical practitioner.

6-4 CLASSIFYING COMPOUNDS

Our discussion concerns the naming of simple ionic and molecular compounds, not complex organic compounds. Different methods of nomenclature are used for ionic and molecular compounds. Consequently, to name a compound, first classify it as ionic or molecular. Fortunately, classification is quite simple. Use the formula of a compound to classify it into one of the following categories. A binary compound is a compound containing only two elements.

1. Ionic Compounds
   (a) Metal–nonmetal binary compounds
       (e.g., sodium chloride, $NaCl$)
   (b) Metal ion–polyatomic ion compounds
       (e.g., potassium nitrate, KNO_3)
   (c) Ammonium ion–polyatomic ion or nonmetal ion compounds [e.g., ammonium sulfate, $(NH_4)_2SO_4$, and ammonium chloride, NH_4Cl]
2. Molecular Compounds
   (a) Nonmetal–nonmetal binary compounds
       (e.g., carbon dioxide, CO_2)
   (b) Metalloid–nonmetal binary compounds
       (e.g., silicon tetrafluoride, SiF_4)
   (c) Compounds including three or more nonmetals
       (e.g., nitric acid, HNO_3, and glucose, $C_6H_{12}O_6$)

Some examples illustrate the classification of compounds.

CaS	Formula starts with a metal, so this binary metal–nonmetal compound is ionic.
$Al_2(SO_4)_3$	Formula starts with a metal, so this metal–polyatomic ion compound is ionic.
NH_4Br	Formula starts with NH_4, so this ammonium ion–containing compound is ionic.
P_4O_{10}	Formula has no metal, so this binary nonmetal–nonmetal compound is molecular.

ACTIVITY

6-1

Classify the following compounds as ionic or molecular.
Ca_3N_2, NH_4Br, $C_{12}H_{22}O_{11}$, H_2S, K_3PO_4, NO

Table 6-3 *Names and Formulas of Some Common Metal Cations*

FORMULA	NAME
Li^+	lithium ion
Na^+	sodium ion
K^+	potassium ion
Mg^{2+}	magnesium ion
Ca^{2+}	calcium ion
Sr^{2+}	strontium ion
Ba^{2+}	barium ion
Cr^{3+}	chromium (III) ion (chromic ion)
Mn^{2+}	manganese (II) ion (manganous ion)
Fe^{3+}	iron(III) ion (ferric ion)
Fe^{2+}	iron(II) ion (ferrous ion)
Co^{2+}	cobalt(II) ion (cobaltous ion)
Ni^{2+}	nickel(II) ion (nickelous ion)
Cu^+	copper(I) ion (cuprous ion)
Cu^{2+}	copper(II) ion (cupric ion)
Ag^+	silver ion
Zn^{2+}	zinc ion
Cd^{2+}	cadmium ion
Hg^{2+}	mercury(II) ion (mercuric ion)
Hg_2^{2+}	mercury(I) ion (mercurous ion)
Al^{3+}	aluminum ion

H_2SO_4 Formula has no metal, so this nonmetal compound is molecular.

$AsCl_3$ Formula has no metal, so this binary metalloid–nonmetal compound is molecular.

6-5 NAMING IONS

The name of an ionic compound comes from the names of the ions that compose it. Metals form positive ions by loss of electrons. Normally, a **metal ion** (a metal cation) uses the name of the metal followed by the word *ion*. Some examples are

Na^+ sodium ion (All group 1 metals form 1+ ions.)
Ca^{2+} calcium ion (All group 2 metals form 2+ ions.)
Zn^{2+} zinc ion (Zn, Cd, and Hg form 2+ ions.)

Those metals that form only one kind of ion include the group 1 (IA) metals, the group 2 (IIA) metals, and zinc (Zn^{2+}), silver (Ag^+), cadmium (Cd^{2+}), and aluminum (Al^{3+}).

Some transition metals and the group 14 (IVA) metals, tin and lead, can form more than one type of metal ion. For example, iron forms the Fe^{2+} and Fe^{3+} ions and lead forms the Pb^{2+} and Pb^{4+} ions. Name such ions by using the name of the metal followed by a Roman numeral in parentheses. The Roman numeral shows the charge.

Fe^{2+} iron(II) ion (read as iron–two–ion)
Fe^{3+} iron(III) ion (read as iron–three–ion)
Pb^{2+} lead(II) ion (read as lead–two–ion)
Pb^{4+} lead(IV) ion (read as lead–four–ion)

Table 6-3 gives the formulas and names of common metal ions. Some have alternate names, for example, Fe^{2+} is the ferrous ion and Fe^{3+} is the ferric ion. Roman numeral names for ions are preferred, but sometimes the older names are used, particularly in medicine, agriculture, and other areas of applied chemistry. You may notice ferrous sulfate as an ingredient in some vitamin–mineral supplements or food products containing iron-fortified flour. In the older nomenclature, the ion of lower charge is the -*ous* ion, and the ion of higher charge is the -*ic* ion. The ferrous and ferric ions illustrate this point.

Nonmetals form simple negative ions that are found in combination with positive metal ions in ionic compounds. Name **simple negative ions** (nonmetal anions) by using the root of the name of the nonmetal with an -*ide* ending.

N^{3-} nitride ion   (Group 15 nonmetals and metalloids form 3− ions.)
S^{2-} sulfide ion   (Group 16 nonmetals and metalloids form 2− ions.)
F^- fluoride ion   (Group 17 elements form 1− ions.)

Table 6-4 lists the formulas and names of some nonmetal ions, and Table 6-5 lists the roots of the names of some nonmetals. Notice how the charges of these simple ions relate to the group numbers of the elements in the periodic table.

Now consider the names of **polyatomic ions.** Table 6-1 and the table inside the front cover lists the formulas and names of the most common polyatomic ions. When dealing with polyatomic ions, you must learn the names, formulas, and charges of the ions. The names of polyatomic ions are common names that do not reveal the formulas. Some names, however, suggest the presence of certain elements. The sulfate ion, SO_4^{2-}, is a polyatomic ion containing sulfur. In fact, you may notice that many polyatomic ions have names that end in *-ate*. (What do these ions have in common?)

The one common positively charged polyatomic ion is the ammonium ion, NH_4^+. In a sense, the mercury(I) ion, Hg_2^{2+}, is a polyatomic ion. We might expect the mercury(I) ion to have the formula of Hg^+, but it is a polyatomic ion that occurs as a combination of two mercury atoms and it has a double positive charge. The polyatomic mercury(I) ion always occurs as Hg_2^{2+} in ionic compounds. Most of the polyatomic ions are negative ions and most contain oxygen and another nonmetal. Some nonmetals form two different ions with oxygen. The name of the ion with the lesser number of oxygen atoms ends in *-ite*, whereas the name of the ion with the greater number of oxygen atoms ends in *-ate*. Nomenclature rules use suffixes with slight variations in endings much like various endings are used in the conjugation of verbs in many languages. The endings used in nomenclature come from Latin. It is interesting that similar endings occur in some common words, such as arsenic, sedate, tripartite, and gracious (Spanish). Examples of ions with different endings in their names are

SO_4^{2-} sulfate ion   and SO_3^{2-} sulfite ion
NO_3^- nitrate ion   and NO_2^- nitrite ion

6-6 FORMULAS OF IONIC COMPOUNDS FROM ION FORMULAS

All ionic compounds are electrically neutral so the total positive charge always equals the total negative charge. This fact is reflected in the formula of an ionic compound. Thus, predicting the formula of an ionic compound is easy if we know the formulas of the ions in the compound. For

Table 6-4 *Names and Formulas of Some Common Nonmetal Anions*

FORMULA	NAME
N^{3-}	nitride ion
P^{3-}	phosphide ion
As^{3-}	arsenide ion
O^{2-}	oxide ion
S^{2-}	sulfide ion
Se^{2-}	selenide ion
Te^{2-}	telluride ion
F^-	fluoride ion
Cl^-	chloride ion
Br^-	bromide ion
I^-	iodide ion

Table 6-5 *Roots of the Names of Some Common Elements*

ELEMENT	ROOT
hydrogen	hydr-
boron	bor-
carbon	carb-
silicon	silic-
nitrogen	nitr-
phosphorus	phosph-
arsenic	arsen-
antimony	antimon-
oxygen	ox-
sulfur	sulf- or sulfur-
selenium	selen-
tellurium	tellur-
fluorine	fluor-
chlorine	chlor-
bromine	brom-
iodine	iod-

example, the formula for the compound containing sodium ions, Na^+, and fluoride ions, F^-, is simply

$$Na^+ \, F^- \qquad NaF \quad \text{(one sodium ion for each fluoride ion)}$$

The formula for the compound containing potassium ions, K^+, and sulfide ions, S^{2-}, is

$$K^+ \, S^{2-} \qquad K_2S$$

The arrow shows that two K^+ ions are needed to balance one S^{2-} ion. In other words, the charge of each ion shows the number of oppositely charged ions needed in the formula.

The formula for the compound containing aluminum ions, Al^{3+}, and oxide ions, O^{2-}, is

$$Al^{3+} \, O^{2-} \qquad Al_2O_3$$

You can deduce the formulas for metal–polyatomic ion compounds in the same way. The formula for the compound containing sodium ions and nitrate ions, NO_3^-, is

$$Na^+ \quad NO_3^- \qquad NaNO_3$$

The formula for the compound of calcium ions, Ca^{2+}, and hydroxide ions, OH^-, is

$$Ca^{2+} \, OH^- \qquad Ca(OH)_2$$

The hydroxide ion is a unit so it is enclosed in parentheses to show two units in the formula. The formula for the compound of magnesium ions, Mg^{2+}, and phosphate ions, PO_4^{3-}, is

$$Mg^{2+} \, PO_4^{3-} \qquad Mg_3(PO_4)_2$$

The phosphate ion is a unit, so it is enclosed in parentheses to show two units in the formula. Use parentheses as needed to show a multiple of any polyatomic ion in a compound. In fact, you typically see some formulas written using parentheses to enclose groups of atoms. Do not, however, use parentheses if there is only one unit of a polyatomic ion in a formula. For instance, write $NaNO_3$ but not $Na(NO_3)$.

6-7 NAMING IONIC COMPOUNDS

As mentioned in Section 6-4, classes of ionic compounds include metal–nonmetal binary, metal–polyatomic ion, and ammonium ion–nonmetal or polyatomic ion compounds. We can recognize from the formula that a compound is ionic. We name it simply by giving the name of the positive ion (cation) followed by the name of the negative ion (anion). Consider some examples of each type of ionic compound.

Metal–Nonmetal Binary

We name metal–nonmetal binary compounds by using the name of the metal followed by the root of the name of the nonmetal adding an *-ide* ending. See Table 6-5 for the roots of nonmetal names. The charge on a simple nonmetal anion is deduced by finding its group in the periodic table. Remember that group 17 elements form singly charged $(-)$ negative ions. Group 16 nonmetals form doubly charged $(2-)$ negative ions. Group 15 nonmetals form triply charged $(3-)$ negative ions.

KBr is binary and contains potassium and bromine; thus, its name is potassium bromide.

Ca_3N_2 is binary and contains calcium and nitrogen; thus, its name is calcium nitride.

If the metal in the compound is an element that forms more than one kind of ion, its charge is deduced from the charge of the nonmetal ion. The positive charge is shown in the ion name by using a Roman numeral.

FeS   The compound contains sulfide, S^{2-}, so the iron must be Fe^{2+} to balance the charge. Its name is iron(II) sulfide.

PbO_2 The compound contains oxide, O^{2-}, so the lead must be Pb^{4+} to balance the charge. Its name is lead(IV) oxide.

Metal–Polyatomic Ion

To name metal–polyatomic compounds, we must recognize which polyatomic ions are in the compound. Furthermore, for metals that form more than one kind of ion, the charge of the metal must be deduced from the charge of the polyatomic ion.

$NaNO_3$ NO_3^- is the nitrate ion, so the compound name is sodium nitrate.

$Al(OH)_3$ OH^- is the hydroxide ion, so the compound name is aluminum hydroxide.

K_2SO_4 SO_4^{2-} is the sulfate ion, so the name is potassium sulfate.

$HgCO_3$ CO_3^{2-} is the carbonate ion; thus, the mercury ion must be Hg^{2+}, which is the mercury(II) ion. The name of the compound is mercury(II) carbonate.

$Cu_3(PO_4)_2$ PO_4^{3-} is the phosphate ion; thus, the copper ion must be Cu^{2+} to balance the charge. That is, since two PO_4^{3-} are in the compound $[2 \times (3-) = 6-]$, there must be three copper(II) ions $[(6+)/3 = 2+]$. The name of the compound is copper(II) phosphate.

Ammonium Ion–Nonmetal or Polyatomic Ion

Name the following compounds.
SnF_2, Ca_3N_2, K_3PO_4, NH_4Br

We name ammonium ion–nonmetal or polyatomic ion compounds "ammonium" followed by the name of the anion. If the anion is a polyatomic ion, we need to recognize it from its formula.

NH_4Cl contains the Cl^-, which is chloride ion, so the compound name is ammonium chloride.

$(NH_4)_2SO_4$, SO_4^{2-} is the sulfate ion, so the name of the compound is ammonium sulfate.

6-8 NAMING NONMETAL–NONMETAL BINARY COMPOUNDS

Glance at a periodic table and recall that the nonmetals populate the upper right region of the table. The metalloids separate the metals and nonmetals. Although there are fewer nonmetals than metals, numerous nonmetal–nonmetal compounds are known. Note that this category of compounds is binary and, therefore, they contain only two different elements. The elements must be nonmetals or metalloids. Some nonmetals can form more than one compound with one another; for instance, nitrogen and oxygen form N_2O, NO, NO_2, N_2O_3, N_2O_4, and N_2O_5.

An example of a binary nonmetal–nonmetal compound is CO_2, which has the name carbon dioxide. We name binary nonmetal–nonmetal and metalloid–nonmetal compounds according to the general pattern that uses the name of the first element followed by the root of the name of the second element with an *-ide* ending. Each part of the name is given an appropriate prefix to denote the number of combined atoms of each element in the compound.

	prefix | name of first element	prefix | root of name of second element | ide suffix		
CO_2	carbon	di	ox	ide

As an example, the compound BF_3 is named boron trifluoride to suggest one boron in combination with three fluorines. Table 6-5 lists the roots used for the nonmetals. Below is a list of common prefixes used to name binary compounds.

mono-   One as in mononucleosis. (This prefix is usually omitted except in carbon monoxide, CO)

di-   Two as in dipole.

tri-	Three as in triangle.
tetra-	Four as in tetravalent.
penta-	Five as in pentathlon.
hexa-	Six as in hexapod.
hepta-	Seven as in heptagon.
octa-	Eight as in octapus.
nona-	Nine as in nonagenerian.
deca-	Ten as in decade.

Which of the nonmetal is named first in a binary compound and which is given the *-ide* ending? The preferred order of elements in nonmetal–nonmetal or metalloid–nonmetal compounds is

B, Ge, Si, C, Sb, As, P, N, H, Te, Se, S, I, Br, Cl, O, and F

Recall that, with the exceptions of hydrogen and oxygen, this order is from bottom to top of each group moving from left to right in the periodic table. By convention hydrogen is written after any element in group 15 or lower and before any element in group 16 or higher (e.g., NH_3 and H_2O). Oxygen comes after all elements except fluorine (e.g., Cl_2O and OF_2). Naming compounds by the prefix method relates directly to the formula of the compound; if we know the formula, the name is obvious; if we know the name, the formula is obvious. Here are some examples. Remember that the basis of the name is always the first element followed by the root of the second element with an *-ide* ending, as illustrated in the first parts of each of the following examples. In the actual name, the prefixes are used to show the relative number of combined atoms.

N_2O	_di_ nitrogen oxide	dinitrogen oxide
NO_2	nitrogen _di_ oxide	nitrogen dioxide
N_2O_5	_di_ nitrogen _pent_ oxide	dinitrogen pentoxide (the "a" of penta is omitted for easier pronunciation)
P_4O_{10}	_Tetra_ phosphorus _dec_ oxide	tetraphosphorus decoxide (the "a" of deca is omitted for easier pronunciation)
OF_2	oxygen _di_ fluoride	oxygen difluoride
P_4S_7	_Tetra_ phosphorus _Hepta_ sulfide	tetraphosphorus heptasulfide
PCl_3	phosphorus _tri_ chloride	phosphorus trichloride
PCl_5	phosphorus _penta_ chloride	phosphorus pentachloride
CO	carbon _mon_ oxide	carbon monoxide
CS_2	carbon _di_ sulfide	carbon disulfide

Name the following compounds.

SiF_4, P_4O_6, N_2O_3, KNO_3, HBr

6-9 NAMING BINARY ACIDS

Acids are an important set of chemicals that have their own unique method of nomenclature. The binary compounds involving hydrogen and the halogens (group 17 or VIIA) are named according to the nonmetal–nonmetal nomenclature rules. As a group they are the hydrogen halides. For example, HCl is hydrogen chloride and HI is hydrogen iodide. When hydrogen halides dissolve in water, the resulting solutions display specific properties called **acidic properties.** Some common properties of acidic solutions are that they taste sour, dissolve many metals, can burn the skin, and turn blue litmus paper red. Acids are described in detail in Chapter 13.

The water solutions of the hydrogen halides are called **acids,** and their solutions are named accordingly. Normally, we give names only to pure compounds. Since these acid solutions are very important in chemistry, however, they are assigned special names. The prefix is *hydro-*, and the root is the name of the halogen with an *-ic* ending. This first part of the name is followed by the word *acid*.

hydro (root of halogen name) ic acid

The names of the four **hydrogen-halide acids** are given in the following table.

HF

HCl

HBr

HI

FORMULA OF PURE COMPOUND	NAME OF PURE COMPOUND	NAME OF WATER SOLUTION
HF	Hydrogen fluoride	Hydrofluoric acid
HCl	Hydrogen chloride	Hydrochloric acid
HBr	Hydrogen bromide	Hydrobromic acid
HI	Hydrogen iodide	Hydroiodic acid

The water solutions of acids are denoted by using the formula of the compound followed by (aq) to suggest aqueous or water solution, for example, $HCl(aq)$, $HF(aq)$. Note that HF is named hydrogen fluoride and $HF(aq)$ is named hydrofluoric acid. We use the same pattern for any of the four hydrogen-halide acids. Hydrochloric acid is the most common of these acids; it is typically found in chemistry laboratories and has many industrial uses. Its trivial name is muriatic acid.

6-10 NAMING OXOANIONS AND OXOACIDS

Many nonmetals form ternary compounds that include hydrogen and oxygen. Ternary compounds contain three elements. Most of these compounds and their corresponding water solutions have acidic properties.

The compounds that contain hydrogen, oxygen, and another element are **oxoacids** or **oxyacids.** For a given oxoacid, the same name is usually used to refer to both the pure oxoacid and its water solutions. For example, H_2SO_4 and $H_2SO_4(aq)$ are both called sulfuric acid. Very often, you can recognize an oxoacid if it has a formula that begins with hydrogen and contains another element and oxygen. Common examples include nitric acid, HNO_3; sulfuric acid, H_2SO_4; and phosphoric acid, H_3PO_4. Molecular models for these acids are shown below.

COMMON OXOACIDS

H_2SO_4	sulfuric acid
HNO_3	nitric acid
H_3PO_4	phosphoric acid
CH_3CO_2H or $HC_2H_3O_2$	acetic acid

HNO₃   H₂SO₄   H₃PO₄

Some oxoacids that include carbon are called **organic acids** or **carboxylic acids.** Two examples are formic acid, $HCHO_2$, and acetic acid, $HC_2H_3O_2$. Molecular models for these acids are given below.

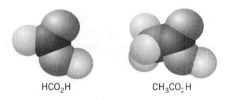

HCO₂H   CH₃CO₂H

The formulas for carboxylic acids are sometimes written to suggest the bonding sequences in the molecules. For instance, formic acid is written as HCO_2H and acetic acid as CH_3CO_2H.

Oxoanions or **oxyanions** are negative polyatomic ions related to the oxoacids. The formulas of these ions follow the same pattern as the formulas of the oxoacids. The difference is that the acidic hydrogen is not present in the ion and the ions carry a corresponding number of negative charges. For instance, if two hydrogens are in the acid, then the corresponding ion carries a double negative charge; H_2SO_4 and SO_4^{2-}. Some nonmetals form more than one oxoanion, so they are distinguished by use of different suffixes or prefixes in their names. The different oxoanions of a given element differ in the number of combined oxygen atoms. Nitrogen forms two oxoanions.

$$NO_3^- \text{ nitrate ion} \qquad NO_2^- \text{ nitrite ion}$$

When there are two oxoanions for an element, the one with the greater number of oxygen atoms has an **-ate** suffix and the one with the lesser number of oxygen atoms has an *-ite* suffix. The name always includes the word *ion*. The oxoanions of group 16 (VIA) elements (sulfur, selenium, and tellurium) are exemplified by those of sulfur.

$$SO_4^{2-} \text{ sulfate ion} \qquad SO_3^{2-} \text{ sulfite ion}$$

All the halogens form four oxoanions except fluorine, which forms none. These ions are exemplified by those of chlorine. Since there are four, special prefixes are needed in their names.

$$ClO_4^- \quad \text{perchlorate ion}$$
$$ClO_3^- \quad \text{chlorate ion}$$
$$ClO_2^- \quad \text{chlorite ion}$$
$$ClO^- \quad \text{hypochlorite ion}$$

ACTIVITY 6-4

Give the names of the BrO^- and IO_3^-.

The formula ClO_3^- represents the chlorate ion, and ClO_4^-, which has one more oxygen, represents the perchlorate ion. The **prefix *per-*** is related to *hyper*, meaning above or elevated (e.g., hyperactive). ClO_2^- is chlorite ion, and ClO^-, which has one less oxygen, is hypochlorite ion. The **prefix *hypo-*** means below or under (e.g., hypodermic means under the skin).

Oxoanions and oxoacids are related in name and formula. The formula of an oxoanion is derived from the formula of an oxoacid by removing hydrogen ions, H^+. The charge of the anion depends on the number of hydrogen ions that are removed. Three examples are shown here.

nitric acid HNO_3 nitrate ion NO_3^-

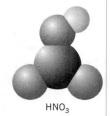

HNO₃

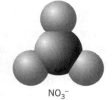

NO₃⁻

sulfuric acid H_2SO_4 sulfate ion SO_4^{2-}

H₂SO₄   SO₄²⁻

acetic acid CH_3CO_2H or $HC_2H_3O_2$ acetate ion $CH_3CO_2^-$ or $C_2H_3O_2^-$

CH_3CO_2H $CH_3CO_2^-$

We name an oxoacid by using the name of the corresponding oxoanion, changing the suffix, and adding the word *acid.* The *-ate* ending of the ion name is changed to *-ic* for the acid name, and the *-ite* ending of the ion name is changed to *-ous* for the acid name. Some examples of polyatomic oxoacids related to oxoanions are

$C_2H_3O_2^-$	acetate ion	$HC_2H_3O_2$	acetic acid
HCO_2^-	formate ion	HCO_2H	formic acid
NO_3^-	nitrate ion	HNO_3	nitric acid
NO_2^-	nitrite ion	HNO_2	nitrous acid
SO_4^{2-}	sulfate ion	H_2SO_4	sulfuric acid
SO_3^{2-}	sulfite ion	H_2SO_3	sulfurous acid
ClO_4^-	perchlorate ion	$HClO_4$	perchloric acid
ClO_3^-	chlorate ion	$HClO_3$	chloric acid
ClO_2^-	chlorite ion	$HClO_2$	chlorous acid
ClO^-	hypochlorite ion	$HClO$	hypochlorous acid

Note that the syllable *-ur-* is added in the names of sulfur acids by convention.

The names of oxoacids do not directly reveal their formulas, so we need to memorize the names and formulas of the common acids. By convention, the formula of an acid begins with hydrogen (i.e., HCl, H_2SO_4). Thus, if we encounter such a formula, we normally classify the compound as an acid; water, H_2O, is a notable exception.

Binary acids and oxoacids with more than one hydrogen can form polyatomic ions that result from the loss of one or more hydrogen ions (H^+) from the parent acid. For instance, when sulfuric acid, H_2SO_4, loses one hydrogen ion, it forms the hydrogen sulfate ion, HSO_4^-. The charges of the ions correspond to the absence of the hydrogens. We name these ions by using the anion name preceded by the word *hydrogen.* If the ion has two hydrogens, the prefix *di-* is used with the word hydrogen to tell that two hydrogens are associated with the ion.

$H_2PO_4^-$	dihydrogen phosphate ion	(H_3PO_4 less one H^+)
HPO_4^{2-}	hydrogen phosphate ion	(H_3PO_4 less two H^+)
HSO_4^-	hydrogen sulfate ion	(H_2SO_4 less one H^+)

ACTIVITY

6-5

Give the names and formulas for the oxoacids of selenium and bromine.

HS^-	hydrogen sulfide ion	(H_2S less one H^+)
HCO_3^-	hydrogen carbonate ion	(H_2CO_3 less one H^+. This ion is often called bicarbonate ion, where "bi" refers to hydrogen, not two of something.)

6-11 NAMES FROM FORMULAS

We have considered the nomenclature of common ionic compounds and some molecular compounds; these include most of the compounds you will encounter in this book. Nevertheless, other compounds are named according to other nomenclature methods. Compounds of carbon, organic compounds, fall into many classes, which follow prescribed methods of nomenclature. Here we will only be concerned with the nomenclature of the types of compounds described in this chapter.

To name a compound given its formula, we first classify the compound by considering the elements in its formula. If it is classed as either ionic, binary molecular, or an acid, we can name it. Be careful to use the prefix method only for binary molecular compounds. Before you use the prefix method, check to see whether or not the compound is binary and contains only nonmetals or a metalloid and a nonmetal. Never use the prefix method to name an ionic compound, an oxoacid, or a complex molecular compound. If a compound does not belong to a class we have considered, we can name it only if we recognize it as a common compound. A flowchart for nomenclature is shown in Focus 6.

Once it is classified, we name a compound using the appropriate method of nomenclature. Refer to the flowchart in Focus 6 to trace the following examples. The compound CaS starts with a metal so it classified as a binary ionic compound. It contains only two elements, the calcium ion and the sulfide ion, so its name is

calcium sulfide

The compound $Cu(OH)_2$ contains a metal and is classified as an ionic compound. Dissecting the formula, we see that it includes the hydroxide ion, OH^-. The cation must have a double positive charge, which makes it the Cu^{2+} or copper(II) ion. The name of the compound is

copper(II) hydroxide

The compound Cl_2O_5 is classified as a binary nonmetal–nonmetal compound, so by the prefix method its name is

____chlorine   ____oxide   dichlorine pentoxide

The compound SiF_4 is classified as a binary metalloid–nonmetal compound, so by the prefix method its name is

silicon   ____fluoride   silicon tetrafluoride

The compound NH_3 is classified as a binary nonmetal–nonmetal compound, but it should be recognized as having the common name

ammonia

The compound $NaHCO_3$ contains a metal and is classified as an ionic compound. Dissecting the formula, we see that it contains the sodium ion and the hydrogen carbonate ion, HCO_3^-. Thus, its name is

sodium hydrogen carbonate

The compound H_3PO_4 is classified as an oxoacid and has the common name

phosphoric acid

ACTIVITY 6-6

Name the following compounds.
Ag_2S, $PbCl_4$, SiC, I_2O_5, HNO_3, H_2O_2

6-12 FORMULAS FROM NAMES

Another aspect of nomenclature is to deduce the formula of a compound from its name. This is easily done for binary nonmetal compounds. For these compounds, the name gives the precise formula. Some examples are

dinitrogen trioxide   Two N and three O: N_2O_3

phosphorus pentafluoride   One P and five F: PF_5

The name of an ionic compound suggests which ions are in the compound. To deduce the formula from the name, we need to start with the precise formula of each ion and the charges of the ions. You need to memorize the formulas of common polyatomic ions or find them in a table of ions. Consulting a periodic table can help you recall the charges of many simple ions. Also remember that the Roman numeral in an ion name tells the charge of the ion. The formula is deduced by writing the ion formulas and deciding on the number of positive and negative ions needed to balance the charges. In other words, find the number of ions needed that make the ionic formula unit neutral. Magnesium fluoride is composed of magnesium ion and fluoride ion:

Mg^{2+} F^- so the formula is MgF_2.

Calcium phosphide is composed of calcium ion and phosphide ion:

Ca^{2+} P^{3-} so the formula is Ca_3P_2.

Iron(II) nitrate is composed of iron(II) ion and nitrate ion:

Fe^{2+} NO_3^- so the formula is $Fe(NO_3)_2$.

Remember that the Roman numeral is the charge.

ACTIVITY 6-7

Give the formulas for the following compounds: ammonium iodide, boron trifluoride, potassium hydrogen sulfate, acetic acid, calcium nitrate

ACTIVITY 6-8

A label from an over-the-counter mineral supplement is shown in Figure 6-2. Using the methods discussed in this chapter, write the formulas of any of the compounds on the label that you recognize as ionic compounds. Don't bother trying to write formulas for the other compounds.

Ammonium phosphate is composed of ammonium ion and phosphate ion:

$$NH_4^+ \quad PO_4^{3-} \qquad \text{so the formula is } (NH_4)_3PO_4.$$

Lead(II) acetate is composed of lead(II) ion and acetate ion:

$$Pb^{2+} \; C_2H_3O_2^- \qquad \text{so the formula is } Pb(C_2H_3O_2)_2.$$

The Roman numeral gives the charge.

Sulfuric acid is one of the common oxoacids. We associate it with the formula H_2SO_4.

Reading Labels

Many over-the-counter medicines, cosmetics, household products, and prepared food products have the ingredients listed on the label. Some products list the ingredients as required by law and some list the ingredients as consumer information. Occasionally, a chemical ingredient is listed by its systematic names. Most ingredients, however, are listed by common names or trade names.

When you read a product label, you may recognize an occasional chemical name. Be content if you can interpret a few chemical names on labels. Keep in mind that even a trained chemist needs to use reference books or the internet to identify many ingredients listed on labels. Figure 6-2 shows the labels from table salt and from an over-the-counter mineral supplement. See if you can recognize any ionic compound names on these labels.

INGREDIENTS: SALT, SODIUM SILICO-ALUMINATE, DEXTROSE, POTASSIUM IODIDE 0.01% AND SODIUM BICARBONATE.

DISTRIBUTED BY UNITED GROCERS LTD. RICHMOND, CA 94804 • PRODUCT OF U.S.A.

FIGURE 6-2 The labels from table salt and from a vitamin–mineral supplement

EACH TABLET CONTAINS	For Adults-Percentage of U.S. Recommended Daily Allowance (U.S. RDA)
Vitamin A (as Acetate and Beta Carotene)	6000 IU (120%)
Vitamin B$_1$ (as Thiamine Mononitrate	1.5 mg (100%)
Vitamin B$_2$ (as Riboflavin)	1.7 mg (100%)
Vitamin B$_6$ (as Pyridoxine Hydrochloride)	3 mg (150%)
Vitamin B$_{12}$ (as Cyanocobalamin)	25 mcg (416%)
Biotin	30 mcg (10%)
Folic Acid	200 mcg (50%)
Niacinamide	20 mg (100%)
Pantothenic Acid (as Calcium Pantothenate)	10 mg (100%)
Vitamin C (as Ascorbic Acid)	60 mg (100%)
Vitamin D	400 IU (100%)
Vitamin E (as di-Alpha Tocopheryl Acetate)	45 IU (150%)
Vitamin K$_1$ (as Phytonadione)	10 mcg*
Calcium (as Dibasic Calcium Phosphate and Calcium Carbonate)	200 mg (20%)
Copper (as Cupric Oxide)	2 mg (100%)
Iodine (as Potassium Iodide)	150 mcg (100%)
Iron (as Ferrous Fumarate)	9 mg (50%)
Magnesium (as Magnesium Oxide)	100 mg (25%)
Phosphorus (as Dibasic Calcium Phosphate)	48 mg (5%)
Zinc (as Zinc Oxide)	15 mg (100%)
Chloride (as Potassium Chloride)	72 mg*
Chromium (as Chromium Chloride)	100 mcg*
Manganese (as Manganese Sulfate)	2.5 mg*
Molybdenum (as Sodium Molybdate)	25 mcg*
Nickel (as Nickelous Sulfate)	5 mcg*
Potassium (as Potassium Chloride)	80 mg*
Selenium (as Sodium Selenate)	25 mcg*
Silicon (as Sodium Metasilicate)	10 mcg*
Vanadium (as Sodium Metavanadate)	10 mcg*

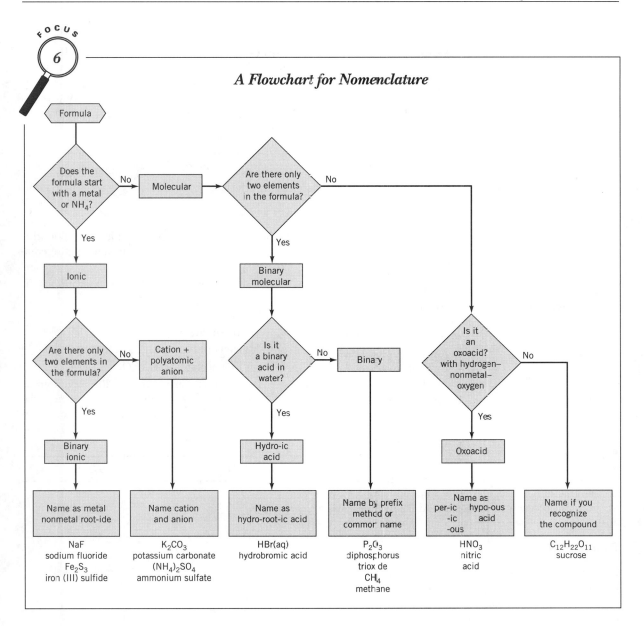

QUESTIONS

1. Match the following terms with the appropriate lettered definitions.

atom

molecule

simple ion

a. a positively charged ion

b. a negatively charged ion

c. chemical names that originated historically and do not reveal the chemical composition of compounds

dichromate (handwritten)

polyatomic ion

d. chemical names of compounds obtained using a nomenclature pattern

common names

e. a group of two or more chemically bound atoms that carries a charge

systematic names

f. an atom that carries a charge resulting from the loss or gain of electrons

cation

g. a chemical particle made of two or more chemically bound atoms

anion

h. water solutions of the binary compounds of hydrogen and the halogens

hydrogen-halide acids

i. acids containing hydrogen, oxygen, and some other element

oxoacids

j. polyatomic ions related to oxyacids through the loss of hydrogen ions

oxoanions

k. the simplest characteristic particle of an element

Section 6-1

2. Describe four kinds of chemical particles; atoms, monatomic ions, molecules, and polyatomic ions. Give an example of each of them, including the formulas and names.

Section 6-2

3. What does the term *generic* mean in naming drugs and medicines?

4. Make a diagram that starts with the title "common compounds." Underneath the title, list the three types of ionic compounds and two types of molecular compounds discussed in this chapter. Next to each type of compound, tell how to recognize a compound of this type.

Section 6-3

5. Give the names of the following ions.

(a) Al^{3+} (b) Br^- (c) CO_3^{2-} (d) Ag^+
(e) S^{2-} (f) Cl^- (g) CrO_4^{2-} (h) Cu^+

(i) Ni^{2+} (j) PO_4^{3-} (k) $Cr_2O_7^{2-}$ (l) Fe^{2+}
(m) HSO_4^- (n) MnO_4^- (o) HCO_3^- (p) OH^-
(q) Mg^{2+} (r) NH_4^+ (s) Ca^{2+} (t) K^+

6. Give the formulas for the following ions.

(a) hydrogen sulfate ion   (b) copper(II) ion Cu^{2+}
(c) chlorate ion ClO_3 (d) fluoride ion F^-
(e) nitrate ion NO_3^- (f) sulfite ion SO_3^{2-}
(g) iron(III) ion Fe^{3+} (h) acetate ion CH_3O_2H
(i) oxalate ion $C_2O_4^{2-}$ (j) cyanide ion CN^-
(k) chlorite ion ClO_2^- (l) chromium(III) ion Cr^{3+}

Section 6-4

7. Explain how the formula of an ionic compound is written to convey the electrical neutrality of the compound.

8. The following compounds contain parentheses. In each case, explain why parentheses are needed.

(a) calcium hydroxide, $Ca(OH)_2$
(b) iron(III) sulfate, $Fe_2(SO_4)_3$
(c) mercury(I) phosphate, $(Hg_2)_3(PO_4)_2$
(d) strontium acetate, $Sr(C_2H_3O_2)_2$

9. Give the correct formulas for the compounds formed by combinations of the positive and negative ions given in the following table. (NaCl is given as an example.)

	Cl^-	CO_3^{2-}	OH^-	NO_3^-	SO_4^{2-}	PO_4^{3-}
Na^+	NaCl					
Ba^{2+}	$BaCl_2$					
Al^{3+}	$AlCl_3$					
Cu^{2+}	$CuCl_2$					
Fe^{2+}						
Fe^{3+}						
K^+						

Sections 6-5 to 6-10

10. *Classify each of the following compounds as metal–nonmetal, metal–polyatomic ion, nonmetal–nonmetal, or acid, and name each by the appropriate nomenclature method. For example, CO_2, nonmetal–nonmetal, carbon dioxide.

(a) PBr_5 (b) ClO_2 (c) Cu_2S
(d) AlN (e) N_2O (f) $Ba(OH)_2$

(g) $Ni(C_2H_3O_2)_2$ (h) $SnCl_4$ (i) BCl_3

(j) Na_2CrO_4 (k) H_3PO_4 (l) $HgCl_2$

(m) Hg_2Br_2 (n) Cr_2O_3 (o) $Fe(CN)_3$

(p) Ca_3P_2 (q) $KClO_3$ (r) $HF(aq)$

(s) $Sn(NO_2)_2$ (t) H_2SeO_4 (u) $NaCN$

(v) N_2O_3 (w) SF_6 (x) HNO_3

(y) CuO (z) $NaOH$ (aa) CCl_4

(bb) I_2O_5 (cc) H_2S (dd) H_2Se

(ee) Na_2CO_3 (ff) $SbBr_5$ (gg) $LiNO_2$

11. Classify each of the following compounds as metal–nonmetal, metal–polyatomic ion, nonmetal–nonmetal, or acid, and name each by the appropriate nomenclature method. For example, CO_2, nonmetal–nonmetal, carbon dioxide.

(a) H_3PO_4 (b) $Cu(NO_3)_2$ (c) $ZnCl_2$

(d) Cr_2S_3 (e) $Cd(CN)_2$ (f) Mg_3P_2

(g) $NaIO_3$ (h) PCl_5 (i) OF_2

(j) Ag_2S (k) $Ca(OH)_2$ (l) $Co(C_2H_3O_2)_2$

(m) $PbCl_4$ (n) $K_2Cr_2O_7$ (o) $HF(aq)$

(p) SnF_2 (q) H_2SO_4 (r) AlN

(s) N_2O_3 (t) $MgCO_3$ (u) SF_4

(v) HNO_2 (w) ZnO (x) $AuBr_3$

(y) I_2O_5 (z) SiC (aa) H_2Te

(bb) CF_4 (cc) PCl_3 (dd) NaH_2PO_4

(ee) $GeCl_4$

12. *Give the formulas for the following compounds.

(a) sodium nitrate $NaNO_3$

(b) tin(II) acetate

(c) manganese(II) sulfate Mn_2SO_4

(d) ammonium acetate $NH_4HC_2H_3O_2$

(e) chlorine trifluoride ClF_3

(f) diarsenic pentasulfide $As SO_3^{-}$

(g) dinitrogen trioxide N_2O_3

(h) calcium fluoride CaF^{-}

(i) iron(II) nitrate $Fe_2NO_3^{-}$

(j) magnesium carbonate $MgCO_3^{2-}$

(k) methane CH_4

(l) copper(I) oxide

(m) silver phosphate

(n) strontium cyanide $SrCN^{-}$

(o) nitric acid HNO_3

(p) ammonium sulfide $NH_3SO_3^{2-}$

(q) phosphoric acid H_3PO_4

13. Give the formulas of the following compounds.

(a) barium carbonate

(b) mercury(I) chloride

(c) chloric acid

(d) copper(I) sulfide

(e) acetic acid $HC_2H_3O_2$

(f) potassium dihydrogen phosphate $KH_2PO_4^{-3}$

(g) carbon monoxide CO

(h) nitrous acid HNO_4

(i) chromium(III) fluoride

(j) sodium cyanide

(k) hydrogen iodide

(l) magnesium hydroxide $MgOH^{-}$

(m) sodium oxalate NaC_2O_4

(n) sodium phosphate $NaPO_4^{2-}$

(o) lead(II) chromate $Pb_2CrO_4^{2-}$

(p) manganese(II) sulfide Mn_2S^{2-}

(q) ammonium carbonate $NH_4CO_3^{2-}$

(r) boron trifluoride BF_3

(s) diantimony pentasulfide

(t) iron(II) oxide Fe_2O

(u) dinitrogen oxide N_2O

(v) potassium hydrogen phosphate $KHPO_4^{2-}$

(w) calcium hydrogen carbonate $CHCO_3$

(x) dichlorine trioxide Cl_2O_3

(y) magnesium chloride $MgCl$

(z) potassium permanganate $KMnO_4$

Sections 6-1 to 6-10

14. Give the chemical name for each of the following compounds, which have the trivial names given.

(a) $NaHCO_3$, baking soda   (b) $CaCO_3$, marble   Calcium Tricarbonate

(c) Hg_2Cl_2, calomel   (d) N_2O, laughing gas   nitrous oxide

---

In the top-right region (problem 10 answers):

(r) magnesium oxalate

(s) zinc chromate

(t) mercury(II) iodide

(u) potassium sulfite

(v) sodium dichromate $NaCr_2O_7^{2-}$

(w) carbon disulfide

(x) dinitrogen pentoxide

(y) magnesium chloride $MgCl$

(z) hydrogen fluoride HF

(e) CaO, lime   (f) NaOH, lye

(g) K_2CO_3, potash   (h) $NaNO_3$, saltpeter

(i) $Ca(OH)_2$, slaked lime   (j) Al_2O_3, alumina

(k) $(NH_4)_2CO_3$, smelling salts

15. List the names and formulas of three compounds with common names that do not describe their chemical compositions.

16. *Classify and name each of the following compounds.

(a) CCl_4 (b) $FeSO_4$ (c) MgO

(d) $HgBr_2$ (e) KI (f) Li_3As

(g) $NaC_2H_3O_2$ (h) $KMnO_4$ (i) PF_3

(j) HCl (k) UF_6 (l) $CoCl_2$

(m) $BaSO_3$ (n) CaF_2 (o) $SnBr_2$

(p) $Zn(OH)_2$ (q) $NaBr$ (r) Al_2S_3

(s) $AgCl$ (t) $Ni(OH)_2$ (u) CrO_3

(v) $HC_2H_3O_2$ (w) NH_4I (x) CaC_2O_4

17. Classify and name each of the following compounds.

(a) $FePO_4$ (b) ZnC_2O_4 (c) $AgNO_3$

(d) SeF_4 (e) NH_3 (f) H_2SO_3

(g) PbO_2 (h) HgS (i) $HClO_2$

(j) Ba_3N_2 (k) $Fe(OH)_3$ (l) Hg_2I_2

(m) KBr (n) Fe_2O_3 (o) $CsC_2H_3O_2$

(p) $CdSO_4$ (q) CrI_3 (r) K_2O

(s) Na_2SO_3 (t) KOH (u) Cl_2O

(v) $Mg(HSO_4)_2$ (w) UF_6 (x) $(NH_4)_3PO_4$

(y) AlP (z) $KHSO_3$

Section 6-8

18. What are oxoacids? What are oxoanions?

19. *Give the names of the missing oxoacids or oxoanions.

(a) HNO_3_____, NO_3^- nitrate ion

(b) HNO_2 nitrous acid, NO_2^- _____

(c) H_3PO_4_____, PO_4^{3-} phosphate ion

(d) $HBrO_3$ bromic acid, BrO_3^- _____

(e) H_2SeO_4_____, SeO_4^{2-} selenate ion

(f) H_3AsO_4 arsenic acid, AsO_4^{3-}_____

(g) H_2SO_3_____, SO_3^{2-} sulfite ion

20. Give the names of the following ions.

(a) HSO_4^- (b) HSO_3^- (c) HCO_3^-

(d) HS^- (e) $H_2PO_4^-$ (f) HPO_4^{2-}

Reading Labels

21. Read the labels on two foods, over-the-counter medicines, or other products. List the names and write the formulas of any compounds that follow nomenclature discussed in this chapter.

22. Read the label on a bottle of commercial soda water and give the formulas of the chemicals it contains.

QUESTION TO PONDER

23. The structural formula for butane given at the end of Section 6-1 represents a form of butane called normal butane. Another compound called isobutane has the formula C_4H_{10} but has a different structural formula than normal butane. Give the structural formula for isobutane. Compounds with the same molecular formulas but different structural formulas are called structural isomers.

CHAPTER

7

CHEMICAL

STOICHIOMETRY

7-1 THE MOLE REVISITED

A cup of water has a mass of about 237 g. Since molecules are so small, we can visualize a cup of water as a huge number of water molecules. How many molecules are in a cup of water? We can't count the molecules one by one, but we do know that each molecule has mass and contributes to the total mass of the sample. As an analogy, imagine a jar containing a pound of jelly beans. Each jelly bean has a specific mass, and the total mass in the jar results from the contribution of each jelly bean. If we know the average mass of a jelly bean, we could easily determine the number of jelly beans in the jar.

In chemistry, weighing samples of chemicals is easy, but when we want to refer to the number of atoms or molecules in a sample of a chemical, **the mole is used most often.** The molar mass of a compound is used as a conversion factor to relate the mass of a compound to the number of moles. Thus, we can easily count the particles in a known mass of a specific chemical. Using the jelly bean analogy, notice that if we know the average mass per dozen jelly beans (# lbs./doz), we can use this as a factor to find the number of dozen jelly beans in the 1 lb jar. To do the calculation, we multiply the pound by the inverse of the mass per dozen factor. In terms of units, this calculation would be

$$1.00 \ \text{lbs.} \left(\frac{1 \ \text{doz}}{\# \ \text{lbs.}} \right) = ? \ \text{doz}$$

The molar mass of a compound is found using its formula as discussed in Section 4-9. A molar mass is used as a factor in the following ways:

1. To find the number of moles in a measured mass of a chemical by multiplying the mass by the inverse of the molar mass.

$$X \text{g} \left(\frac{1 \ \text{mol}}{\# \ \text{g}} \right) = ? \ \text{mol}$$

Mole

The amount of a chemical that contains Avogadro's number of chemical species. The species may be atoms, molecules, ions, or formula units.

175

2. To find the mass of a given number of moles by multiplying the number of moles by the molar mass.

$$Y \text{ mol} \left(\frac{\# \text{ g}}{1 \text{ mol}} \right) = ? \text{ g}$$

EXAMPLE 7-1

How many moles of water, H_2O, are in a 237 g sample?

To relate mass to moles, we need the molar mass of water. Use the periodic table to find the necessary molar masses of the elements in the compound. Round them to convenient numbers of digits.

$$16.00 + 2(1.008) = 18.02 \quad \text{or} \quad 18.02 \text{ g/mol } H_2O$$

To find the number of moles in a given mass, we multiply the sample mass by the inverted molar mass.

$$X \text{ mol } H_2O = 237 \text{ g} \left(\frac{1 \text{ mol } H_2O}{18.02 \text{ g}} \right) = 13.2 \text{ mol } H_2O$$

Remember that a mole is Avogadro's number of particles. The sample has 13.2 times Avogadro's number of water molecules. A cup of water has 7,920,000,000,000,000,000,000,000 water molecules (7.92 million billion billion molecules). Obviously, it is more convenient to refer to this amount of water as 13.2 moles of H_2O.

EXAMPLE 7-2

A sugar cube contains 5.15 g of sucrose, $C_{12}H_{22}O_{11}$. How many moles of sucrose are in this sample?

To relate mass to moles, we need the molar mass of sucrose. From the formula, the molar mass is

$$12(12.01) + 22(1.008) + 11(16.00) = 342.30 \quad \text{or} \quad 342.3 \text{ g/mol } C_{12}H_{22}O_{11}$$

To find the moles multiply the mass by the inverted molar mass

$$X \text{ mole } C_{12}H_{22}O_{11} = 5.15 \text{ g} \left(\frac{1 \text{ mol } C_{12}H_{22}O_{11}}{342.3 \text{ g}} \right) = 1.50 \times 10^{-2} \text{ mol } C_{12}H_{22}O_{11}$$

ACTIVITY 7-1

How many moles of ammonia, NH_3, are in a 525 g sample? How many moles of $NaHCO_3$ are in a 26.6 g sample?

Since the molar mass of a compound is a conversion factor, it is also used to find the mass of a given number of moles. Like any conversion factor, the molar mass is used so that the unwanted units cancel and the desired units remain.

EXAMPLE 7-3

What is the mass in grams of 0.555 mole of H_2O?

To find the mass in a given number of moles of a compound, multiply by the molar mass. See Example 7-1 for the molar mass of water.

$$X \text{ g } H_2O = 0.555 \text{ mol } H_2O \left(\frac{18.02 \text{ g}}{1 \text{ mol } H_2O} \right) = 10.0 \text{ g}$$

EXAMPLE 7-4

What is the mass in grams of 0.10 mole of NaCl?

To relate moles to mass, we need the molar mass of NaCl.

$$22.99 + 35.45 = 58.44 \quad \text{or} \quad 58.44 \text{ g/mol NaCl}$$

To find the mass in grams, multiply the given number of moles by the molar mass.

$$X \text{ g NaCl} = 0.10 \text{ mol NaCl} \left(\frac{58.44 \text{ g}}{1 \text{ mol NaCl}} \right) = 5.8 \text{ g}$$

What is the mass of 202 moles of ammonia, NH_3? What is the mass of 0.049 mole of $NaHCO_3$?

7-2 CONSERVATION OF MASS IN CHEMICAL REACTIONS

Imagine what happens when a candle burns. The candle contains a variety of hydrocarbon compounds. Hydrocarbons are compounds that contain only hydrogen and carbon. Combustion of these compounds involves their vigorous reaction with oxygen in the air to form gaseous carbon dioxide and water. Although release of energy as heat and light accompanies this reaction, the mass is the same before and after the reaction. Suppose we place a candle in a container of air and weigh the container. Then we light the candle, seal the container, and let the candle burn until it goes out. When we reweigh the container we'll find that no detectable mass is lost or gained in the reaction (see Figure 7-1). Experimental measurements done on a variety of reactions reveal that mass is conserved in chemical reactions. Conservation of mass means that no detectable mass is lost or gained in a reaction.

That mass is conserved in a chemical reaction is an expression of the law of conservation of matter as discussed in Section 3-8. What significance does the law have when we work with equations? Mass is conserved in a reaction because the atoms are conserved. In other words, as you know, the number of combined atoms in the products must be the same as the number of combined atoms in the reactants. A balanced equation for a chemical reaction reflects the conservation of atoms. Since atoms are not

Write the balanced equation for the reaction between hydrogen gas and oxygen gas to give water. To confirm the mass balance in the reaction, add the molar masses of the reactants multiplied by their coefficients in the balanced equation, and compare with the molar mass of the product multiplied by its coefficient from the equation.

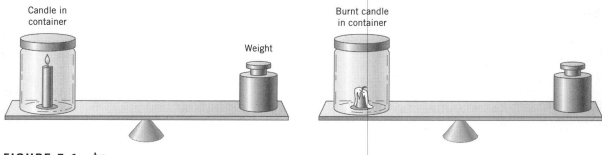

Candle in container

Weight

Burnt candle in container

FIGURE 7-1 An experimental confirmation of the law of conservation of mass.

destroyed or created in a chemical reaction, it is reasonable that the total mass of the chemicals that react must equal the total mass of the products formed, which suggests another meaning for a balanced equation. The total mass of the products formed in a reaction balances the total mass of reactants used.

7-3 CHEMICAL STOICHIOMETRY

Recall from Chapter 4 that molar ratios from formulas are used to relate number of moles and masses in compounds. Similar methods are used to relate moles and masses of chemicals involved in reactions. A chemical equation contains a wealth of information. An equation gives the formulas of the reactants and products, and the coefficients reveal the relative proportions of the reactants and products. Using an equation as a guide, we can deduce the mass relations between individual reactants and products. For instance, we can use an equation to predict amounts of products formed from specified amounts of reactants.

Calculations involving mass relations are **stoichiometric** (pronounced stoy-key-oh-met′-rik) calculations. Stoichiometric is an interesting word. The *-metric* part refers to the measuring of masses. The term *stoichio-* refers to elements or parts of compounds; thus, stoichiometric, taking measure of the masses of elements or compounds in chemical reactions. In other words, stoichiometric refers to how masses of elements and compounds involved in a reaction relate. The methods of stoichiometric calculations are called **stoichiometry**. This word relates to the ancient Greek and Roman school of philosophy called Stoicism. Stoics worshiped the Greek Goddess Nemesis (Fate) who ruled the constant combining, separating, and recombining of the elements. Of course, in the view of the ancient Greeks, everything was made of combinations of the four elements: earth, air, water and fire.

Ammonia, NH_3, is an important industrial and agricultural chemical. Its main uses are in the manufacture of fertilizers, plastics, and explosives.

The industrial synthesis of ammonia is accomplished by a process in which diatomic molecules of hydrogen and nitrogen combine using an iron catalyst.

$$3H_2 + N_2 \xrightarrow{\text{Fe}} 2NH_3$$

This process is named the Haber process for Fritz Haber, a German chemist who developed it in the early 1900s. Air is the source of nitrogen used in the Haber process. The hydrogen typically comes from the high-temperature reaction of methane and water.

$$CH_4 + 2H_2O \longrightarrow 4H_2 + CO_2$$

Ammonia

The raw materials used in the industrial synthesis of ammonia are air, natural gas, and purified water.

Typically, for a reaction to take place it is necessary to mix the reactants and then adjust the reaction conditions of temperature and pressure. Some reactions require the addition of a catalyst. For instance, finely divided iron is used as a catalyst in the Haber process. See Section 14-3 for more about catalysts. Some possible stoichiometric questions relating to the equation for formation of ammonia are:

1. How many grams of ammonia can be produced from specific masses of the reactants?
2. What masses of hydrogen or nitrogen are needed to produce a specific mass of ammonia?
3. What mass of hydrogen is needed to react with a specific mass of nitrogen?

Similar stoichiometric questions can be asked about any reaction.

7-4 MOLAR INTERPRETATION OF AN EQUATION

To represent a chemical reaction, we write an equation. When writing an equation, be sure to use the proper formulas for all chemicals involved in the reaction. Once a balanced equation is written, the equation is useful for stoichiometric calculations. For example, the equation for the combustion of methane is

$$1CH_4 + 2O_2 \longrightarrow 1CO_2 + 2H_2O$$

Coefficients of 1 are shown for emphasis. They are normally not written in an equation. How can we interpret such an equation? We could say that the equation shows that 1 molecule of methane reacts with 2 molecules of oxygen to produce 1 molecule of carbon dioxide and 2 molecules of water. Naturally if this is true for a few molecules, it is also true for any number, even Avogadro's number.

$$1 \text{ molecule } CH_4 + 2 \text{ molecules } O_2 \longrightarrow 1 \text{ molecule } CO_2 + 2 \text{ molecules } H_2O$$

$$1 \text{ dozen } CH_4 + 2 \text{ dozen } O_2 \longrightarrow 1 \text{ dozen } CO_2 + 2 \text{ dozen } H_2O$$

$$1 \text{ gross } CH_4 + 2 \text{ gross } O_2 \longrightarrow 1 \text{ gross } CO_2 + 2 \text{ gross } H_2O$$

$$1 \text{ million } CH_4 + 2 \text{ million } O_2 \longrightarrow 1 \text{ million } CO_2 + 2 \text{ million } H_2O$$

$$1 \text{ billion } CH_4 + 2 \text{ billion } O_2 \longrightarrow 1 \text{ billion } CO_2 + 2 \text{ billion } H_2O$$

$$1 \text{ mole } CH_4 + 2 \text{ moles } O_2 \longrightarrow 1 \text{ mole } CO_2 + 2 \text{ moles } H_2O$$

$$(6.022 \times 10^{23}) \, CH_4 + 2(6.022 \times 10^{23}) \, O_2 \longrightarrow (6.022 \times 10^{23}) CO_2 + 2(6.022 \times 10^{23}) H_2O$$

For stoichiometric purposes, it is best to interpret an equation in terms of Avogadro's number of particles. In other words, an equation shows the relative number of moles of the reactants and products. Such an interpretation is called the **molar interpretation** (see Figure 7-2).

FIGURE 7-2 The molar interpretation of an equation.

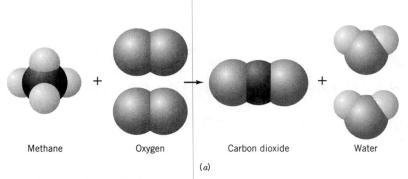

Methane   Oxygen   Carbon dioxide   Water

(a)

In this reaction one methane molecule reacts with two oxygen molecules to give one carbon dioxide molecule and two water molecules.

To interpret this reaction from a molar point of view it is assumed that the reaction that occurs with individual molecules will be the same if Avogadro's number of molecules are involved. The same reaction from a molar point of view is

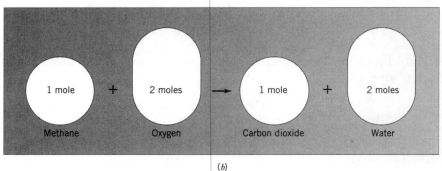

(b)

One mole of methane reacts with two moles of oxygen to give one mole of carbon dioxide and two moles of water.

From a molar point of view, the equation for the burning of methane shows that 1 mole of methane reacts with 2 moles of oxygen to produce 1 mole of carbon dioxide and 2 moles of water. The relative molar amounts of reactants and products must correspond to the coefficients. Thus, it follows that 0.5 mole of methane reacts with 1 mole of oxygen, or 2 moles of methane reacts with 4 moles of oxygen, and so on for any other amounts that have a 1:2 molar ratio. The coefficients reveal the molar ratios relating reactants and products. Understanding this kind of molar interpretation is very important because such a view is essential to stoichiometric calculations.

7-5 MOLAR RATIOS

In an equation, the coefficient in front of each formula shows the relative number of moles of each chemical involved in the reaction. When no numerical coefficient is shown, a 1 is understood. Expressing the relation between any two chemicals involved in a reaction as a molar ratio is very useful. A **molar ratio** expresses the relation between chemicals as a proportion. For the reaction of methane and oxygen to give carbon dioxide and water, molar ratios relating the various reactants and products are

$$\left(\frac{1 \text{ mol } CH_4}{2 \text{ mol } O_2}\right) \quad \left(\frac{1 \text{ mol } CH_4}{1 \text{ mol } CO_2}\right) \quad \left(\frac{1 \text{ mol } CH_4}{2 \text{ mol } H_2O}\right)$$

$$\left(\frac{2 \text{ mol } O_2}{1 \text{ mol } CO_2}\right) \quad \left(\frac{2 \text{ mol } O_2}{2 \text{ mol } H_2O}\right) \quad \left(\frac{1 \text{ mol } CO_2}{2 \text{ mol } H_2O}\right)$$

Expressing any of these ratios in the inverse or reciprocal form is possible. For instance, the molar ratio relating methane and oxygen is written as

$$\left(\frac{1 \text{ mol } CH_4}{2 \text{ mol } O_2}\right) \quad \text{or} \quad \left(\frac{2 \text{ mol } O_2}{1 \text{ mol } CH_4}\right)$$

EXAMPLE 7-5

Give the molar ratios for the reaction corresponding to the equation

$$3H_2 + N_2 \longrightarrow 2NH_3$$

Six ratios are possible. (Why?) We can start with the ratios relating the reactants.

$$\left(\frac{3 \text{ mol } H_2}{1 \text{ mol } N_2}\right) \quad \text{and} \quad \left(\frac{1 \text{ mol } N_2}{3 \text{ mol } H_2}\right)$$

Next, we can write the ratios relating the reactants and product.

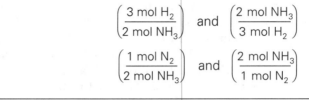

$$\left(\frac{3 \text{ mol } H_2}{2 \text{ mol } NH_3}\right) \quad \text{and} \quad \left(\frac{2 \text{ mol } NH_3}{3 \text{ mol } H_2}\right)$$

$$\left(\frac{1 \text{ mol } N_2}{2 \text{ mol } NH_3}\right) \quad \text{and} \quad \left(\frac{2 \text{ mol } NH_3}{1 \text{ mol } N_2}\right)$$

7-6 MOLAR RATIOS AS CONVERSION FACTORS

Molar ratios come from the coefficients in the balanced equation. They apply only to the specific reaction and are used as conversion factors to relate the number of moles of one chemical to the number of moles of another chemical. As an example, suppose we want to know how many moles of oxygen gas react with 426 moles of methane gas if equation for the reaction is

$$CH_4 + 2O_2 \longrightarrow CO_2 + 2H_2O$$

We know from the coefficients of the equation that these chemicals react in a 2:1 molar ratio. We are given the number of moles of methane and want to find the corresponding number of moles of oxygen gas. Thus, to change to moles of oxygen, we need to multiply the given number of moles of methane by a molar ratio having moles of O_2 in the numerator and moles of CH_4 in the denominator.

$$426 \text{ mol } CH_4 \left(\frac{2 \text{ mol } O_2}{1 \text{ mol } CH_4}\right) = 852 \text{ mol } O_2$$

Molar ratios are used as **mole-to-mole factors** in stoichiometric calculations. Molar ratios serve as a bridge or connection between the number of moles of any two chemicals involved in a reaction.

EXAMPLE 7-6

How many moles of oxygen gas can react with 3.27 moles of methane gas, the major component of natural gas? Use the equation for the combustion of methane.

A summary of the question is

$$\begin{array}{ccc} CH_4 & + & 2O_2 \longrightarrow CO_2 + 2H_2O \\ 3.27 \text{ mol } CH_4 & & ? \text{ mol } O_2 \end{array}$$

In this summary, we write the given number of moles below CH_4 and show that we want to find the moles of O_2 using "? mol" under its formula. The appropriate molar ratio relating methane and oxygen comes from the balanced equation.

$$X \text{ mol O}_2 = 3.27 \text{ mol CH}_4 \left(\frac{2 \text{ mol O}_2}{1 \text{ mol CH}_4}\right) = 6.54 \text{ mol O}_2$$

As a double check, we note that using the appropriate molar ratio, the moles of methane cancel to give moles of oxygen.

EXAMPLE 7-7

How many moles of ammonia can form when 153 moles of hydrogen gas react with sufficient nitrogen gas? The equation for the reaction is

$$N_2 + 3H_2 \longrightarrow 2NH_3$$
$$153 \text{ mol H}_2 \quad ? \text{ mol NH}_3$$

To find the moles of one chemical corresponding to the moles of another chemical, we use the appropriate molar ratio from the balanced equation.

$$X \text{ mol NH}_3 = 153 \text{ mol H}_2 \left(\frac{2 \text{ mol NH}_3}{3 \text{ mol H}_2}\right) = 102 \text{ mol NH}_3$$

ACTIVITY 7-4

Find how many moles of hydrogen gas are formed from 895 moles of methane, CH_4, if the equation for the reaction is

$$CH_4 + 2H_2O \longrightarrow 4H_2 + CO_2$$

As we will see, the molar ratios from a chemical equation serve as the basis for many types of stoichiometric calculations. Consequently, you need to have a correctly balanced equation to do calculations.

7-7 STOICHIOMETRIC MASS-TO-MOLE CALCULATIONS

If we know the formula of a chemical involved in a reaction, we can easily determine the molar mass of that chemical. The molar mass is useful for calculating the number of moles contained in a known mass of the chemical or vice versa. This technique is important in **mass-to-mole stoichiometric calculations.** See Figure 7-3 on page 184 for an illustration.

Consider an example of a mass-to-mole type of stoichiometric calculation. Iron ore is the industrial source of iron metal. The overall reaction in the refining of iron includes iron(III) oxide from iron ore reacting with carbon monoxide to give iron metal and carbon dioxide.

$$Fe_2O_3 + 3CO \longrightarrow 2Fe + 3CO_2$$

How many moles of iron form when 385 g of carbon monoxide reacts with sufficient iron(III) oxide? First, we decide what is wanted (the unknown). This question asks for how much iron in moles. Next, we note what is given (the knowns). In this question, it is given that 385 g of carbon monoxide

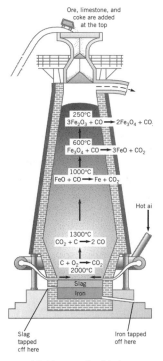

Ore, limestone, and coke are added at the top

250°C
$3Fe_2O_3 + CO \longrightarrow 2Fe_3O_4 + CO_2$

600°C
$Fe_3O_4 + CO \longrightarrow 3FeO + CO_2$

1000°C
$FeO + CO \longrightarrow Fe + CO_2$

Hot air

1300°C
$CO_2 + C \longrightarrow 2 CO$

$C + O_2 \longrightarrow CO_2$
2000°C

Slag
Iron

Slag tapped off here

Iron tapped off here

A blast furnace used in refining iron

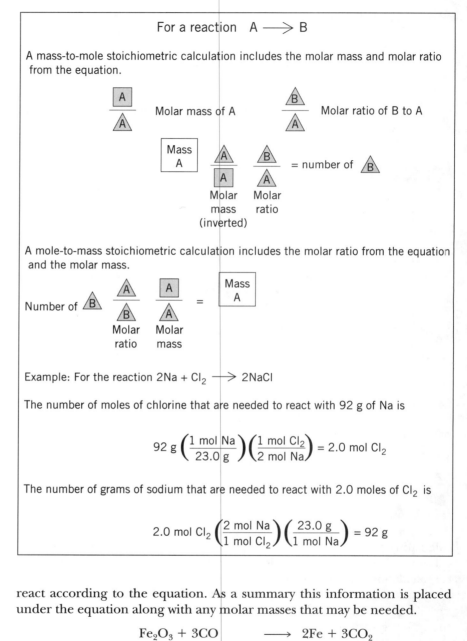

FIGURE 7-3 The conversion of mass to mole and mole to mass.

react according to the equation. As a summary this information is placed under the equation along with any molar masses that may be needed.

$$
\begin{array}{lccc}
& Fe_2O_3 + 3CO & \longrightarrow & 2Fe + 3CO_2 \\
\text{mass} & 385\ g & & \\
\text{molar mass} & 28.01\ g/mol & & \\
\text{moles} & mol\ CO & \Longrightarrow & ?\ mol\ Fe
\end{array}
$$

In such a summary the top row is for mass and the bottom row is for moles. Since molar masses relate masses and moles, the middle row is for any molar masses. We enter the unknown as "?" including its units. Fill the table with any given data, placing each item below the formula of the appropriate reactant or product in the equation. Determine any molar masses from the formulas of the reactants and products. Amounts of chemicals in reactions connect only through moles, so in the row for moles we draw an arrow connecting carbon monoxide to iron; this emphasizes the connection.

Finding the number of moles of iron directly from the mass of carbon monoxide is not possible. Moles of chemicals relate through molar ratios, so we must work in moles. In other words, the equation gives the molar relations and the molar ratios serve as the bridge between the amounts of the two chemicals. The calculation sequence is

$$\text{g CO} \xRightarrow{\substack{\text{molar}\\\text{mass}}} \text{mol CO} \xRightarrow{\substack{\text{molar}\\\text{ratio}}} \text{? mol Fe}$$

We find the moles of carbon monoxide from the grams of carbon monoxide and then, using the appropriate molar ratios, we use the moles of carbon monoxide to find the moles of iron.

To begin, first calculate the molar mass of carbon monoxide.

$$12.011 + 15.999 = 28.010 \quad \text{or} \quad 28.01$$

Then, use it as a factor to find the number of moles of CO. Since the grams, not moles, of CO are given, we use the molar mass in the form in which grams cancel and moles remain. The number of moles of CO is

$$385 \, \text{g} \left(\frac{1 \, \text{mol CO}}{28.01 \, \text{g}} \right)$$

To calculate the moles of iron, multiply the moles of CO by the molar ratio from the balanced equation. Choose the ratio that cancels moles of CO and gives moles of Fe.

$$385 \, \text{g} \left(\frac{1 \, \text{mol CO}}{28.01 \, \text{g}} \right) \left(\frac{2 \, \text{mol Fe}}{3 \, \text{mol CO}} \right) = 9.16 \, \text{mol Fe}$$

When working a problem of this type, first devise a calculation sequence and then carry out the calculation in steps. We do not have to do any numerical calculations until the final setup of the calculation sequence is written. We need to find any necessary molar masses as side calculations. In the preceding discussion, we did not calculate the numerical values for the number of moles of carbon monoxide in the first step; instead, we just multiplied the product by the molar ratio. Once we have set up the calculations, including all factors, a numerical answer is calculated and rounded to the correct number of digits. As a double check, the units should cancel properly; hence, using units is important.

If you want, you can calculate an answer for each step in the calculation sequence. If you do, always carry more digits than needed in each part and round to the correct number of digits in the final answer. Carrying more digits prevents incorrect answers, which result from rounding intermediate answers. As an example of step-by-step calculation, let us use the same question as given above: How many moles of iron form when 385 g of carbon monoxide react with sufficient iron(III) oxide? First, the molar mass of carbon monoxide is used to find the number of moles of CO.

$$385 \cancel{g} \left(\frac{1 \text{ mol CO}}{28.01 \cancel{g}}\right) = 13.745 \text{ mol CO}$$

To calculate the moles of iron, the above result is multipled by the molar ratio from the equation and then rounded to the correct number of significant digits.

$$13.745 \cancel{\text{mol CO}} \left(\frac{2 \text{ mol Fe}}{3 \cancel{\text{mol CO}}}\right) = 9.16 \text{ mol Fe}$$

EXAMPLE 7-8

How many moles of ammonia can form when 56.0 g of nitrogen react with excess hydrogen? The equation for the reaction is

$$N_2 + 3H_2 \longrightarrow 2NH_3$$

Find any necessary molar masses from the formulas and the atomic weights. The question is summarized below. The molar mass of N_2 is (2 × 14.007 = 28.01).

$$
\begin{array}{llll}
& N_2 + 3H_2 & \longrightarrow & 2NH_3 \\
\text{mass} & 56.0 \text{ g} & & \\
\text{molar mass} & 28.01 \text{ g/mol} & & \\
\text{moles} & \text{mol } N_2 & \Longrightarrow & \text{? mol } NH_3
\end{array}
$$

Work on a molar basis. First, use the molar mass to find the number of moles of N_2.

$$56.0 \cancel{g} \left(\frac{1 \text{ mol } N_2}{28.01 \cancel{g}}\right)$$

To find the moles of ammonia multiply by the molar ratio from the equation. The ratio is chosen so that the moles of N_2 cancel to give the answer in moles of NH_3.

$$X \text{ mol } NH_3 = 56.0 \cancel{g} \left(\frac{1 \cancel{\text{mol } N_2}}{28.01 \cancel{g}}\right) \left(\frac{2 \text{ mol } NH_3}{1 \cancel{\text{mol } N_2}}\right) = 4.00 \text{ mol } NH_3$$

The reverse problem of calculating the number of grams of a reactant or product corresponding to a given number of moles of another reactant or product (mole-to-mass calculation) is accomplished using a pattern similar to the above examples. The difference is that the calculation sequence is reversed.

EXAMPLE 7-9

Ethane gas, C_2H_6, is obtained from natural gas and used to make plastics like polyethylene. Ethane can react with oxygen to give carbon dioxide and water. How many grams of ethane react with 14 moles of oxygen gas? Placing the information beneath the balanced equation for the reaction gives

$$2C_2H_6 \quad + \quad 7O_2 \longrightarrow 4CO_2 + 6H_2O$$

mass	? g
molar mass	30.1 g/mol
moles	? mol C_2H_6 $\Longleftarrow$ 14 mol O_2

The calculation sequence is

$$\text{mol } O_2 \xRightarrow[\text{ratio}]{\text{molar}} \text{mol } C_2H_6 \xRightarrow[\text{mass}]{\text{molar}} \text{g } C_2H_6$$

The appropriate molar ratio is used to change moles of O_2 to moles of C_2H_6. Since we want mass of ethane, we need its molar mass [C_2H_6 2(12.01) + 6(1.008) = 30.1]. The molar mass of C_2H_6 is used to cancel moles and gives the answer in grams.

$$X \text{ g } C_2H_6 = 14 \text{ mol } O_2 \left(\frac{2 \text{ mol } C_2H_6}{7 \text{ mol } O_2}\right)\left(\frac{30.1 \text{ g}}{1 \text{ mol } C_2H_6}\right) = 120 \text{ g or } 1.2 \times 10^2 \text{ g}$$

ACTIVITY 7-5

Find how many moles of hydrogen gas are formed from 825 grams of methane, CH_4, if the equation for the reaction is

$$CH_4 + 2H_2O \longrightarrow 4H_2 + CO_2$$

Also calculate the number of grams of CO_2 produced when 51.4 moles of CH_4 react.

7-8 STOICHIOMETRIC MASS-TO-MASS CALCULATIONS

Since we often work with masses of reactants and products in the laboratory or on an industrial scale, the three most common stoichiometric questions asked are

1. How many grams of a product form from a given number of grams of a reactant?

2. How many grams of a reactant are needed to form a given number of grams of a product?

3. How many grams of a reactant are needed to react with a given number of grams of another reactant?

Mass-to-mass calculations are stoichiometric calculations that provide solutions to these questions. Just use combinations of mass-to-mole and

Aluminum

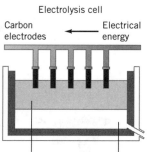

Electrolysis cell

Carbon electrodes ← Electrical energy

Molten cryolite with dissolved aluminum oxide

Molten aluminum metal

An electrolytic cell used to obtain aluminum from alumium ore.

mole-to-mole calculations for mass-to-mass conversions. Mass-to-mass calculations are very important in chemistry because we usually work with masses of materials. Since an equation relates chemicals on a molar basis, we need to change mass to the number of moles.

Bauxite ore is the major source of natural aluminum oxide. Aluminum metal is made by dissolving aluminum oxide in a high temperature vat of melted cryolite (a mineral). Then carbon rods are immersed into the vat and electrical current is passed through the molten mixture. The electrical energy encourages a reaction in which the aluminum oxide combines with carbon to form aluminum metal and carbon dioxide gas. See Section 15-12 for a further discussion of the process. The equation for the overall reaction is

$$2Al_2O_3 + 3C \xrightarrow{\text{electrical energy}} 4Al + 3CO_2$$

As an example of a mass-to-mass question, suppose we have a 204 g sample of aluminum oxide. How many grams of aluminum metal are formed from this mass? The information is summarized as

	$2Al_2O_3$	$+ 3C \longrightarrow$	$4Al$	$+ 3CO_2$
mass	204 g		? g	
molar mass	102.0 g/mol		26.98 g/mol	
moles	mol Al_2O_3	$\Longrightarrow$	? mol Al	

Converting directly from the mass of one chemical to the mass of another is not possible. The conversion must be done on a molar basis. Thus, the necessary calculation sequence is

$$\text{g } Al_2O_3 \overset{\substack{\text{molar} \\ \text{mass}}}{\Longrightarrow} \text{mol } Al_2O_3 \overset{\substack{\text{molar} \\ \text{ratio}}}{\Longrightarrow} \text{mol Al} \overset{\substack{\text{molar} \\ \text{mass}}}{\Longrightarrow} \text{g Al}$$

According to this sequence, we need two molar masses and a molar ratio from the equation. The calculation involves multiplying the grams of aluminum oxide by the three appropriate factors. First, convert the grams of aluminum oxide to moles using its molar mass [2(26.98) + 3(16.00) = 101.96 or 102.0].

$$204 \text{ g} \left(\frac{1 \text{ mol } Al_2O_3}{102.0 \text{ g}} \right)$$

Multiplying by the molar ratio, from the equation, gives the corresponding number of moles of aluminum.

$$204 \text{ g} \left(\frac{1 \text{ mol } Al_2O_3}{102.0 \text{ g}} \right) \left(\frac{4 \text{ mol Al}}{2 \text{ mol } Al_2O_3} \right)$$

Finally, the molar mass of aluminum is used to find the mass of aluminum.

$$204 \, \cancel{g} \left(\frac{1 \, \cancel{mol \, Al_2O_3}}{102.0 \, \cancel{g}} \right) \left(\frac{4 \, \cancel{mol \, Al}}{2 \, \cancel{mol \, Al_2O_3}} \right) \left(\frac{26.98 \, g}{1 \, \cancel{mol \, Al}} \right) = 108 \, g$$

No numerical calculations need be done until the final setup is written. Check the setup by making sure the various units cancel and the answer has the desired unit. Note that in the final setup the grams of aluminum oxide cancel, the moles of aluminum oxide cancel, and the moles of aluminum cancel, leaving grams of aluminum.

You can calculate an intermediate answer for each step and then use that answer for the next step. Remember, however, to include enough digits, and do not round until the final answer. The number of moles of aluminum oxide is found in the first step.

$$204 \, \cancel{g} \left(\frac{1 \, mol \, Al_2O_3}{102.0 \, \cancel{g}} \right) = 2.000 \, mol \, Al_2O_3$$

The number of moles of aluminum is found in the second step.

$$2.000 \, \cancel{mol \, Al_2O_3} \left(\frac{4 \, mol \, Al}{2 \, \cancel{mol \, Al_2O_3}} \right) = 4.000 \, mol \, Al$$

The mass of aluminum is found in the last step.

$$4.000 \, \cancel{mol \, Al} \left(\frac{26.98 \, g}{1 \, \cancel{mol \, Al}} \right) = 108 \, g$$

This problem illustrates the general approach to a mass-to-mass stoichiometric calculation. We can recognize such problems by the context. When the mass of one chemical is given and we want to find the corresponding mass of a second chemical, we have a mass-to-mass question. Problems like this always involve multiplying the given mass by three factors: the molar mass of the given chemical, the molar ratio from the equation, and the molar mass of the second chemical. See Figure 7-4 on page 190. A summary written below the equation allows us to visualize what is given and what is wanted. In addition, the summary suggests a calculation path for the solution to the problem.

EXAMPLE 7-10

The element silicon is made by a reaction represented by the equation

$$SiCl_4 + 2Mg \longrightarrow 2MgCl_2 + Si$$

How many grams of magnesium are needed to react completely with 6.25×10^3 grams of $SiCl_4$?

The information is summarized as

$$SiCl_4 \quad + \quad 2Mg \longrightarrow 2MgCl_2 - Si$$

mass	6.25×10^3 g	? g
molar mass	169.9 g/mol	24.31 g/mol
moles	mol $SiCl_4$ $\Longrightarrow$	? mol Mg

This problem is a mass-to-mass calculation. First, find the molar mass of $SiCl_4$ (28.09 + 4(35.45) = 169.89 or 169.9). Then use it to find the moles of $SiCl_4$.

$$6.25 \times 10^3 \, g \left(\frac{1 \text{ mol } SiCl_4}{169.9 \, g} \right)$$

Next, use the proper molar ratio to find the corresponding number of moles of Mg.

$$6.25 \times 10^3 \, g \left(\frac{1 \text{ mol } SiCl_4}{169.9 \, g} \right) \left(\frac{2 \text{ mol Mg}}{1 \text{ mol } SiCl_4} \right)$$

Finally, the mass of magnesium is found using its molar mass.

$$6.25 \times 10^3 \, g \left(\frac{1 \text{ mol } SiCl_4}{169.9 \, g} \right) \left(\frac{2 \text{ mol Mg}}{1 \text{ mol } SiCl_4} \right) \left(\frac{24.31 \, g}{1 \text{ mol Mg}} \right) = 1.79 \times 10^3 \, g$$

FIGURE 7-4 The conversion of mass to mass.

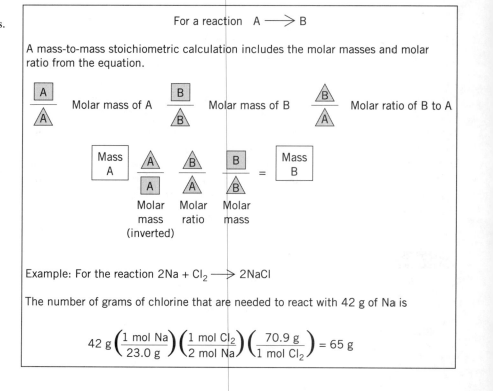

For a reaction   A $\longrightarrow$ B

A mass-to-mass stoichiometric calculation includes the molar masses and molar ratio from the equation.

$\frac{A}{A}$ Molar mass of A $\frac{B}{B}$ Molar mass of B $\frac{B}{A}$ Molar ratio of B to A

$$\text{Mass A} \quad \frac{A}{A} \quad \frac{B}{A} \quad \frac{B}{B} = \text{Mass B}$$

Molar   Molar   Molar
mass   ratio   mass
(inverted)

Example: For the reaction $2Na + Cl_2 \longrightarrow 2NaCl$

The number of grams of chlorine that are needed to react with 42 g of Na is

$$42 \, g \left(\frac{1 \text{ mol Na}}{23.0 \, g} \right) \left(\frac{1 \text{ mol } Cl_2}{2 \text{ mol Na}} \right) \left(\frac{70.9 \, g}{1 \text{ mol } Cl_2} \right) = 65 \, g$$

The three common types of stoichiometric calculations are mole to mole, mass to mole, and mass to mass. A stoichiometry map, as described in Focus 7, summarizes these calculation types. Notice that the solution to any stoichiometry problem is given by the map. According to the map, to go from the mass of one chemical to the mass of another we need to use moles and molar ratios. The molar ratio is the bridge or connecting point of any stoichiometric calculation.

Find how many grams of hydrogen gas are produced from 825 grams of methane if the equation for the reaction is

$$CH_4 + 2H_2O \longrightarrow 4H_2 + CO_2$$

7-9 LIMITING REACTANT

Gasoline is a complex mixture of organic chemicals, most of which are hydrocarbons. The chemical reactions that occur in an automobile engine involve these chemicals reacting with oxygen from the air. With an adequate supply of air, an engine runs until all the gasoline is used. The amount of gasoline is the limiting factor. On the other hand, if the air supply is cut off, the engine stops when the supply of oxygen is depleted. Here, the amount of oxygen limits the reaction.

When reactants are mixed in reactions used in the laboratory or industrial processes, the more expensive or the least available reactant is normally the limiting factor. Measuring reactants in correct molar proportions is usually not convenient or necessary. More commonly, reactants are mixed and the reaction occurs until one reactant is depleted. One reactant limits the reaction leaving an unreacted portion of the other reactant. The amount of the reactant that limits the reaction dictates the amount of product formed. We call this reactant the **limiting reactant.** The other reactant is in excess. Some of it is used, but an unused portion remains after the reaction has taken place. As an analogy, suppose we are making cheese sandwiches and we have 10 pieces of cheese and 24 slices of bread. Since the "formula" of a sandwich is "cheese(bread)$_2$," we could make 10 sandwiches and have 4 slices of bread left over. The cheese is the limiting component. See Figure 7-5 on page 192.

As a chemical example, suppose we want to find the mass of ammonia formed when 56.0 g of nitrogen gas and 12.0 g of hydrogen gas react

$$N_2 + 3H_2 \longrightarrow 2NH_3$$

This requires a mass-to-mass calculation, but which of the reactants is the limiting reactant? To find out, we need to compare the ratio of the moles of the reactants with the molar ratio needed as given by the equation. First, note the molar ratio of hydrogen to nitrogen as given in the balanced equation.

$$\left(\frac{3 \text{ mol H}_2}{1 \text{ mol N}_2}\right)$$

To compare this with the ratio of the numbers of moles given in the question, we first calculate the number of moles of each reactant using each

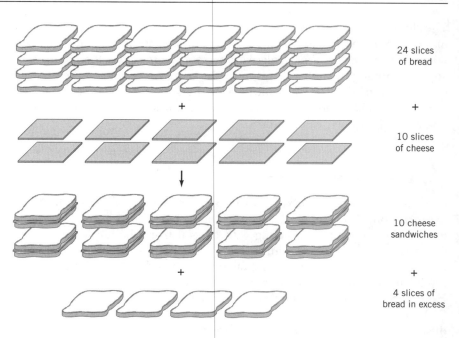

24 slices
of bread

+

10 slices
of cheese

10 cheese
sandwiches

+

4 slices of
bread in excess

FIGURE 7-5 Limiting
reactant using bread and
cheese.

given mass and the corresponding molar mass. In the example, we need
the molar masses of H_2 [2(1.0079) = 2.016] and N_2 [2(14.007) = 28.01].
The number of moles of hydrogen is (carry extra digits in the calculation)

$$12.0 \, g \left(\frac{1 \text{ mol } H_2}{2.106 \, g} \right) = 5.952 \text{ mol } H_2$$

The number of moles of nitrogen is (carry extra digits in the calculation)

$$56.0 \, g \left(\frac{1 \text{ mol } N_2}{28.01 \, g} \right) = 1.999 \text{ mol } N_2$$

The ratio of these numbers of moles is found by dividing the moles of H_2
by the moles of N_2 and compared with the required ratio.

$$\left(\frac{5.952 \text{ mol } H_2}{1.999 \text{ mol } N_2} \right) = \left(\frac{2.997 \text{ mol } H_2}{1 \text{ mol } N_2} \right)$$

This ratio is less than the ratio needed as given by the balanced equation.
Therefore, we conclude that hydrogen is the limiting reactant since there
are only 2.997 moles of hydrogen per mole of nitrogen. If this ratio turned
out to be greater than 3 to 1, then hydrogen would be in excess and nitrogen would be the limiting reactant. Since hydrogen is the limiting reactant, we use the number of moles of hydrogen in a mole-to-mass
stoichiometric calculation to find the amount of product formed. In the

calculation, we need the molar mass of NH_3 $[14.007 + 3(1.008) = 17.03]$. The final answer is rounded to the correct number of digits.

$$X \text{ g } NH_3 = 5.952 \text{ mol } H_2 \left(\frac{2 \text{ mol } NH_3}{3 \text{ mol } H_2} \right) \left(\frac{17.03 \text{ g}}{1 \text{ mol } NH_3} \right) = 67.6 \text{ g } NH_3$$

Notice that carrying enough digits in these calculations is very important so that a rounding error does not obscure the answer.

Since hydrogen is the limiting reactant, the nitrogen is the reactant in excess. How many grams of nitrogen are used and how many grams remain unused? Since we know that the number of moles of hydrogen limits the reaction, we use this number to find the mass of nitrogen that reacted. This step requires another mole-to-mass stoichiometric calculation changing from moles of hydrogen to mass of nitrogen. The amount of nitrogen used is

$$X \text{ g } NH_3 = 5.952 \text{ mol } H_2 \left(\frac{1 \text{ mol } N_2}{3 \text{ mol } H_2} \right) \left(\frac{28.10 \text{ g}}{1 \text{ mol } N_2} \right) = 55.6 \text{ g } N_2$$

To find the amount of nitrogen unused, after the reaction has taken place, we subtract the grams used from the number of grams present at the outset.

$$56.0 \text{ g} - 55.6 \text{ g} = 0.4 \text{ g } N_2$$

How do you recognize a limiting reactant question? Typically, such a question gives the masses of two reactants and asks for the mass of product formed. When working limiting reactant problems, use the following approach.

1. Which reactant is the limiting reactant and how much product is formed?
   (a) We need the required molar ratio of reactants corresponding to the coefficients in the balanced equation. Find this ratio by dividing the larger coefficient by the smaller coefficient.
   (b) Calculate the number of moles of each reactant using its mass and molar mass. Carry extra digits.
   (c) Find the molar ratio of the given numbers of moles of reactants by division.
   (d) Compare the calculated ratio with the ratio of reactants found from the equation. If the calculated ratio is less than the coefficient ratio, the element in the numerator is the limiting reactant. If the calculated ratio is greater than the coefficient ratio, the element in the denominator is the limiting reactant.
   (e) Use the number of moles of the limiting reactant in a mole-to-mass calculation to find the mass of the product formed.

2. What mass of the reactant in excess was used and what mass remains?
   (a) Use the calculated number of moles of the limiting reactant to find the mass of the other reactant used in the reaction by using a mole-to-mass calculation.
   (b) Find the remaining mass of the reactant in excess by subtracting the mass used from the initial starting mass.

EXAMPLE 7-11

One step in the manufacture of methanol, CH_3OH, involves a reaction represented as

$$CO + 2H_2 \longrightarrow CH_3OH$$

How many grams of methanol can form when 72.8 g of CO and 10.6 g of H_2 react? How many grams of the reactant in excess are used and how many grams remain?

Since the amounts of both reactants are given, we have a limiting reactant question. We will need molar masses, so a summary of the question is useful.

	CO	+	$2H_2$	$\longrightarrow$	CH_3OH
mass	72.8 g		10.6 g		? g
molar mass	28.01 g/mol		2.106 g/mol		32.04 g/mol
mol	? mol CO		? mol H_2		? mol CH_3OH

First, note the required molar ratio given by the coefficients in the equation.

$$\left(\frac{2 \text{ mol } H_2}{1 \text{ mol } CO}\right)$$

Next, calculate the number of moles of each reactant using the given masses and the molar masses.

$$X \text{ mol CO} = 72.8 \text{ g} \left(\frac{1 \text{ mol CO}}{28.01 \text{ g}}\right) = 2.599 \text{ mol CO}$$

$$X \text{ mol } H_2 = 10.6 \text{ g} \left(\frac{1 \text{ mol } H_2}{2.016 \text{ g}}\right) = 5.258 \text{ mol } H_2$$

Use these results to find the ratio of the moles of H_2 to the moles of CO.

$$\left(\frac{5.258 \text{ mol } H_2}{2.599 \text{ mol CO}}\right) = \left(\frac{2.023 \text{ mol } H_2}{1 \text{ mol CO}}\right)$$

Compare this ratio with the ratio found from the coefficients. Since the calculated ratio is greater, then H_2 is in excess and CO is the limiting reactant. The number of moles of CO is then used to find the mass of product formed in the reaction using a mole of CO to mass of CH_3OH calculation. Thus, the number of grams of methanol formed in the reaction is

$$X \text{g } CH_3OH = 2.599 \text{ mol CO} \left(\frac{1 \text{ mol } CH_3OH}{1 \text{ mol CO}} \right) \left(\frac{32.04 \text{ g}}{1 \text{ mol } CH_3OH} \right) = 83.3 \text{ g}$$

For the second part of the question, we use the number of moles of the limiting reactant to calculate the mass of the other reactant used in the reaction. This step involves a mole of CO to mass of H_2 calculation using the equation for the reaction.

$$X \text{g } H_2 = 2.599 \text{ mol CO} \left(\frac{2 \text{ mol } H_2}{1 \text{ mol CO}} \right) \left(\frac{2.016 \text{ g}}{1 \text{ mol } H_2} \right) = 10.5 \text{ g}$$

Finally, to find the mass of the reactant in excess that remains after the reaction, we subtract the mass of hydrogen used from the mass at the beginning.

$$10.6 \text{ g} - 10.5 \text{ g} = 0.1 \text{ g } H_2$$

The reaction that occurs when baking soda and vinegar are mixed is

$$NaHCO_3 + HC_2H_3O_2 \longrightarrow$$
$$H_2O + CO_2 + NaC_2H_3O_2$$

How many grams of carbon dioxide are formed when 4.28 g of $NaHCO_3$ and 3.00 g of $HC_2H_3O_2$ react? Which reactant is in excess and how many grams of it remain after the reaction?

7-10 THEORETICAL YIELD, ACTUAL YIELD, AND PERCENT YIELD

Often we carry out a stoichiometric calculation to find the amount of product formed by a specified amount of a reactant. Such a calculation gives the potential or theoretical amount of product that could be formed. This amount is called the theoretical yield of the reaction. Typically, in an actual synthesis reaction the product is not highly pure. It may be contaminated by some reactant or dissolved in some solvent used in the process, or it may contain some impurities formed by side reactions that accompany the synthesis. Some product is lost in any purification process, and some original reactant may be used in side reactions. The result is that the actual yield of product may be less than the theoretical (stoichiometric) yield. The percent yield for the product of any reaction is simply

$$\text{percent yield} = \frac{\text{actual yield}}{\text{theoretical yield}} \times 100$$

If we know the given amount of a reactant in a reaction, we can use it to find the theoretical yield. The actual yield is then divided by the theoretical yield and the ratio is multiplied by 100 to give the percent yield.

EXAMPLE 7-12

The fertilizer urea is made by a reaction represented by the equation

$$2NH_3 + CO_2 \longrightarrow (NH_2)_2CO + H_2O$$

When 2.5×10^3 g of NH_3 reacted with sufficient CO_2, the actual yield of urea was 3.2×10^3 g. Calculate the percent yield for the reaction.

First, we calculate the theoretical yield based on the starting amount of the reactant using mass-to-mass stoichiometry. The calculation sequence is

$$g\,NH_3 \xrightarrow[\text{mass}]{\text{molar}} mol\,NH_3 \xrightarrow[\text{ratio}]{\text{molar}} mol\,(NH_2)_2CO \xrightarrow[\text{mass}]{\text{molar}} g\,(NH_2)_2CO$$

$$X\,g\,(NH_2)_2CO = 2.5 \times 10^3\,g \left(\frac{1\,mol\,NH_3}{17.04\,g}\right)\left(\frac{1\,mol\,(NH_2)_2CO}{2\,mol\,NH_3}\right)\left(\frac{46.05\,g}{1\,mol\,(NH_2)_2CO}\right)$$

$$= 3.4 \times 10^3\,g$$

The percent yield is then found from the actual yield and the theoretical yield.

$$percent\,yield = \left(\frac{3.2 \times 10^3\,g}{3.4 \times 10^3\,g}\right) \times 100 = 94\%$$

EPA
Toxic Releases

Conservation of Matter and Society

Chemical reactions follow the law of conservation of matter, which means that the mass of the reactants used in a reaction must equal the mass of the products formed. This statement is sometimes called the mass balance or material balance. Mass balance is related to pollution control in chemical manufacturing processes. Any difference between the mass of the useful product and the masses of the raw materials represents lost materials or byproducts of the process. Losses in the masses of the raw materials must be accounted for. Thus, any mass that is not part of the products must be given off as a gas, washed down the drain, or accumulated as a byproduct. Such mass balance reasoning can also apply to the use of fuels used to supply energy for the manufacturing process or to any solvents used in the process. No chemicals can disappear, so all materials that go into a process and come out of a process must balance in terms of mass. Any solvents, unused reactants, byproducts that are not recycled, must go into the atmosphere, into wastewater, or end up as solid wastes. There is no other possible fate.

The U.S. Environmental Protection Agency (EPA) is a federal agency that includes various subagencies involved with environmental laws involving air pollution, water pollution, solid wastes, noise pollution, pesticide control, radiation health, and toxic wastes. Among its numerous activities, the EPA is charged with the implementation of the federal Clean Water Act, the Clean Air Act, and the Emergency Planning and Community Right to Know Act, which includes the Toxic Release Inventory Law. By law, some industries and companies have to give an annual report of specified wastes and toxic materials that they produce. Known amounts of materials, solvents, and cooling waters are used in industrial and manufacturing processes. As discussed above, the masses of substances used must balance with the masses of substances produced. For example, liquid solvents used can be accounted for, at the end of a process, as liquids or evaporated gases. Furthermore, any wastes generated must be disposed of in an acceptable way. Wastes are recycled, stored in waste sites, carried away in wastewater, or possibly incinerated. In times past, industries often disposed of wastes in ways we now find unacceptable. Strict environmental laws in the United States and many industrialized nations now require industries to be responsible for their wastes. Unfortunately, some illegal waste disposal still occurs.

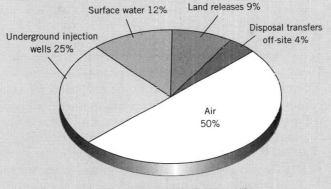

Disposal of approximately 750 million
pounds of toxic wastes each year

The major sources of energy in the world come from the combustion of the fossil fuels: petroleum (gasoline, diesel, and fuel oil), natural gas, and coal. The major products of fuel combustion are gaseous carbon dioxide and water. For example, the combustion of octane, a compound found in gasoline, involves the reaction

$$2C_8H_{18} + 25O_2 \longrightarrow 18H_2O + 16CO_2$$

All carbon-based fuels burned in the environment release carbon dioxide into the atmosphere. Since the annual amounts of fossil fuels used in the world are known, it is easy to estimate the amount of carbon dioxide generated each year. Essentially, all the carbon in fuels that are burned ends up as CO_2. Consequently, the principle of mass balance allows us to know how much carbon dioxide is being produced. Based on the amounts of fuels used, the annual worldwide production of carbon dioxide from fossil fuel use is about 22 billion metric tons (1 metric ton is 1000 kg). Some of this carbon dioxide ends up dissolved in the oceans, but the use of fossil fuels by humans has increased and continues to increase the total concentration of carbon dioxide in the atmosphere. See Section 10-19 for further discussion of atmospheric carbon dioxide. The law of conservation of matter tells us that the materials we use end up somewhere in the environment. When wastes are thrown away, they are not gone but continue to exist somewhere in the environment.

7-11 ENERGY AND REACTIONS

Recall that energy is exchanged during a chemical reaction. Exothermic reactions release energy and endothermic reactions require energy. It is possible to measure amounts of heat exchanged in chemical processes. An exothermic reaction warms the surroundings and an endothermic reaction absorbs heat from the surroundings. Heat is thermal energy, so its units are units of energy. A common unit of energy is the calorie (cal). One calorie is the amount of heat needed to change 1 g of liquid water by 1°C. Water is simply a convenient reference compound; the calorie can be used as a unit of heat with any kind of material or as a unit to express amounts of any kind of energy. For example, it takes 0.2 calorie to heat 1 g of aluminum by 1°C. Recall that 1 food calorie is equal to 1000 cal or 1 kilocalorie (kcal). The calorie is often used to express the amounts of energy exchanged in chemical reactions.

A unit of energy used in the metric or SI system is the joule (abbreviated as J and pronounced jewel). To understand this unit, imagine the physical work or energy expended in picking up and moving an object. To pick it up, you have to apply a force, and to move it, the force needs to be exerted over a distance. For instance, you have to use force to pick up a book and continue applying the force if you want to raise it above your head. Exerting a force through a distance requires work or the expenditure of energy. Forces are measured in the SI system using units of newtons, N. Newtons have SI units of kg m/s². One newton of force exerted over a distance of 1 m requires 1 J of energy. That is, 1 N m = 1 J = 1 kg m²/s². The joule is used as a unit for measuring any kind of energy. In this text, we'll seldom have to use the detailed units for the joule, so don't worry about them. Both the calorie and joule are used to express amounts of heat exchanged in chemical processes. The calorie and the joule are different units, and the relation between them is **1 cal = 4.184 J.** You may be familiar with some other units of energy like the kilowatt-hour (kwhr) or the British thermal unit (Btu). These units are seldom used in chemistry.

How much energy is exchanged in a reaction? The answer depends on the reaction, the reaction conditions, and the amounts of reactants. The amount of energy exchanged in a reaction is measured by allowing the reaction to occur under controlled conditions. The characteristic energy exchanged in a chemical reaction is called the **heat of reaction** or the **enthalpy change for the reaction.** The word *enthalpy* comes from the Greek *thalpein* (to warm or heat). The conventional symbol used to represent heat of reaction or enthalpy change is ΔH. The symbol Δ is the Greek letter delta, so the enthalpy change is pronounced delta H.

Imagine that we place 1 mole of carbon and 1 mole of oxygen gas in a reaction vessel at specific conditions. We initiate the reaction and 1 mole of carbon dioxide is produced. By appropriate design of the experiment, the total amount of heat released is measured in terms of kilojoules or kilocalories. Heats of reaction are usually thousands of joules or calories, so they are normally measured in kilojoules or kilocalories. The measured heat of reaction or enthalpy change for the reaction of solid carbon and oxygen gas is 393 kJ/mol C, which is shown as

$$C(s) + O_2(g) \longrightarrow CO_2(g) \qquad \Delta H = -393 \text{ kJ/mol C}$$

By convention, the enthalpy change of an exothermic reaction is given a negative sign, which corresponds to heat being released from the reaction system as the reaction occurs. The enthalpy change of an endothermic reaction is given a positive sign, which corresponds to heat being gained by the reaction system as the reaction occurs.

When 1 mole of carbon and 1 mole of oxygen gas react to form 1 mole of carbon dioxide under specified conditions, the reaction releases 393 kJ of heat. The negative enthalpy change reveals the heat released per mole of C that reacts. This fact is shown by including the energy on the product side of the equation.

$$C(s) + O_2(g) \longrightarrow CO_2(g) + 393 \text{ kJ}$$

Each reaction has a unique enthalpy change that is measured by experiment. Heats of reaction or enthalpy changes for many different reactions are listed in chemical references.

An example of an endothermic reaction is the reaction of gaseous hydrogen and iodine to give hydrogen iodide.

$$\text{energy} + H_2(g) + I_2(g) \longrightarrow 2HI(g)$$

Heat energy is absorbed from the surroundings when this reaction occurs. That is, the surroundings lose an amount of energy equal to the absorbed energy. Experimentally, the enthalpy change is $\Delta H = +25.9 \text{ kJ/mol HI}$. The plus sign reveals that the reaction is endothermic; note that the value given is per mole of HI. For an endothermic reaction, the enthalpy change is included on the reactant side, but in this case, the value of the enthalpy change is multiplied by 2 since the equation includes 2 moles of HI.

$$51.8 \text{ kJ} + H_2(g) + I_2(g) \longrightarrow 2HI(g)$$

7-12 STOICHIOMETRY AND ENERGY EXCHANGES

The numerical value of the heat of reaction is sometimes shown as a reactant or product in an equation. Relating the heat of reaction to the molar amounts of each chemical in the reaction is possible. The equation for the combustion of methane is expressed as

$$CH_4 + 2O_2 \longrightarrow 2H_2O + CO_2 \qquad \Delta H = -803 \text{ kJ/mol } CH_4$$

The heat of reaction is related to each chemical by expressing the following stoichiometric ratios or heat ratios. When writing heat ratios note the number of moles of the specific species as it relates to the enthalpy change.

$$\left(\frac{803 \text{ kJ}}{1 \text{ mol } CH_4} \right) \quad \text{or} \quad \left(\frac{803 \text{ kJ}}{2 \text{ mol } O_2} \right) \quad \text{or} \quad \left(\frac{803 \text{ kJ}}{2 \text{ mol } H_2O} \right) \quad \text{or} \quad \left(\frac{803 \text{ kJ}}{1 \text{ mol } CO_2} \right)$$

These ratios are used as stoichiometric factors to relate heat to the amount of any chemical in the reaction. Focus 7 shows a stoichiometry map that includes heat.

EXAMPLE 7-13

Find how many kilojoules of heat are produced when 15.3 g of oxygen gas react with sufficient methane according to the equation

$$CH_4 + 2O_2 \longrightarrow 2H_2O + CO_2 \qquad \Delta H = -803 \text{ kJ/mol } CH_4$$

The calculation sequence for this type of stoichiometric question is

$$\begin{array}{ccc} \text{molar} & \text{heat} \\ \text{mass} & \text{ratio} \\ \text{g } O_2 \Longrightarrow & \text{mol } O_2 \Longrightarrow & ? \text{ kJ} \end{array}$$

This type of question is similar to a mass-to-mass calculation since going directly from the mass of a chemical to heat is not possible. The heat of reaction relates kilojoules and moles, so we need to work on a molar basis. First, the number of moles of oxygen gas is found from the mass using its molar mass. Then, the ratio relating the heat of reaction and moles of oxygen,

$$\left(\frac{803 \text{ kJ}}{2 \text{ mol } O_2} \right)$$

is used as a factor to find the heat released.

$$X \text{ kJ} = 15.3 \cancel{g} \left(\frac{1 \cancel{\text{ mol } O_2}}{32.00 \cancel{g}} \right) \left(\frac{803 \text{ kJ}}{2 \cancel{\text{ mol } O_2}} \right) = 192 \text{ kJ released}$$

EXAMPLE 7-14

A Btu or British thermal unit is a unit of heat equivalent to 1.054 kilojoules. Find how many grams of methane are needed to give 1.054 kilojoules of heat by the reaction shown as

$$CH_4 + 2O_2 \longrightarrow 2H_2O + CO_2 \qquad \Delta H = -803 \text{ kJ/mol } CH_4$$

The calculation sequence for this heat to mass question is

$$\begin{array}{ccc} \text{heat} & \text{molar} \\ \text{ratio} & \text{mass} \\ \text{kJ} \Longrightarrow & \text{mol } CH_4 \Longrightarrow & \text{g } CH_4 \end{array}$$

Since we are given the number of kilojoules of heat, the appropriate heat ratio is used to find the number of moles of methane that produce this amount of heat. The molar mass of methane is then used to change the moles of methane to mass. Using the factors so that the kilojoules and moles cancel gives

$$X \text{ g } CH_4 = 1.054 \cancel{\text{kJ}} \left(\frac{1 \cancel{\text{ mol } CH_4}}{803 \cancel{\text{ kJ}}} \right) \left(\frac{16.01 \text{ g}}{1 \cancel{\text{ mol } CH_4}} \right) = 2.10 \times 10^{-2} \text{ g}$$

Energy

ACTIVITY 7-8

The overall reaction for the metabolism of sucrose in the body is

$$C_{12}H_{22}O_{11} + 12O_2 \longrightarrow$$
$$12CO_2 + 11H_2O$$
$$\Delta H = -1350 \text{ kcal/1 mol}$$
$$C_{12}H_{22}O_{11}$$

Calculate the number of dietetic calories or kilocalories released by the metabolism of 1 g of sucrose. Since sucrose is a carbohydrate, what do you notice about your answer? See Section 5-12 on Food and Calories.

Energy and Society

Fossil fuels (petroleum, coal, and natural gas) are the main source of energy in industrial societies. Figure 7-6 shows the sources and uses of energy in the United States. The percentages are approximations since the energy mix changes yearly. Notice in Figure 7-6 that the fossil fuels provide about 84% of our energy. Coal provides 23%. More than one-half of this is used to produce electricity. Natural gas accounts for about 23% of our energy. Most of it is used directly as heating fuel. Petroleum and natural gas liquids (butane and propane) contribute about 38%. Of all petroleum used in the United States, around 55% is imported. Another 7% of our energy comes from nuclear reactors and 7% from renewable sources such as hydroelectric and wind farms.

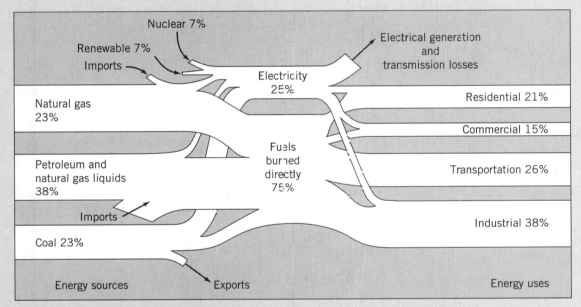

FIGURE 7-6   Approximate uses of energy and sources of energy in the United States.

Overall, around one-fourth of energy resources are used to produce electricity, and three-fourths is consumed directly as fuels. Some is used to make petrochemical products, such as plastics and chemicals, but most of it is used to produce energy. Industry uses 38% of the energy in producing consumer goods. The largest fraction of this energy is used in steel furnaces and metalworks to produce metals, such as iron and aluminum. Other large industrial energy consumers include oil refineries, mines, chemical plants, glass factories, food processing plants, paper plants, and the construction industry. Commercial establishments (stores, offices, hotels, etc.) account for about 15% of our energy use. One-half the energy used commercially is for space heating (heaters) and air conditioning. Transportation uses 26% of our energy. Most of this comes from the burning of gasoline and diesel oil in cars, buses, and trucks. Trains, ships, and airplanes are additional energy-consuming transportation modes. Household uses consume around 21% of our energy. Around one-half of this energy is for space heating and cooling using natural gas and heating oil. Other home uses include electricity for lighting, cooking, and various appliances.

Stoichiometry Map

The stoichiometry map applies to any balanced equation. The letters A and B refer to any chemicals. The factors shown are needed to move from one part of the map to another part. To find the mass of B related to the mass of A requires the use of the molar mass of A, the molar ratio from the equation, and the molar mass of B.

Example: Hydrogen and oxygen gas react to give water, $2H_2 + O_2 \longrightarrow 2H_2O$. The process needed to find the number of grams of hydrogen needed to react with 64 g of oxygen is found by following this mass to mass question on the stoichiometry map. Mass of oxygen converts to moles of oxygen. Then the moles of oxygen is changed to moles of hydrogen and the moles of hydrogen are converted to grams of hydrogen.

$$64 \text{ g} \left(\frac{1 \text{ mol } O_2}{32.00 \text{ g}} \right) \left(\frac{2 \text{ mol } H_2}{1 \text{ mol } O_2} \right) \left(\frac{2.016 \text{ g}}{1 \text{ mol } H_2} \right) = 8.1 \text{ g}$$

The enthalpy change of a reaction is used to express appropriate heat ratios that are used to relate amounts of heat to the number of moles or mass of any species involved in a reaction.

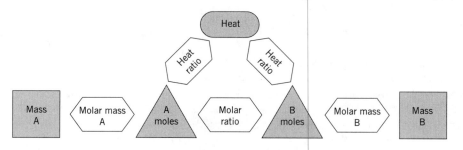

Example: For the reaction $2H_2 + O_2 \longrightarrow 2H_2O$ the enthalpy change is $\Delta H = -121$ kJ/mol H_2O. The number of grams of oxygen that is needed to react with sufficient hydrogen to give 484 kJ of heat is found using the appropriate heat ratio obtained from the enthalpy change.

$$\left(\frac{121 \text{ kJ}}{1 \text{ mol } H_2O} \right) \left(\frac{2 \text{ mol } H_2O}{1 \text{ mol } O_2} \right) = \left(\frac{242 \text{ kJ}}{1 \text{ mol } O_2} \right)$$

Following the stoichiometry map from heat to the mass of oxygen gives

$$484 \text{ kJ} \left(\frac{1 \text{ mol } O_2}{242 \text{ kJ}} \right) \left(\frac{32.00 \text{ g}}{1 \text{ mol } O_2} \right) = 64.0 \text{ g}$$

QUESTIONS

1. Match the following terms with the appropriate lettered definitions.

mole	a. the SI or metric unit of energy
molar mass	b. 4.184 J
stoichiometry	c. the reactant that controls the amount of product formed in a reaction
Haber process	d. the industrial method used to make ammonia from hydrogen and nitrogen
molar ratio	e. the amount of a chemical that contains as many chemical particles as there are atoms in 12 grams of carbon-12
limiting reactant	f. methods used to relate the amounts of chemicals in a reaction
percent yield	g. the heat exchanged in a chemical reaction
calorie	h. the yield of reaction expressed as the relation between the actual yield and the theoretical yield
joule	i. a factor relating two chemicals involved in a reaction in terms of their coefficients in the balanced equation
enthalphy change	j. the number of grams per mole of a chemical

See Chapter 4 for a review of the use of the mole.

Section 7-1

2. *Liquid mercury has a density of 13.6 g/mL. How many moles of Hg atoms are in a thermometer that contains 0.055 mL of Hg?

3. Liquid bromine, Br_2, has a density of 2.90 g/mL. How many moles of Br_2 are in a 25 mL sample?

4. *A quartz crystal has a density of 2.75 g/cm^3. How many moles of SiO_2 are in a 37.5 cm^3 crystal?

5. A typical aspirin tablet contains 5.0 grains of acetylsalicylic acid, $C_9H_8O_4$ (180 g/mole). How many moles of acetylsalicylic acid are in two tablets? (0.0648 g = 1 grain)

6. *Thyroxine is a hormone produced in the thyroid gland. It has the formula $C_{15}H_{11}O_4I_4N$. How many moles of combined iodine atoms are contained in 3.15 micrograms of thyroxine? (1 microgram = 10^{-6} g)

7. Lithium carbonate, Li_2CO_3, is used in the treatment of bipolar mental disorders. How many moles of lithium ion, Li^+, are contained in a 500 mg dose of lithium carbonate?

8. *Green vegetables are an important source of dietary magnesium because they contain chlorophyll, $C_{55}H_{72}MgN_4O_5$. If the recommended dietary allowance (RDA) for magnesium for adults is 280 mg/day, how many grams of chlorophyll meet this need?

9. Sodium fluoride is used as an anticavity agent in some toothpaste. How many moles of fluoride ion are in a 4.5 mg sample of sodium fluoride?

10. *Baking soda, $NaHCO_3$, is often used as a leavening agent in baked goods. If a recipe calls for 2 tsp of baking soda, how many formula units of $NaHCO_3$ is this? (1 tsp = $\frac{1}{8}$ oz; give your answer to two significant digits)

11. *A sample of moon rock is found to contain 12.5% Al_2O_3. How many moles of Al_2O_3 are contained in 2.25 kg of the rock? (Hint: The rock contains 12.5 g Al_2O_3/100 g rock.) How many moles of oxide ion does the 2.25 kg sample contain?

12. Determine the mass or the number of moles for each of the following.

   (a) the number of moles of O_2 in a 25.0 g sample

   (b) the mass of 3.94 moles of fructose, $C_6H_{12}O_6$

   (c) the number of moles of calcium carbonate, $CaCO_3$, in a 125 mg sample

   (d) the mass of 0.445 mole of acetic acid, $HC_2H_3O_2$

13. Give a general definition of the mole.

14. How many grams of NaCl contain 2400 mg of sodium ion?

Section 7-2

15. Why is mass conserved in a chemical reaction?

Sections 7-3 to 7-6

16. *Acetylene, C_2H_2, burns in oxygen according to the unbalanced equation

$$C_2H_2(g) + O_2(g) \longrightarrow CO_2(g) + H_2O(g)$$

Balance the equation. Give the molar interpretation of the equation and give the molar ratios that relate the following pairs of reactants and products.

    (a) C_2H_2 and O_2

    (b) C_2H_2 and CO_2

    (c) C_2H_2 and H_2O

    (d) O_2 and CO_2

17. *Refer to the balanced equation in Question 16. Find the number of moles of the first chemical listed in each of the following.

    (a) O_2, to react with 0.570 mole of C_2H_2

    (b) C_2H_2, to produce 0.400 mole of CO_2

    (c) H_2O, formed when 17.6 moles of C_2H_2 react

    (d) CO_2, produced when 3.89 moles of O_2 react with acetylene

18. Butane gas burns in air to form carbon dioxide and water as shown by the unbalanced equation

$$C_4H_{10}(g) + O_2(g) \longrightarrow CO_2(g) + H_2O(g)$$

Balance the equation. Give the molar interpretation of this equation and give the molar ratios that relate the following pairs of reactants and products.

    (a) C_4H_{10} and O_2

    (b) C_4H_{10} and CO_2

    (c) C_4H_{10} and H_2O

    (d) O_2 and CO_2

19. Refer to the balanced equation in Question 18. Find the number of moles of the first chemical listed in each of the following.

    (a) O_2, to react with 8.75 moles of C_4H_{10}

    (b) C_4H_{10}, to produce 14.9 moles of CO_2

    (c) H_2O, formed when 6.05×10^3 moles of C_4H_{10} react

    (d) CO_2, produced when 0.075 mole of O_2 reacts with butane

Sections 7-7 to 7-9

20. *The ethanol or ethyl alcohol component of gasohol burns according to the unbalanced equation

$$C_2H_5OH + O_2 \longrightarrow CO_2 + H_2O$$

Balance the equation. Find the amount of the first chemical listed in each of the following.

    (a) Oxygen, to react with 52.6 g of C_2H_5OH; answer in moles

    (b) Oxygen, to react with 52.6 g of C_2H_5OH; answer in grams

    (c) CO_2, formed when 52.6 g of C_2H_5OH react; answer in grams

    (d) CO_2, produced when 52.6 g of C_2H_5OH and 75.0 g of O_2 are mixed; answer in grams (limiting reactant problem)

21. Nitrogen dioxide can form nitric acid by reaction with water as shown by the unbalanced equation

$$NO_2 + H_2O \longrightarrow HNO_3 + NO$$

Balance the equation. Find the number of grams of the first chemical listed in each of the following.

    (a) HNO_3, formed from 50 moles of NO_2

    (b) H_2O, to form 500 g of HNO_3

    (c) NO_2, to form 250 g of HNO_3

    (d) HNO_3, formed when 125.0 g of NO_2 and 95.0 g of water are mixed (limiting reactant problem)

22. *Antacids containing $CaCO_3$ react with "stomach acid" according to the unbalanced equation

$$CaCO_3(s) + HCl(aq) \longrightarrow CaCl_2(aq) + CO_2(g) + H_2O$$

Balance the equation. Find the amount of the first chemical listed in each of the following.

    (a) CO_2, formed from 500 mg of $CaCO_3$; answer in grams

    (b) HCl, to react with 1.00 g of $CaCO_3$; answer in moles

    (c) $CaCO_3$, to produce 1200 mg of calcium ion; answer in grams (some people use $CaCO_3$ as a supplemental calcium source)

    (d) CO_2, formed when 9.45 g of HCl are mixed with 28.4 g of $CaCO_3$; answer in grams (limiting reactant problem)

23. The fermentation of glucose to form ethyl alcohol occurs according to the unbalanced equation

$$C_6H_{12}O_6 \longrightarrow C_2H_5OH + CO_2$$

Balance the equation. Find the amount of the first chemical listed in each of the following.

    (a) $C_6H_{12}O_6$, to form 500 g of ethyl alcohol; answer in grams

(b) CO_2, produced when 10.0 g of $C_6H_{12}O_6$ react; answer in moles

(c) C_2H_5OH, formed from 1.00 kg of $C_6H_{12}O_6$; answer in grams

24. *Calcium cyanamide, $CaCN_2$, is used as a fertilizer. It reacts with water to form $CaCO_3$, which counteracts excess acidity in the soil, and ammonia, NH_3, which fertilizes the soil, according to the equation

$$CaCN_2 + 3H_2O \longrightarrow CaCO_3 + 2NH_3$$

(a) How many grams of NH_3 are formed from 1.00 moles of $CaCN_2$?

(b) How many grams of water are needed to produce 625 g of NH_3?

(c) How many grams of $CaCN_2$ are needed to produce 500 kg of NH_3?

(d) If the $CaCN_2$ used in the reaction is only 70% pure, the rest being inert impurities, how many kilograms of NH_3 are formed from 500 kg of impure $CaCN_2$?

25. Nitroglycerine, $C_3H_5(NO_3)_3$, is a powerful chemical explosive. Since this oily liquid can explode (even in the absence of atmospheric oxygen) by the slightest shock, it is very dangerous to handle. The equation for the reaction is

$$4C_3H_5(NO_3)_3 \longrightarrow$$

$$6\,N_2(g) + O_2(g) + 12\,CO_2(g) + 10\,H_2O(g) + energy$$

(a) How many moles of nitrogen are produced from 15.5 moles of $C_3H_5(NO_3)_3$?

(b) How many grams of oxygen are produced from 475 g of $C_3H_5(NO_3)_3$?

(c) If a vial of nitroglycerine contains 2.86 g of $C_3H_5(NO_3)_3$, how many grams of CO_2 are formed?

(d) Dynamite is made by mixing nitroglycerine with inert diatomaceous earth. How many grams of water can form when 1.00 kg of dynamite, containing 50% nitroglycerine, explodes?

26. *Titanium metal is used to make relatively light-weight but high-strength alloys used in aircraft. It is obtained as the metal by the reaction

$$TiCl_4 + 4Na \longrightarrow 4NaCl + Ti$$

(a) How many moles of Ti are formed when 11.6 g of Na react?

(b) How many grams of Ti are formed from 385 g of $TiCl_4$?

(c) If you want to produce 5.0 kg of Ti, how many grams of Na are needed?

(d) How many grams of NaCl are formed when 0.353 g of $TiCl_4$ react?

(e) How many grams of titanium are formed when 965 g of $TiCl_4$ are mixed with 480 g of Na? (limiting reactant problem)

27. An important reaction in a blast furnace used to make iron is

$$Fe_2O_3 - 3CO \longrightarrow 2Fe + 3CO_2$$

(a) How many grams of Fe_2O_3 are needed to produce 887 g of Fe?

(b) How many kilograms of CO are needed to produce 750 kg of Fe?

(c) If the iron ore is 76% Fe_2O_3, how many grams of Fe are formed from 945 g of ore?

(d) How many kilograms of Fe are formed when 1.63×10^4 kg of Fe_2O_3 are mixed with 779 kg of CO? (limiting reactant problem)

28. *An important enzyme in your body, sometimes called catalase, converts hydrogen peroxide, H_2O_2, to oxygen and water.

$$\overset{\text{catalase}}{2H_2O_2 \longrightarrow O_2 + 2H_2O}$$

(a) How many grams of H_2O_2 are converted if 1.36×10^{-6} mole of O_2 is formed?

(b) A single cell in your body produces 4.61×10^{-12} g of hydrogen peroxide. How many grams of water are produced if all the H_2O_2 reacts?

(c) How many moles of H_2O_2 are required to produce 1.00 mg of water?

(d) How many grams of O_2 are formed when 3.32×10^{-8} g of H_2O_2 reacts?

29. The enzyme urease, isolated from jack beans, catalyzes the decomposition of urea according to the reaction

$$\overset{\text{urease}}{CN_2H_4O + H_2O \longrightarrow 2NH_3 + CO_2}$$

(a) How many grams of urea are needed to produce 56.7 g of NH_3?

(b) How many moles of water are required to react with 118 mg of urea?

(c) How many grams of NH_3 are produced from 7.62 g of urea?

(d) How many grams of urea are needed to react with 1.006 g of water?

30. *Wine is made by fermenting a sugar, such as glucose, by the action of enzymes found in yeast. The fermentation forms ethanol and carbon dioxide.

$$C_6H_{12}O_6 \xrightarrow{\text{yeast}} 2C_2H_5OH + 2CO_2$$

(a) How many grams of ethanol are produced from 525 g of glucose?

(b) How many moles of CO_2 can be produced from 81.5 moles of glucose?

(c) If 255 kg of ethanol is produced, how many kilograms of carbon dioxide are also produced?

31. An astronaut excretes about 2.61×10^3 g of water a day. If lithium oxide is used in the spacecraft to absorb the water, how many kilograms of Li_2O must be included for a 28-day space trip? The reaction involved is given by the equation

$$Li_2O + H_2O \longrightarrow 2LiOH$$

32. A 8.15 g sample of salt water is analyzed by carrying out the reaction

$$Ag^+(aq) + Cl^-(aq) \longrightarrow AgCl(s)$$

If 0.910 g of AgCl is formed from the sample, what is the percent of chloride ion in the salt water?

33. *A quick bread recipe uses a mixture of baking soda, $NaHCO_3$, and vinegar, which contains acetic acid, $HC_2H_3O_2$. This mixture serves as a leavening agent according to the equation

$$NaHCO_3 + HC_2H_3O_2 \longrightarrow NaC_2H_3O_2 + CO_2 + H_2O$$

If 49.5 g of $NaHCO_3$ and 42.6 g of $HC_2H_3O_2$ are mixed, find

(a) how many grams of CO_2 are formed

(b) which reactant is in excess and how many grams of this reactant remain after the reaction is complete

34. Sodium nitrite, $NaNO_2$, is added to packaged meats to prevent the growth of *Clostridium botulinum*, the organism that causes botulism poisoning. Nitrites, however, are potentially dangerous since they are converted in the stomach to nitrous acid, HNO_2, which can lead to the formation of carcinogenic nitrosamines.

$$NaNO_2 + HCl \longrightarrow HNO_2 + NaCl$$

If 3.75 mg of $NaNO_2$ is added to 15.0 mg HCl, find

(a) how many milligrams of HNO_2 are formed

(b) which reactant is in excess and how many milligrams remain after the reaction is complete

35. *Aspirin, $C_9H_8O_4$, is synthesized from salicylic acid, $C_7H_6O_3$, and acetic anhydride, $C_4H_6O_3$, in a reaction represented by the equation

$$C_7H_6O_3 + C_4H_6O_3 \longrightarrow C_9H_8O_4 + HC_2H_3O_2$$

If 219 g of salicylic acid and 210 g of acetic anhydride are mixed, find

(a) how many grams of aspirin are formed

(b) which reactant is in excess and how many grams remain after the reaction is complete

36. A method used to make uranium metal involves the reaction

$$U_3O_8 + 4C \longrightarrow 3U + 4CO_2$$

If 11.16 kg of U_3O_8 and 0.0900 kg of C are mixed, find

(a) how many grams of uranium are formed

(b) which reactant is in excess and how many grams remain after the reaction is complete

37. *Lime or calcium oxide, CaO, reacts with water to form calcium hydroxide, $Ca(OH)_2$. Write the balanced equation for the reaction. If 25.0 g of CaO is added to 21.0 g of water, find

(a) how many grams of $Ca(OH)_2$ are formed

(b) which reactant is in excess and how many grams remain after the reaction is complete

38. Tungsten metal is used in light bulb filaments. One method used to make tungsten metal involves the reaction

$$WO_3 + 3H_2 \longrightarrow W + 3H_2O$$

If 69.8 g of WO_3 and 3.90 g of H_2 are mixed, find

(a) how many grams of tungsten are formed

(b) which reactant is in excess and how many grams remain after the reaction is complete

Section 7-10

39. Define the terms *theoretical yield* and *percent yield*.

40. *Aspirin, $C_9H_8O_4$, is made from the reaction of salicylic acid, $C_7H_6O_3$, and acetic anhydride, $C_4H_6O_3$:

$$C_7H_6O_3 + C_4H_6O_3 \longrightarrow C_9H_8O_4 + HC_2H_3O_2$$

When 69.2 g of salicylic acid react with acetic anhydride, 82.5 g of aspirin are obtained. What is the percent yield of aspirin?

41. The fermentation of glucose forms ethyl alcohol:

$$C_6H_{12}O_6 \longrightarrow 2C_2H_5OH + 2CO_2$$

When 90 g of glucose react, 42 g of ethyl alcohol are obtained. What is the percent yield of ethyl alcohol?

Sections 7-11 and 7-12

42. Define the following terms.

(a) joule

(b) calorie

43. How do the calorie and joule compare?

44. A watt (W) is a unit of power and has units of joule per seconds (J/s). Determine the number of joules corresponding to 1 kW-hr. (Hint: Substitute joules per second for the watt and convert the hour to seconds.)

45. A British thermal unit is defined as the amount of heat needed to change 1 lb of water by 1°F. Determine the number of calories corresponding to 1 Btu. [Hint: Change the units of the (1°F 1 lb/Btu) to (1°C 1 g/ Btu).]

46. *Sucrose (table sugar) burns in oxygen to form carbon dioxide and water.

$$C_{12}H_{22}O_{11}(s) + 12O_2(g) \longrightarrow 11H_2O(\ell) + 12CO_2(g)$$
$$\Delta H = -5649 \text{ kJ/1 mol } C_{12}H_{22}O_{11}$$

(a) How many grams of sucrose need to react to give 1.00 kcal of heat? (1 kcal = 4.184 kJ.

(b) When sucrose is metabolized in your body, the overall process is represented by the above equation. How many kilojoules of heat could be produced by the metabolism of 1 cup, 220 g, of sugar? How many kilocalories or dietetic calories is this?

(c) Sugar is a carbohydrate. Calculate the number of kcal produced per gram of sugar metabolized. (1 kcal = 4.184 kJ)

47. Glucose sugar burns in oxygen to form carbon dioxide and water.

$$C_6H_{12}O_6(s) + 6O_2(g) \longrightarrow 6H_2O(\ell) + 6CO_2(g)$$
$$\Delta H = -2816 \text{ kJ/1 mol } C_6H_{12}O_6$$

(a) How many grams of glucose need to react to give 1.00 kcal of heat? (1 kcal = 4.184 kJ)

(b) When glucose is metabolized in your body, the overall process is represented by the above equation. How many kilojoules of heat could be produced by the metabolism of 1 oz of glucose (1 oz = 28.3 g) and how many kcal or dietetic calories is this?

(c) Glucose is a carbohydrate. Calculate the number of kcal produced per gram of glucose metabolized. (1 kcal = 4.184 kJ)

48. *When 1 mole of nitrogen gas and 1 mole of oxygen gas react to give 2 moles of nitrogen oxide gas, NO, 181 kJ of heat are absorbed. Write a balanced

equation for the reaction that includes the heat of reaction. If 175.2 g of nitrogen react with sufficient oxygen, how many kilojoules of heat are required?

49. When magnesium metal, Mg, reacts with oxygen, magnesium oxide, MgO, is formed and intense heat and a bright flash of light accompany the reaction. When 1 mole of oxygen reacts with magnesium, 2420 kJ of heat are released. Write a balanced equation for the reaction; include the heat of reaction. How many kilojoules of heat are released when 1.00 g of magnesium reacts with oxygen?

50. *The combustion reactions and the enthalpy changes for methane, propane, and butane are

$$CH_4(g) + 2O_2(g) \longrightarrow CO_2(g) + 2H_2O(g)$$
$$\Delta H = -803 \text{ kJ/1 mol } CH_4$$

$$C_3H_8(g) + 5O_2(g) \longrightarrow 3CO_2(g) + 4H_2O(g)$$
$$\Delta H = -2046 \text{ kJ/1 mol } C_3H_8$$

$$2C_4H_{10}(g) + 13O_2(g) \longrightarrow 8CO_2(g) + 10H_2O(g)$$
$$\Delta H = -2261 \text{ kJ/1 mol } C_4H_{10}$$

Determine which is the best fuel in terms of the number of kilojoules produced per gram of fuel. The number of kilojoules per gram is called the fuel value.

51. *Refer to the equations and enthalpy changes in Question 50 to answer the following.

(a) How many grams of methane, CH_4, are needed to produce 1.00×10^3 kJ of heat?

(b) A Btu is 1.05 kJ. How many Btu are produced from the combustion of 65.0 g of propane, C_3H_8?

(c) How many grams of butane, C_4H_{10}, are needed to produce 35.0 Btu of heat?

(d) If a methane heater produces 30,000 Btu of heat per hour (30×10^3 Btu/hr) and 1 Btu = 1.05 kJ, how many grams of methane are burned per hour?

52. Gasohol is a mixture of ethanol and gasoline that is found in some gas stations. The heats of reaction for the combustion of ethanol and octane (one component of gasoline) are

$$C_2H_6O(\ell) + 3O_2(g) \longrightarrow 2CO_2(g) + 3H_2O$$
$$\Delta H = -1235 \text{ kJ/1 mol } C_2H_6O$$

$$2C_8H_{18}(\ell) + 25O_2(g) \longrightarrow 16CO_2(g) + 18H_2O(g)$$
$$\Delta H = -5080 \text{ kJ/1 mol } C_8H_{18}$$

(a) Calculate how much heat is produced by burning 1.00 gallons of ethanol (density = 0.789 g/mL).

(b) Calculate how much heat is produced by burning 1.00 gallons of octane (density = 0.702 g/mL).

(c) Compare octane and ethanol in terms of the number of kilojoules of heat produced per gram of each.

53. Refer to the equations and enthalpy changes in Question 52 to answer the following.

(a) How many grams of ethanol are needed to produce 7.50×10^3 kcal of heat?

(b) How many grams of octane are needed to produce 1.00 kcal of heat?

(c) When 1.00 liters of ethanol are burned how many kilojoules of heat are released [see Question 52(a)]?

(d) How many kilocalories of heat are released when 47.1 g of ethanol and 39.8 g of oxygen react?

54. Hydrogen is a potential fuel of the future. The combustion reaction of hydrogen and the heat of reaction are

$$2H_2(g) + O_2(g) \longrightarrow 2H_2O(g)$$

$$\Delta H = -121 \text{ kJ}/1 \text{ mol } H_2O$$

(a) How many grams of hydrogen are needed to produce 1.20×10^3 kJ of heat?

(b) How many kilojoules of heat are produced when 675 g of oxygen react with sufficient hydrogen?

(c) How many kilojoules of heat are produced per gram of hydrogen? How does hydrogen compare with methane in terms of kilojoules per gram of fuel? See Question 50.

55. What are the main sources and uses of energy in our society?

56. What fuels are commonly used as sources of energy?

ATOMIC

STRUCTURE

8-1 THE EVOLUTION OF THE ATOM

Ideas about the structure of atoms have evolved since John Dalton first proposed the atomic theory. Dalton viewed atoms as characteristic, indestructible particles that somehow combined to form compounds.

The discovery of electrons and protons produced a view of the atom as some sort of collection of electrons and positive particles.

The atom, as suggested by Lord Rutherford in 1911, was visualized as electrons in motion about a positively charged nucleus. This view was called the nuclear model or the solar system atom.

In 1913, Niels Bohr, a Danish physicist, theorized that an atom has electrons in motion about the nucleus in definite, fixed orbits.

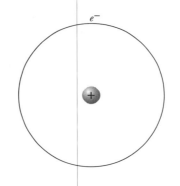

Scientists

The modern model of the atom was developed in the 1920s by the efforts of many scientists; among the most notable are European scientists Erwin Schrödinger, Paul A. M. Dirac, and Werner Heisenberg. The quantum mechanical atom views electrons in atoms as three-dimensional clouds surrounding the nucleus.

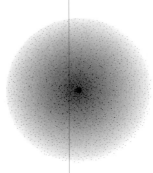

Atomic structure refers to the ways that electrons are distributed in atoms. Experimental evidence suggests that the chemical behavior of atoms is directly related to their atomic structures. Knowing something about atomic structure makes it easier to understand the similarities and differences in the chemistry of various elements.

As you know, each element has a unique atomic number, which is the number of protons and electrons in atoms of that element. Hydrogen, atomic number 1, has atoms containing one proton and one electron. Helium, atomic number 2, has atoms with two protons and two electrons. This pattern continues through the elements, with each subsequent element having atoms with one more proton and one more electron than the previous element of lower atomic number. Elements are arranged from left to right in the periodic table by increasing atomic number. To get the atomic number of an element just find its location in the table.

8-2 CATHODE-RAY TUBES

Much experimental evidence of atomic structure came from interesting scientific devices first constructed in the mid-1800s. These devices are sealed glass tubes that have the air pumped out and a metal electrode attached at each end (see Figure 8-1). A high-voltage source is connected to the two electrodes. When the voltage is high enough, electricity passes through the tube. This situation is similar to a spark of electricity or a bolt of lightning passing through air except that the flow in an evacuated tube is not visible. Within the tube, electricity flows from one electrode to the other. That is, it flows from the cathode to the anode. The flow of electricity in the tube is called a cathode ray, so the tube is known as a **cathode-ray tube (CRT).**

CRT Spectra

When a CRT tube is placed between electrically charged plates, as illustrated in Figure 8-2 on page 212, the cathode rays are attracted toward the positive plate. Furthermore, when the tube is placed between the poles of a strong magnet, the cathode-ray beam deflects from a straight-line path. It was theorized that cathode rays were negatively charged particles that were the same as the characteristic particles of electricity. Of course, the characteristic particles of electricity are electrons. As we know, electrons are subatomic particles found in atoms. A **cathode-ray** beam is a free-moving beam of electrons. That is, electrons in the beam are not bound tightly to atoms but are free to move under the influence of voltage.

Cathode-ray tubes provide evidence for the existence of electrons and provide ways in which the properties of electrons are observed. Incidentally, tubes in television sets and computer terminals are cathode-ray tubes in which the cathode-ray beam is moved about by electromagnetic fields. The electron beam is focused on and scanned across a screen coated with phosphors. Phosphors are chemicals that glow temporarily when energized by the beam.

FIGURE 8-1 A simple discharge tube.

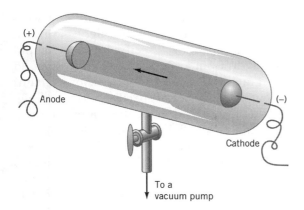

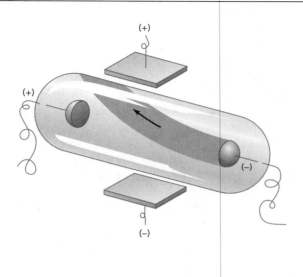

FIGURE 8-2 Cathode-ray beam is deflected toward a positively charged plate.

8-3 GAS DISCHARGE TUBES

Construction of a cathode-ray tube involves pumping the air out of the glass tube to which the electrodes are attached. Cathode-ray tubes are vacuum tubes; that is why they implode when they are broken. A device, known as a **gas discharge tube,** is made by placing a small amount of a gas in the cathode-ray tube after the air is removed. A normal cathode-ray tube has a partial vacuum in it, whereas a gas discharge tube has some kind of gas in it. When electricity flows through a gas discharge tube, a distinct light is produced. The light produced is of a unique color that depends on the gas used in the tube. That is when electricity is passed through a specific gas in a discharge tube, it gives off light of a characteristic color. Hydrogen emits a distinct reddish purple light and neon emits a reddish orange light. Discharge tubes are used in neon signs and in mercury vapor and sodium vapor street lights. You probably have seen a hydrogen discharge tube as a "black" light source or as a sunlamp.

Sunlight or white light from an incandescent light bulb passing through a prism or diffraction grating is dispersed into light having all the colors of the rainbow; red, orange, yellow, green, blue, indigo, and violet **(ROY G. BIV).** When the light from a gas discharge tube is passed through a prism or diffraction grating, however, the dispersion contains not all colors of the rainbow, but a unique set of specific colors (see Figure 8-3). For instance, the light from a hydrogen discharge tube is a distinct set of red, bluish green, and purple light. No other element gives the same colors. To the human eye, the mixture of the color set appears as reddish purple light; we only see the separate colors when the light is diffracted. The characteristic pattern of light from a discharge tube, called its emission spectrum, serves as a fingerprint of an element. The

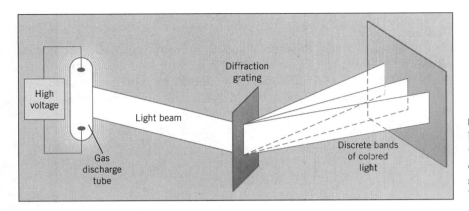

FIGURE 8-3 A light beam from a discharge tube passes through a diffraction grating and is separated into discrete bands of colored light.

emission spectrum pattern suggests an element's unique atomic structure and the arrangement of its electrons, a message from the inner spaces of atoms. Experimental evidence obtained from gas discharge tubes was very important to the development of models or theories of atomic structure.

Seeing inside an atom is not possible. It is possible to develop models of atomic structure using experimental evidence about the behavior of collections of atoms. A good model provides answers to many questions and supplies useful mental images of atoms. How are the electrons distributed about the nucleus? How does atomic structure explain the properties of the various elements? How is atomic structure related to the ways in which atoms chemically combine?

Somewhere over the Rainbow

You can use a compact disc to separate white light into the colors of the rainbow. The closely spaced lines on the disc diffract white light into light of various colors.

1. Carefully cut a piece of dark paper or cardboard in half, making as straight a cut as you can.
2. Lay the two halves flat. Leave a few millimeters of space between them and use a length of clear tape, the entire length of the cardboard, to tape the halves together. Then turn over the paper and use another length of tape on the other side. This space is your light slit.
3. Obtain a CD and be careful not to touch the shiny side.
4. Hold the paper slit between a light bulb and the CD and move the CD around until you see a vivid spectrum of light. Describe your observations. You'll get more vivid results if you work in a dark room with only one source of light.

Bohr

8-4 THE BOHR MODEL

The modern model of the atom has evolved from the ideas of the nuclear atom. Niels Bohr developed a model to explain how electrons orbit the nucleus. He developed his model in an attempt to explain why a hydrogen discharge tube produces colored light that is characteristic of hydrogen. According to the **Bohr model,** a hydrogen atom has a nucleus with a 1+ charge, around which an electron is moving in an orbit.

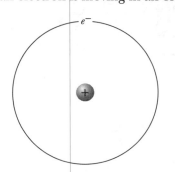

The electron has energy that results from its motion and position. This energy is somewhat analogous to the energy of a ball tied to a string and twirled in a circular motion. The ball has energy of motion (kinetic energy) and energy of position (potential energy) as it is held by the string. The electron has kinetic energy of motion and potential energy arising from attraction between the positively charged nucleus and the negative charges of the electron.

Bohr made the revolutionary assumption that an electron in an atom has quantized energy. In this view an electron can only have a fixed energy or some multiple of that energy, but it cannot have any intermediate energies. In terms of orbits, an electron in a hydrogen atom is found only at specific distances from the nucleus in specific orbits corresponding to the allowable quantized energies.

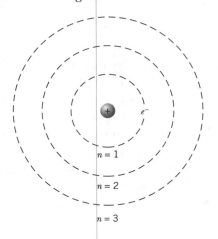

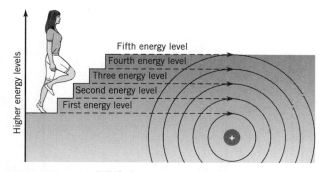

FIGURE 8-4   While hopping up a staircase, a person may take on only specific "quantized" energy positions. The steps represent energy levels. The person cannot stop between "energy level" steps. Likewise, electrons occupy specific quantized energy levels; they cannot exist between levels. An electron occupies a specific level corresponding to its energy.

The idea of quantized energy needs some further explanation. As an analogy to the quantized energy of an electron, consider climbing a staircase by hopping on one leg. When hopping, you can take on only "quantized" positions as you climb, as shown in Figure 8-4. That is, you cannot remain at any position between steps. As you climb the staircase, your energy (potential energy with respect to the bottom of the staircase) is of a certain value at the first step. Your energy increases by some fixed amount as you hop to higher steps.

The Bohr model gives a view of a hydrogen atom as a central nucleus orbited by an electron in an orbit that depends on its energy. The electron can only have a specific energy out of a set of possible energy values. Each allowed energy corresponds to a specific orbit. Orbits corresponding to different energies are found at different distances from the nucleus. The possible quantized orbits of the electron are the **energy states** or **energy levels.** An electron is found in one of these possible orbits, depending on its energy. It cannot be found anywhere between these orbits since the orbits represent the allowed quantized energy states.

8-5 QUANTUM JUMPS

In the Bohr view, the single electron in a hydrogen atom normally occupies the lowest energy level. The lowest level is called the **ground state.** This level is the one closest to the nucleus. If the electron gains energy from some outside sources, such as a flame or electricity, it can jump from a lower energy level to a higher level. Such a change in energy levels is a **quantum jump** or **quantum leap.** Similar to the stair-hopping analogy, the electron can hop from level to level but cannot stop anywhere between levels.

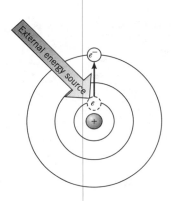

Since the development of modern models of atomic structure, the terms *quantum jump* and *quantum leap* have become literary terms meaning dramatic or profound change.

When an electron jumps to a higher energy level, it is considered excited or in an **excited state.** If an electron gains enough energy, it breaks away from the atom, leaving a positive ion. Positive ions or cations are formed by the loss of electrons from atoms. Those excited electrons that do not break away from atoms occupy various higher energy levels. These excited electrons spontaneously fall back to lower energy levels. They fall to various lower levels, ultimately reaching the ground state (see Figure 8-5). As an excited electron falls from a higher energy level to a lower energy level, it releases its excess energy, like what happens when you leap down a staircase and give off the potential energy as kinetic energy or energy of motion. As an electron jumps from a higher to a lower energy level, it releases an amount of energy that corresponds to the difference in energy between the two levels.

The energy released by electrons falling to lower energy levels in atoms very often takes the form of radiant energy or light. Some light you see in a flame or the colored light given off by fireworks comes from the loss of energy of excited electrons. Since atoms of various elements differ in the number of electrons and the energies of the electron levels, each element emits different characteristic colors of light when energetically excited. In fireworks, compounds of specific elements are added to produce desired colors: sodium compounds for yellow, strontium compounds for red, copper compounds for blue, and barium compounds for green.

Bohr's model is used to explain why a hydrogen discharge tube gives off specific colors of light. Ground-state hydrogen atoms have their electrons in the lowest possible energy level. In the discharge tube, electrons within the hydrogen atoms can gain energy from the electron beam in the tube and become excited to higher energy levels. The excited electrons spontaneously undergo quantum jumps in which they move from the

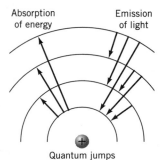

FIGURE 8-5 Some possible quantum jumps in a hydrogen atom according to the Bohr model.

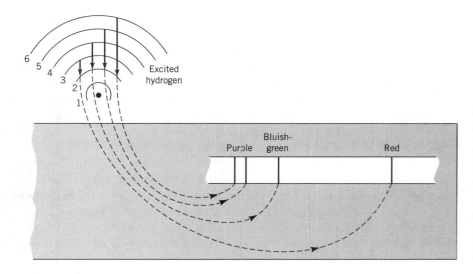

FIGURE 8-6 An interpretation of the quantum jumps in excited hydrogen atoms. Jumps from higher levels to level 2 produce red, blue-green, and purple light. Jumps from and to other levels produce ultraviolet and infrared light, which is not visible.

higher energy levels to the lower levels. As the electrons drop to lower energy levels, their excess energy is released as light energy. The hydrogen atom has a specific set of allowable energy levels for its electron. Thus, only a fixed set of quantum jumps is possible for electrons in hydrogen atoms. Consequently, specific colors of light corresponding to the possible quantum jumps are expected. Bohr used his model to explain why hydrogen gas in a discharge tube emits the specific colors of light as seen in its emission spectrum (see Figure 8-6).

In neon discharge tubes, the neon gas is energetically excited by passing electricity through the tube. Some electrical energy is absorbed by the neon atoms as the electrons are excited. As the excited electrons fall back to lower energy levels, the characteristic reddish orange light of a neon tube is emitted. As long as the electricity flows, the atoms become excited, and as they relax and lose their energy, the tube emits light. Different gases give different colors. Helium gives a yellowish pink light, argon a lavender light, krypton a silvery white light, and xenon a blue light. Hydrogen gives the distinctive reddish purple light. Each color is characteristic of a particular gas. A hydrogen discharge tube also emits invisible ultraviolet light similar to the ultraviolet light in sunlight, which is why hydrogen discharge tubes are used in sunlamps.

8-6 THE QUANTUM MECHANICAL MODEL

The Bohr model gives a description of a simple hydrogen atom. It was soon found, however, that it was not a good model for complex atoms with many electrons. To encourage further development of theories of atomic structure, Bohr established a scientific institute in Copenhagen, Denmark. In the 1920s

Quantum

and 1930s, many scientists contributed to the work of this institute. As a result, one of the most important scientific models of the twentieth century developed: the **quantum mechanical model** of atomic structure.

In the Bohr model, an electron follows a determined circular orbit. The quantum mechanical model views electrons in a significantly different way. An electron in an atom has energy because of its motion and position. In the quantum mechanical model, an electron of a specific energy takes the form of a wave called an electron orbital. As you know, energy is carried through water as waves. When a pebble is dropped into a pool of still water, some energy forms waves that move out from the point of impact. Electron orbitals are wave patterns that surround the nucleus of an atom. The electron energy waves remain in place around the nucleus; they do not move and dissipate like energy waves in water. Different electron energies correspond to specific fixed wave patterns called standing waves since they do not move away like water waves. Using a different analogy, a vibrating string on a guitar has energy that is dissipated by moving the air around it and creating a sound wave. As a standing wave, a vibrating string is imagined as one that continually vibrates and whose energy never dissipates.

The mathematical methods used in the model are called quantum mechanics or wave mechanics. Using quantum mechanics, a complex set of equations predicts the number and nature of the quantized energy levels of electrons in atoms. The equations are sometimes called wave equations because they relate to electron waves. To illustrate how an electron is visualized in the quantum mechanical model, consider the bicycle wheel shown in Figure 8-7. At rest, you can clearly see all the spokes and you can easily stick your finger between the spokes. When the wheel is moving rapidly, the images of the spokes appear to blur and you will hurt yourself if you try to put your finger between them. Thus, the moving spokes occupy or fill the entire volume of the wheel. Just as the moving spokes define a volume in which there is a high probability of finding a spoke, an electron defines a certain wave pattern in which it occupies or

FIGURE 8-7 A bicycle wheel cloud and an electron cloud.

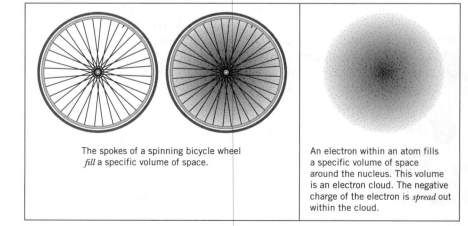

The spokes of a spinning bicycle wheel *fill* a specific volume of space.

An electron within an atom fills a specific volume of space around the nucleus. This volume is an electron cloud. The negative charge of the electron is *spread* out within the cloud.

fills the entire volume of the pattern. The electron orbitals represent three-dimensional probability volumes of electrons.

The electron is not considered as a particle in a specific orbit following a determined path, but rather as a cloud filling a volume about the nucleus just as the spokes fill a "spoke cloud" in the moving wheel. The term *electron cloud* comes from the idea that the negatively charged electron is filling the volume as a wave. It spreads over a volume of space. The shapes of these orbitals correspond to the regions of space "filled" by the rapid motion of the electron. The term *electron orbital* emphasizes that these are not orbits like those of the Bohr model but rather are electron clouds. The wave equations of the model define a set of possible wave patterns that an electron may occupy. Thus, within an atom, a set of electron orbitals represents the quantized energies of the various electrons in the atom. Electrons filling the orbitals give shape and substance to atoms. Using the guitar string analogy again, note that a string may vibrate at one energy level,

Orbitals

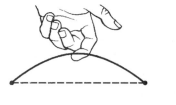
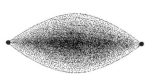

or at some higher energy level,

The two vibrations have different energies and different wave patterns.

The quantum mechanical model is more abstract than the Bohr model. It does, however, allow the visualization of atoms in three-dimensional space. Atoms are visualized as diffuse fuzzy balls of electron clouds surrounding tiny nuclei. Furthermore, the model also defines a pattern for the electron distribution in the various energy levels of a hydrogen atom. It is assumed that this pattern is the same for all atoms, and it is used to describe how electrons distribute in atoms with many electrons.

8-7 ENERGY STATES

According to the quantum mechanical model, there are several possible **main or principal energy states** in which the electrons are found. These main energy states are the energy levels or energy shells. Each main energy level has one or more energy states called **energy sublevels or energy subshells.** Furthermore, each energy sublevel consists of one or more specific

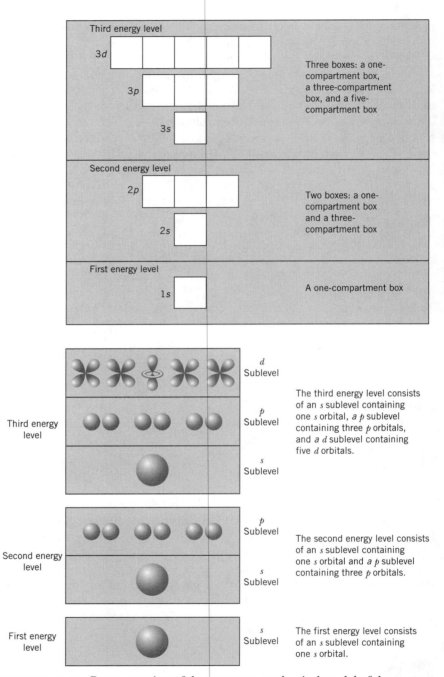

FIGURE 8-8   Representation of the quantum mechanical model of the atom.

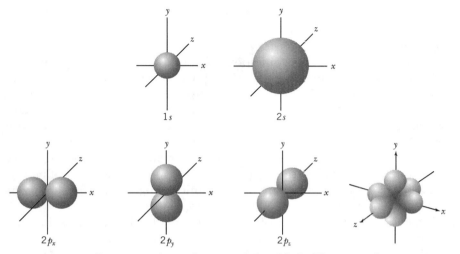

FIGURE 8-9   Representations of some atomic orbitals. The x, y, and z axes are given to show the three-dimensional spatial distribution of the orbitals. The first energy level contains only the $1s$ orbital; the second energy level contains the $2s$ orbital and three $2p$ orbitals. The three p orbitals are sometimes distinguished as p_x, p_y, and p_z to emphasize their different spatial orientations.

electron energy states. These specific energy states within the energy sublevels are the **electron orbitals.** Electrons can occupy these orbitals.

We can think of the main energy levels as a set of large boxes, one box for each level. Opening the boxes, we find that each contains a certain number of smaller boxes, the energy sublevels. Inside the smaller boxes, we find a certain number of compartments that serve as positions for the electrons. These compartments are the electron orbitals. Of course, energy levels are not boxes but are shells of electron clouds of varying complexity. See Figure 8-8 for a summary of the quantum mechanical view of the energy states in an atom. The description of an atom in terms of energy states does not give us a picture or mental image of the atom. As shown in Figure 8-9, however, the theory does provide good images of atoms using three-dimensional shapes and sizes of the electron clouds or electron orbitals.

8-8 ENERGY LEVEL STRUCTURE

An atom of a given element consists of a nucleus, with a charge corresponding to the atomic number, surrounded by electron clouds that overlap and extend outward from the nucleus. The atomic number of an element tells us how many electrons are in its atoms. The electrons occupy the lowest available energy states. Several questions arise at this point, however. How many electrons can fit in each main energy level and in each sublevel? How many electrons can fit in each electron orbital? How do the energies of the various orbitals and sublevels differ with respect to one another?

With answers to these questions, figuring out the arrangement of electrons in the atom of an element is possible. Since we know the number of electrons in the atom, we can hypothetically build an atom by placing electrons in the orbitals starting at the lowest energy position and working up to higher positions. Using the box analogy, if we know the relative energies of the compartments in the boxes, we can distribute electrons in the compartments starting a low energy and working up as the compartments fill. The electronic arrangement of an atom of an element is its electronic structure or **electronic configuration.** Each element has a unique configuration.

Since there are a variety of electron energy levels, distinguishing one from another by numbers is useful. The numbers used to designate the main energy levels are

1   First energy level
2   Second energy level
3   Third energy level
4   Fourth energy level, and so on.

The energy-level number is called the **principal quantum number.** Higher numbers correspond to higher energies and increased distances from the nucleus.

$$\text{energy level} \quad 1 \quad 2 \quad 3 \quad 4 \quad 5 \quad 6 \quad 7 \ldots$$

distance from nucleus increases
-------------------->
increasing energy

Each energy level consists of a specific number of sublevels. The number of sublevels in an energy level is the same as the energy level number. That is, the first energy level has one sublevel, the second energy level has two sublevels, and so on. Instead of numbers, the letters *s, p, d,* and *f* are used to reference various sublevels. Sublevel structure becomes more complex as we move to higher energy levels.

1   The first energy level has one sublevel, an *s* sublevel.
2   The second energy level has two sublevels, an *s* sublevel and a *p* sublevel.
3   The third energy level has three sublevels, an *s* sublevel, a *p* sublevel, and a *d* sublevel.
4   The fourth energy level has four sublevels, an *s* sublevel, a *p* sublevel, a *d* sublevel, and an *f* sublevel.

Within a given energy level the sublevel energies differ. The order of sub-level energies is

$$\text{sublevel} \quad s \quad p \quad d \quad f$$

increasing energy
------------→

Within an energy level, an *f* sublevel is of higher energy than a *d* sublevel, a *d* sublevel is of higher energy than a *p* sublevel, and a *p* sublevel is of higher energy than an *s* sublevel.

A specific sublevel is identified by giving the main energy-level number followed by the letter corresponding to the sublevel. So we refer, for instance, to the 1*s* (pronounced one–S) sublevel, the 2*p* (pronounced two–P) sublevel, or the 3*d* (pronounced three–D) sublevel. The sublevel structure of the first four energy levels is

ENERGY LEVEL	SUBLEVELS
First	1*s*
Second	2*s*, 2*p*
Third	3*s*, 3*p*, 3*d*
Fourth	4*s*, 4*p*, 4*d*, 4*f*

Notice, as shown in Table 8-1, that as we go to higher energy levels they become larger and more complex. Higher levels can hold more electrons. Each sublevel consists of a certain number of electron orbitals. See Figure 8-9 for illustrations of the shapes of some orbitals.

An *s* sublevel has one orbital, the *s* orbital.

A *p* sublevel has three orbitals, the *p* orbitals.

Table 8-1   *The Sublevel Structures of the First Four Energy Levels*

ENERGY LEVEL	TYPE OF SUBLEVEL	NUMBER OF ORBITALS OF GIVEN TYPE	ELECTRON STATE NOTATION
First	*s*	One *s*	1*s*
Second	*s*	One *s*	2*s*
	p	Three *p*	2*p*
Third	*s*	One *s*	3*s*
	p	Three *p*	3*p*
	d	Five *d*	3*d*
Fourth	*s*	One *s*	4*s*
	p	Three *p*	4*p*
	d	Five *d*	4*d*
	f	Seven *f*	4*f*

A *d* sublevel has five orbitals, the *d* orbitals.

An *f* sublevel has seven orbitals, the *f* orbitals.

This information is summarized in Table 8-1. Notice that the pattern is that the number of orbitals in a sublevel increases as a sequence of odd numbers (1, 3, 5, 7, . . .).

Orbitals within a given sublevel have the same electron energies. For example, the three *p* orbitals of a *p* sublevel are equivalent in electron energy but the orbitals have different orientations in three-dimensional space, as shown in Figure 8-9.

8-9 ELECTRON SPIN

The important question at this point is, How many electrons can a single orbital hold? Electrons have negative charges and repel. Thus, we might expect that an orbital can only hold one electron. Electrons have another property, however, that makes it possible for two electrons to occupy a given orbital. Electrons have a property called spin that is analogous to the spinning of a top. A top can spin in one direction or in the opposite direction. Similarly, a given electron can have only one of two possible spin states that we can represent symbolically as spin up or spin down arrows (↑ or ↓).

The spin of electrons gives them properties somewhat like tiny bar magnets. If two bar magnets are forced together so that their like poles align, they strongly repel. When they arrange with their opposite poles aligned, they do not repel. In a similar fashion, two electrons cannot occupy the same orbital if they have the same spin but can occupy the same orbital if they have opposite spins. Wolfgang Pauli, a German scientist involved in the development of the quantum mechanical model, first expressed this idea, which is now called the Pauli exclusion principle. See the margin.

An orbital that contains two electrons ↑↓ is a **filled** orbital, whereas an orbital with one electron ↑ is a **half-filled** orbital. Electrons repel and occupy single orbitals whenever possible. This becomes important when there are several equivalent orbitals within a sublevel. Friedrich Hund, another German scientist, described the pattern electrons follow in such cases. See the margin.

This pattern is called Hund's rule, and it has significance when deciding where the electrons are within the equivalent energy sublevels of atoms.

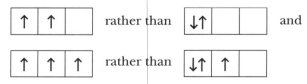

Pauli Exclusion Principle

An orbital can contain a maximum of two electrons and those electrons must have opposite spin states.

Why is it not possible to walk through a brick wall?

Hund's Rule

Electrons in a set of orbitals within a sublevel tend to occupy empty orbitals and pair only when all orbitals have one electron.

These somewhat abstract ideas of electron spin and the Pauli exclusion principle do have practical significance. The magnetic properties of spinning electrons are responsible for the magnetic properties of matter. This property is most notable with iron. Pieces of iron are made into permanent magnets that take advantage of electron spins in iron atoms. Furthermore, moving electrons in a wire produce a magnetic field used to make electric motors work. An obvious but notable illustration of the Pauli exclusion principle is that two solids cannot occupy the same space. When the solids are forced together, the electrons repel with great force. (Why can two gas samples occupy the same volume but not two solids?)

8-10 RELATIVE ENERGIES OF THE SUBLEVELS

The maximum capacity of an orbital is two electrons. Since we know how many orbitals are in the various sublevels, we can easily predict the maximum electron capacity of the sublevels as shown on the next page.

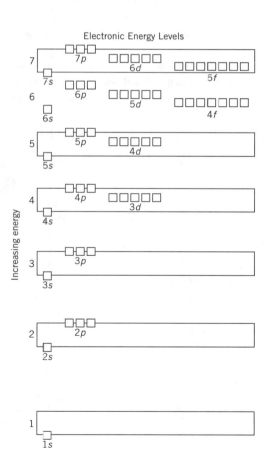

FIGURE 8-10 The relative energies of the various electron orbitals.

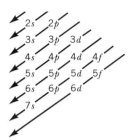

(a) The sublevels are written as shown. The direction of the arrows froom upper right to lower left indicates the general energies of the energy sublevels.

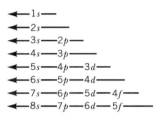

(b) The sublevels are written in columns as shown. The relative energies are given in the direction of horizontal arrows from right to left.

FIGURE 8-11

Mnemonic devices for remembering the order of energy sublevels.

s sublevel	one s orbital	2 electrons	an s sublevel can hold 1 or 2 electrons
p sublevel	three p orbitals	6 electrons	a p sublevel can hold 1 to 6 electrons
d sublevel	five d orbitals	10 electrons	a d sublevel can hold 1 to 10 electrons
f sublevel	seven f orbitals	14 electrons	an f sublevel can hold 1 to 14 electrons

To figure out how the electrons are distributed in an atom of an element, we hypothetically place the electrons of the atom in the various energy sublevel positions. To do this, we start at the lowest position and move to a higher position when the lower level is full. Using the box analogy, the orbitals are the compartments within the sublevel boxes. Each main energy-level box contains one or more sublevel boxes. The electrons are distributed in compartments of the boxes with the lowest sublevel boxes filled first. Since the boxes have limited capacities, they fill with electrons, and any additional electrons must be placed in the boxes corresponding to the next-higher sublevel. The process continues from level to level until all the electrons are placed. Figure 8-10 on page 225 shows the order of sublevel energies.

To figure out the electronic configurations of the atoms of elements, we need to know the order of **increasing energy of the sublevels.** That is, we need to know which sublevel is the lowest energy, which is the next lowest, and so on. This order is

$$1s\ 2s\ 2p\ 3s\ 3p\ 4s\ 3d\ 4p\ 5s\ 4d\ 5p\ 6s\ 4f\ 5d\ 6p\ 7s\ 5f\ 6d$$

The order follows a pattern, and Figure 8-11 shows a way to remember it.

8-11 ENERGY OVERLAP

Note that, according to the sublevel energy pattern, the 4s sublevel is of lower energy than the 3d sublevel. This example shows that in some cases a sublevel of a higher main energy level has lower energy than a sublevel of a lower main energy level. **Energy overlap** is the term used for this situation. There is no way to draw actual pictures of energy levels in terms of energy. To understand energy overlap better, imagine the energy levels as a series of ladders extending upwards, as shown in Figure 8-12. Where one ladder ends the next begins. The ladders, representing various energy levels, have a specific number of rungs corresponding to the sublevels. Energy overlap occurs when the higher rungs of a ladder extend beyond the lower rungs of the next ladder. The higher rungs of the first ladder are of higher energy than the lower rungs of the second ladder and correspond to an energy overlap.

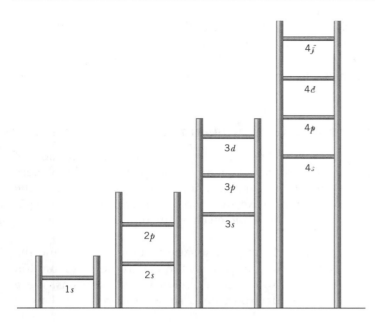

FIGURE 8-12 Ladder
analogy of the sublevels
showing energy overlap.

As a simple explanation of energy overlap consider that the sublevels
within a main level increase in energy in the order

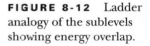

and the levels increase in energy in the order

$$\text{-------------}\rightarrow$$
$$1\quad 2\quad 3\quad 4\quad 5\quad 6\quad 7$$

Sometimes the increase in energies of sublevels causes the higher energy
sublevels to exceed the lower energy sublevels of the next main energy
level. For example, the sublevel energy pattern for the third and fourth
main energy level are

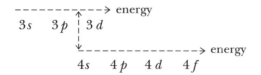

which means that the energy of the $3d$ sublevel is higher than the energy
of the $4s$ sublevel; we say that the $3d$ sublevel overlaps the $4s$ sublevel. Figure 8-10 shows that there are several energy sublevel overlaps between
various main levels. Energy overlap does not mean that the higher
energy sublevel is physically farther from the nucleus than the lower
energy sublevel. Distance and energy are not the same. That is, the $4s$
orbital is not necessarily located closer to the nucleus than the orbitals

The $3d$ sublevel overlaps the
$4s$ sublevel. List three other
energy sublevel overlaps by
indicating which sublevels
are involved.

of the $3d$ sublevel, although the $4s$ sublevel is lower in energy than the $3d$ sublevel. The point is that typically the $4s$ sublevel is farther from the nucleus than the $3d$ sublevel, but the $4s$ fills with electrons before the $3d$. This situation happens because the $4s$ sublevel is lower energy than the $3d$.

8-12 THE ELECTRONIC CONFIGURATIONS OF THE ELEMENTS

We know that the atomic number of an element is equal to the number of electrons in neutral atoms of that element. Elements are listed in the periodic table by increasing atomic numbers, and atoms of an element in the table have one more electron than the atoms of the preceding element. Hydrogen, atomic number 1, has one electron; helium, atomic number 2, has two electrons; and so on.

By using the number of electrons that various sublevels hold and the order of sublevel energies, we can predict the electronic configuration of an atom of an element. **An s sublevel holds 1 or 2 electrons, the p sublevel holds from 1 to 6 electrons, the d sublevel holds from 1 to 10 electrons, and the f sublevel holds from 1 to 14 electrons.** The electronic structure of an element is represented by the sublevel distribution of the electrons. To show the structure, we distribute electrons in the lowest energy sublevel first, and when it is full, we use the next-higher energy sublevel. We continue until all the electrons are placed in sublevels. To illustrate the process, we can use the ladder analogy shown in Figure 8-13.

FIGURE 8-13 Bird roost analogy for electrons in sublevels.

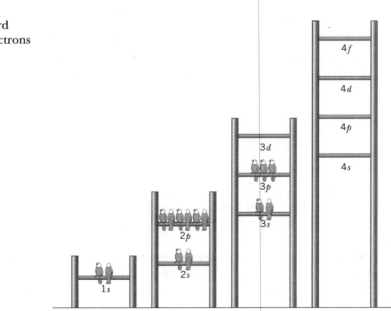

Imagine birds landing on the rungs of the ladders. Each rung can hold a specific number of birds. Suppose the birds prefer the lower rungs (lower energy sublevels). They occupy these rungs first. As the lower rungs are filled, the birds locate in the next-higher rungs. Also, when sublevel energy overlap occurs, the birds prefer the lower energy rung even when it is on the next-higher ladder (next-higher main energy level).

The one electron in a hydrogen atom locates in the *s* sublevel of the first main energy level. The standard method of showing the **electronic configuration** is

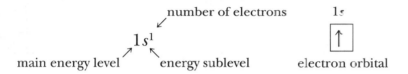

This sublevel or orbital notation shows one electron in the *s* sublevel of the first main energy level. The number on the left is the main energy level, and the letter is the energy sublevel. The superscript digit shows the number of electrons in that sublevel. The electronic configuration of hydrogen is read as one–S–one. Note on the right above that the electron in the 1*s* orbital is represented as an arrow in a box. This box is given for comparison and clarity. Such boxes are not typically shown as part of an electronic configuration. The two electrons in a helium atom are found in the lowest energy position; that is, they are found in the *s* sublevel of the first main energy level. The electronic configuration of helium is one–S–two.

The next element, lithium, has atoms with three electrons. Two of these electrons occupy the lowest energy position, the 1*s* sublevel, which fills this sublevel. The third electron must occupy the next-lowest-energy position, the 2*s* sublevel. Thus, the electronic structure of lithium is one–S–two, two–S–one.

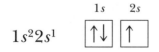

To figure out the electronic configuration of an element, all we have to do is decide how many electrons are in an atom and then distribute these electrons; we start in the lowest energy sublevel and continue until all the electrons are distributed. The next page shows elements 4 to 10.

ELEMENT	ATOMIC NUMBER	ELECTRONIC CONFIGURATION	ELECTRON ORBITAL BOX NOTATION		
			1s	2s	2p
Beryllium, Be	4	$1s^2 2s^2$	[↑↓]	[↑↓]	[ ][ ][ ]
Boron, B	5	$1s^2 2s^2 2p^1$	[↑↓]	[↑↓]	[↑][ ][ ]
Carbon, C	6	$1s^2 2s^2 2p^2$	[↑↓]	[↑↓]	[↑][↑][ ]
Nitrogen, N	7	$1s^2 2s^2 2p^3$	[↑↓]	[↑↓]	[↑][↑][↑]
Oxygen, O	8	$1s^2 2s^2 2p^4$	[↑↓]	[↑↓]	[↑↓][↑][↑]
Fluorine, F	9	$1s^2 2s^2 2p^5$	[↑↓]	[↑↓]	[↑↓][↑↓][↑]
Neon, Ne	10	$1s^2 2s^2 2p^6$	[↑↓]	[↑↓]	[↑↓][↑↓][↑↓]

ACTIVITY 8-3

The maximum electron capacity of an energy level is given by the expression $2n^2$, where n is the energy level number. Confirm this expression by filling the maximum number of electrons for the various sublevels and counting the total electrons in each level.

Level 1   1s $n = 1$, so $2n^2 = 2$

Level 2   2s2p $n = 2$, so $2n^2 = 8$

Level 3   3s3p $n = 3$,
              3d   so $2n^2 = 18$

Level 4   4s4p $n = 4$,
              4d4f   so $2n^2 = 32$

The box notation shows that, as the 2p sublevel fills with electrons, one electron is placed in each of the three p orbitals according to Hund's rule. Again, we can use the ladder analogy of Figure 8-13. The ladder rungs have specific numbers of perching spots. An s rung has one, a p rung three, a d rung five, and an f rung seven. There is room for two birds on each spot. As the birds occupy a rung, they prefer their own spot if possible, so they occupy empty spots on a rung. If all spots are used, one bird pairs up with another. This pairing continues until the rung is filled, then the next-higher rung is used.

The boxes shown along with the electronic configurations represent the electrons in the orbitals. Notice where the electrons are found in the p orbitals of a p sublevel. Nitrogen has three unpaired electrons in the 2p sublevel, one in each orbital. As more electrons are added, they must be placed in half-filled orbitals, which results in neon having a completely filled 2p sublevel with three filled p orbitals. To write the electronic configuration of an element, distribute its electrons within the sublevels using the order of their energies and their electron capacities (an s sublevel holds 2 electrons, a p sublevel 6 electrons, a d sublevel 10 electrons, and an f sublevel 14 electrons).

EXAMPLE 8-1

What is the electronic configuration of phosphorus, atomic number 15? Phosphorus atoms have 15 electrons. The order of energy of the sublevels is

$$1s \ 2s \ 2p \ 3s \ 3p \ldots$$

Distributing the 15 electrons in these sublevels so that the lower energy sublevels are filled gives the electronic configuration

$$1s^2 \ 2s^2 \ 2p^6 \ 3s^2 \ 3p^3$$ read as one-S-two, two-S-two, two-P-six, three-S-two, three-P-three

8-13 OUTER-LEVEL ELECTRONS

Vanadium has an atomic number of 23. The electronic configuration of vanadium is $1s^2 2s^2 2p^6 3s^2 3p^6 4s^2 3d^3$. Note that the $4s$ sublevel comes before the $3d$ sublevel. When writing electronic configurations, however, distributions are more often written by grouping all the sublevels of each energy level in numerical order. To do this, the configuration of vanadium is rewritten so that the $4s$ level is last.

$$1s^2 2s^2 2p^6 3s^2 3p^6 3d^3 4s^2$$

Such notation emphasizes the number of outer-energy-level electrons. **Outer-energy-level electrons** are those in the highest-numbered energy levels. For instance, vanadium has two outer-energy-level electrons, the $4s$ electrons. Energy level 4 is the highest-numbered level used in a vanadium atom. We will soon see why the notation that shows the number of outer-level electrons is useful. Just remember that outer-level electrons are those electrons with the highest-numbered energy level. To emphasize the number of outer-level-electrons, first write the electronic configuration using the relative energies of the sublevels. Next, rearrange the configuration so that the sublevels are in increasing numerical order of the energy levels.

What is the electronic configuration of zinc, atomic number 30?

Useful Quantum Jumps: Noble Gases, Starlight, Advertising, and Motor Oil

When atoms of an element are energetically excited, they give off a characteristic signature light called the emission spectrum. Emission spectra consist of specific colors of light, and when passed through a spectroscope, this light appears as a series of lines. A spectroscope is a device that separates a light beam into a distinct set of colors. Since each element has unique atoms, the spectrum of an element consists of a characteristic set of lines. No two elements give the same emission line spectrum. As a result, the emission lines in a source of light are used to identify elements in the light source. Line spectra have been used to identify and characterize newly discovered elements contained in discharge tubes.

In 1894, William Ramsay, a Scottish chemist, was studying a new gas obtained from air. This gas was very chemically inert and would not form compounds with any other element. The emission spectrum of the gas revealed that it was a previously unknown element. Ramsay named this element argon, from the Greek word for lazy because it did not form compounds. In the following years, an entire set of colorless, chemically inert gaseous elements was discovered by use of emission spectra. This group is the noble gases: helium, neon, argon, krypton, xenon, and radon. Since they are all colorless gases, we can only "see" them by their emission spectra.

What makes the noble gases chemically inactive? Notice that as a family these gases occupy the far right column (18 or VIIIA) of the periodic table. All these elements have eight outer-level electrons except helium, which has two. Since helium is actually a noble gas, it is placed in the same column as the others even though it has only two outer-level electrons. For helium, the two electrons represent a filled first energy level. Apparently, eight electrons in the outer level of an atom is a particularly stable arrangement, since the noble gases do not readily form compounds. There are a few known compounds of the noble gases, but they only form under extreme conditions and are quite rare. In any case, an octet of outer-level electrons or a duo for helium is a very stable electronic arrangement.

All we know about stars comes from starlight. When starlight is passed through a spectroscope, it is possible to detect the presence of specific elements as reflected in their characteristic spectral lines. Numerous elements have been shown, in this way, to be present in stars. The light of our own star, the sun, reveals numerous elements ranging from hydrogen and helium to iron.

The characteristic light emitted by elements in discharge tubes finds many uses in advertisement, art, and lighting. Neon gas in a discharge tube gives off an intense red-orange light. Of course, this is the familiar color of a neon sign. Other gases are used to give different colors of light. A discharge tube can even contain a solid like sodium. When the tube heats up, some sodium vaporizes and produces an intense yellow light. Sodium vapor lamps are used in some streetlights, identified by the characteristic yellow color. A mercury vapor lamp produces a silvery blue light. These lamps are also used for outdoor lighting.

A sample of motor oil from your car engine can give some chemical information. Part of the oil is energized in an electric arc and the light emitted is subjected to spectral analysis. The presence of specific elements in the oil reveals undue wear in the pistons, valves, rings, bearings, and other metallic parts.

8-14 THE PERIODIC TABLE AND ELECTRONIC CONFIGURATIONS

As discussed in Section 5-1, the periodic table lists the elements according to increasing atomic number and its groups contain families of elements having similar properties. Not until the properties of the elements were related to electronic configurations did an explanation of the unique shape and appearance of the periodic table become apparent. It was found that the elements having similar electronic configurations in their outermost energy levels are those with similar properties. To illustrate, the electronic configurations of lithium, sodium, potassium, rubidium, cesium, and francium are

ATOMIC NUMBER		
3	Li	$1s^2 2s^1$
11	Na	$1s^2 2s^2 2p^6 3s^1$
19	K	$1s^2 2s^2 2p^6 3s^2 3p^6 4s^1$
37	Rb	$1s^2 2s^2 2p^6 3s^2 3p^6 3d^{10} 4s^2 4p^6 5s^1$
55	Cs	$1s^2 2s^2 2p^6 3s^2 3p^6 3d^{10} 4s^2 4p^6 4d^{10} 5s^2 5p^6 6s^1$
87	Fr	$1s^2 2s^2 2p^6 3s^2 3p^6 3d^{10} 4s^2 4p^6 4d^{10} 4f^{14} 5s^2 5p^6 5d^{10} 6s^2 6p^6 7s^1$

These six elements have very similar chemical properties. Also, note that each has one electron in the outermost or highest numbered main energy level. This similarity in configuration is shown as ns^1, where n refers to the number of the outer main energy level. These elements share a similar ns^1 outer-level configuration with n ranging from 2 to 7. As expected, these elements make up a group of the periodic table.

Elements can be arranged by increasing atomic number so that all elements of similar electronic configurations are in groups. Such an arrangement gives the **periodic table.** Ordering elements by similarities in electronic configurations gives essentially the same table that Mendeleev created more than 130 years ago. It is impressive that he did it long before any understanding of atomic structure existed. His creative arrangement anticipated and now confirms the quantum mechanical model of the atom. In turn, the model predicts the shape of the table as a reflection of the sublevel structures of atoms. Figure 8-14 shows the electronic configurations of most of the elements and emphasizes that elements of similar configurations are found in the same group of the table. Since the table is based on the electronic configuration of elements, its shape corresponds to the filling of the energy sublevels. Various parts of the table relate to the filling of specific sublevels.

Recall the sublevel filling pattern in Figure 8-11. An s sublevel can hold 2 electrons, so the s-sublevel groups are 2 elements wide. A p sublevel can hold 6 electrons, so the p-sublevel groups are 6 elements wide. A d sublevel can hold 10 electrons, so the d-sublevel part of the table is 10 elements wide. An f sublevel can hold 14 electrons, so the f-sublevel part of the table is 14 elements wide. Once we realize that the elements are arranged in the table according to the filling pattern of the various sublevels, we have an

Periodic Table

FIGURE 8-14
Similarities in outer-level configurations of elements. Members of the same group in the table have similar electronic configurations. This table includes only elements up to actinium.

Period	1	2	3	4	5	6	7	8	9	10	11	12	13	14	15	16	17	18 NOBLE GASES
1	H $1s^1$																	He $1s^2$
2	7 $2s^1$	Be $2s^2$											B $2s^22p^1$	C $2s^22p^2$	N $2s^22p^3$	O $2s^22p^4$	F $2s^22p^5$	Ne $2s^22p^6$
3	Na $3s^1$	Mg $3s^2$											Al $3s^23p^1$	Si $3s^23p^2$	P $3s^23p^3$	S $3s^23p^4$	Cl $3s^23p^5$	Ar $3s^23p^6$
4	K $4s^1$	Ca $4s^2$	Sc $3d^14s^2$	Ti $3d^24s^2$	V $3d^34s^2$	Cr $3d^54s^1$	Mn $3d^54s^2$	Fe $3d^64s^2$	Co $3d^74s^2$	Ni $3d^84s^2$	Cu $3d^{10}4s^1$	Zn $3d^{10}4s^2$	Ga $4s^24p^1$	Ge $4s^24p^2$	As $4s^24p^3$	Se $4s^24p^4$	Br $4s^24p^5$	Kr $4s^24p^6$
5	Rb $5s^1$	Sr $5s^2$	Y $4d^15s^2$	Zr $4d^25s^2$	Nb $4d^45s^1$	Mo $4d^55s^1$	Tc $4d^55s^2$	Ru $4d^75s^1$	Rh $4d^85s^1$	Pd $4d^{10}$	Ag $4d^{10}5s^1$	Cd $4d^{10}5s^2$	In $5s^45p^1$	Sn $5s^25p^2$	Sb $5s^25p^3$	Te $5s^25p^4$	I $5s^25p^5$	Xe $5s^25p^6$
6	Cs $6s^1$	Ba $6s^2$	La $5d^16s^2$	Hf $5d^26s^2$	Ta $5d^36s^2$	W $5d^46s^2$	Re $5d^56s^2$	Os $5d^66s^2$	Ir $5c^76s^2$	Pt $5d^96s^1$	Au $5d^{10}6s^1$	Hg $5d^{10}6s^2$	Tl $6s^26p^1$	Pb $6s^26p^2$	Bi $6s^26p^3$	Po $6s^26p^4$	At $6s^26p^5$	Rn $6s^26p^6$
7	Fr $7s^1$	Ra $7s^2$	Ac $6d^17s^2$															

The Periodic Table

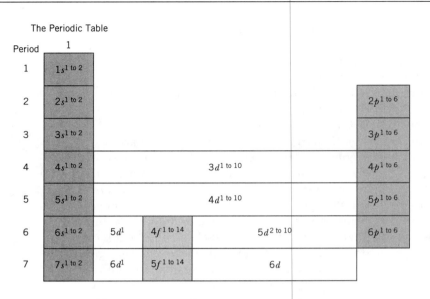

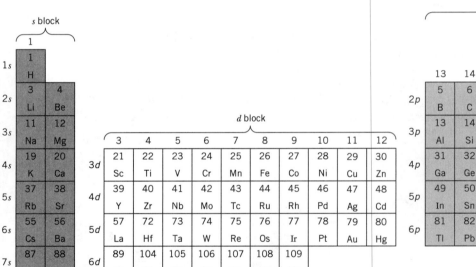

FIGURE 8-15   The shape of the periodic table corresponds to the filling of sublevels.

explanation for the curious shape of the periodic table. In summary, the periodic table has its unique shape because the electron energy sublevels in atoms have specific filling orders and capacities. See Figure 8-15.

8-15 PARTS OF THE PERIODIC TABLE

The elements are sometimes classified according to the part of the periodic table in which they are found. Figure 8-15 shows such a classification. The elements are classified into four blocks, the *s*-**block elements,** the *p*-**block elements,** the *d*-**block elements,** and the *f*-**block elements.** Note that, as expected, the *s* block is 2 elements wide, the *p* block 6 elements wide, the *d* block 10 elements wide, and the *f* block 14 elements wide. Confirm this by counting them yourself. Strictly speaking, the *f*-block elements should be located within the *d* block, but this makes the table inconveniently wide. To understand what this means, take a close look at the atomic numbers of the elements in the *f* block and those in the lower parts of the *d* block. To shorten the space occupied by the periodic table, the *f*-block elements are normally placed below the *d*-block elements.

The far right-hand group of the periodic table contains the unique collection of chemically inactive **noble gases.** Helium has the configuration $1s^2$, and all the other noble gases have eight outer-level electrons corresponding to the configuration ns^2np^6. Confirm this for yourself by writing the electronic configurations for Ne (atomic number 10), Ar (18), and Kr (36). For years, chemists viewed these gases as chemically inert, so they called them inert gases. But after the successful synthesis of a few compounds of xenon and krypton, chemists realized that they were not completely inert and renamed them noble gases, suggesting a reluctance to form compounds. The lack of much chemical activity for the noble gases suggests that the electron configuration of eight outer-level electrons is quite stable. Note that ns^2np^6 totals eight electrons, two *s* and six *p* electrons.

The elements that occupy the *s* and *p* blocks are called the **representative elements.** The *s*-block representative elements have the outer-energy-level configuration ns^1 or ns^2, and the *p*-block representative elements have an outer-energy-level configuration ranging from ns^2np^1 to ns^2np^6. The *d*-block elements have electronic configurations that correspond to the filling of the *d* sublevels. These elements are called the **transition elements.** They represent a transition between the *s*-block and the *p*-block elements in a period. The *f*-block elements have electronic configurations that correspond to the filling of the *f* sublevels; these elements are called the **inner transition elements** or the rare earths. The inner transition elements involving the $4f$ sublevel start after lanthanum and are called the lanthanide series; those that involve the $5f$ sublevel start after actinium and are called the actinide series.

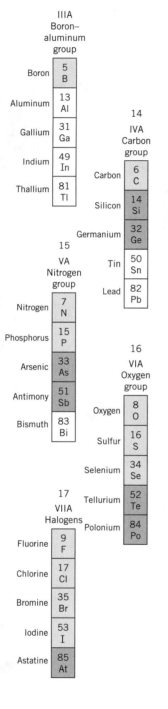

8-16 GROUPS AND CLASSES OF ELEMENTS

Not counting the *f* block there are 18 columns or **groups** in the periodic table, simply numbered left to right as 1 to 18. The representative elements are groups 1, 2, 13, 14, 15, 16, 17, and 18. Using the alternate notation, these groups correspond to IA, IIA, IIIA, IVA, VA, VIA, VIIA, and VIIIA; VIIIA is sometimes labeled "noble gases." Group 1 (IA) is the family of **alkali metals**, and group 17 (VIIA) is the family of **halogens.** Alkali comes from the ancient Arabic name for potash, and potassium is one member of this group. Halogen comes from the Greek *hals* (salt) and *genes* (born or generate) because all members of the family form saltlike compounds with sodium. The transition metal groups are sometimes numbered in sequence as IIIB, IVB, VB, VIB, VIIB, VIII, VIII, VIII, IB, and IIB. The B designations suggest a slight similarity in properties to the corresponding A groups of representative elements. The three transition metal groups designated as VIII have no corresponding representative element groups.

The left-to-right horizontal sequences or rows of elements in the table are the **periods.** The six complete periods contain 2, 8, 8, 18, 18, and 32 elements; the seventh period is incomplete and contains 26 elements. The periods correspond to the filling of the electronic sublevels, and the last period is incomplete because currently there are not enough known elements to fill the $6d$ and $7p$ sublevels.

In the periodic table, the **nonmetals** include hydrogen and those elements in the upper right portion of the periodic table. The **metalloids** separate the metals and nonmetals in a steplike fashion. The **metals** occupy the left portion of the table including the *s* block, the *d* block (transition metals), and the *f* block. There are even some metals found in the *p* block, such as the group IIIA metals and tin, lead, and bismuth. Generally, metals

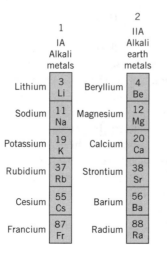

are those elements with few (one, two, or three) outer-level electrons. Tin, lead, and bismuth have atoms with four or five outer-level electrons, but they are classified as metals because they have metallic properties. Nonmetals include those elements with greater numbers (4, 5, 6, 7, and 8) of outer-level electrons. As we will see, the chemical behaviors and similarities of elements are directly related to the number of outer-level electrons.

8-17 SHORTHAND ELECTRONIC CONFIGURATIONS

For most elements, reading the electronic configuration directly from the periodic table is possible (see Figure 8-15). Since the table reflects electronic configurations, it makes sense that reading configurations directly from the table should be possible. Do this by noting the pattern of sublevel filling up to the position of an element in the table. For the element boron, we note in the table that the five electrons in boron involve the filling of the $1s$ sublevel, the $2s$ sublevel, and one electron in the $2p$ sublevel.

Therefore, the configuration of boron is $1s^2 2s^2 2p^1$. For zinc (atomic number 30), we fill the $1s$, $2s$, $2p$, $3s$, $3p$, $4s$, and $3d$ sublevels to reach zinc. So the configuration is $1s^2 2s^2 2p^6 3s^2 3p^6 4s^2 3d^{10}$, or $1s^2 2s^2 2p^6 3s^2 3p^6 3d^{10} 4s^2$ as we normally write it. Using this counting method, the periodic table is a quick source of the electronic configuration of an element.

Reading the electronic configuration of an element directly from the table is often convenient, and its configuration is typically written in the abbreviated or shorthand fashion. Look at a periodic table. Each period ends with a noble gas. The next period that begins after a noble gas represents the s level of the next-higher energy level. For instance, the third period ends with argon (18) and the next period begins with potassium (19). The outer-level electron in an atom of potassium begins the $4s$ sublevel of the fourth energy level. This pattern repeats from period to period.

To write the **shorthand electronic configuration** of an element, find the element in the table and trace back to the previous noble gas. The electronic configuration of the element is the same as that noble gas plus the additional electrons the element has beyond it. Write an abbreviated configuration by giving the symbol of the noble gas, enclosed in square brackets, and then the various extra sublevels with the proper numbers of electrons. To do this, pay attention to the parts of the table that correspond to the filling of various sublevels (see Figure 8-15). For boron, we move back to He, the previous noble gas, and fill the $2s$ followed by one electron in the $2p$. Thus, the shorthand configuration is [He]$2s^2 2p^1$. As another example, find sodium, Na, in the table and note that Ne is the previous noble gas. Sodium has one more electron than neon, and this electron is in the $3s$ sublevel, so the shorthand electronic configuration of sodium is $[Ne]3s^1$. In the same fashion, the electronic configuration of zinc includes the structure of argon plus electrons in the $4s$ and $3d$ levels. We count back 12 elements from zinc to argon; that represents 12 electrons. Starting with argon $[Ar]$, we add the 12 electrons, 2 in the $4s$ sublevel and 10 in the $3d$ sublevel. Thus, the shorthand configuration of zinc is

$$[Ar]4s^2 3d^{10} \quad or \quad [Ar]3d^{10}4s^2$$

The shorthand method is useful for any of the elements. Consider lead, Pb, atomic number 82. The preceding noble gas is xenon, atomic number 54. So, we need $82 - 54 = 28$ more electrons. We place these electrons in the sublevels corresponding to the period of the element, filling each sublevel then moving to new sublevels as needed. Thus, the shorthand configuration of lead is

$$[Xe]6s^2 4f^{14}5d^{10}6p^2 \quad or \quad [Xe]4f^{14}5d^{10}6s^2 6p^2$$

This example shows that we must remember to fill the f sublevels when writing electronic configurations of elements in this part of the table. Working with f-block elements, the pattern is that one electron goes in the $5d$ (or $6d$), then the $4f$ (or $5f$) gets electrons. After the f sublevel fills, continue filling the d sublevel. You can always remember this pattern by noting that La (element 57) is in the d block followed by elements 58 to 71 in the f block, and Ac (element 89) is in the d block followed by elements 90 to 103 in the f block. Consult a table to confirm this pattern. Following the pattern, the electronic configuration of curium, Cm, atomic number 96 is

$$[Rn]7s^2 6d^1 5f^7 \quad or \quad [Rn]5f^7 6d^1 7s^2$$

ACTIVITY 8-6

Write the shorthand electronic configurations for

(a) mercury, Hg
(b) barium, Ba
(c) iodine, I.

The counting method works quite well for most elements, but a close look at Figure 8-14 shows that some d-block configurations do not fit this method exactly. The exceptions are only a few, however, and we need not bother with them.

8-18 IONIZATION ENERGY

The term *periodic* comes from elements of higher atomic number having properties similar to those elements of lower atomic number. These similarities follow a pattern that repeats several times through the periods of elements. Two properties that follow this **periodic behavior** are discussed in the next few sections.

It requires energy to remove an electron from an atom to form a positive ion. The process of forming an ion from an atom is called ionization. Measuring the amount of energy required to remove one electron from a neutral atom is possible. This energy is called the first **ionization energy.** A specially designed instrument using a device similar to a cathode-ray tube measures ionization energies. An ionization energy is measured by bombarding atoms of an element in the gaseous state with a beam of free electrons. If the beam of electrons is of sufficient energy, bombarding electrons cause the loss of the most loosely bound outer-energy-level electron of the atom. This results in the formation of a singly charged ion (see Figure 8-16). The general ionization process is shown as

$$\text{energy} + X(g) \longrightarrow X^+(g) + e^-$$

where X represents an atom of a specific element and e^- represents the electron removed to form the X^+ ion. Since the neutral atom loses an electron, the resulting ion has a positive charge. In the positive ion there is one less electron, which results in an excess of one proton and a net positive charge.

The ionization energies of the elements are related to their atomic structures. Figure 8-17 gives a plot of these ionization energies versus atomic number. The plot illustrates the up-and-down periodic behavior of this property quite well. The sequence is a series of increasing ionization energies followed by sharp decreases in a repeating, periodic way. The ionization energies of the elements generally decrease from the top to the bottom of a given group and increase from left to right within a given period. This pattern is reasonable if we consider the factors that affect the ionization energies. Two important generalizations are that metals lose electrons more readily than nonmetals and that the noble gases do not readily lose electrons.

Two factors can account for the general decrease in the ionization energies of the elements within a group. One factor is the increased distance of the outer-energy-level electrons from the nucleus. In addition, the inner core of the lower-energy-level electrons shield the outer-energy-level electrons

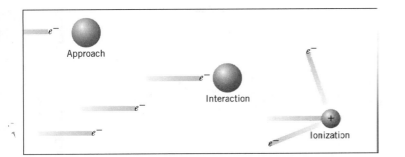

FIGURE 8-16
Ionization of an atom by an electron beam. An electron is a cathode-ray beam that interacts with the orbital electron, causing it to gain energy and break away from the atom, leaving a positive ion.

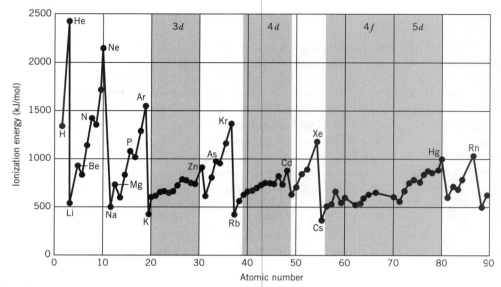

FIGURE 8-17 Plot of the ionization energies of elements versus their atomic numbers.

from the positive nucleus. The **shielding effect** decreases the attraction of the nucleus for the outer-energy-level electrons, and they are more loosely held (see Figure 8-18). As we move down a given group, the outer-energy-level electrons in the atoms of the elements are farther from the nucleus, and the attraction by the nucleus is diminished by shielding. The result is that the electrons are less tightly bound to the atoms and thus more easily removed.

The increase in ionization energy from left to right within a given period of the table occurs for a different reason. This increase results from an increase in nuclear charge without an increase in shielding. No increase in shielding occurs because electrons are added to the same outer energy level and, thus, have little shielding effect on one another. The increase in nuclear charge, as we go from left to right in a given period, usually causes an increase in the attraction between the nucleus and the outer-level electrons. The result is that these electrons are more tightly bound and removing one requires more energy.

ACTIVITY 8-7

Read the plot of ionization energies and record the approximate values for He, Ne, Ar, Kr, Xe, and Rn.

FIGURE 8-18 Factors that influence ionization energies.

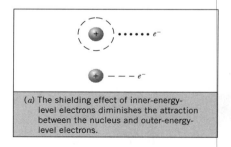

(a) The shielding effect of inner-energy-level electrons diminishes the attraction between the nucleus and outer-energy-level electrons.

(b) Higher-energy-level electrons that are further from the nucleus are less strongly attracted to the nucleus.

8-19 ATOMIC SIZES

In the light of the quantum mechanical model, how do we visualize an atom? Imagine slicing into one. Inside is a tiny central nucleus with surrounding layers of electron clouds extending outward, starting with the lower energy sublevels and ending with the outer energy sublevels. Of course, we never deal with such atomic slices. How can we imagine an atom if we look at it from the outside? Generally, we view an atom as essentially a spherical distribution of electron clouds. That is, we imagine them as small spheres or possibly bumpy-surfaced spheres since the outer level may consist of several orbitals.

As part of our chemical vision, we picture atoms as various-sized fuzzy balls with the outer-level electrons near the outer surfaces. As we will see, these outer-level electrons are normally responsible for the chemical behavior of an atom. Average atomic sizes are found indirectly from experimental information concerning molecules. One method used to find atomic size is X-ray diffraction (see Section 3-15). X rays are reflected from a sample of an element or compound. The pattern of X-ray reflection gives information about the distances between atoms and, therefore, atomic sizes. As an analogy, if we know the distance between the centers of two balls that are touching we know the radius of each ball.

If we assume atoms to have essentially spherical shapes, we can express sizes in terms of atomic radii. The radius of an atom is the distance from the center of the nucleus to the outer electrons. Atomic radii are in the range of about 10^{-10} meters. Since atomic radii are so small, they are expressed in units of nanometers; 1 nm = 10^{-9} m. Figure 8-19 on page 242 shows atomic radii of most of the representative elements. Hydrogen atoms, which have only one electron, are the smallest atoms of all elements.

As shown in Figure 8-19, the sizes of atoms within a group generally increase down the group from top to bottom. The sizes of atoms within a period generally become smaller from left to right. The periodic relations between sizes are those we might expect, since the factors affecting the sizes are essentially the same as those discussed previously for ionization energies. Atomic sizes increase down a given group because higher-energy-level electrons are found farther from the nucleus. The shielding effect decreases the attraction between outer-energy-level electrons and the nucleus. The result is that these loosely held electrons occupy greater volumes about the nucleus. The decrease in atomic sizes from left to right

The relative sizes of some common atoms

FIGURE 8-19
Variations in the sizes of atoms. The atomic radius of each element is given in nanometers. The noble gases, the transition elements, and the inner transition elements are not included in this table.

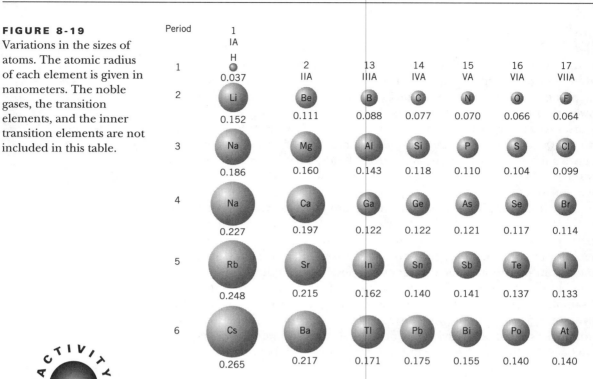

Using the radii values given in Figure 8-19, draw circles to show the relative sizes of atoms of hydrogen, oxygen, sodium, and chlorine.

within a period results from an increase in nuclear charge in that period. Each subsequent element in a period has one more proton and one more electron than the previous element. Since there is an increase in nuclear charge with no increase in shielding, the result is a greater attraction for the outer-energy-level electrons and, thus, smaller atoms. Figure 8-20 illustrates these factors.

FIGURE 8-20   Effect of number of electrons, shielding, and nuclear charge on atomic size.

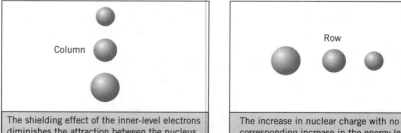

The shielding effect of the inner-level electrons diminishes the attraction between the nucleus and outer-energy-level electrons. As the number of inner-energy-level electrons increases the outer-energy-level electrons occupy greater volumes. This accounts for the increase in atomic sizes within a group of elements.

The increase in nuclear charge with no corresponding increase in the energy levels of the electrons, within a row of representative elements, results in the outer-energy-level electrons being more strongly attracted to the nucleus. Thus, since the nuclear charge increases from left to right, the atomic sizes will decrease from left to right.

8-20 FAMILY RELATIONSHIPS

The periodic table is a valuable source of information about the elements. The idea that the location of an element in the table corresponds to its electronic configuration is insightful. The families of elements that occupy the various columns of the table have similar properties because they have **similar outer-level electronic configurations.** As discussed in Section 5-1, elements in families form compounds that have similar formulas. (See Section 5-1 to review the formulas of the oxides of some families of elements.) This does not mean, however, that elements in a family form the same number and kinds of compounds. Each element has its unique and characteristic chemistry. The chemistries of the transition and inner transition elements are related to their outer-level electronic configurations and sometimes even to their next-to-outer-level electronic configurations. As a result, there are similarities and distinct differences between the elements, and the periodic table is a fundamental guide to making sense of the variety of chemical compounds they form. For example, when compounds containing calcium are found in nature, they often contain small amounts of magnesium, which is another common element of Group 2. A radioactive isotope of the group 2 element strontium is one product of nuclear fission. When nuclear accidents contaminate environments with low levels of this isotope, some of it replaces calcium in its compounds. Humans and other animals have calcium compounds in bones. As a result, it is possible for the radioactive strontium to replace some calcium in bones. Such radioactivity in the bones may cause radiation-induced cancer, as discussed in Section 16-11.

FOCUS
8

Reading the Periodic Table

Much information about an element can be read from the periodic table. The box of a given element in the table usually contains the atomic number, the symbol, and the atomic weight. The position of an element in the table reveals its group or family, whether it belongs to the s block, p block, d block, or f block and whether it is a representative, transition, or inner transition element. Its position also shows an element as a metal, nonmetal, or metalloid. For the representative elements, the last digit of the group number reveals the number of outer-level electrons. For example, group 2 elements have two outer-level electrons, group 13 elements have three outer-level electrons, and group 17 elements have seven outer-level electrons. Roman numeral group numbers for the A groups also reveal the number of outer-level electrons within a group.

As an example of the use of the table, let us locate the element boron, B. Finding B in the table reveals the following:

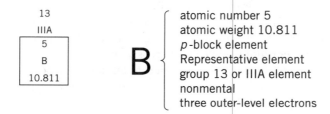

$$\begin{array}{c} 13 \\ \text{IIIA} \\ \boxed{\begin{array}{c} 5 \\ B \\ 10.811 \end{array}} \end{array} \quad B \left\{ \begin{array}{l} \text{atomic number 5} \\ \text{atomic weight 10.811} \\ p\text{-block element} \\ \text{Representative element} \\ \text{group 13 or IIIA element} \\ \text{nonmental} \\ \text{three outer-level electrons} \end{array} \right.$$

The position of an element in the table reveals its electronic configuration. The shorthand configuration is deduced by tracing back to the previous noble gas and then counting to the element's location in the table. Thus, the shorthand configuration of boron is $[He]2s^22p^1$. The counting method does not give the correct configuration for some transition and inner transition elements because their atoms exhibit such small differences in the relative energies of their sublevels. These exceptions are rare and of little concern to us.

The position of a representative element in the table also reveals how it differs in atomic size from neighbors in its period or group. For example, we predict that boron has smaller atoms than beryllium and larger atoms than carbon. In addition, we predict that boron has smaller atoms than aluminum or any of the other elements in its group. Furthermore, we predict that boron has a higher ionization energy than aluminum or any other members of its group.

QUESTIONS

1. Match the following terms with the appropriate lettered definitions.

cathode-ray tube

gas discharge tube

visible light

Bohr model

a. an electron moving from one energy level to another level

b. electron clouds that electrons occupy according to their energies

c. electrons in atoms occupy quantized energy levels or orbits

d. electrons in atoms occupy electron orbitals contained in energy sublevels that make up various quantized energy levels

quantum mechanical model

energy levels

quantum jump

electron orbital

electronic configuration

Pauli exclusion principle

noble gases

representative elements

e. quantized positions that electrons occupy according to their energies

f. ROY G. BIV

g. CRT

h. elements in the far right column of the periodic table

i. s- and p-block elements

j. d-block elements

k. f-block elements

l. energy needed to remove one electron from a neutral atom

transition elements

m. inner-level electrons shield outer-level electrons from the nucleus

inner transition elements

n. cathode ray tube filled with an element in the gaseous state

ionization energy

o. an orbital can contain only two electrons of opposite spin

shielding effect

p. the location of electrons in an atom in terms of the energy sublevels that they occupy

Sections 8-1 to 8-3

2. What information about an atom of an element is given by its atomic number?

3. Describe a cathode-ray tube. Describe a gas discharge tube.

Sections 8-4 and 8-5

4. What factors contribute to the energy of an electron in an atom?

5. Give a brief description of the Bohr model of the atom.

6. Explain what is meant by the quantized energy of electrons.

7. What is an energy state or energy level of an electron in an atom?

8. What is a quantum jump or a quantum leap?

Sections 8-6 to 8-13

9. Describe the quantum mechanical model of the atom and include the sublevel and orbital structure of energy levels.

10. What is an electron orbital? What is an electron cloud?

11. What is meant by the energy overlap of energy levels? Give an example of an energy overlap.

12. What is electron spin?

13. Using the pattern shown below as a beginning, construct a diagram of the electron energy sublevels in increasing order of energy. Be sure to indicate the

positions of energy overlap (e.g., the $3d$ sublevel overlaps the $4s$ sublevel).

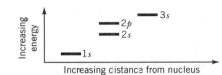

14. *Which of the following electron states do not exist? $2d$, $3p$, $3f$, $2p$, $7s$, $4d$, $1p$

15. Which of the following electron states do not exist? $1p$, $2s$, $2f$, $3f$, $2p$, $3s$, $4f$

16. What is the maximum number of electrons possible in

(a) any electron orbital   (b) a p orbital

(c) an s orbital   (d) an s sublevel   (e) a p sublevel

(f) a d sublevel   (g) an f sublevel   (h) an f orbital

17. In what way does the second main energy level of an atom differ from the fourth main energy level?

18. Consider the capacities of the sublevels. Given these capacities, determine the maximum number of electrons that can be contained in

(a) the first main energy level

(b) the second main energy level

(c) the third main energy level

(d) the fourth main energy level

19. *Give the electronic configuration in sublevel notation for the following elements for which the atomic numbers are given in parentheses.

(a) H (1)   (b) Al (13)   (c) Ga (31)

(d) Se (34)   (e) Kr (36)   (f) Zr (40)

(g) Sn (50)   (h) O (8)   (i) Pt (78)

(j) P (15)   (k) I (53)

20. Give the electronic configuration in sublevel notation for the following elements for which the atomic numbers are given in parentheses.

(a) C (6)   (b) F (9)   (c) Ca (20)

(d) As (33)   (e) Ne (10)   (f) Ba (56)

(g) Fr (87)   (h) Ni (28)   (i) Pb (82)

(j) Br (35)   (k) Cl (17)

21. What are outer-energy-level electrons? How is the number of outer-level electrons determined for a specific element?

Sections 8-14 to 8-16

22. What is the periodic table of elements and why is it useful?

23. What is meant by the term *periodic* in the periodic table?

24. Describe the shape of the periodic table and indicate how many elements comprise the various parts of the table. Relate the shape of the table, and the number of elements in each part of the table, to the electronic configuration of the elements.

25. How many elements wide are each of the following?

   (a) the *s* block   (b) the *p* block

   (c) the *d* block   (d) the *f* block

26. Sketch an outline of the periodic table and indicate the following.

   (a) *s* block, *p* block, *d* block, and *f* block

   (b) Label the portions of the table corresponding to the filling of the 2*s* sublevel, the 2*p* sublevel, the 4*d* sublevel, the 5*p* sublevel, and the 4*f* sublevel.

27. What is a period in the periodic table? What are rows in the periodic table? What is a group in the periodic table? Relate periods and groups in the periodic table to the electronic configurations of the elements.

28. Explain why elements in the same group have similarities in chemical behavior.

29. Explain why the numbers of elements in the periods of the periodic table follow the pattern 2, 8, 8, 18, 18, 32, 26.

30. Write the electronic configurations of carbon (6), silicon (14), germanium (32), and tin (50). What do you notice about the number of electrons in the outer energy level of each of these atoms? The outer energy level has the highest energy-level number.

31. Write the electronic configurations for the following (the atomic number is given in parentheses).

   (a) Na (11) (b) K (19) (c) Rb (37) (d) Cs (55)

What do the electrons of the outermost energy level of each of these elements have in common? What do these configurations reveal about the positions of these elements in the periodic table? Would you expect them to have chemical properties similar to one another? Why?

32. Write the electronic configurations for the following (the atomic number is given in parentheses).

   (a) O (8)   (b) S (16) (c) Se (34) (d) Te (52)

What do the electrons of the outermost energy level of each of these elements have in common? What do these configurations reveal about the positions of these elements in the periodic table? Would you expect them to have chemical properties similar to one another? Why?

33. Write the electronic configurations of the elements having atomic numbers 29, 47, and 79. What do these configurations tell you about the positions of these elements in the periodic table?

34. Suppose that a new element, given the name paulingium, had recently been made by nuclear scientists. If the new element has an atomic number of 116, to what group in the periodic table would you assign the element?

35. List some properties associated with metals. Where are the metals located in the periodic table?

36. Describe nonmetals and metalloids. Where are the nonmetals located in the periodic table?

37. Sketch an outline of the periodic table and indicate the positions of the following.

   (a) the noble gases, the representative elements, the transition elements, and the inner transition elements

   (b) the alkali metals, the halogens, the lanthanides, and the actinides

38. *Refer to the periodic table inside the front cover of this book and classify the following elements as representative element, transition element, or inner transition element, and metal, nonmetal, metalloid or noble gas.

   (a) scandium, Sc   (b) argon, Ar

   (c) radium, Ra   (d) radon, Rn

   (e) cadmium, Cd   (f) antimony, Sb

   (g) thorium, Th   (h) chlorine, Cl

   (i) osmium, Os   (j) indium, In

39. Refer to the periodic table inside the front cover of this book and classify the following elements as representative element, transition element, or inner transition element, and metal, nonmetal, metalloid or noble gas.

   (a) molybdenum, Mo (b) helium, He

   (c) rubidium, Rb   (d) iodine, I

   (e) zirconium, Zr   (f) gallium, Ga

   (g) neptunium, Np (h) astatine, At

   (i) lead, Pb   (j) phosphorus, P

Section 8-17

40. *Answer Question 19, but give the shorthand electronic configurations.

41. Answer Question 20, but give the shorthand electronic configurations.

Section 8-18

42. Define ionization energy.

43. Sketch an outline of the main portion of the periodic table (s, p, and d blocks) and indicate with arrows how the ionization energies of the elements vary from left to right in periods and from top to bottom in groups.

44. Describe the shielding effect and how it influences the attraction of a nucleus for outer-level electrons.

45. For each of the following sets of three elements, indicate which has the lowest and which has the highest ionization energy.

   (a) barium, Ba; calcium, Ca; magnesium, Mg

   (b) carbon, C; nitrogen, N; fluorine, F

   (c) chlorine, Cl; argon, Ar; potassium, K

   (d) neon, Ne; copper, Cu; cesium, Cs

Section 8-19

46. Explain how the sizes of the atoms vary within the periods and groups of the periodic table. In which part of the table are the elements with the smallest atoms found? In which part of the table are the elements with the largest atoms found?

47. What is a nanometer? Express the relation between centimeters and nanometers in terms of the number of centimeters per one nanometer.

48. The element with the largest atoms is francium. Explain why.

49. *A sulfur atom has a diameter of 0.208 nm. Calculate the diameter of the sulfur atom in inches. (1 in. = 2.54 cm)

50. An old copper penny is 1.0 mm thick. Assuming that the copper atoms are stacked on top of one another, calculate how many copper atoms make up the thickness of the penny. The diameter of a copper atom is 0.256 nm.

51. *A carbon atom has a diameter of about 0.154 nm. If Avogadro's number of carbon atoms were stacked one on top of the other, how many meters high would the stack be? For comparison, note that the sun is about 1×10^{11} meters from Earth. (It is not possible to stack atoms in this way.)

52. The volume of a sphere is given by the formula $V = \pi d^3 / 6$, where d is the diameter of the sphere. Calculate the volume of a single calcium atom in cubic centimeters if its diameter is 0.394 nm and $\pi = 3.142$.

53. For each of the following sets of elements, indicate which element has the smallest atoms.

   (a) calcium, Ca; bromine, Br; germanium, Ge

   (b) carbon, C; silicon, Si; tin, Sn

   (c) rubidium, Rb; sodium, Na; francium, Fr

   (d) fluorine, F; nitrogen, N; boron, B

54. *Gold is the most malleable metal known. It can be pressed or beaten into very thin sheets called gold leaf. The diameter of a gold atom is 0.29 nm. Assuming the atoms are stacked one on top of the other, calculate the number of atoms of gold that make up the thickness of a 0.000005-in.-thick gold leaf. (1 in. = 2.54 cm)

Section 8-20

55. *Using the periodic table, give the atomic number, atomic weight, block, type of element, shorthand electronic configuration, and number of electrons in the outer level for the following elements.

   (a) nitrogen, N   (b) sodium, Na

   (c) iron, Fe   (d) nickel, Ni

   (e) silicon, Si   (f) nobelium, No

   (g) barium, Ba   (h) neon, Ne

   (i) phosphorus, P (j) oxygen, O

56. Using the periodic table, give the atomic number, atomic weight, block, type of element, shorthand electronic configuration, and number of electrons in the outer level for the following elements.

   (a) titanium, Ti   (b) tantalum, Ta

   (c) magnesium, Mg (d) hydrogen, H

   (e) boron, B   (f) tin, Sn

   (g) lawrencium, Lr (h) fluorine, F

   (i) tellurium, Te   (j) tungsten, W

57. Why are groups in the periodic table sometimes called families?

Questions to Ponder

58. Consult some reference books or the internet and give a brief description of how a laser works at the atomic level.

59. Calculate the number of electrons in 1.0 tons of aluminum metal. What percent of the total mass of 1.0 tons of aluminum is represented by electrons?

60. How many electrons are in the head of a pin?

CHEMICAL
BONDS

9-1 THE NATURE OF THE CHEMICAL BOND

Hundreds of years ago, before scientists had knowledge of atoms and atomic structure, Isaac Newton, famous English scientist, made the following prescient observation:

> Have not the small particles of bodies certain powers, virtues or forces, by which they act at a distance, not only upon the rays of light for reflecting, refracting and inflecting them, but also upon one another for producing a great part of the phenomena of Nature?

Newton anticipated that atoms, as particles of matter, associate with one another to make compounds and that those compounds are involved in natural processes. Today we recognize that compounds contain atoms held in chemical combination by bonds. **Chemical bonds** are forces that hold atoms in combination. Bonds result from attractions between opposite charges, the same forces that hold electrons in atoms. Recall that particles that carry electrical charges attract and repel according to Coulomb's law, described in Section 1-3. Electrical forces of attraction are somewhat like magnetic forces. You know that when two magnets are moved closer and closer, attractive forces increase until they strongly bond and stick together. Of course, this happens only when the opposite poles are aligned. In a like manner, forces of attraction occur between objects of positive and negative electrical charge.

When atoms form compounds, the outer-level electrons are involved. If atoms did not interact with one another to form compounds, all elements would be monatomic gases. It is interesting that the noble gases (group 18 elements) all occur as monatomic gases at normal earth conditions. For example, a sample of helium gas is just a collection of separate helium atoms. In contrast, elemental gases such as hydrogen, nitrogen, oxygen, fluorine, and chlorine occur as diatomic molecules. In diatomic

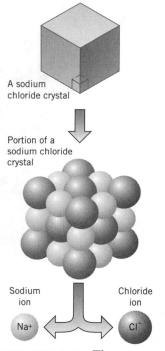

A sodium
chloride crystal

Portion of a
sodium chloride
crystal

Sodium
ion

Chloride
ion

Na+

Cl⁻

FIGURE 9-1 The
chemical particles in
sodium chloride are
sodium ions and chloride
ions.

molecules, the atoms are held in combination by chemical bonds. Such groups of bonded atoms are more stable than separate individual atoms.

Salt and sugar are examples of common chemical compounds that occur as crystalline solids. Chemical investigations reveal a distinct difference in the combination of sodium and chlorine in salt and the combination of carbon, hydrogen, and oxygen in sucrose. Figure 9-1 shows the fundamental particles found in sodium chloride are not ordinary atoms. The particles are ions having positive and negative charges. Positively charged sodium ions and negatively charged chloride ions compose salt. In sugar, as illustrated in Figure 9-2, the fundamental particles are not independent atoms but groups of atoms linked by the sharing of electrons. These bonded groups of atoms are sucrose molecules. Sugar and salt are examples of the two kinds of chemically combined elements found in compounds. One type involves ion formation, the other involves molecule formation. Atoms interact through their outer-level electrons to form ions or molecules. The **number of outer-level electrons** an atom holds dictates the kinds of chemical compounds it forms.

9-2 ELECTRON DOT SYMBOLS

When atoms of elements form compounds, the outer-energy-level electrons are normally the only electrons involved. These outer-energy-level electrons are called **valence electrons.** The word *valence* comes from the Latin *valentia* (capacity). For atoms, the valence refers to the capacity to form bonds. Rather than write the electronic configurations for elements to show the number of valence electrons, special symbols are used. The valence electrons of each element are shown by an electron dot symbol. To write an **electron dot symbol,** give the usual symbol of the element surrounded by dots representing the valence electrons. For example, the dot symbol for hydrogen, with its one electron, is

H•

Electron dot symbols are normally used for the representative elements. For these elements, the electron dot symbol is the symbol of the element surrounded by a number of dots corresponding to the number of valence electrons. The symbol represents the nucleus with the core electrons, and the dots represent the outer-level electrons. When writing a dot symbol, imagine a square around the symbol of the element and put a dot on each side until all the valence electrons are used. After one dot is placed on each side, dots are doubled up as necessary. For example, symbols for the elements sodium, magnesium, boron, carbon, nitrogen, oxygen, fluorine, and neon are

Group	1	2	13	14	15	16	17	18
Number	IA	IIA	IIIA	IVA	VA	VIA	VIIA	noble gas
	Na·	·Mg·	·Ḃ·	·Ċ·	·N̈·	:Ö·	:F̈·	:N̈e:

The side used for pairing dots is not important, so the dot symbol for oxygen could be written

$$\ddot{O}\cdot \quad \cdot\ddot{O}: \quad :\dot{O}: \quad :\dot{O}: \quad \cdot\ddot{O}\cdot$$

You can interpret a dot symbol for groups 15, 16, 17, and 18 elements by imagining that the four sides around the element symbol represent the *s* orbital and the three *p* orbitals of the outer energy level.

$$
\begin{array}{c}
s \\
p\,X\,p \\
p
\end{array}
$$

The outer-level electrons are found within these orbitals; the *s* orbital is filled first, then the *p* orbitals. According to Hund's rule, whenever possible each *p* orbital contains no more than one electron. For elements with more than three *p* electrons, all three *p* orbitals do contain one electron, then each additional electron fills a *p* orbital until the noble gas configuration is reached.

Table 9-1 gives the dot symbols for most of the representative elements. Note carefully that the dot symbols for all the elements in a family with the same outer-level configurations are the same, since the number of valence electrons is the same. Although these dot symbols are simplistic representations of the distributions of valence electrons, they are very useful in the discussion of chemical bonding. Remember that the number of outer-level electrons for a representative element is revealed by its group number.

9-3 METALLIC BONDING

Metals characteristically have atoms with few valence electrons; most of them have two, some have one, and a few have more than two. Metals are typically solids that are good conductors of electricity and heat. Normally, they are easily stretched into wires, pressed into sheets, and cast or molded into various shapes.

Table 9-1 *Electron Dot Symbols for Some Representative Elements*

			GROUP				
	IA 1	IIA 2	IIIA 13	VIA 14	VA 15	VIA 16	VIIA 17
	ns^1	ns^2	ns^2np^1	ns^2np^2	ns^2np^3	ns^2np^4	ns^2np^5
H·	Li·	·Be·	·Ḃ·	·Ċ·	·N̈·	:Ö·	:Ḟ·
	Na·	·Mg·	·Al·	·Ṡi·	·P̈·	:S̈·	:Cl̈·
	K·	·Ca·		·Ġe·	·Äs·	:S̈e·	:Br̈·
	Rb·	·Sr·				:T̈e·	:Ï·

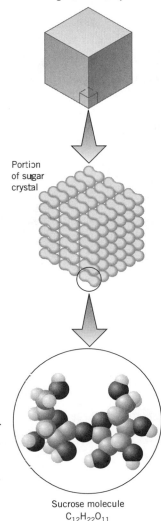

A sugar (sucrose) crystal

Portion of sugar crystal

Sucrose molecule $C_{12}H_{22}O_{11}$

FIGURE 9-2 The chemical particles in table sugar are $C_{12}H_{22}O_{11}$ molecules.

Consult a periodic table and list those metals that have one valence electron and those that have more than two valence electrons.

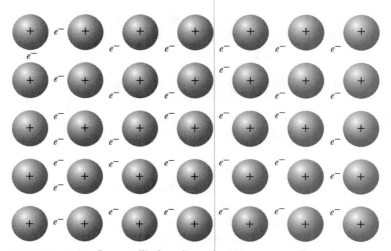

FIGURE 9-3   In metallic bonding, the inner cores of metal atoms are immersed in a "sea" of outer-level electrons.

How are metal atoms held in combination in a metallic solid made of a pure metal? A simple picture is that positively charged atomic cores are surrounded by a "sea" of loosely held valence electrons. See Figure 9-3. A simple vision of sodium metal is

Na	•	Na	•	Na	•	Na	•	Na	•	Na	•
•	Na	•	Na	•	Na	•	Na	•	Na
Na	•	Na	•	Na	•	Na	•	Na	•	Na
•	Na	•	Na	•	Na	•	Na	•	Na

Electrons can easily move within a sample of a solid metal, which is why metals are good conductors. The various atoms in a metallic solid are held in place by relatively strong electrostatic forces of attraction. The positive atomic cores and outer sea of electrons mutually attract. Such attractions are nondirectional and act in three dimensions. Consequently, metals are typically crystalline solids in which the atoms are stacked in a three-dimensional pattern. The relative positions of atoms in such solids are readily changed. Consequently, metals typically are flexible and malleable so they are easily bent, stretched, shaped, molded, and cast.

9-4 THE OCTET RULE

Chemists developed ideas about chemical bonding by observing the nature and behavior of many chemical compounds. The ability of an atom to form chemical bonds relates to the distribution of electrons in that atom.

Knowledge of the electronic configurations of elements is the basis of understanding chemical bonding between elements. In the 1920s, an American chemist, Gilbert Lewis, proposed the theory that elements combine chemically by the loss, gain, or sharing of electrons. Studies of the properties of many compounds of the representative elements have produced an important generalization called the **octet rule.** This rule states that atoms typically lose, gain, or share electrons so that they attain a total of **eight (an octet) outer-energy-level electrons.** To show an octet of electrons, eight dots are written around the symbol. For example, the octet of electrons in argon corresponding to its outer-energy level is shown as

$$:\!\ddot{A}r\!: \qquad [Ne]\ 3s^2\ 3p^6$$

Notice that an octet of electrons corresponds to filled s and p sublevels (ns^2np^6) in the outer level. An octet of electrons in the outer level of an atom is a particularly stable arrangement. Hydrogen has only one valence electron. It gains stability when it has two outer-level electrons. These two electrons correspond to a filled first energy level ($1s^2$). Hydrogen follows a duet rule rather than the octet rule since there is no room in the first energy level for eight electrons. The noble gases are chemically stable and form very few compounds. Unlike other elements, all the noble gases except one have an octet of outer-level electrons. The exception is helium, which has atoms with two outer-level electrons corresponding to a filled $1s$ sublevel ($1s^2$).

Noble gases have atoms that are stabilized by an octet of outer-level electrons. Atoms of other elements have fewer than eight outer-level electrons and typically combine in ways through which they attain eight outer-level electrons. The octet rule suggests that atoms combine not only in ways governed by electrostatic attractive forces but also in ways that they become surrounded by a stable arrangement of electrons. Apparently, the attainment of a stable arrangement of outer-level electrons allows chemical bonding.

The octet pattern is called a rule rather than a law since it is not followed for all cases of compound formation. In fact, when noble gases form compounds they do not follow the rule. Nevertheless, this rule does apply to a great number of compounds and serves as a useful guide to the way in which atoms of elements combine. Two basic types of chemical bonds are common. An **ionic bond** results when electrons transfer from one atom to another creating ions. Here the octet rule is followed by the loss and gain of electrons. A **covalent bond** results when atoms share electrons between them. Here the octet rule is followed by mutual sharing of electrons.

9-5 ION FORMATION

When sodium metal and chlorine gas mix, they readily react to form salt. Imagine a reaction between sodium atoms and chlorine atoms. The

Atoms   Ions

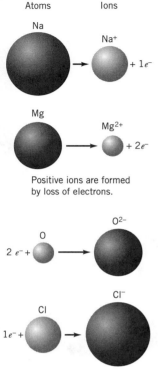

Positive ions are formed
by loss of electrons.

Negative ions are formed
by gain of electrons.

FIGURE 9-4 Ion
formation by the loss and
gain of electrons.

sodium atoms each have one outer-level electron. In the reaction, they lose these electrons to form positive sodium ions.

$$\text{Na} \longrightarrow \text{Na}^+ + 1e^-$$
$$[\text{Ne}]3s^1 \longrightarrow [\text{Ne}] \quad \text{or} \quad [\text{He}]2s^2 2p^6$$

The electron lost by a sodium atom is gained by a chlorine atom to form a negative chloride ion.

$$\text{Cl} + 1e^- \longrightarrow \text{Cl}^-$$
$$[\text{Ne}]3s^2 3p^5 \longrightarrow [\text{Ar}] \quad \text{or} \quad [\text{Ne}]3s^2 3p^6$$

The transfer of the electron allows both kinds of atoms to have an octet of outer-energy-level electrons. The sodium atom loses an electron and ends up with an octet in the next-lowest energy level. The chlorine atom gains an electron to complete an octet in the outer-energy level.

When an atom loses electrons, negative charges are removed (each electron carries a single negative charge), leaving behind a positively charged particle consisting of a nucleus and the remaining electrons. The result is a **positive ion or a cation**. See Figure 9-4. The sodium atom loses an electron to form a singly charged sodium cation.

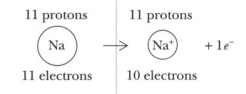

An atom that gains electrons takes on the negative charges of the electrons. Such a negatively charged particle is also an ion, but in this case, a **negative ion or an anion**. See Figure 9-4. The chlorine atom gains an electron to form a singly charged anion.

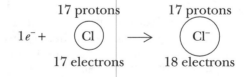

Ions are fundamental chemical species or particles found in compounds that are appropriately called ionic compounds. When ionic compounds form from elements, the atoms lose and gain electrons to form ions. The number of protons in an atom does not change when it forms an ion, so the charge of the ion is a result of excess or deficient electrons. Ions are very different chemical particles compared with the neutral atoms from which they form. We sprinkle salt on our food and eat it. In fact, we need it in small amounts to keep our bodies functioning correctly. But no one dares eat either sodium metal or chlorine gas. Both are highly toxic and dangerous. Chemists and professionals sometimes use terminology that may be misleading to a beginner. For instance, a doctor may refer to a low-sodium diet or a nutritionist may claim that we need a certain amount of calcium in our diets. These professionals know that they are referring to sodium ions and calcium ions, not sodium or calcium metal. As you learn more chemistry, you'll become more aware of the use of this terminology.

9-6 ION FORMULAS

Ions formed by the loss or gain of electrons from atoms are called **monatomic ions** or **simple ions.** A simple ion is represented by a formula showing the symbol of the element from which the ion is formed with the charge of the ion as a superscript. The formula for a sodium ion is Na^+ (read N–A–plus), and the formula for the chloride ion is Cl^- (read C–L–minus). A simple positive ion has a name that is the same as the element name. The formula for the calcium ion, Ca^{2+} (read C–A–two–plus), shows that calcium ion carries a double positive charge. The formula for the oxide ion, O^{2-} (read O–two–minus), shows that the oxide ion has a double negative charge.

Metals with few (one, two, or three) outer-level electrons typically lose electrons. Metals generally have lower ionization energies and so they form cations readily. The nonmetals, with many outer-level electrons (five, six, or seven), typically gain electrons. Nonmetals generally have higher ionization energies and do not lose electrons readily. Nonmetals can gain electrons to form anions.

Table 9-2 shows the ions formed by common metals and nonmetals. There is a pattern in the way that metals form ions. The pattern relates to

Consult a periodic table and predict the cation formed by the element barium and the anion formed by the element sulfur.

Ionic

Table 9-2   *Names and Formulas for Some Common Cations and Anions*

CATIONS		ANIONS	
Group 1 (IA)		Group 15 (VA)	
Na^+	sodium ion	N^{3-}	nitride ion
K^+	potassium ion	P^{3-}	phosphide ion
Group 2 (IIA)		Group 16 (VIA)	
Ca^{2+}	calcium ion	O^{2-}	oxide ion
Mg^{2+}	magnesium ion	S^{2-}	sulfide ion
Sr^{2+}	strontium ion	Se^{2-}	selenide ion
Ba^{2+}	barium ion	Te^{2-}	telluride ion
Group 13 (IIIA)		Group 17 (VIIA)	
Al^{3+}	aluminum ion	F^-	fluoride ion
		Cl^-	chloride ion
		Br^-	bromide ion
		I^-	iodide ion

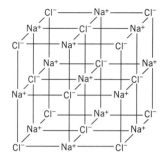

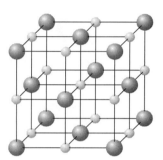

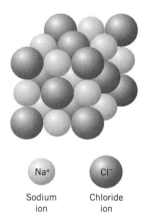

Sodium
ion

Chloride
ion

FIGURE 9-5 Sodium ions and chloride ions are held in chemical combination by ionic bonds, the electrostatic attraction between ions of opposite charge.

Ionic Bond

The electrostatic force of attraction between oppositely charged ions.

the periodic table and the number of valence electrons. The group number from the periodic table reveals the number of valence electrons. For representative metals, the charge on the positive ions they form is the number of valence electrons. Sodium (group 1 or IA) forms Na^+, magnesium (group 2 or IIA) forms Mg^{2+}, and aluminum (group 13 or IIIA) forms Al^{3+}. For nonmetals, the number of electrons needed to make an octet is found by subtracting the number of valence electrons from 8. Thus, nitrogen (group 15 or VA) forms N^{3-}, nitride ion; oxygen (group 16 or VIA) forms O^{2-}, oxide ion; and fluorine (group 17 or VIIA) forms F^-, fluoride ion. As you know, a simple anion always has a name that includes the root of the element name with an *-ide* ending.

9-7 THE IONIC BOND

We return to the reaction between sodium metal and chlorine gas, which forms positive sodium ions and negative chloride ions. The sodium atoms lose one electron each and the chlorine atoms gain one electron each. In the resulting compound, the number of positive ions equals the number of negative ions. As these ions form, they are in close proximity and intermingle. Since the ions are of opposite charge, they are attracted to one another by electrostatic attractive forces. The attractive forces cause the ions to group into a three-dimensional pattern of alternating ions, as illustrated in Figure 9-5. Even a tiny crystal of salt contains a vast number of sodium ions and chloride ions stacked in three-dimensional space. This three-dimensional accumulation of ions forms a crystal. The resulting compound is crystalline. In the reaction of sodium and chlorine, the reactants change from the soft metal sodium and gaseous chlorine to a product that is a colorless crystalline solid. The force of attraction between the ions that allows a compound to form is the ionic bond as defined in the margin.

Compounds containing ions held together by ionic bonds are called **ionic compounds.** When a metal reacts with a nonmetal, a transfer of electrons occurs, which results in the formation of ions that combine through ionic bonds. Hence, a binary or two-element compound involving a metal and a nonmetal is typically an ionic compound containing metal cations and nonmetal anions. Figure 9-6 illustrates a few examples of ionic compound formation. Ionic compounds are **crystalline solids** and have ions stacked in three-dimensional space. The three-dimensional pattern of stacked ions is called a **crystal lattice.** Because of differences in sizes, shapes, and charges of ions, not all ionic compounds have cubic crystals like sodium chloride.

Ionic bonds are relatively strong, making ionic compounds hard solids. Ionic bonds are nondirectional because negative ions attract positive ions in all directions and a positive ion attracts negative ions in all directions. This attraction in ions arranged in space as a specific three-dimensional crystalline solid. Typically, ionic solids easily crack or fracture. As shown in Figure 9-7, when part of an ionic solid is pushed out of position, like charges can align and their mutual repulsion pushes the crystal apart, causing it to crack.

9-8 PREDICTING FORMULAS OF BINARY IONIC COMPOUNDS

An ionic compound contains oppositely charged ions. All ionic compounds, however, are overall electrically neutral because, as mentioned in Section 6-6, an ionic compound always has as many positive charges as negative charges. The positive ions balance the negative ions. The formulas of ionic compounds reflect this balance. The formula of sodium chloride is simply NaCl, since a sodium ion carries one positive charge and a chloride ion carries one negative charge. The formula could be written as $(Na^+)(Cl^-)$ or Na^+Cl^-. The charges are not normally included in the formulas of ionic compounds, however; they are implied by the number of combined atoms shown.

How do we know how to write the likely formula for a **binary ionic compound**? To make the amount of positive charge equal the amount of negative charge, the appropriate number of ions of each type is needed. Take, as an example, the compound involving calcium and chlorine. Calcium is a group 2 element, so it forms Ca^{2+} ions; chlorine is a group 17 element, so it forms Cl^- ions. Group numbers and, thus, the expected formulas of the ions are predicted from a periodic table. Two chloride ions balance

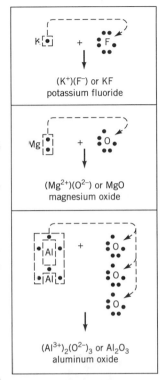

$(K^+)(F^-)$ or KF
potassium fluoride

$(Mg^{2+})(O^{2-})$ or MgO
magnesium oxide

$(Al^{3+})_2(O^{2-})_3$ or Al_2O_3
aluminum oxide

FIGURE 9-6 The idealized formation of some ionic compounds.

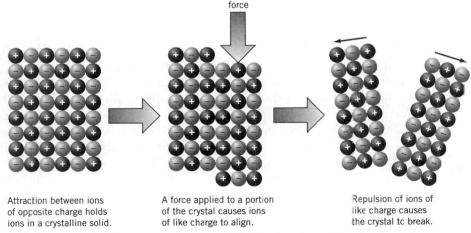

Attraction between ions of opposite charge holds ions in a crystalline solid.

A force applied to a portion of the crystal causes ions of like charge to align.

Repulsion of ions of like charge causes the crystal to break.

FIGURE 9-7 Ionic compounds are typically brittle; when a shearing force is exerted on a portion of an ionic crystal, ions of the same charge come into contact and repel, causing the crystal to break.

one calcium ion since the number of negative charges must equal the number of positive charges as

$$(Cl)^- (Ca^{2+}) (Cl^-) \quad \text{giving the formula } CaCl_2$$

By convention, the metal always comes first in the formula of an ionic compound.

EXAMPLE 9-1

What is the likely formula for a binary compound of potassium and sulfur ions?

Potassium (group 1) forms the ion K^+ and sulfur (group 16) forms the sulfide ion, S^{2-}. Thus, the formula contains two K^+ for each S^{2-}.

$$K^+ \quad S^{2-} \qquad \text{gives the formula } K_2S$$

EXAMPLE 9-2

What is the likely formula for a binary compound of aluminum and oxygen?

Aluminum (group 13) forms the ion Al^{3+} and oxygen (group 16) forms the oxide ion, O^{2-}. Thus, the formula contains two aluminum ions (to give a total charge of 6+) and three oxide ions (to give a total charge of 6−).

$$Al^{3+} \quad O^{2-} \qquad \text{gives the formula } Al_2O_3$$

ACTIVITY 9-3

What is the likely formula of the binary compound of magnesium and nitrogen?

9-9 THE COVALENT BOND

Many compounds contain molecules rather than ions. Molecules involve the sharing of electrons between bonded atoms. To illustrate how this sharing occurs, we will use the formation of a compound of hydrogen and chlorine. Suppose an atom of hydrogen and an atom of chlorine approach one another, as shown in Figure 9-8. The stable electron arrangement for a hydrogen atom is two electrons. A chlorine atom is stabilized with eight electrons in the outer level. As shown in the figure, the atoms attain stability by sharing electrons. When an electron pair locates between the nuclei, the electrostatic forces of the electron–nucleus attractions hold the atoms in combination.

$$H\cdot \; + \; \cdot\ddot{\underset{\cdot\cdot}{Cl}}: \; \longrightarrow \; H\!:\!\ddot{\underset{\cdot\cdot}{Cl}}:$$

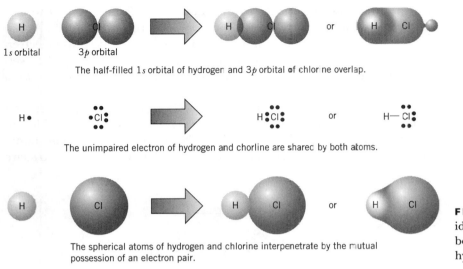

The half-filled $1s$ orbital of hydrogen and $3p$ orbital of chlorine overlap.

The unimpaired electron of hydrogen and chorline are shared by both atoms.

The spherical atoms of hydrogen and chlorine interpenetrate by the mutual possession of an electron pair.

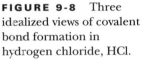

FIGURE 9-8 Three idealized views of covalent bond formation in hydrogen chloride, HCl.

The result of sharing is that both atoms attain a stable electronic configuration by mutual possession of electrons. Each nucleus attracts the shared pair of electrons and no transfer of electrons occurs. The sharing produces a molecule.

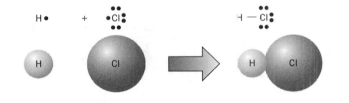

The sharing of an electron pair gives hydrogen two electrons and chlorine an octet of outer-level electrons. A duet of electrons for hydrogen and an octet for chlorine are attained when they share electrons rather than lose or gain them. Notice that the shared electrons "belong" to both the hydrogen and the chlorine atoms. The shared pairs are included in the count of valence electrons for each atom.

When nonmetals or metalloids form compounds with other nonmetals, the atoms attain stability by sharing outer-level electrons. The sharing of an electron pair between atoms acts as a bonding force to link the atoms. This type of chemical bond is the **covalent bond** (co- as in cooperation and -valent from valence or combining capacity). See the margin for a definition.

Covalently bonded atoms form molecules. By definition, a **molecule** is a group of two or more covalently bonded atoms. A simple way to picture the formation of a covalent bond is to imagine the valence electron or-

Molecules

Covalent Bond
The force of attraction between two atoms resulting from the sharing of a pair of electrons.

bitals of the bonded atoms. Recall that an orbital is filled when it contains two electrons and an orbital containing one electron is half-filled. As two atoms covalently bond, the half-filled orbitals overlap so that the electron pair is shared. In other words, we can picture a covalent bond as the **overlap of half-filled atomic orbitals** that allows the electrons to become a shared pair.

Figure 9-8 illustrates the formation of a compound of hydrogen and chlorine. In this case, the overlap of the half-filled $1s$ atomic orbital of hydrogen and the half-filled $3p$ orbital of chlorine is the covalent bond. The overlap of half-filled orbitals makes it possible for the atoms to share the electrons.

Ethyl Alcohol

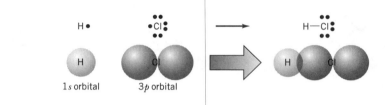

$1s$ orbital $3p$ orbital

Ethyl Alcohol CH₃CH₂OH

Ethyl alcohol or ethanol was most likely one of the earliest molecular compounds synthesized by humans since it is the active ingredient in alcoholic beverages. Ethyl alcohol is made by the yeast controlled fermentation of carbohydrates. Fermentation is a complex process in which enzymes from the yeast encourage the break down of starches to glucose. Other enzymes encourage the reaction of glucose to form ethyl alcohol and carbon dioxide gas:

$$C_6H_{12}O_6 \text{ (glucose)} \xrightarrow[\text{from yeast}]{\text{enzymes}} 2CO_2 + 2CH_3CH_2OH$$

The initial starch may be from wheat, barley, rice (beer), honey (mead), sugar cane (rum), grapes (wine and brandy), rye, barley, corn (whiskeys), rice (sake), potatoes (vodka, now usually made from grain), cactus pulp (tequila), or almost any fruit or vegetable sugars. Ethyl alcohol is the psychoactive drug in alcoholic beverages. In all such beverages, ethyl alcohol is the active agent but it varies in concentration, as shown in Table 9-3.

Different alcoholic beverages get their distinct tastes from the presence of small amounts of impurities. The aftereffects (hangover) of excessive use of alcoholic beverages come from toxic byproducts of the metabolism of ethyl alcohol and the presence of toxic impurities in alcoholic beverages called fusel oils. The term *fusel* comes from the German word for evil spirit. Ethyl alcohol is the most commonly used and abused drug in our society. Drunkenness is, of course, characterized by the loss of motor control and the slurring of speech. Imbibing in too much ethyl alcohol over a short time causes alcohol poisoning. Such poisoning results in sickness, a coma, and sometimes death. Ethyl alcohol is known to be an addictive drug, especially for those with a

Table 9-3   *The Ethyl Alcohol Content of Alcoholic Beverages*

BEVERAGE	PERCENT BY VOLUME ALCOHOL	PROOF (DOUBLE THE PERCENT)
Light beers	3% to 4%	6 to 8
Lager beers	4% to 5%	8 to 10
Malt liquors and ales	5% to 6%	10 to 12
Wines	Up to 12%	Up to 24
Fortified wines	Up to 20%	Up to 40
Brandy (distilled wine)	40% to 45%	80 to 90
Rum	40% and higher	80 and up
Whiskey	40% to 55%	80 to 110
Tequila	40% to 55%	80 to 110
Vodka	40% to 55%	80 to 110

genetic penchant toward alcoholism. Besides its use as a mood-altering drug, ethyl alcohol finds use in industry as a solvent and as a starting material to make other molecular chemicals. Furthermore, ethyl alcohol is used as a fuel and when mixed with gasoline it is used as gasohol, an alternate automotive fuel. Industrial alcohol is made by the reaction of ethylene with water:

$$C_2H_4 + H_2O \longrightarrow C_2H_5OH$$

Some alcohol for industrial and fuel use is made by the fermentation of cornstarch by using enzymes that decompose the starch to glucose which then ferments using yeast enzymes. Methods are currently being researched that will allow the production of ethyl alcohol from cellulose containing wastes from agricultural or waste paper. These methods involve the decomposition of plant cellulose to various simple sugars and the fermentation of these sugars using genetically engineered enzymes.

$$\text{cellulose} \xrightarrow{\substack{\text{cellulase} \\ \text{enzyme or} \\ \text{sulfuric acid}}} \text{simple sugars} \xrightarrow{\substack{\text{recombinant} \\ \text{yeast or} \\ \text{bacteria}}} \text{ethanol}$$

Since such ethyl alcohol comes from biomass wastes, it is viewed as a renewable fuel and may be an important addition to fuels of the future.

9-10 POLYATOMIC IONS

By definition, a **polyatomic ion** is a group of two or more covalently bonded atoms that carries an electrical charge. As you know, polyatomic ions occur in ionic compounds found in nature and some manufactured chemicals. Some common polyatomic ions are found dissolved in natural waters and cellular fluids along with simple ions. Table 6-1 and the table inside the cover of this book list common polyatomic ions.

Although polyatomic ions contain several combined atoms, they are components of ionic compounds just like simple ions. Negative polyatomic ions are found in combination with simple positive ions or with positive polyatomic ions, such as the ammonium ion. The formula of a compound containing polyatomic ions reflects a balance of positive and negative charges. When you write the formula of a polyatomic ion, the charge is part of the formula. The sulfate ion is not SO_4 but SO_4^{2-}. These charges are important when we want to know the formula of a compound containing ions. As you know, however, the charges are not written as part of the compound formula.

EXAMPLE 9-3

What is the likely formula of the compound containing potassium ions and carbonate ions?

Potassium is a group 1 metal, so it forms the K^+ ion. The carbonate ion is CO_3^{2-}. It requires two K^+ to balance its charge.

$$K^+ \quad CO_3^{2-} \quad \text{gives the formula } K_2CO_3$$

EXAMPLE 9-4

What is the predicted formula of the compound of aluminum ions and sulfate ions?

Aluminum is a group 13 metal that forms the Al^{3+} ion. The sulfate ion is SO_4^{2-}. To have the same amount of positive and negative charge, the compound must contain two aluminum cations (charge 6+) balanced with three sulfate anions (charge 6−).

$$Al^{3+} \quad SO_4^{2-} \quad \text{gives the formula } Al_2(SO_4)_3$$

Parentheses are used to show that the formula has three SO_4^{2-} units.

ACTIVITY 9.4

Give the formulas of the compounds made up of the following sets of ions.

(a) ammonium ion and phosphate ion

(b) calcium ion and nitrate ion

(c) sodium ion and oxalate ion

(d) zinc ion and acetate ion

9-11 MOLECULES AND IONS COMPARED

An ionic compound consists of a three-dimensional stack of ions held in place by the electrostatic attractive forces between the ions. The ionic bond is the force holding the ions in place. Positive and negative ions are the characteristic particles that compose ionic compounds, which are typically crystalline solids. Many ionic solids are water soluble. Because ionic compounds have saltlike crystalline forms, they are sometimes collectively called salts. In other words, the generic meaning of **salt** is an ionic compound. Compounds involving covalent bonds are quite different. Covalent compounds have molecules as fundamental particles. The covalent bond is the force holding atoms in molecules.

Appreciating the differences between ions and molecules is important. Ions carry charges, and ionic compounds consist of positive ions balanced with negative ions. Positive and negative ions always exist in proximity. Having a sample of a pure negative ion or a pure positive ion is not possible. Ions are found as components of ionic compounds. Since many ionic compounds are soluble in water, however, ions are also found dissolved in natural waters, such as ocean water, drinking water, and body fluids. Ocean water is salty due to dissolved ionic compounds. Body electrolytes refer to various ions dissolved in the fluids of our bodies. Ions can exist in solution, and such solutions contain as many negative charges as positive charges.

The **formula unit of an ionic compound** is the simplest combination of ions that satisfies the charges. **Ionic compounds** normally occur as pieces of crystalline solid and sometimes as beautifully formed crystals, but they occur more often as irregular chunks or even finely divided or powdered solids. Some are found concentrated in nature as mineral deposits. Other ionic compounds are manufactured by industrial chemical processes.

A molecule is a unique chemical species or particle that is electrically neutral. All molecules contain two or more atoms in close contact held in combination by covalent bonds. A **molecular compound** is a collection of molecules each containing atoms of elements shown in the formula (e.g., H_2O, CH_4, and $C_{12}H_{22}O_{11}$). In a sample of a pure covalent compound, all the molecules are alike, and the molecule is a representative particle of the compound. Molecules are stable particles, and it requires energy to break the covalent bonds within molecules.

Covalent or molecular compounds occur as gases, liquids, and solids. Generally, those molecular compounds with relatively low molar mass and molecules with few atoms are gases at room temperature. For instance, carbon dioxide, CO_2, and chlorine, Cl_2, are gases. Covalent compounds with more complex molecules or higher molar masses are liquids. Bromine, Br_2, and carbon tetrachloride, CCl_4, are examples of molecular liquids. Those molecular compounds with relatively high molar masses or many atoms in their molecules are usually solids. Iodine, I_2, and sucrose, $C_{12}H_{22}O_{11}$, are solids.

Figure 9-9 shows that when we have a sample of an ionic compound, it is imagined to contain cations and anions. If a compound is molecular, it

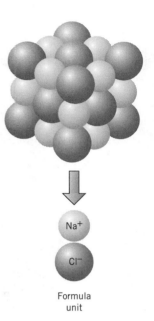

Formula unit

Ionic Compound

A compound that has a formula starting with a metal.

Molecular Compound

A compound containing only nonmetals or metalloids and nonmetals.

ACTIVITY 9-5

Classify the following compounds as ionic or molecular. Explain why in each case. KBr, CH_4, FeO, H_2S, MgS, SiF_4, C_3H_8, $ZnCl_2$

Solid iodine I_2

Powdered (ground up crystals)
calcium oxide

CCl_4

Liquid
carbon tetrachloride

$O = C = O$

CO_2

Gaseous
carbon dioxide

KNO_3

Crystalline
potassium nitrate

FIGURE 9-9
Visualizations of some ionic
and molecular compounds.

is imagined to contain molecules. This concept is an important part of a chemical vision. How do we know whether a compound is ionic or molecular? Generally, we can use the guidelines shown in the margin on page 263, although there are exceptions.

9-12 BONDING ABILITIES OF NONMETALS

Molecular compounds normally involve nonmetals bonded to nonmetals. Recall the locations of nonmetals in the periodic table. How many covalent bonds do the atoms of the various nonmetals commonly form? A hydrogen atom, with only one electron to share, forms one covalent bond. The other nonmetals typically form the number of covalent bonds corresponding to single or unpaired electrons in the dot symbol (see Table 9-4). Some examples are

$$-\overset{|}{\underset{|}{C}}- \qquad -\overset{..}{N}- \qquad :\overset{..}{O}- \qquad :\overset{..}{\underset{..}{F}}-$$

4 bonds   3 bonds   2 bonds   1 bond

Table 9-4 *Typical Numbers of Covalent Bonds Formed by Nonmetals*
(In certain compounds some elements can form more or less bonds than
the typical number.)

GROUP					
IIIA 13	IVA 14	VA 15	VIA 16	VIIA 17	IA 1
$-\overset{|}{B}-$	$-\overset{|}{\underset{|}{C}}-$	$-\overset{..}{N}-$	$:\overset{..}{O}-$	$:\overset{..}{\underset{..}{F}}:$	$H-$
3 bonds	4 bonds	3 bonds	2 bonds	1 bond	1 bond
	$-\overset{|}{\underset{|}{Si}}-$	$-\overset{..}{P}-$	$:\overset{..}{S}-$	$:\overset{..}{\underset{..}{Cl}}-$	
	4 bonds	3 bonds	2 bonds	1 bond	
			$:\overset{..}{Se}-$	$:\overset{..}{\underset{..}{Br}}-$	
			2 bonds	1 bond	
				$:\overset{..}{\underset{..}{I}}-$	
				1 bond	

Since hydrogen forms one bond and oxygen two, the compound water involves the combination of two hydrogen atoms and one oxygen atom.

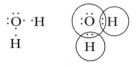

Notice that oxygen follows the octet rule by sharing two pairs of electrons and that each hydrogen has two shared electrons.

A molecule of water is sometimes represented by a **structural formula** such as

$$:\ddot{O}-H$$
$$|$$
$$H$$

where the lines represent the electron pairs of covalent bonds. Structural formulas show the sequences of covalently bonded atoms in molecules. More commonly, the **molecular formula** H_2O is used for water. Molecular formulas take up less room, but they do not show the bonding sequence. The formula H_2O suggests that water consists of molecules in which two atoms of hydrogen bond to one oxygen atom. To show how the atoms are bonded, we use a structural formula or a molecular model.

Molecular compounds may involve two, three, or more elements. **Binary molecular compounds** are those that contain only two elements. Many molecular compounds involve a few atoms bonded to form simple molecules. However, molecules can consist of several atoms, tens of atoms, hundreds of atoms, or thousands of covalently bonded atoms. Millions of molecular compounds are known. Learning chemistry, however, does not mean that we have to learn about millions of different compounds. The bonding patterns that apply to simple molecules also apply to complex molecules.

9-13 ELECTRON DOT OR LEWIS STRUCTURES OF MOLECULES

The formula of a molecule shows the number of atoms of each element but does not show the bonding arrangement of the atoms. To represent the bonding sequences in a molecule, the electron dot symbols of the elements are arranged so that the shared pairs are shown and the octet rule (or duet for hydrogen) is satisfied. For instance, a molecule of fluorine, F_2, is shown as

Dots

$$:\ddot{F}:\ddot{F}: \quad \text{or} \quad :\ddot{F}-\ddot{F}:$$

and a molecule of hydrogen fluoride, HF, is shown as

$$H \!:\! \ddot{\underset{..}{F}} \!:\! \quad \text{or} \quad H \!-\! \ddot{\underset{..}{F}} \!:\!$$

Arrangements of dot symbols used to represent molecules are called **Lewis structures** in honor of Gilbert Lewis, who first suggested their use. These Lewis structures do not convey any information concerning the shapes of the electron clouds nor do they suggest that electrons are dots. They merely serve as convenient representations of molecules. Usually, the **shared pairs** of electrons are represented by lines between atoms. Any unshared electron pairs are shown as dot pairs. These **unshared pairs** are also called **lone pairs** or **nonbonding pairs** since they are not directly involved in covalent bonds.

Lewis structures are written by fitting the element dot symbols together to show shared electron pairs and to satisfy the octet rule. Some examples are water, ammonia, and carbon tetrachloride.

1. Water, H_2O, two H· and one $:\!\ddot{\underset{.}{O}}\!\cdot$, fit together as

$$:\!\ddot{\underset{..}{O}} \!-\! H$$
$$|$$
$$H$$

2. Ammonia, NH_3, three H· and one $\cdot\ddot{\underset{.}{N}}\!\cdot$, fit together, with the three hydrogens bonded to the nitrogen, as

$$\overset{..}{H \!-\! N \!-\! H}$$
$$|$$
$$H$$

3. Carbon tetrachloride, CCl_4, four $:\!\ddot{\underset{.}{Cl}}\!\cdot$ and one $\cdot\dot{\underset{.}{C}}\!\cdot$, fit together, with the four chlorines bonded to the carbon, as

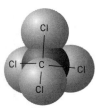

$$:\!\ddot{\underset{..}{Cl}}\!:$$
$$|$$
$$:\!\ddot{\underset{..}{Cl}} \!-\! C \!-\! \ddot{\underset{..}{Cl}}\!:$$
$$|$$
$$:\!\ddot{\underset{..}{Cl}}\!:$$

When writing dot structures for molecules, first try to fit the dot symbols together and try to satisfy the octet rule. Remember hydrogen forms one bond, oxygen forms two bonds, and carbon forms four bonds. If the octet rule cannot be satisfied, another approach is used as discussed in the next section. Keep in mind, however, that there are exceptions to the octet rule. For some molecules, the dot structures do not satisfy the rule. In other cases, writing a simple dot structure for a molecule is not possible. These exceptions are not important to us. We'll consider molecules containing elements that follow the octet rule.

EXAMPLE 9-5

What is the Lewis structure of hydrogen fluoride, HF?

Fitting a hydrogen, H•, and fluorine, $\cdot\ddot{\text{F}}\!:$, gives the Lewis structure.

$$\text{H}\!-\!\ddot{\text{F}}\!:$$

EXAMPLE 9-6

What is the Lewis structure of sulfur dichloride, SCl_2?

Using the dot symbol of sulfur, $:\!\dot{\ddot{\text{S}}}\cdot$, with two chlorines, $\cdot\ddot{\text{C}}\text{l}\!:$, gives a possible structure of

$$:\!\text{S}\!-\!\ddot{\text{C}}\text{l}\!:$$
$$|$$
$$:\!\ddot{\text{C}}\text{l}\!:$$

We can check the structure by noting that each atom has an octet of electrons when the shared pairs are counted for both atoms sharing.

EXAMPLE 9-7

Draw the Lewis structure for iodoform, CHI_3.

This compound contains one carbon, $\cdot\dot{\text{C}}\cdot$, one hydrogen, H•, and three iodines, $\cdot\ddot{\text{I}}\!:$. As noted in Table 9-4, carbon commonly forms four bonds and hydrogen and iodine each commonly form one bond. We can merge the dot symbols using the carbon in the center.

It does not matter which position we use for the hydrogen. We could just as easily draw the Lewis structure as

Write Lewis structures for the following molecules. PBr_3 , H_2S, CF_4, OF_2

9-14 MULTIPLE BONDS

When we attempt to write the Lewis structures of some molecules and try to obey the octet rule, we find that moving electron pairs around is necessary so that more than one pair is shared between atoms. For instance, the electron dot symbol for nitrogen is :N̈·. Attempting to write the Lewis structure for molecular nitrogen, N_2, by just merging two dot symbols gives

$$:\dot{N}—\dot{N}:$$

This structure provides for only six electrons around each nitrogen. Remember that the shared pair counts for both atoms. If we move one electron from each atom to make another shared pair, we have

$$:\dot{N}=\dot{N}:$$

This structure gives only seven electrons around each nitrogen. But if we move one more electron from each atom to make a third shared pair, we have a Lewis structure in which each nitrogen has an octet.

$$:N{\equiv}N: :N{\equiv}N:$$

Note that each nitrogen has an octet when the three shared pairs are counted for each nitrogen. Five valence electrons from each nitrogen give a total of 10 outer-energy-level or valence electrons in a diatomic nitrogen. To satisfy the octet rule, the two combined nitrogen atoms must share three pairs of electrons.

As another example, when we try to write the Lewis structure of carbon dioxide, CO_2, by simply connecting dot symbols, we obtain the incorrect structure of

$$:\dot{O}—\dot{C}—\dot{O}:$$

We achieve a satisfactory structure, however, by having each oxygen share two pairs of electrons with the carbon.

$$:\ddot{O}{=}C{=}\ddot{O}:   :O{=}C{=}O:$$

There are six valence electrons from each oxygen and four from the carbon, for a total of 16 valence electrons. To satisfy the octet rule for all three atoms, two pairs of electrons are placed between the carbon and each oxygen.

The sharing of more than one pair of electrons by atoms is entirely possible. This is called **multiple bonding.** The sharing of two pairs of electrons between two atoms is a **double bond,** and the sharing of three pairs of electrons between two atoms is a **triple bond.**

9-15 WRITING LEWIS STRUCTURES

To predict a possible Lewis structure for a molecule, we need to know which atoms are bonded in sequence. Which atoms bond to one another is sometimes obvious, but we often need more information besides the formula to establish the bonding sequence. Next, we write the dot symbols of each atom and bond the atoms in the sequence by single covalent bonds. Finally, we move any extra electrons about to form multiple bonds needed to satisfy the octet rule and the bonding tendencies of the atoms. There are four steps involved in this pattern.

1. Select the appropriate dot symbols for each element. If you are not sure, consult a periodic table to see in which groups the elements are found.

2. Fit the dot symbols together according to the bonding sequence so that there is one single bond between the bonded atoms.

3. If the octet rule is not satisfied for any atoms, move electrons about to make multiple bonds as needed. Hydrogen is satisfied with two electrons and always forms a single bond.

4. For some molecules writing simple Lewis structures that obey the octet rule is not possible.

EXAMPLE 9-8

What is the Lewis structure of formaldehyde, H_2CO, in which the carbon bonds to the hydrogens and the oxygen?

The bonding sequence gives this arrangement of dot symbols:

Forming single covalent bonds gives the structure

$$
\begin{array}{c}
\text{H} \\
| \\
\cdot\text{C}-\ddot{\text{O}}\!: \\
| \\
\text{H}
\end{array}
$$

The unpaired electrons of carbon and oxygen are moved so that carbon and oxygen are double bonded. This gives a structure in which the carbon and oxygen have octets of electrons.

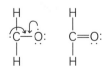

ACTIVITY 9-7

What is the Lewis structure of hydrogen cyanide, HCN, in which a hydrogen bonds to carbon that, in turn, is bonded to nitrogen? What is the Lewis structure of carbon disulfide, CS_2, in which two sulfur atoms are bonded to one carbon atom?

EXAMPLE 9-9

What is the Lewis structure of acetylene, C_2H_2, in which a hydrogen is bonded to each carbon and the carbons are bonded to one another?

The dot symbols are written in the bonding sequence so that each atom is bonded to the next by single covalent bonds, which gives

$$H{-}\dot{C}{-}\dot{C}{-}H$$

To have an octet around each carbon, the unpaired electrons are moved to form a triple bond between the carbons

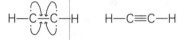

9-16 LEWIS STRUCTURES OF POLYATOMIC IONS

Since polyatomic ions are covalently bonded groups of atoms, writing Lewis structures for them is often possible. The same rules for Lewis structures of molecules are used for writing Lewis structures of polyatomic ions except that the charge of the ion must be considered. The charge of a polyatomic ion changes the way we count the valence electrons. A negatively charged polyatomic ion has excess valence electrons corresponding to the charge (i.e., a 1− ion has one extra valence electron and a 2− ion has two extra valence electrons). Consider, for example, the chlorate ion, ClO_3^-, in which the chlorine has three oxygen atoms bonded to it.

$$
\begin{array}{c}
O \\
O \; Cl \; O
\end{array}
$$

First, put in the valence electrons according to the dot symbols of the elements.

Since the charge is 1−, add one more valence electron, usually on the central atom, which is chlorine in this case.

added electron

Next, move the electrons about to make at least one single covalent bond to each oxygen. When writing Lewis structures, any of the electrons of the dot symbols are moved about as needed to satisfy the octet rule.

$$
\begin{array}{c}
:\!\overset{..}{\underset{..}{O}}\!: \quad - \\[4pt]
:\!\overset{..}{O}\!:\overset{..}{\underset{..}{Cl}}\!:\overset{..}{\underset{..}{O}}\!: \quad \text{gives} \quad
\end{array}
\qquad
\begin{array}{c}
:\!\overset{..}{\underset{}{O}}\!: \quad - \\[2pt]
| \\
:\!\overset{}{\underset{..}{O}}\!\!-\!\!\overset{}{\underset{}{Cl}}\!\!-\!\!\overset{}{\underset{..}{O}}\!:
\end{array}
$$

Note that since an extra electron was added, the group of atoms has a charge (one more electron than protons) and the octet rule is satisfied for each atom. Show the charge as part of the Lewis structure to emphasize that the species is an ion, not a molecule.

EXAMPLE 9-10

Give a Lewis structure for the ammonium ion, NH_4^+, in which the nitrogen is bonded to four hydrogens.

First, write each atom in the bonding sequence as its dot symbol.

$$
\begin{array}{c}
H \\
\cdot \\
H \cdot \cdot \overset{}{N} \cdot \cdot H \\
\cdot \\
H
\end{array}
$$

When we bond each atom of hydrogen to the nitrogen with a single bond, we notice that there is one too many electrons. Since the charge of the ion is $1+$, we remove one valence electron to give the Lewis structure.

$$
\begin{array}{cc}
H & \quad + \\
| & \\
H\!-\!N\!-\!H & \\
| & \\
H &
\end{array}
$$

This positive ion has 8 valence electrons. Nitrogen (group 15) has 5 valence electrons and the four hydrogens contribute one each. Subtract one from the total since the ion has a $1+$ charge $(5 + 4 - 1 = 8)$.

EXAMPLE 9-11

Give a Lewis structure for the cyanide ion, CN^-.

First, write the dot symbols of the elements and add one extra electron for the $1-$ charge.

$$
\cdot\overset{}{\underset{.}{C}}\cdot\cdot\overset{}{\underset{.}{N}}: \qquad :\overset{}{\underset{.}{C}}\cdot\cdot\overset{}{\underset{.}{N}}:
$$

$\qquad\qquad\qquad$ └── added electron

Next, try a single bond:

$$:\ddot{C}—\ddot{N}:^-$$

This bond does not satisfy the octet rule, so we try a double bond.

$$:\ddot{C}=\ddot{N}:^-$$

In this case, we need one more bond to satisfy the octet rule, which gives the Lewis structure

$$:C\equiv N:^-$$

ACTIVITY 9-8

What is the likely arrangement of four pairs of electrons about a central atom? You can answer this question yourself by using four small balloons of the same size. Inflate each balloon to the same volume and tie the stems. Tie or twist the necks of all the balloons together and note that they arrange as far apart as possible. Instead of balloons, you could use four straight pins and a small eraser removed from the end of a pencil. Push the ends of the pins into the eraser and try to arrange them in a three-dimensional pattern so that each pin is at an equal distance from the others. In other words, try to find the arrangement in which the angle between any two pins is the same and the pins are as far apart as possible. Remember to work in three-dimensional space; you should not make a flat square shape.

9-17 SHAPES OF MOLECULES

Unlike ionic bonds, covalent bonds are directional. Covalent bonds occur between atoms, and when atoms have more than one single bond, the bonds point in different directions in three-dimensional space. As a result, molecules have three-dimensional shapes, which makes it possible to show the spatial shapes of molecules using models. A molecule is visualized as a group of atoms bonded through the merging of their electron orbitals. The orbitals have specific spatial orientations. Thus, a water molecule has two hydrogen atoms bonded to an oxygen atom in an arrangement that gives a specific shape to the molecule. We can describe the shape of a molecule by imagining straight lines connecting the centers of the bonded atoms. These lines define a skeletal structure for a molecule that allows a description of the molecular shape. In a water molecule, the hydrogen atoms bond to the oxygen atom so that the skeletal structure is angular or bent. Thus, the shape of a water molecule is angular, bent, or V-shaped.

The Lewis structure of a molecule shows the number of electrons distributed about each atom. Within a given molecule in which the octet rule is satisfied, a specific atom occupies a somewhat central position. This atom is the atom to which other atoms are bonded; we call it the **central atom.** For example, in methane, the carbon is the central atom; around it are four pairs of electrons involved in bonding. If we figure out how the electron pairs arrange in space, we can visualize the shape of the molecule. What is the likely arrangement of four pairs of electrons about the central atom? The **valence shell electron pair repulsion theory (VSEPR theory)** gives an answer. The basis of this theory is that electron pairs repel each other and are arranged about the central atom so that they are as far apart as possible.

The likely arrangement of four pairs of electrons around a central atom is based on the idea that the pairs arrange as far from each other as possible, which maximizes the space the orbitals can occupy and ensures they do not crowd each other. The arrangement of four electron pairs is **tetrahedral,** with the pairs directed toward the corners of a **tetrahedron,** as shown in Table 9-5. In a tetrahedral distribution, the angle between two pairs of electrons is about 109.5°. This angle is the internal tetrahedral

VSEPR

Table 9-5   *Common Molecular Shapes*

Formula (A is central atom)	Electron dot structure	Pair Arrangement	Shape
AX$_4$ Four bonding pairs No lone pairs			Tetrahedral
AX$_3$ Three bonding pairs One lone pair			Triangular pyramid
AX$_2$ Two bonding pairs Two lone pairs			Angular

angle. A **tetrahedral distribution** is represented in two dimensions by sketches such as

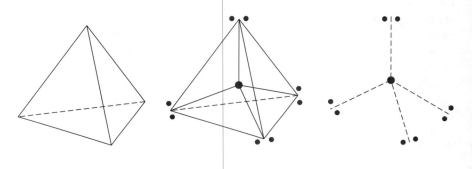

Tetrahedral Molecules

If we know that four pairs of electrons have a tetrahedral distribution, we can predict the likely shape of many molecules, as shown in Table 9-5. Accordingly, we predict that a molecule of methane has a tetrahedral shape that is sketched as

$$
\begin{array}{c}
H \\
| \\
C \\
H \diagup \ \diagdown \ H \\
H
\end{array}
$$

Triangular Pyramid Molecules

The Lewis structure for ammonia, NH_3, is

$$
H\!-\!\overset{\displaystyle\cdot\cdot}{N}\!-\!H
$$
$$
\begin{array}{c}
| \\
H
\end{array}
$$

The central nitrogen has four pairs of valence electrons, which arrange tetrahedrally. The shape of ammonia, however, relates to the locations of the centers of the various atoms in the molecule. How do the four atoms in ammonia arrange in space? The tetrahedral arrangement of the valence shell electron pairs in ammonia results in the molecule having a **triangular pyramid** shape (see Table 9-5). A triangular pyramid refers to the nitrogen atom locating at the peak of a pyramid and the hydrogen atoms having a triangular arrangement at the base of the pyramid.

$$
\begin{array}{c}
\overset{\displaystyle\cdot\cdot}{N} \\
H \diagup \ \diagdown \\
H \ \ H
\end{array}
$$

Angular or Bent Molecules

The Lewis structure for water is

$$:\ddot{O}-H$$
$$|$$
$$H$$

The central oxygen has four valence electron pairs, which arrange tetrahedrally. The shape of water, however, depends on how the hydrogen atoms locate about the central oxygen. As shown in Table 9-5, the water molecule is expected to have an **angular shape**, also called a V-shape or bent shape. Within the molecule are two bonding pairs of electrons and two lone pairs of valence electrons. The shape is represented as

$$\overset{..}{O}$$
$$H \diagup \diagdown H$$

Linear Molecules

When a molecule consists of only two atoms with one bonding electron pair and three lone pairs around the "central" atom, the molecule is **linear** in shape. All diatomic molecules are linear. Linear means that all the atoms are arranged along a straight line. For example, a molecule of hydrogen fluoride, HF, is linear although the four pairs of electrons are arranged in a tetrahedral fashion around the fluorine.

$$H-\ddot{F}:$$

Some Other Shapes

The atoms involved in a double bond or a triple bond arrange in a linear sequence. Thus, a molecule of carbon dioxide with two double bonds is linear.

$$:\underset{..}{O}=C=\underset{..}{O}:$$

A molecule of acetylene with a triple bond is also linear.

$$H-C\equiv C-H$$

Consider a central atom that has one double bond and two single bonds. In such a case, the central atom and the three other atoms to which it is bonded locate in the same plane and arrange in a **triangular shape** around the center. Thus, in a molecule of formaldehyde, H_2CO, all the atoms are in the same plane and the molecule has a triangular planar shape. Triangular planar means that all the atoms lie in a plane with the outer atoms on the corners of a triangle forming angles of 120° apart.

$$\overset{..}{O}$$
$$120°\underset{H}{\diagup}\overset{||}{\underset{\diagdown H}{C}}$$

Predict the shapes and make drawings or sketches of molecules for the following compounds. $AsBr_3$, $GeCl_4$, OF_2, CS_2

A variation of this case occurs when a central atom has a single bond, a double bond, and one lone pair. For example, a Lewis dot structure for ozone is $:\ddot{O}=\ddot{O}-\ddot{O}:$. The electrons around the central atom arrange in a triangular fashion, which gives the molecule an angular shape.

Models   For more information and examples of VSEPR theory see Appendix 3.

Models in the Chemical Imagination

Atoms cannot be seen, but they can be imagined as the structural units of molecules. Models are very useful for visualizing the shapes and forms of molecules and have played an important role in the evolution of chemical thought. Models are used for simple molecules and complex molecules. John Dalton, who developed the atomic theory in the early 1800s, used a set of wooden balls to represent atoms and atomic combinations. The American chemist Linus Pauling used models to aid his thinking about the structures of molecules. He once told a story about how he discovered the structure of certain protein molecules while in bed with a cold. He drew on sheets of paper and folded the paper to help him envision the three-dimensional structures of complex molecules. In 1953, Francis Crick and James Watson constructed an elaborate ball-and-stick model to represent their theory of the structure of DNA. This model served as the basis of the now-famous double-helix structure of DNA molecules.

In 1985, American chemist Richard Smalley proposed a structure for the newly discovered form of carbon composed of molecules containing 60 carbon atoms. He used paper hexagons and pentagons to fit together a three-dimensional soccer-ball shape as a possible structure. This polygon had 60 vertices, each of which corresponded to the position of a carbon atom in the molecule. The structure had symmetrical integrity. This new type of molecule was given the chemical name of buckministerfullerene, in honor of the American inventor of the geodesic dome, Buckminister Fuller. Since these molecules have a hollow soccer-ball shape, they are informally called buckyballs. In 1996 Smalley, Robert Curl and Harold Kroto shared the Nobel Prize in Chemistry for their discovery of fullerenes.

In modern times, the structures of complex molecules are modeled on computers. Computer imagery allows for the testing of various possible structures. The structures can be rotated and investigated from various perspectives to give insight to the properties and behaviors of the molecules. For instance, a molecule with possible physiological effects is modeled on a computer. Other computer programs are used to predict possible pharmaceutical uses of the compound represented by the molecule.

FOCUS
9

Ions, Molecules, and Bonding

Octet rule: Atoms tend to lose, gain, or share electrons so that they attain a total of eight outer-energy-level electrons.

Ion: A chemical particle that carries an electrical charge.

Cation: An ion having a positive charge.

Anion: An ion having a negative charge.

Simple ion: An atom of an element that has an electrical charge resulting from the loss or gain of electrons. They are also known as monatomic ions.

Ionic bond: The electrostatic force of attraction between ions that holds them in combination.

Ionic compounds: Compounds containing cations and anions. Typical ionic compounds contain metals in combination with nonmetals and have formulas that start with a metal. Formulas of ionic compounds reflect the balance of positive and negative charges.

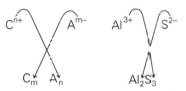

Covalent bond: The force of attraction between atoms resulting from the sharing of an electron pair.

Molecule: A chemical particle composed of two or more covalently bonded atoms.

Molecular compound: A compound that consists of molecules. Typical molecular compounds contain only nonmetals or metalloids and nonmetals.

Polyatomic ion: A charged particle composed of two or more covalently bonded atoms. Many ionic compounds contain polyatomic anions in combination with metallic cations.

Molecules and polyatomic ions are represented by Lewis dot structures that show which atoms are bonded. The pattern for writing a simple Lewis dot structure is as follows:

1. Select the appropriate dot symbols for each element.

2. Fit the dot symbols together according to the bonding sequence so that there is one single bond between each of the bonded atoms. Use lines to represent single covalent bonds and pairs of dots to represent electron pairs not involved in covalent bonds.

3. If the octet rule is not satisfied for an atom, move electrons about to make double bonds or triple bonds as needed. Hydrogen does not obey the octet rule. It forms one single bond.

4. For some molecules, it is not possible to write simple Lewis structures that obey the octet rule.

The shapes of simple molecules and polyatomic ions can be predicted by using their Lewis dot structures. Identify the central atom or atoms in the structure and count the number of bonding pairs (BP) of electrons and nonbonding or lone pairs (LP) of electrons around the central atom. Typical shapes are 4 BP = tetrahedral,

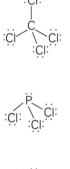

3 BP and 1 LP = triangular pyramid,

2 BP and 2 LP = bent, V-shape or angular,

1 double bond and 2 single bonds = triangular planar,

H
 \\
  C=Ö
 /
H

1 double bond, 1 single bond, 1 LP = angular,

2 double bonds = linear,

:Ö=C=Ö:

and 1 single bond and 1 triple bond = linear,

H—C≡N:

QUESTIONS

1. Match the following terms with the appropriate lettered definitions.

chemical bond _b_

valence electrons_ _)_

octet rule _m_

ionic bond _k_

covalent bond _l_

a. the sharing of one pair of electrons between atoms

b. a particle containing two or more covalently bonded atoms

c. a particle containing two or more covalently bonded atoms that carries a positive or negative charge

d. a simple ion

e. the sharing of two pairs of electrons between atoms

monatomic ion *d*   f.   the sharing of three pairs of electrons between atoms

molecule *k*   g.   electron pairs around the central atom in a molecule or polyatomic ion are located as far apart as possible around the center

polyatomic ion *g c*   h.   representation of a molecule or polyatomic ion showing bonding and nonbonding electron pairs

Lewis structure *h*   i.   the force of attraction that holds atoms in combination

single bond *A*   j.   the electrostatic force of attraction between oppositely charged ions

double bond *e*   k.   the force of attraction between two atoms resulting from the sharing of a pair of electrons

triple bond *f*   l.   outer-level electrons

VSEPR theory *m*   m.   the tendency of atoms of elements to lose, gain or share electrons so as to attain eight outer-level electrons

Section 9-1

2. What is the nature of the force that holds atoms in combination?

Section 9-2

3. What are valence electrons? What is an electron dot symbol? Give an example.

4. Give the electron dot symbols for these elements (refer to a periodic table).

   (a) Be (b) Ne (c) He (d) S (e) Al
   (f) N (g) Te (h) Rb (i) I (j) Xe (k) C

5. Select two elements from the list in Question 4 that you suspect might have similar chemical behavior based on their electron dot symbols.

Section 9-3

6. Describe how metal atoms bond in metallic solids.

Section 9-4

7. Give a statement of the octet rule.

Section 9-5

8. What is an anion? What is a cation?

9. What is the difference between an atom and an ion?

10. A negative ion is larger than the atom from which it is formed. Why? A positive ion is smaller than the atom from which it is formed. Why?

Section 9-6

11. *Give the formulas of the simple cations formed by the following elements (refer to a periodic table for group numbers).

   (a) H (b) K (c) Be (d) Sr (e) Ga
   (f) Fr (g) Ca (h) Rb

12. Give the formulas of the simple cations formed by the following elements (refer to a periodic table for group numbers).

   (a) Li (b) Mg (c) Na (d) Cs (e) Ba
   (f) Al (g) Zn

13. *Give the formulas of the simple anions formed by the following elements (refer to a periodic table for group numbers).

   (a) S (b) I (c) P (d) Br (e) Se (f) As

14. Give the formulas of the simple anions formed by the following elements (refer to a periodic table for group numbers).

   (a) F (b) Te (c) O (d) N (e) Cl

15. Describe the kinds of simple ions formed by elements of the following groups.

   (a) group 1 (b) group 2 (c) group 13
   (d) group 15 (e) group 16 (f) group 17

Section 9-7

16. Describe the ionic bond. How does it differ from a covalent bond?

17. What physical form do ionic compounds usually have? Why?

18. Give definitions or descriptions of the following.

   (a) ionic compound   (b) crystal lattice
   (c) electrostatic force

Section 9-8

19. *Give the formulas of the ionic compounds containing the following ions.

(a) Na^+ and S^{2-} (b) Li^+ and Br^-

(c) Ca^{2+} and O^{2-} (d) Mg^{2+} and F^-

(e) Al^{3+} and O^{2-} (f) Al^{3+} and Br^-

20. Give the formulas of the ionic compounds containing the following ions.

(a) K^+ and S^{2-} (b) Na^+ and F^-

(c) Ba^{2+} and Cl^- (d) Be^{2+} and O^{2-}

(e) Al^{3+} and P^{3-} (f) Al^{3+} and S^{2-}

21. *Give the formulas of the ionic compounds formed between the following metals and nonmetals. Use the expected positive ions for the metals and the expected negative ions for the nonmetals (refer to a periodic table for group numbers).

(a) lithium and chlorine

(b) magnesium and nitrogen

(c) barium and chlorine

(d) strontium and oxygen

(e) sodium and sulfur

(f) aluminum and sulfur

(g) calcium and arsenic

(h) sodium and phosphorus

(i) potassium and oxygen

22. Give the formulas of the ionic compounds formed between the following metals and nonmetals. Use the expected positive ions for the metals and the expected negative ions for the nonmetals (refer to a periodic table for group numbers).

(a) aluminum and bromine

(b) magnesium and phosphorus

(c) calcium and sulfur

(d) strontium and fluorine

(e) sodium and nitrogen

(f) gallium and oxygen

(g) barium and arsenic

(h) lithium and nitrogen

(i) potassium and iodine

Section 9-9

23. Describe the covalent bond. How does it differ from an ionic bond?

24. Which elements tend to form covalent bonds and molecules? Which elements tend to form binary ionic compounds?

25. Give definitions or descriptions of the following.

(a) molecule

(b) shared electron pair

(c) molecular compound

26. Explain the sharing of electrons between two atoms of hydrogen in terms of electron orbitals.

Section 9-10

27. What is a polyatomic ion?

28. Give the names of the following anions.

(a) $C_2H_3O_2^-$ (b) NO_2^- (c) Br^- (d) CO_3^{2-}

(e) CrO_4^{2-} (f) HSO_4^- (g) I^- (h) OH^-

(i) $Cr_2O_7^{2-}$ (j) SO_3^{2-}

29. Give the formulas of the following polyatomic ions. Where they are commonly found is shown in parentheses.

(a) sulfate ion (ammonium sulfate fertilizer)

(b) hydroxide ion (sodium hydroxide)

(c) bicarbonate ion (baking soda)

(d) nitrate ion (saltpeter)

(e) oxalate ion (spinach and rhubarb)

(f) nitrite ion (sodium nitrite, used to preserve processed meats)

(g) chlorate ion (fireworks)

(h) phosphate ion (cola beverages and nondairy coffee creamers)

(i) dichromate ion (safety matches)

(j) cyanide ion (sodium cyanide, a violent poison)

(k) permanganate ion (some antifungal medicines)

30. The formula for the ionic compound calcium phosphate is $Ca_3(PO_4)_2$. Explain why the parentheses are needed in this formula.

31. *Give the formulas for the ionic compounds formed between the following pairs of ions.

(a) K^+ and SO_3^{2-} (b) Al^{3+} and NO_2^-

(c) Mg^{2+} and PO_4^{3-} (d) K^+ and $C_2H_3O_2^-$

(e) Ca^{2+} and HCO_3^- (f) Ag^+ and I^-

(g) Hg^{2+} and Br^- (h) NH_4^+ and SO_4^{2-}

32. Give the formulas for the ionic compounds formed between the following pairs of ions.

    (a) Na^+ and SO_4^{2-} (b) Al^{3+} and SO_4^{2-}

    (c) Ba^{2+} and NO_3^- (d) K^+ and ClO_3^-

    (e) Na^+ and $C_2O_4^{2-}$ (f) Cu^+ and O^{2-}

    (g) Zn^{2+} and Br^- (h) NH_4^+ and PO_4^{3-}

Section 9-11

33. The following are among the most common chemicals manufactured in the United States. Indicate which of them are ionic and which are molecular.

    (a) sulfuric acid, H_2SO_4

    (b) ammonia, NH_3

    (c) lime, CaO

    (d) oxygen, O_2

    (e) nitrogen, N_2

    (f) ethylene, C_2H_4

    (g) sodium hydroxide, $NaOH$

    (h) chlorine, Cl_2

    (i) phosphoric acid, H_3PO_4

    (j) ammonium nitrate, NH_4NO_3

    (k) sodium carbonate, Na_2CO_3

    (l) urea, CH_4N_2O

    (m) propylene, C_3H_6

    (n) toluene, C_7H_8

    (o) benzene, C_6H_6

    (p) ethylene dichloride, $C_2H_4Cl_2$

    (q) ethyl benzene, C_8H_{10}

    (r) carbon dioxide, CO_2

    (s) methanol, CH_3OH

    (t) nitric acid, HNO_3

34. For all the ionic compounds listed in Question 33, write the formulas with charges of the individual ions that make up the compounds.

Section 9-12

35. How does the bonding tendency of a nonmetal correspond to its group position in the periodic table?

Section 9-13

36. What does it mean to say that an element is diatomic?

37. List the names and formulas of those elements that occur in the form of diatomic molecules when they are not combined with other elements.

Sections 9-14 to 9-17

38. *Give Lewis structures for the following.

    (a) chlorine, Cl_2

    (b) hydrogen iodide, HI

    (c) chloroform, $CHCl_3$ (1 H and 3 Cl bonded to a central C)

    (d) water, H_2O

    (e) hydrogen telluride, H_2Te

    (f) dichlorine oxide, Cl_2O (2 Cl bonded to a central O)

    (g) perchloric acid, $HClO_4$ (4 O bonded to Cl; H bonded to one of the O)

    (h) hydrazine, N_2H_4 (2 H bonded to each N, which are bonded to one another)

    (i) phosphorus trichloride, PCl_3

    (j) boron trifluoride, BF_3 (exception to octet rule)

    (k) ethane, C_2H_6 (3 H bonded to each C, which are bonded to one another)

    (l) cyclobutane, C_4H_8 (4 C bonded in a square; 2 H bonded to each C)

    (m) carbon tetrachloride, CCl_4

    (n) hydrogen peroxide, H_2O_2 (1 H bonded to each O, which are bonded to one another)

    (o) hydroxide ion, OH^-

    (p) ammonium ion, NH_4^+

    (q) phosphate ion, PO_4^{3-}

    (r) nitrate ion, NO_3^-

39. Give Lewis structures for the following.

    (a) hydrogen selenide, H_2Se

    (b) dibromine oxide, Br_2O (2 Br bonded to a central O)

    (c) iodine, I_2

    (d) hydrogen chloride, HCl

    (e) methyl bromide, CH_3Br

    (f) silicon tetrachloride, $SiCl_4$

    (g) nitrogen trifluoride, NF_3

    (h) stibine, SbH_3

    (i) chloric acid, $HClO_3$ (3 O bonded to Cl; H bonded to one of the O)

(j) hydroxyl amine, NH_2OH (2 H and O bonded to N; 1 H bonded to O)

(k) chloroethane, C_2H_5Cl (2 C bonded; 3 H bonded to one C; 2 H and 1 Cl bonded to the other C)

(l) cyclopropane, C_3H_6 (3 C bonded in a triangle; 2 H bonded to each C)

(m) carbon tetrafluoride, CF_4

(n) methyl alcohol, CH_3OH (3 H and 1 O bonded to C; 1 H bonded to O)

(o) permanganate ion, MnO_4^- Use $:\overset{..}{\underset{.}{Mn}}\cdot$ as dot symbol.

(p) chlorate ion, ClO_3^-

(q) peroxide ion, O_2^{2-}

(r) hydronium ion, H_3O^+

40. Define a multiple covalent bond. Give an example of a molecule that has such a bond.

41. What is the difference between a double bond and a triple bond?

42. *Give the Lewis structures for the following molecules; all contain multiple bonds.

(a) phosgene, a nerve gas, $COCl_2$ (2 Cl and 1 O bonded to a central C)

(b) acetylene, a fuel, C_2H_2 (1 H bonded to each C, which are bonded to one another)

(c) carbon dioxide, used in cold storage of produce, CO_2 (2 O bonded to C)

(d) hydrogen cyanide, a poison, HCN (H and N bonded to a central C)

(e) urea, a product of protein metabolism, CN_2H_4O (1 O and 2 N bonded to a central C; 2 H bonded to each N)

(f) acetone, a solvent, C_3H_6O (2 C and 1 O bonded to a central C; 3 H bonded to each of the outer C's)

(g) tetrabromoethene, C_2Br_4 (2 Br bonded to each C)

(h) vinyl chloride, C_2H_3Cl (2 H bonded to one C; 1 H and 1 Cl bonded to the other C)

43. Give the Lewis structures for the following molecules; all contain multiple bonds.

(a) silicon dioxide, SiO_2 (2 O bonded to Si)

(b) cyanogen, C_2N_2 (1 N bonded to each C, which are bonded to one another)

(c) nitrous acid, HNO_2 (2 O bonded to N; H bonded to an O)

(d) trichloroethylene, C_2HCl_3 (2 C bonded and 2 Cl bonded to one C; 1 H and 1 Cl bonded to the other C)

(e) dichloroethyne, C_2Cl_2 (1 Cl bonded to each C, which are bonded to one another)

(f) allene, C_3H_4 (3 C bonded; 2 H bonded to each of the end C's)

(g) propyne, C_3H_4 (3 C bonded; 3 H bonded to one end C and 1 H bonded to the other end C)

(h) formaldehyde, CH_2O (2 H and 1 O bonded to C)

44. Write Lewis structures for (b), (e), (f), (h), (r), and (s) in Question 33. In methanol, 3 H and 1 O are bonded to C; 1 H is bonded to O.

Section 9-18

45. *Describe the shapes and give Lewis structures that suggest the shapes of the following (the central atoms are underlined).

(a) $\underline{Si}Cl_4$, silicon tetrachloride

(b) $\underline{P}Cl_3$, phosphorus trichloride

(c) $H_2\underline{S}$, hydrogen sulfide

(d) $\underline{O}F_2$, oxygen difluoride

(e) $\underline{C}H_3Br$, methyl bromide

(f) $Br_2\underline{O}$, dibromine oxide

(g) $\underline{C}H_2F_2$, difluoromethane

(h) $\underline{N}H_3$, ammonia

(i) $H\underline{I}$, hydrogen iodide

(j) H_2O_2, hydrogen peroxide (both oxygens are viewed as central atoms)

(k) C_2H_6, ethane (both carbons are viewed as central atoms)

(l) $\underline{N}O_2^-$, nitrite ion

46. Describe the shapes and give Lewis structures that suggest the shapes of the following (the central atoms are underlined).

(a) $\underline{C}Cl_2F_2$, dichlorodifluoromethane

(b) $\underline{N}F_3$, nitrogen trifluoride

(c) $Cl_2\underline{O}$, dichlorine oxide

(d) $\underline{C}F_4$, carbon tetrafluoride

(e) $H_2\underline{Se}$, hydrogen selenide

(f) $\underline{Si}H_4$, silane

(g) $\underline{As}H_3$, arsine

(h) $H\underline{F}$, hydrogen fluoride

(i) $\underline{Cl}O_2^-$, chlorite ion

(j) NH_4^+, ammonium ion

(k) H_3O^+, hydronium ion

(l) PO_4^{3-}, phosphate ion

47. *Describe the shapes of the molecules or ions in parts (b), (c), (d), (e), (f), (i), (j), (m), (o), (p), and (r) of Question 38.

48. Describe the shapes of the molecules or ions in parts (a), (b), (e), (f), (g), (h), (o), (p), (q), and (r) of Question 39.

Questions to Ponder

49. Draw a picture to show how the half-filled $2p$ electron orbitals in an oxygen atom and the half-filled $1s$ electron orbitals in two hydrogen atoms overlap to form a water molecule.

50. Use some creative materials to make a model of a tetrahedron.

GASES AND
GAS LAWS

10-1 PROPERTIES OF GASES

When riding a bicycle you notice that the rear tire needs some air. Sitting on the bicycle increases the pressure on the tire and it becomes flatter as its volume decreases. You add air with a pump and you notice that the volume increases until the tire becomes rigid, then the pressure increases as you add more air. When riding the bicycle on a hot day, you notice that as the tires heat up, they increase in volume and become more rigid, reflecting an increase in pressure. Your experiences with the bicycle are confirmations of the kinetic molecular theory of gases and the behavior of gases as described by the gas laws.

The curious behaviors of gases have interested and stimulated the imaginations of scientists for centuries. Gases are fascinating because we can experience them without seeing them. A gas occupies the entire volume of any container in which it is placed. To prevent a gas from escaping into the atmosphere, we have to keep it in a sealed container. From experience with gas-filled balloons, we know that a gas is very flexible and that the volume of a balloon changes when it is squeezed or heated. Furthermore, we know that a gas readily leaks from a balloon, even through very tiny holes.

Sometimes gases are called vapors. Actually, the terms *gas* and *vapor* have different meanings. The term *gas* refers to any substance in the gaseous state. A **vapor** is a gas that is easily changed to a liquid by a slight change in its pressure or temperature. Air is a mixture of gases that includes some water vapor. The water vapor in the air can easily condense to form liquid water, which happens when dew forms. It is not easy, however, to liquefy the nitrogen and oxygen gases in air. It takes a combination of very low temperature and high pressure to cause these components of air to liquefy. Nevertheless, liquefaction of air is done on an industrial scale and is the major source of industrial nitrogen and oxygen.

285

Some compounds occur as gases at typical earth temperatures and pressures. The noble gases are minor components of the atmosphere; they exist as monatomic gases. All other gases occur as collections of molecules; hydrogen, nitrogen, oxygen, fluorine, and chlorine gases are collections of diatomic molecules. A sample of hydrogen gas, for example, contains H_2 molecules; a sample of chlorine gas contains Cl_2 molecules. Some other common gases are carbon dioxide, CO_2; ammonia, NH_3; nitrogen dioxide, NO_2; and the hydrocarbons: methane CH_4; ethane, C_2H_6; propane, C_3H_8; and butane, C_4H_{10}. Molecular models of nitrogen dioxide, ethane, and butane are shown here.

NO$_2$ C$_2$H$_6$ or CH$_3$CH$_3$ C$_4$H$_{10}$ or
                                                CH$_3$CH$_2$CH$_2$CH$_3$

10-2 KINETIC MOLECULAR THEORY

KMT

The theory used to explain the properties and dynamic behavior of gases is known as the kinetic molecular theory. The term *molecular* refers to the important chemical vision that all matter is composed of minute particles much too small to be seen even with microscopes. The characteristic particles of gases are molecules. The term *kinetic* refers to motion or movement. A moving object has energy of motion or **kinetic energy.** A moving ball, for instance, has kinetic energy, and this energy is exchanged with objects that contact the ball. This is why a catcher's mitt gets warm when a ball hits it. Simply stated, the kinetic molecular theory provides a picture of a gas as a collection of molecules in motion. The **kinetic molecular theory** is stated as follows.

1. A gas is a collection of molecules that are on average very far apart.

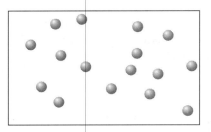

2. The molecules of a gas are in rapid, random, straight-line motion and are constantly colliding with one another and any object in their envi-

ronment. The molecules of a gas travel at various speeds, and the motion of the molecules gives them kinetic energy.

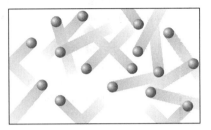

3. The attractive forces between molecules are negligible, so they do not stick together but instead rebound upon collision. Molecules exchange kinetic energies when they collide.

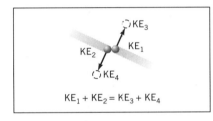

This theory provides a dynamic view of a gas as a collection of molecules rapidly darting about and constantly colliding with one another and other objects. The random motion of the molecules causes them to "fill" any container. The molecules of a gas sample are somewhat like minute balls on a three-dimensional pool table. A given molecule moves randomly and rapidly about until it collides with another molecule or the walls of the container. The molecules travel at various speeds, and the motion of molecules gives them kinetic energy. Thus, some molecules have relatively high kinetic energies and some have relatively low kinetic energies. Molecules exchange kinetic energy when they collide. Most molecules have similar kinetic energies, which fall in the range of the average or typical kinetic energy of the collection of molecules. As an example of rapid motion, consider that, in air, a typical oxygen molecule travels at about 1000 mi/hr. At such a speed, it does not take long for a given molecule to collide with another molecule or some object. A typical molecule averages about four billion collisions each second. Therefore, the molecules travel relatively short distances between collisions. Upon collision, they rebound, like pool balls, and continue their "cosmic dance." The difference between molecules and pool balls is that the molecules keep moving and do not roll to a stop.

The kinetic molecular theory provides images of gases that allow us to understand their behaviors. We envision a gas sample as a collection of minute particles darting about at high speeds; it is this motion that contributes to the interesting properties of a gas. This picture allows us to imagine the particles of a gas even if we cannot see them. Most gases are colorless, so it is interesting that some are colored. Chlorine gas, for example, has a pale greenish yellow color that fills the sample space. We can imagine that the molecules of chlorine gas have a greenish yellow color and that they "fill" the entire space of the sample by their continuous motion. The result is a gas sample we can see through but that has a distinct color. Another common colored gas is nitrogen dioxide, a reddish brown gas that is one component of smog. This gas gives a reddish brown color to smog as it "fills" a region of the atmosphere.

A theory allows us to understand observations and behaviors, and it might also suggest some novel applications. In addition, a theory allows us to understand natural phenomena of environmental interest. As an air mass or pocket of air near Earth's surface is warmed, it behaves somewhat like a flexible container and increases in volume. The increase in volume makes the pocket less dense than the air above it, so it moves up and mixes with the cooler air. This process is how hot air rises. If we heat air in a lightweight balloon, the air expands and becomes less dense than the surrounding cooler air. The warmer air rises and takes the balloon with it. Hot-air balloons work nicely, but it is a good idea to take an air heater along on the ride. Similarly, a fire heats the air around it; the hot air rises and carries the smoke with it. For this reason, smoke detectors are placed on ceilings or high on walls. Furthermore, during a structural fire, the cooler, less smoky air is usually near the floor.

10-3 PRESSURE

If we capture a gas in a container and keep the container closed, it remains as a gas indefinitely. As long as the sample container is not opened or the gas does not change chemically, the amount of gas in terms of the mass or number of moles does not change. We can describe a gas sample by stating the volume of the container and the temperature. Since a gas occupies the entire container, the volume of the container is the volume of the gas. A gas sample also exerts a pressure on the container. A pressure-measuring device attached to the container registers a specific pressure.

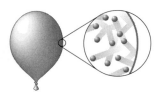

Even though you do not feel it, you are constantly being bombarded by the molecules of the air, as are all objects exposed to the atmosphere. The pressure exerted by a gas comes from this constant bombardment. To illustrate the idea of pressure, place the tip of a pen or pencil on a magazine or notebook and push down to make an indentation. Now jab the pen or pencil repeatedly into the paper. The repeated jabbing applies a pressure to the paper. This pressure is similar to what happens when gas molecules collide with objects. The continuous collisions of gas molecules with objects in their environment cause the pressure. The molecules continually "jab" or "push" on the surface of objects. The pressure exerted by a gas depends on how many collisions are occurring and how fast the molecules are moving.

By definition, **pressure** is force per unit area. According to the kinetic molecular theory, the pressure exerted on objects by a gas comes from the molecules of the gas constantly colliding with surfaces. A continuous series of collisions exert a specific force per unit area on any object.

10-4 MEASUREMENT OF GAS PRESSURES

Sometimes when you dive deeply in water, you can feel the pressure on your eardrums. In deep water, you feel the pressure of the layer of water above. The atmosphere is the layer of air contained about Earth by gravitational forces. The molecules and atoms present in the atmosphere exert pressures on all objects in the atmosphere; this pressure is the **air pressure** or **atmospheric pressure**. It arises from the continuous bombardment of objects by molecules and atoms in the air. Since gravitational forces decrease with altitude, the atmosphere is more dense near the lower surfaces of Earth and becomes less dense at distances removed from Earth's surface. Atmospheric pressure decreases with the altitude because the air is less dense at higher altitudes and fewer molecules are available to exert pressure.

Atmospheric pressure is measured with a device known as a barometer, invented by Evangelista Torricelli, an Italian scientist, in the 1600s. A **barometer** is constructed by first filling a glass tube, one end of which is sealed, with liquid mercury. This tube is then inverted and supported in a container of mercury that is open to the atmosphere (see Figure 10-1). The pressure exerted on the surface of the mercury by the atmosphere supports a column of mercury in the tube. The tube must be long enough so that some space remains at the top of the tube. The atmospheric pressure supports the mercury column, and the space at the top of the column is essentially a vacuum. The pressure results from the continuous bombardment of the surface of the mercury by air molecules. The height of the mercury column is directly proportional to the atmospheric pressure. As the atmospheric pressure changes due to changes in temperature and weather conditions, the height of the mercury column fluctuates accordingly. A falling barometer—that is, a decreasing barometer reading—usually foretells an approaching storm.

10-5 UNITS FOR PRESSURE

The height of a column of mercury supported by a gas is directly proportional to the pressure of the gas. Consequently, gas pressures are often expressed in terms of the height of a mercury column as measured with a

Explain why a decrease in barometric pressure (a falling barometer) is often associated with a change in the weather from fair to foul.

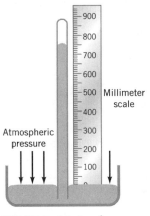

FIGURE 10-1 A mercury barometer.

barometer. This height is usually expressed in **millimeters of mercury** or **mm Hg.** An alternate unit for pressure is the torr, named in honor of Torricelli. One **torr** is defined as 1 mm of mercury. In U.S. weather reports, atmospheric pressures are usually given in inches of mercury. Since 1 in. = 25.4 mm, it then follows that 1 in. Hg = 25.4 torr.

Another unit of pressure is the **atmosphere, atm,** defined as the pressure equal to 760 torr. The term *atmosphere* comes from the atmospheric pressure around sea level being typically about 760 torr. The atmosphere as a unit of pressure is often used to express pressures common to industrial and experimental work. For instance, the process used to make synthetic diamonds includes subjecting graphite to temperatures of about 2000°C and pressures around 70,000 atm.

Torr and atmosphere units are used to express values of air pressure. They are also used to express the pressure of any sample of gas in a closed container. The relation between the torr and the atmosphere is written as a conversion factor.

$$\left(\frac{760 \text{ torr}}{1 \text{ atm}}\right)$$

1 torr = 1 mm Hg

1 atm = 760 torr

$$\frac{1 \text{ atm}}{760 \text{ torr}}$$

EXAMPLE 10-1

The measured pressure in the eye of a hurricane was 669 torr. What is this pressure in atmospheres?

Use the factor relating torr and atm as a conversion factor.

$$669 \text{ torr}\left(\frac{1 \text{ atm}}{760 \text{ torr}}\right) = 0.880 \text{ atm}$$

The SI unit of force is the newton, N. Gas pressures are sometimes expressed in terms of newtons per square meter (N/m^2), which is a force per unit area. The SI unit for pressure is the pascal (Pa). One pascal is equivalent to 1 N/m^2. The pascal represents a relatively small unit of pressure. One torr is 133 Pa. Consequently, typical pressures in SI units are expressed in units of kilopascals, kPa. Another SI pressure unit is the bar. One bar is defined as 10^5 Pa. Thus, the relation between the bar and the atmosphere is 1.01 bar = 1 atm.

Gas pressures in chemistry are typically measured and expressed in torr and atmospheres and, occasionally, in units of kilopascals or bars. You may be familiar with barometric pressures expressed in inches of mercury (1 atm = 29.92 in. Hg) and gas pressures expressed in pounds per inch (1 atm = 14.696 psi) used by engineers, but these units are seldom used in

ACTIVITY 10-2

Use the values 1 bar = 10^5 Pa, 133 Pa = 1 torr, and 760 torr = 1 atm to calculate the number of bar per atm.

ACTIVITY 10-3

One day in Death Valley, California, the atmospheric pressure was 772 torr. What is this pressure in atmospheres? What is this pressure in inches of mercury?

chemistry. Incidently, about 34 ft of water is equivalent to 1 atm. So when you scuba dive in water to 34 ft, you are subjected to a pressure of about 1 atm over the prevailing atmospheric pressure.

10-6 OBSERVING THE BEHAVIOR OF GASES

Because gases represented an illusive form of matter, they fascinated scientists for centuries. Using your insight into the kinetic molecular theory, reread the portion of the Lucretian poem given in Section 3-1. Most gases are not visible, but they do have properties that can be observed and measured. A gas sample has measurable volume, pressure, temperature, and mass. The mass of a gas sample, of course, relates to number of moles of chemical particles that make up the sample. If a gas sample is composed of only one chemical, the molar mass of that chemical is used to find the number of moles of particles from the mass of the sample. Pressure and temperature are intensive properties since they have the same value any place in a sample and do not depend on the sample size. In contrast, volume is an extensive property that depends on the sample size as it relates to the temperature and pressure.

Historically, the behaviors of gases were observed by noting how the various properties of volume, V, pressure, P, temperature, T, and number of moles, n, relate. These relations revealed consistent patterns of gas behavior and came to be called gas laws. Recall that a scientific law is a statement of some consistent behavior or pattern in nature.

Boyle's Law

In 1662, Robert Boyle, an English scientist, experimented with the compressibility of air. Boyle, using an apparatus similar to that shown in Figure 10-2, measured the relation between the volume and the pressure of a sample of gas. He observed that if a sample of gas is kept at a constant temperature, the volume of the gas decreases with increasing pressure. Boyle observed that if the pressure of the gas doubles, the gas compresses to occupy one-half its original volume; if the pressure triples, the gas compresses to occupy one-third its original volume. Boyle also noted that relieving the pressure increases the volume of the gas sample. For instance, if the pressure decreases by one-half, the gas expands to occupy twice its original volume. Today, after many confirmations of Boyle's work and additional experiments, a generalization concerning the relation between the volume and the pressure of a gas is known as Boyle's law, as shown in the margin.

This law relates to simple observations of gas samples in flexible containers. Squeezing a balloon filled with a gas increases the pressure and decreases the volume. Releasing the balloon decreases the pressure and

Boyle's Law

The volume of a gas sample at constant temperature is inversely proportional to the pressure.

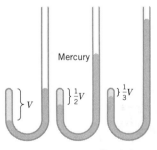

FIGURE 10-2 Boyle's law apparatus.

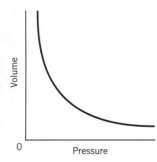

Volume

0   Pressure

$V = k/P$ where k depends upon the mass and temperature of the gas

FIGURE 10-3 Boyle's law plot.

increases the volume. In a chamber with a movable piston a gas sample can be compressed or expanded by moving the piston. Figure 10-3 is a graph that shows that as the pressure of a gas sample increases, the volume decreases proportionally; as the pressure decreases, the volume increases. Two quantities are **inversely proportional** when one increases as the other decreases or vice versa. The volume and pressure of a gas are inversely proportional.

Symbolically, the inverse proportion between volume and pressure is shown as

$$V \propto \frac{1}{P}$$

where $\propto$ is read as "proportional to." Thus, this expression reads that the volume is proportional to the inverse of the pressure.

Charles' Law

Charles' Law

The volume of a gas sample at a constant pressure is directly proportional to the temperature.

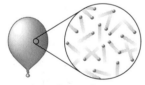

Lower temperature

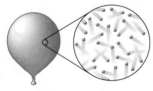

Higher temperature

In 1887, Jacques Charles carefully measured the relation between the volume and temperature of a sample of a gas at constant pressure. When the pressure is constant, the volume is affected only by the temperature change. Charles observed that when the temperature of such a gas sample increases, the volume increases; when the temperature decreases, the volume decreases. Specifically, he noted a direct linear relation between the volume and temperature. His observation has become known as Charles' law, as stated in the margin.

A **direct proportion** between two quantities means that when one is increased the other increases and as one is decreased the other decreases. A way to remember Charles' law is to recall that a filled balloon increases in volume when warmed and decreases in volume when cooled. Symbolically, the direct proportionality between volume and temperature is expressed as

$$V \propto T$$

The Pressure–Temperature Law

Pressure–Temperature Law

The pressure of a gas sample at constant volume is directly proportional to the temperature.

When you use a pressure cooker, the inside pressure increases as the temperature increases. Experimental observations of gas samples in rigid, inflexible containers reveal that pressure varies directly with the temperature. When the temperature of a gas sample of fixed volume increases, the pressure increases; when the temperature decreases, the pressure decreases. This behavior is stated in the pressure–temperature law, as defined in the margin. The pressure-temperature law is expressed symbolically as

$$P \propto T$$

10-7 RELATING *P, V, T,* and *n* for Gases

We can picture a gas sample using the kinetic molecular theory. The theory describes an idealized gas, referred to as an ideal gas, that behaves according to the theory. The theory states that there are no significant attractive forces between the molecules of a gas. Actually, weak attractive forces do exist between molecules. This situation makes real gases deviate from ideal behavior, but most real gases behave as though they are ideal gases within certain ranges of temperature and pressure. Thus, for our purposes it is reasonable to treat real gases as ideal and expect them to obey gas laws.

The properties of pressure, volume, and temperature of a sample of gas relate to the dynamic nature of the gas. These properties are not independent and interrelate as expressed by the gas laws. Boyle's law states that the volume of a gas sample is inversely proportional to the pressure.

$$V \propto \frac{1}{P}$$

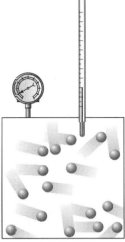

Lower temperature

According to Charles' law, the volume of a gas sample is directly proportional to its Kelvin temperature.

$$V \propto T$$

Another relevant proportionality is how the volume of a gas sample relates to the number of moles of gas. The number of moles, n, is an expression of the number of particles in a sample. Imagine a balloon filled with a gas at a constant temperature and pressure. If we blow air into the balloon, we increase the number of moles of gas and the volume increases. Allowing some of the gas to escape decreases the number of moles. The result is a decrease in volume. Generally, as expected, the volume of a gas at constant temperature and pressure is directly proportional to the number of moles of gas.

$$V \propto n$$

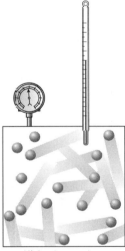

Higher temperature

It is possible to combine the three proportions into a general form that relates volume to any of the other three factors.

$$V \propto \frac{nT}{P}$$

This combined relation shows how the volume is directly proportional to the number of moles and the temperature and is inversely proportional to the pressure. Another way to express this relation is to rearrange it so that pressure and volume are on one side and the number of moles and temperature are on the other side; to do this, we multiply both sides by the pressure, *P,* giving the proportionality

$$PV \propto nT$$

ACTIVITY 10-4

Under what conditions of temperature and pressure do you think gases deviate the most from ideal behavior? Why?

This relation states that the product of the pressure and volume of a gas sample is directly proportional to the product of the number of moles and temperature of the sample. The next section shows how this proportion is written as an algebraic equation using a proportionality constant.

 THE IDEAL GAS LAW

The proportionality between the pressure, volume, number of moles, and temperature of an ideal gas written in an algebraic form using a proportionality constant is

$$PV = knT$$

This relation can also be expressed with the proportionality constant on one side and P, V, n, and T on the other side.

$$k = \frac{PV}{nT}$$

In this form, the relation reveals that for any ideal gas the product of the pressure and volume divided by the product of the number of moles and temperature is always constant. This situation is true for any ideal gas and any set of conditions. A value for the proportionality constant is found by using a sample containing a known number of moles of gas and measuring its pressure, volume, and temperature. The proportionality constant is known as the universal gas constant or ideal gas constant and is normally written as R. Using the symbol R in the proportion gives the expression shown in the margin. By tradition the constant is placed between n and T.

$$PV = nRT$$

This important expression relates the pressure, volume, temperature, and number of moles of a gas and is called the **ideal gas law.** R is a universal constant, and the same numerical value of R applies to any ideal gas. We can use the ideal gas law and the constant value of R with any gas as long as we assume it is ideal.

The ideal gas law is very useful. Since R is a known constant, if we know any three of the properties associated with a gas sample, the fourth can easily be calculated. For example, if we measure the pressure, volume, and temperature of a sample of a gas, the ideal gas law is used to find the number of moles of gas in the sample. In other words, we can count the molecules in a sample of a gas by just measuring the volume, pressure, and temperature; $n = PV/RT$.

Typically, volume is measured in liters or milliliters and pressure in atmospheres or torrs. When working with the gas laws, gas temperatures must be expressed on the Kelvin scale. In the laboratory, the temperature of a gas sample may be measured in Celsius degrees. To be useful

$PV = nRT$

in the gas laws, a Celsius temperature needs to be converted to the Kelvin scale. This conversion is done by adding 273: $T\,K = T\,°C + 273$. We need to use Kelvin temperatures since the Kelvin scale has a starting point of 0 K, the lowest possible temperature. In contrast, the Celsius scale uses an arbitrary zero point, so a Celsius temperature cannot be used in gas law calculations. Never use a Celsius temperature directly in gas law calculations.

10-9 THE GAS CONSTANT, *R*

The numerical value and units of the gas constant, R, depend on the units used to measure the pressure, volume, and temperature of the gas sample. The temperature used in the ideal gas law is always the Kelvin temperature. The volume of a sample of a gas is usually expressed in liters and sometimes in milliliters. Its pressure is usually expressed in atmospheres and sometimes in torr or kilopascals.

The value of R is found by experiment, measuring the pressure, volume, temperature, and number of moles for gas samples. Suppose we find that 1.54 mol of gas occupy 37.9 L at 298 K and 0.994 atm. To calculate R, we first solve $PV = nRT$ for R by dividing both sides by n and T.

$$R = \frac{VP}{Tn}$$

The values $V = 37.9$ L, $P = 0.994$ atm, $T = 298$ K, and $n = 1.54$ mol are substituted into the expression to find R.

$$R = \frac{(37.9\ \text{L})(0.994\ \text{atm})}{(298\ \text{K})(1.54\ \text{mol})} = \left(\frac{0.0821\ \text{L atm}}{\text{K mol}}\right)$$

$$R = \frac{0.0821\ \text{L atm}}{\text{K mol}}$$

The same value of R is found using any set of related P, V, n, and T values for an ideal gas. R serves as a known constant in all ideal gas law computations. Different units for the volume or pressure give different numerical values for the gas constant, R. The value of R given above is convenient for most calculations. When using R in calculations, be sure to include the units.

Various Values of *R*

0.0821 L atm/K mol
82.1 mL atm/K mol
62.4 L torr/K mol
62,400 mL torr/K mol
8.31×10^3 L Pa/K mol
8.31 kPa/K mol

10-10 CALCULATIONS WITH THE IDEAL GAS LAW

The ideal gas law is used in calculations for a specific gas sample having a constant pressure, volume, temperature, and number of moles. Calculations with the ideal gas law usually involve finding P, V, T, or n for a gas. To calculate any one of these, we need values for the other three. The gas law is rearranged algebraically to solve for any one of the four terms P, V, n, or T.

EXAMPLE 10-2

A weather balloon contains 4.73 moles of helium gas. What volume does the gas occupy at an altitude of 4300 m if the temperature is 0°C and the pressure is 0.595 atm?

First, change the temperature to the Kelvin scale (0°C + 273 = 273 K) and tabulate the data.

$$
\begin{array}{cccc}
P & V & n & T \\
0.595 \text{ atm} & ? & 4.73 \text{ mol} & 273 \text{ K}
\end{array}
$$

The ideal gas law is solved for volume by dividing both sides by P.

$$\frac{PV}{P} = \frac{nRT}{P} \quad \text{which gives } V = \frac{nRT}{P}$$

Find the volume using the data and the constant (0.0821 L atm/K mol) in the expression.

$$V = \frac{nRT}{P} = \left(\frac{4.73 \text{ mol}}{0.595 \text{ atm}}\right)\left(\frac{0.0821 \text{ L atm}}{\text{K mol}}\right) 273 \text{ K} = 178 \text{ L}$$

Notice how the units of R cancel the other units, giving liters as the final unit.

EXAMPLE 10-3

A used aerosol can contains 0.0104 mole of gas and has a volume of 255 mL. Calculate the pressure of the gas in the can if the can is accidentally heated to 400°C. **DANGER: Never heat an aerosol can! It may explode.**

First, tabulate the data.

$$
\begin{array}{cccc}
P & V & n & T \\
? & 255 \text{ mL} & 0.0104 \text{ mol} & 673 \text{ K } (400 + 273)
\end{array}
$$

To use the volume in the ideal gas law, we must change the original volume to units of liters to match the units used in R.

$$255 \text{ mL} \left(\frac{1 \text{ L}}{1000 \text{ mL}}\right) = 0.255 \text{ L}$$

Next, we solve the ideal gas law for pressure by dividing both sides by V.

$$\frac{PV}{V} = \frac{nRT}{V} \quad \text{which gives } P = \frac{nRT}{V}$$

We use the data and the constant (0.0821 L atm/K mol) in this relation to calculate the pressure.

$$P = \frac{nRT}{V} = \left(\frac{0.0104 \text{ mol}}{0.255 \text{ L}}\right)\left(\frac{0.0821 \text{ L atm}}{\text{K mol}}\right) 673 \text{ K} = 2.25 \text{ atm}$$

EXAMPLE 10-4

A tire with an interior volume of 3.60 L contains 0.367 mole of air at a pressure of 2.49 atm. What is the temperature of the air in the tire in kelvins?

First, tabulate the data.

P	V	n	T
2.49 atm	3.60 L	0.367 mol	?

Next, solve the ideal gas law for temperature by dividing both sides by nR.

$$\frac{PV}{nR} = \frac{nRT}{nR} \quad \text{which gives } T = \frac{PV}{nR}$$

Use the relation and the constant (0.0821 L atm/K mol) to calculate the temperature.

$$T = \frac{PV}{nR} = \left(\frac{2.49 \,\text{atm}}{0.367 \,\text{mol}}\right)\left(\frac{3.60 \,\cancel{L}}{0.0821 \,\text{L atm}, \text{K mol}}\right) = 298 \text{ K}$$

Note that when the units of R cancel other units, the result is

$$\frac{1}{1/K} \quad \text{or} \quad \frac{1}{\frac{1}{K}} = 1\frac{K}{1} = K$$

To divide 1 by 1/K, we invert the fraction and multiply, which gives the unit of K.

ACTIVITY 10-5

How many moles of gas are in 1.00 ft³ of natural gas at 273 K and 1.00 atm?

10-11 THE COMBINED GAS LAW

The ideal gas law is quite useful for a gas sample with fixed pressure, volume, and temperature. Let's consider how it can be used to deal with a gas sample in which the pressure, volume, and temperature are changed. Consider a gas at pressure P_1, volume V_1, and Kelvin temperature T_1. Using these factors in the ideal gas law gives

$$P_1 V_1 = nRT_1$$

Note that the number of moles and R are constant for such a sample. Rearranging so that the pressure, volume and temperature are set equal to the constants gives

$$\frac{P_1 V_1}{T_1} = nR$$

Suppose the gas sample is changed by heating, cooling, expanding, or compressing it such that it has pressure P_2, volume V_2, and temperature T_2. The number of moles of gas and R do not change and are constant. Thus,

$$\frac{P_2 V_2}{T_2} = nR$$

To represent how the pressure, volume, and temperature relate when the conditions of a gas sample change, we use the principle that things equal to the same thing are equal to one another. This fact results in an expression called the **combined gas law**, commonly written as

$$\frac{V_2 P_2}{T_2} = \frac{V_1 P_1}{T_1}$$

This expression is used to calculate a new volume, pressure, or temperature corresponding to specific changes in the other terms. In other words, if you know any five terms, it is possible to calculate the sixth term.

The new volume of a gas sample that results from a pressure change and a temperature change is found by multiplying the initial volume by both a pressure ratio and a temperature ratio.

$$V_2 = V_1 \frac{P_1}{P_2} \frac{T_2}{T_1}$$

where subscript 1 indicates an initial value and subscript 2 indicates a final value. Suppose we had a balloon of gas that occupies 2.0 L of volume at 20°C, typical room temperature. What volume does the balloon occupy at a temperature of 50°C if the pressure remains constant? When using temperatures in gas laws, they must always be expressed in kelvins rather than degrees Celsius. That's easily done since we just have to add 273 to each Celsius temperature:

$$20°C + 273 = 293 \text{ K} \quad \text{and} \quad 50°C + 273 = 323 \text{ K}$$

The information given in the question is summarized as

V_1	P_1	T_1		V_2	P_2	T_2
2.0 L	constant	293 K		?	constant	323 K

The new volume is found by multiplying the initial volume by the appropriate ratio of temperatures. Since the temperature increases (293 K to 323 K), the ratio of temperatures needs to be greater than 1 to give an increased volume:

$$V_2 = 2.0 \text{ L} \left(\frac{323 \text{ K}}{203 \text{ K}} \right) = 2.2 \text{ L}$$

The pressure is constant, so the volume changes with a change in temperature. We know that the volume and temperature are directly proportional. Since volume is directly proportional to the temperature, the temperature ratio should be greater than 1 for an increase in temperature and less than 1 for a decrease in temperature. In this example, the temperature increases so the volume must increase. The new volume is the initial volume multiplied by a ratio of the temperatures, which is greater than 1. Thus, we just use the ratios deduced from the proportions corresponding to the changes. Using this kind of reasoning we do not have to substitute into an algebraic equation. We get the same result, however, if we substitute the given values into the combined gas law equation.

As another example, suppose we have a 4.5 L sample of a gas at a pressure of 760 torr. What is the new volume when the pressure changes to 790 torr and the temperature remains constant? A summary of the data is

V_1	P_1	T_1		V_2	P_2	T_2
4.5 L	760 torr	constant		?	790 torr	constant

The gas sample has a constant temperature. The new volume is found by multiplying the initial volume by the appropriate ratio of pressures:

$$V_2 = 4.5 \text{ L} \left(\frac{760 \text{ torr}}{790 \text{ torr}} \right) = 4.3 \text{ L}$$

The pressure increases (760 torr to 790 torr), so the original volume decreases. To find the new volume, we multiply the original volume by a ratio of the pressures, which is less than 1. On the other hand, if the pressure decreases, the original volume increases. In such a case, find the new volume by multiplying the original volume by a ratio of the pressures, which is greater than 1. We can use any units of pressure as long as we express both pressures in the same unit. The units cancel in the ratio. Again, we get the same result if we substitute the given values into the combined gas law equation.

ACTIVITY 10-6

A 50.5 mL sample of a gas is in a syringe with a pressure gauge attached. Initially, the gauge shows a pressure of 1.00 atm. The plunger of the syringe is pushed so that the pressure reads 1.45 atm. What is the new volume of the gas if the temperature is constant?

EXAMPLE 10-5

A 325 mL sample of a gas in a balloon is maintained at 25°C and 750 torr pressure. The temperature changes to 20°C and the pressure changes to 760 torr. What is the new volume?

First, summarize the data:

V_1	P_1	T_1		V_2	P_2	T_2
325 mL	750 torr	298 K (25 + 273)		?	760 torr	293 K (20 + 273)

Both the temperature and pressure are changing. The new volume is found by multiplying the initial volume by both pressure and temperature ratios. A pressure increase (750 torr to 760 torr) gives a smaller volume, so the pressure ratio should be less than 1. A temperature decrease (298 K to 293 K) gives a smaller volume, so the temperature ratio should also be less than 1:

$$V_2 = 325 \text{ mL} \left(\frac{750 \text{ torr}}{760 \text{ torr}} \right) \left(\frac{293 \text{ K}}{298 \text{ K}} \right) = 315 \text{ mL}$$

The same result is obtained by substituting the data in the combined gas law.

EXAMPLE 10-6

A balloon occupies a volume of 625 mL at 25°C. It is placed in a freezer and the volume decreases to 552 mL. What is the temperature of the balloon in the freezer in degrees Celsius if the pressure remains constant?

Tabulate the data as

V_1	P_1	T_1		V_2	P_2	T_2
625 mL	constant	298 K (25 + 273)		552 mL	constant	?

Note that we want to find the final temperature in degrees Celsius. Nevertheless, we must use Kelvin temperatures in the calculations. The final Kelvin temperature is changed to give the final temperature in degrees Celsius. Since the volume decreases (625 mL to 552 mL), the temperature must decrease, so to find the final temperature, we multiply the initial temperature by a volume ratio of less than 1.

$$T_2 = 298 \text{ K} \left(\frac{552 \text{ mL}}{625 \text{ mL}} \right) = 263 \text{ K}$$

To find the temperature in Celsius, we need to subtract 273 K from this Kelvin temperature.

$$263 \text{ K} - 273 \text{ K} = -10°C$$

EXAMPLE 10-7

A 4.25 L sample of a gas is contained in a piston chamber at 264 K and 0.989 atm pressure. If the piston moves to compress the gas sample to 1.32 L and the temperature changes to 395 K, what is the new pressure of the gas sample?

First, tabulate the data.

V_1	P_1	T_1		V_2	P_2	T_2
4.25 L	0.989 atm	264 K		1.32 L	?	395 K

To find the new pressure, multiply the initial pressure by a volume ratio and a temperature ratio. A decrease in volume (4.25 L to 1.32 L) increases the pressure, so the volume ratio is greater than 1. An increase in temperature (264 K to 395 K) increases the pressure so the temperature ratio is greater than 1.

$$P_2 = 0.989 \text{ atm} \left(\frac{4.25 \text{ L}}{1.32 \text{ L}} \right) \left(\frac{395 \text{ K}}{264 \text{ K}} \right) = 4.76 \text{ atm}$$

The same result is obtained by substituting the data into the combined gas law.

ACTIVITY 10-7

A potato chip bag occupies a volume of 1.4 L at 760 torr and 20°C. If the bag is transported to the mountains, what volume does it occupy at 725 torr and 18°C?

10-12 STP AND MOLAR VOLUME

Suppose you read that the annual commercial production of nitrogen gas in the United States is 950 billion cubic feet or 27 trillion liters. What is wrong with this statement? We have learned that the volume of a gas sample depends on the pressure and temperature. So the statement has no meaning unless it is accompanied by some specific temperature and pressure conditions.

Specific values of temperature and pressure are defined as standard reference conditions for gases. These defined conditions are

0°C or 273 K   and   **760 torr or 1 atm**

These values are arbitrarily chosen as reference conditions of gases and are called **standard temperature and pressure, abbreviated STP.**

What volume does a gas occupy at STP? This question is not meaningful because, according to the ideal gas law, it is not possible to calculate the volume of a gas sample at a given temperature and pressure unless we know the number of moles of gas. A more meaningful question is, What volume does 1 mole of gas occupy at STP? This volume is easily calculated by substituting a P of 1.00 atm, a T of 273 K, an n of 1.00 mol, and R into the relation $V = nRT/P$.

$$V = \left(\frac{1.00 \text{ mol}}{1.00 \text{ atm}}\right)\left(\frac{0.0821 \text{ L atm}}{K \text{ mol}}\right) 273 K = 22.4 \text{ L}$$

Thus, 1 mole of an ideal gas occupies a volume of 22.4 L at STP, which is expressed as the factor

$$\left(\frac{22.4 \text{ L}}{1 \text{ mol}}\right)_{STP}$$

The subscript denotes the conditions (1 atm and 273 K) at which the relation is valid. This factor is called the **molar volume** of a gas and is valid only at STP. A volume of 22.4 L is equal to the volume of a cube measuring about 28 cm or 11 in. on an edge. The value of 22.4 L/mol applies only to ideal gases, not to liquids or solids. Use the molar volume as a simple conversion factor to calculate the number of moles of a gas in a given volume at STP or vice versa, but remember that it only applies to gases at STP.

STP

0°C or 273 K and
760 torr or 1.00 atm

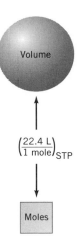

$$\left(\frac{22.4 \text{ L}}{1 \text{ mole}}\right)_{STP}$$

A C T I V I T Y

10-8

The molar mass of a gas has units of grams per mole. The molar volume of a gas has units of moles per liter. How can the molar mass and molar volume be used to calculate the density of a gas at STP? To figure this out, pay attention to units. Use your answer to calculate the density of methane gas, CH_4, at STP and the density of carbon dioxide gas, CO_2, at STP.

EXAMPLE 10-8

One cubic foot (ft³) of volume equals 28.3 L. How many moles of nitrogen are in 1.00 ft³ of nitrogen gas at STP?

Use the molar volume in the inverted form to convert the volume of a gas at STP to the number of moles.

$$X \text{ mol } N_2 = 28.3 \text{ L} \left(\frac{1 \text{ mol } N_2}{22.4 \text{ L}}\right) = 1.26 \text{ mol } N_2$$

10-13 DALTON'S LAW OF PARTIAL PRESSURES

So far, we have discussed samples of gases without being concerned with their chemical composition. Now we consider the relation that exists between the components of a mixture of gases. If a mixture of three gases is placed in a container of fixed volume, V, each of the gases is viewed as occupying the entire volume (see Figure 10-4). Likewise, all three gases must be at the same temperature, T.

The pressure exerted by a component of the mixture, if it alone were to occupy the container, is called the **partial pressure.** Since the particles of each component behave independently, the total pressure exerted by the mixture is a result of all the particles in the mixture. Consequently, the total pressure of the mixture, P_t, is simply the sum of the partial pressures of the components of the mixture. Expressed algebraically, the total pressure is

$$P_t = P_a + P_b + P_c + \cdots$$

where P_a, P_b, P_c, and so on refer to the partial pressures of the components of a mixture. This relation is **Dalton's law of partial pressures,** named for John Dalton of atomic theory fame. Dalton's law is quite convenient when we want to relate the partial pressures of the components of a mixture to the total pressure.

In the laboratory, samples of gases are often collected by displacement of water. As the gas forms, it is bubbled into an inverted container of water. The accumulating gas displaces the liquid water. As the gas bubbles through the water, however, it becomes saturated with gaseous water vapor. Thus, a gas collected by water displacement is actually a mixture of the gas and water vapor. The total pressure of the sample is the sum of the partial pressure of the gas and the partial pressure of the water vapor. According to Dalton's law, $P_t = P_g + P_{water}$, so to find the actual pressure of the gas we subtract the partial pressure of water from the total pressure of the mixture: $P_g = P_t - P_{water}$.

Liquid water evaporates to form water vapor. Any enclosed sample of liquid water with space above it contains water vapor. Since water vapor is a gas,

FIGURE 10-4 Dalton's law of partial pressures. Each gas exerts a pressure according to its amount. When the gases are in the same container, each contributes to the total pressure.

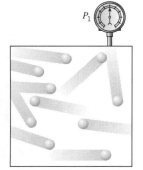
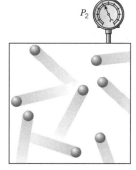
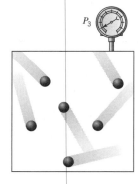
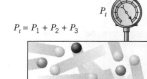

it exerts a pressure called the vapor pressure of water. At higher temperatures, water evaporates more readily, so the vapor pressure increases with an increase in temperature. The vapor pressure of water at various temperatures is measured using a container with a pressure-measuring device. Reference tables of the vapor pressures of water at various temperatures are readily available. Thus, to find the actual pressure of a gas collected by water displacement, we simply find the vapor pressure of water at the temperature of the sample and subtract it from the total pressure of the sample.

EXAMPLE 10-9

A sample of hydrogen gas is collected by water displacement. The container of gas collected has a volume of 645 mL at 25°C and a total pressure of 758 torr. How many moles of hydrogen are in the sample?

We know that we can use the ideal gas law to find the number of moles of gas from the measured volume, temperature, and pressure.

V	T	P_t	n
645 mL	25°C	758 torr	?

In this sample, however, water vapor contributes to the total pressure of the "wet" hydrogen. The vapor pressure of water at the measured temperature is the partial pressure of water in the sample. Since the gas sample is a mixture, the partial pressure of hydrogen is calculated using Dalton's law of partial pressures. The total pressure is the result of two components.

$$P_t = P_{H_2} + P_{H_2O}$$

A table of vapor pressures of water shows that the vapor pressure of water at 25°C is 24 torr.

$$P_t = 758 \text{ torr} \quad \text{and} \quad P_{H_2O} = 24 \text{ torr}$$

Thus,

$$P_{H_2} = P_t - P_{H_2O} = 758 \text{ torr} - 24 \text{ torr} = 734 \text{ torr}$$

The actual pressure of the hydrogen gas is used with the volume and temperature of the sample to find the number of moles of hydrogen gas in the sample. The units of the data need to be changed to make them compatible with the units of the gas constant R. We need pressure in atmospheres, volume in liters, and temperature in kelvin.

$$P = 734 \text{ torr} \left(\frac{1 \text{ atm}}{760 \text{ torr}}\right) = 0.9658 \text{ atm}, \quad V = 645 \text{ mL or } 0.645 \text{ L},$$

$$T = 25°C \text{ or } 298 \text{ K}$$

$$n = \frac{PV}{RT} = \left(\frac{0.9658 \text{ atm}}{0.0821 \text{ L atm}/K \text{ mol } H_2}\right)\left(\frac{0.645 \text{ L}}{298 \text{ K}}\right) = 0.0255 \text{ mol } H_2$$

Note that after canceling, the unit that results is mol H_2.

$$\frac{1}{1/\text{mol } H_2} \quad \text{or} \quad \frac{1}{\dfrac{1}{\text{mol } H_2}} = \text{mol } H_2$$

10-14 THE MOLAR MASS OF A GAS

Suppose we had a sample of a pure gas and wanted to know the **number of grams per mole** or the **molar mass** of the gas. Of course, if we already know the formula of the gas, we can easily find its molar mass. It is possible, however, to find the molar mass of a pure gas by a simple experiment without knowing its formula by measuring the pressure, volume, temperature, and mass of a sample of the gas. Furthermore, the molar mass of a volatile liquid is found by boiling a sample of the liquid and determining the volume, pressure, and temperature of a known mass of vapor (see Figure 10-5). The key to this process is that it is possible to find the number of moles of a gas from the ideal gas law by simply measuring the volume, pressure, and temperature of the sample.

To calculate the number of grams per mole or molar mass of a gas from its properties, we use the ideal gas law, $PV = nRT$. The number of moles of gas is found from the pressure, volume, and temperature using $n = PV/RT$. Then, the mass of the sample is divided by the number of moles to find the number of grams per 1 mole. The resulting ratio is the molar mass.

FIGURE 10-5
Experimental determination of the number of grams per mole of a liquid.

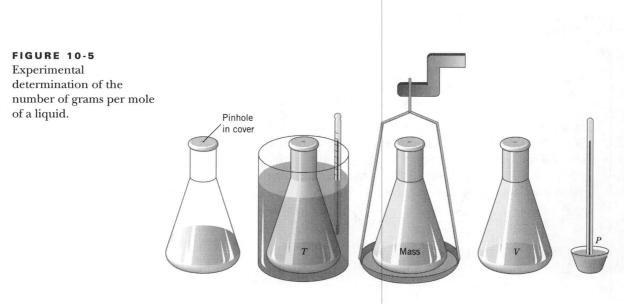

EXAMPLE 10-10

What is the molar mass of a 0.935 g sample of gas that occupies 0.515 L at 20°C and 776 torr?

First, tabulate the data with mass in grams, pressure in atmospheres, volume in liters, and temperature in kelvin.

$$\text{mass} \qquad\qquad P \qquad\qquad\qquad\qquad V \qquad\qquad T$$

$$0.935 \text{ g} \quad 766 \text{ torr}\left(\frac{1 \text{ atm}}{760 \text{ torr}}\right) = 1.021 \text{ atm} \quad 0.515 \text{ L} \quad 293 \text{ K } (20 + 273)$$

The number of moles in the sample is found from the ideal gas law.

$$n = \frac{PV}{RT} = \left(\frac{1.021 \text{ atm}}{0.0821 \text{ L atm/K mol}}\right)\left(\frac{0.515 \text{ L}}{293 \text{ K}}\right) = 0.02186 \text{ mol}$$

Carry an extra digit in the pressure and number of moles. Do not round until the final molar mass calculation. Finally, divide the mass by the number of moles, which gives the molar mass.

$$\left(\frac{0.935 \text{ g}}{0.02186 \text{ mol}}\right) = \left(\frac{42.8 \text{ g}}{1 \text{ mol}}\right) \quad \text{or} \quad 42.8 \text{ g/mol}$$

There is a general pattern for finding the molar mass of a gas from the mass, pressure, volume, and temperature of a sample. The molar mass is the ratio of the sample mass and the number of moles,

$$\text{molar mass} = \frac{\text{mass}}{n}$$

and the number of moles from the gas law is $n = PV/RT$. Substituting PV/RT for n gives a general expression for calculating the molar mass of a gas.

$$\text{molar mass} = \frac{\text{mass}}{PV/RT} \quad \text{or} \quad \frac{\text{mass } RT}{PV}$$

ACTIVITY 10-9

Use the general expression for the molar mass of a gas to calculate the molar mass of the gas sample in Example 10-10. Use the data given in the example.

10-15 STOICHIOMETRY AND THE GAS LAWS

Amounts of gases are expressed in terms of the volume they occupy at a specific pressure and temperature. Consider, for example, the decomposition of potassium chlorate, which gives oxygen gas.

$$2KClO_3 \xrightarrow{\text{heat}} 2KCl + 3O_2$$

Some possible questions related to this equation are

1. What volume of oxygen at specific P and T is formed when a given mass of potassium chlorate reacts?
2. How many grams of potassium chlorate need to react to give a specified volume of oxygen at specific P and T?

Answers to questions like these require the use of gas laws and stoichiometric reasoning. A gas in a reaction is related to other species through the combined use of the gas laws and appropriate molar ratios from the balanced equation. The gas laws relate the volume of gas at specific conditions to the number of moles of the gas. The number of moles provides an entry to stoichiometric calculations. Figure 10-6 shows a stoichiometry map illustrating how the volumes of gas samples are integrated into stoichiometric calculations.

We can use the **molar volume, (22.4 L/mol)$_{STP}$**, to relate the volume of a gas at STP to the number of moles. Alternatively, we can use the **ideal gas law** to relate the volume of a gas at any conditions to the number of moles of the gas.

As an example, let us calculate the number of grams of methane needed to react with 265 L of O_2 contained at 298 K and 1.06 atm pressure. The equation for the reaction is

$$CH_4 + 2O_2 \longrightarrow CO_2 + 2H_2O$$

This problem is a gas law and stoichiometry question since the volume of a gaseous reactant at specific conditions is given and we want to find the mass of other reactant. By consulting the stoichiometry map in Figure 10-6, we can see that a calculation sequence is

$$
\begin{array}{ccc}
 & n = PV/RT & \text{molar ratio} & \text{molar mass} \\
\text{265 L } O_2 \text{ at 298 K} \xrightarrow{\hspace{1cm}} & \text{mol } O_2 \xrightarrow{\hspace{1cm}} & \text{mol } CH_4 \xrightarrow{\hspace{1cm}} & \text{g } CH_4 \\
\text{and 1.06 atm} & & &
\end{array}
$$

FIGURE 10-6
Stoichiometry map showing how gas laws relate to stoichiometric calculations.

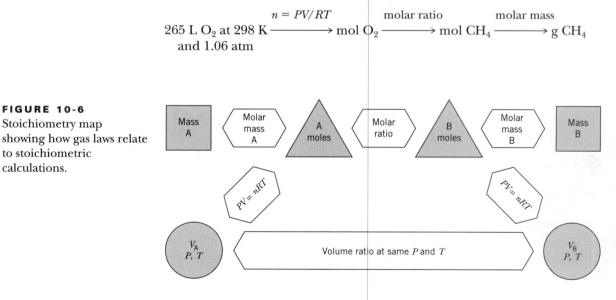

First, find the number of moles of O_2 using the ideal gas law. This step is the gas law part of the question.

$$n = \frac{PV}{RT} = \frac{1.06 \text{ atm } 265 \text{ L}}{(0.0821 \text{ L atm/K mol } O_2) \, 298 \text{ K}}$$

Next, find the number of moles of CH_4 using the molar ratio from the equation. This step is the stoichiometry part.

$$\frac{1.06 \text{ atm } 265 \text{ L}}{(0.0821 \text{ L atm/K mol } O_2) \, 298 \text{ K}} \left(\frac{1 \text{ mol } CH_4}{2 \text{ mol } O_2} \right)$$

Finally, find the mass of CH_4 from the number of moles using its molar mass.

$$X \text{ g } CH_4 = \frac{1.06 \text{ atm } 265 \text{ L}}{(0.0821 \text{ L atm/K mol } O_2) \, 298 \text{ K}} \left(\frac{1 \text{ mol } CH_4}{2 \text{ mol } O_2} \right) \left(\frac{16.04 \text{ g}}{1 \text{ mol } CH_4} \right) = 92.1 \text{ g}$$

EXAMPLE 10-11

How many liters of oxygen gas collected at STP can form when a 246 g sample of potassium chlorate is heated to give potassium chloride and oxygen gas? The balanced equation for the reaction is

$$\overset{heat}{2KClO_3 \longrightarrow 2KCl + 3O_2}$$

A summary of this stoichiometric question is

	heat	
	$2KClO_3 \longrightarrow$	$2KCl + 3O_2$
mass	246 g	
molar mass	122.5 g/mol	
moles	mol $KClO_3$ $\Longrightarrow$	mol O_2
volume		? L (STP)

Notice that we include mass, molar mass, moles, and volume. The first step involves a mass-to-mole stoichiometric calculation. We include volume since the question asks for the volume of a gas from the mass of a reactant. By consulting the stoichiometry map in Figure 10-6, we see that the liters of oxygen gas produced at STP are found by using the following calculation sequence.

$$\text{g } KClO_3 \xrightarrow{\text{molar mass}} \text{mol } KClO_3 \xrightarrow{\text{molar ratio}} \text{mol } O_2 \xrightarrow{22.4 \text{ L/mol}} \text{L } O_2 \text{ at STP}$$

First, we use the molar mass of potassium chlorate to find the number of moles of $KClO_3$ [39.099 + 35.453 + 3(15.999) = 122.5].

$$246 \text{ g} \left(\frac{1 \text{ mol } KClO_3}{122.5 \text{ g}} \right)$$

Second, we multiply the moles of potassium chlorate by the molar ratio to find the moles of oxygen.

$$246\,g \left(\frac{1\ \text{mol KClO}_3}{122.5\,g}\right)\left(\frac{3\ \text{mol O}_2}{2\ \text{mol KClO}_3}\right)$$

Finally, we find the volume of oxygen gas at STP by multiplying the moles of oxygen by the molar volume.

$$L\ O_2\ \text{at STP} = 246\,g\left(\frac{1\ \text{mol KClO}_3}{122.5\,g}\right)\left(\frac{3\ \text{mol O}_2}{2\ \text{mol KClO}_3}\right)\left(\frac{22.4\ L}{1\ \text{mol O}_2}\right)$$

$$= 67.5\ L\ O_2\ \text{at STP}$$

10-16 VOLUME-TO-VOLUME CALCULATIONS

For reactions that involve two or more gaseous reactants or products, it is sometimes necessary to figure out how the volume of one species relates to the volume of another. Often this kind of calculation is solved quite readily if we know how the volumes of gases relate to the numbers of moles. The ideal gas law, $PV = nRT$, reveals that the volume of a gas directly relates to the number of moles if the pressure and temperature are constant; $V = nRT/P$. Furthermore, the ideal gas law shows that two gas samples having the same volume, pressure, and temperature must contain the same number of moles of gas. This idea was first suggested by Amedeo Avogadro in 1811, long before the ideal gas law was known. The relation is often stated in a form called Avogadro's law, as shown in the margin and in Figure 10-7.

Avogadro's Law
Equal volumes of gases at the same temperature and pressure contain the same number of moles.

Any samples of gases that have the same volumes at the same temperature and pressure must have the same number of moles. Another way to state this is that the number of moles of a gas relates directly to the volume. Thus, what is true for the number of moles of a gas in a reaction is also true for the number of liters of the gas. Therefore, it is possible to interpret a reaction involving gases at a constant temperature and pressure on a volume basis rather than a molar basis. The volumes of gases involved in a reaction relate through volume ratios that are equal to molar ratios, which is true as long as the gases are at the same temperature and pressure. Consider the reaction of hydrogen gas and chlorine gas to give hydrogen chloride gas.

$$H_2(g) + Cl_2(g) \longrightarrow 2HCl(g)$$

In this reaction, 1 mole of hydrogen reacts with 1 mole of chlorine to give 2 moles of hydrogen chloride. The ratios of the volumes of the gases involved are the same as the ratio of the number of moles. Remember that

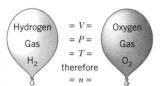

FIGURE 10-7
Avogardro's law.

this is true only if the gases have the same temperature and pressure. **Volume ratios** are expressed in the same manner as molar ratios.

$$\left(\frac{1\ L\ H_2}{1\ L\ Cl_2}\right)\qquad\left(\frac{1\ L\ H_2}{2\ L\ HCl}\right)\qquad\left(\frac{1\ L\ Cl_2}{1\ L\ H_2}\right)$$

$$\left(\frac{2\ L\ HCl}{1\ L\ H_2}\right)\qquad\left(\frac{1\ L\ Cl_2}{2\ L\ HCl}\right)\qquad\left(\frac{2\ L\ HCl}{1\ L\ Cl_2}\right)$$

Volume ratios are written for any reaction involving two or more gases. Volume ratios do not apply to chemicals that are not in the gaseous state. These ratios are conversion factors used to relate the volume of one gas to the volume of another gas in stoichiometric calculations. They are used in a way that is similar to molar ratios. See the stoichiometry map in Figure 10-6.

EXAMPLE 10-12

How many liters of hydrogen chloride gas are produced at 780 torr and 25°C when 32 L of chlorine gas at 780 torr and 25°C react with sufficient hydrogen? The balanced equation is

$$H_2(g)\quad+\quad Cl_2(g)\longrightarrow\quad 2HCl(g)$$
$$\text{volume}\qquad 32\ L\Longrightarrow\ ?\ L$$

Since the two gases are at the same conditions of temperature and pressure, the volume ratio is used as a factor.

$$\text{L HCl}\ =\ 32\ \cancel{L\ Cl_2}\left(\frac{2\ L\ HCl}{1\ \cancel{L\ Cl_2}}\right)=64\ L\ HCl$$

Methane gas reacts with oxygen gas to give carbon dioxide and water. A cubic foot of methane has a volume of 28.3 L at STP. How many liters of oxygen gas at STP are needed to react with 28.3 L of methane at STP?

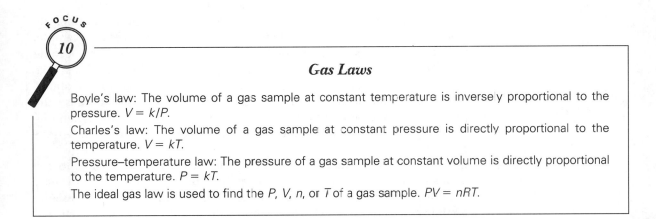

FOCUS
10

Gas Laws

Boyle's law: The volume of a gas sample at constant temperature is inversely proportional to the pressure. $V = k/P$.

Charles's law: The volume of a gas sample at constant pressure is directly proportional to the temperature. $V = kT$.

Pressure–temperature law: The pressure of a gas sample at constant volume is directly proportional to the temperature. $P = kT$.

The ideal gas law is used to find the P, V, n, or T of a gas sample. $PV = nRT$.

The combined gas law is used for changes in the volume, temperature, and pressure of a gas sample.

$$V_2 = V_1 \frac{P_1 T_2}{P_2 T_1} \quad \text{or} \quad \frac{V_2 P_2}{T_2} = \frac{V_1 P_1}{T_1}$$

The universal gas constant R has a value of 0.0821 L atm/K mol.

STP or standard temperature and pressure are

$$0°C \quad \text{or} \quad 273\,K \quad \text{and} \quad 760\,torr \quad \text{or} \quad 1\,atm$$

The molar volume expresses the volume of a mole of an ideal gas at STP. Molar volume = 22.4 L/mol.

Dalton's law of partial pressures: The total pressure of a mixture of gases is equal to the sum of the partial pressures of the components. $P_t = P_a + P_b + P_c + \cdots$.

The molar mass of a gas is found by measuring the mass, volume, temperature, and pressure of a sample of the gas.

$$\text{molar mass} = (\text{mass}) \frac{RT}{PV}$$

Avogadro's law: Equal volumes of gases at the same temperature and pressure contain the same number of moles.

In stoichiometric calculations, gas laws are used to related moles of gases to volumes at specific pressures and temperatures. See the stoichiometry map in Figure 10-6.

EPA

10-17 PRIMARY AIR POLLUTANTS

As you know, air is a mixture of gases of which nitrogen and oxygen are the major components. Industrial processes and the burning of trash produce waste gases, and smoke and cars and trucks produce exhaust gases. There is a difference between gases and smoke. Smoke consists of finely divided solids and liquids suspended in air. In cigarette smoke, for example, you can see these suspended solid and liquid particulates as they become dispersed in the air. You cannot see the gaseous products in the smoke even though they are present. When gases and smoke mix with the atmosphere, they can become semipermanent components. Just because the products are released into the air does not mean they are gone. When pollutants accumulate in specific geographical areas, they can cause serious air pollution.

Waste gases produced by an industrial society and released into the atmosphere are called **primary air pollutants.** The five major primary pollutants are **carbon monoxide, sulfur dioxide, nitrogen oxides, volatile organic compounds (VOCs),** and **particulates.** Table 10-1 describes the nature and origin of each of these primary air pollutants. Consider a burning

Table 10-1   *Sources of Primary Air Pollutants*

Carbon monoxide is formed in the incomplete combustion of fossil fuels.

$$\text{carbon containing compounds} + O_2 \longrightarrow CO + H_2O$$

Sulfur dioxide is formed when sulfur-containing impurities in fossil fuels are burned.

$$\text{sulfur-containing compounds} + O_2 \longrightarrow SO_2$$

Nitrogen oxides are formed when nitrogen and oxygen combine at higher temperatures of fossil fuel combustion.

$$N_2 + O_2 \longrightarrow 2NO$$
$$2NO + O_2 \longrightarrow 2NO_2$$

Volatile organic compounds result when fossil fuels and solvents evaporate or are not completely burned.

$$\text{solvents or fossil fuels} \longrightarrow \text{HC and others VOC}$$

Hydrocarbons include a variety of hydrogen–carbon volatile organic compounds represented collectively as HC.

Particulates consist of liquid droplets and tiny pieces of solids that are formed during the combustion process.

cigarette as a source of primary air pollutants. As the cigarette smolders, the incomplete combustion produces carbon monoxide gas. Any sulfur-containing impurities in the tobacco burn to form sulfur dioxide gas. The high temperature of the burning tobacco forms nitrogen oxides. The hot tobacco decomposes to produce a variety of volatile organic compounds, including nicotine. Cigarette smoke contains liquid and solid particulates that deposit a variety of chemicals in the lungs. That is why the lungs of a smoker have black tar-like deposits. In the atmosphere, primary air pollu-

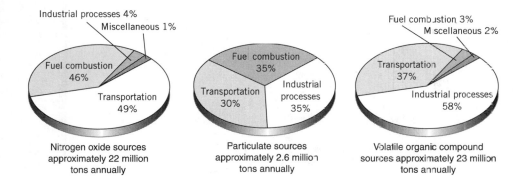

Industrial processes 4%
Miscellaneous 1%
Fuel combustion 46%
Transportation 49%

Nitrogen oxide sources approximately 22 million tons annually

Fuel combustion 35%
Transportation 30%
Industrial processes 35%

Particulate sources approximately 2.6 million tons annually

Fuel combustion 3%
Miscellaneous 2%
Transportation 37%
Industrial processes 58%

Volatile organic compound sources approximately 23 million tons annually

tants come mainly from automobiles and trucks, electrical power plants, industrial and manufacturing processes, and solid waste incineration.

10-18 TEMPERATURE INVERSIONS

The accumulation of air pollutants within a restricted geographical area can cause an air pollution episode. Because of weather fluctuations and winds, air masses are swept horizontally from one region to another region of the atmosphere. Horizontal air movement can disperse and dilute air pollutants. Air near Earth's surface is warmed by the water and rocks that have been heated by sunlight. The warmer air becomes less dense and rises. It is replaced by cooler air above. Figure 10-8 illustrates this normal vertical circulation or convection of air. Vertical movement or circulation of the surface air can disperse air pollutants into higher regions of the atmosphere.

Normally, the temperature of air decreases with an increase in altitude, as shown in Figure 10-9(a). In certain geographical regions, air masses get temporarily trapped in the atmosphere when weather conditions allow a cooler air mass to move below the warmer air mass. This situation is shown in Figure 10-9(b). The warmer air mass becomes trapped between the cooler air below and above it. Normal vertical circulation of air does not happen under these conditions. This phenomenon is called a temperature inversion. In a **temperature inversion,** the temperature of the air decreases with altitude until the warm air mass is reached. The temperature of this air mass increases in altitude until the cooler air mass above is reached. In other words, the normal temperature variation is inverted. The warmer air layer lies between the cooler air layers above and below. A temperature inversion temporarily traps the warmer air mass. Environmental problems result when humanmade air pollutants accumulate in the inversion layer and make smog. When this happens, you can often see a distinct hazy layer in the sky in the region where the warm layer contacts the cooler layer above it.

FIGURE 10-8 The vertical circulation of air.

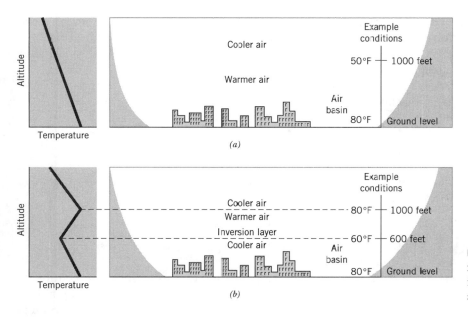

FIGURE 10-9 The formation of a temperature inversion in the atmosphere.

10-19 SMOG

A temperature inversion can result in a trapped and immobile air mass in which pollutants accumulate. Often, temperature inversions break up at night, but a new inversion forms the next day if the weather is favorable. Sometimes an inversion may persist for days. Temperature inversions often occur during cloudless, sunny days, which complicates air pollution episodes.

Once a temperature inversion has formed, primary air pollutants become trapped and accumulate in localized areas. If the inversion occurs on a warm, cloudless day, the primary air pollutants can react to form **secondary air pollutants.** The sunlight-induced chemical reactions that produce secondary pollutants are **photochemical reactions.** The inversion layer acts as a large reaction vessel in which photochemical reactions and subsequent reactions produce a variety of secondary pollutants known collectively as smog. Such **photochemical smog** is typical of air pollution episodes in various regions of the United States, especially in California. The hazy brown appearance of smog is a result of accumulated particulates and nitrogen dioxide, but many components of smog are colorless gases. Photochemical smog is a complex mixture of chemical compounds produced by reactions between nitrogen oxides, hydrocarbons, particulates, water vapor, and oxygen.

An important example of a secondary pollutant is **ozone, O_3.** Ozone is not normally found in the lower atmosphere, but it forms by the following set of reactions. Nitrogen oxide, NO, is produced by the combination of nitrogen and oxygen at the high temperatures of internal combustion engines, incinerators, and power plants. Nitrogen oxide reacts with oxygen in the air to form reddish brown nitrogen dioxide sometimes called nitric oxide.

$$2NO + O_2 \longrightarrow 2NO_2$$

Atomic oxygen forms when nitrogen dioxide is exposed to the ultraviolet (UV) light of sunshine.

$$NO_2 \xrightarrow{\text{UV}} NO + O$$

Individual atoms of oxygen are very reactive and, when they collide with molecules of oxygen adsorbed on the surface of particulates, they form ozone gas.

$$O + O_2(\text{particulates}) \longrightarrow O_3 + \text{particulates}$$

Ozone

Ozone is a toxic gas that can damage plants and cause respiratory problems in humans.

The Clean Air Act is a federal law intended to establish procedures to reduce air pollution and protect air quality. The U.S. Environmental Protection Agency (EPA) is the federal agency that coordinates and monitors efforts to control air pollution.

A Hole in the Ozone Layer

Ozone is an undesirable secondary air pollutant in lower regions of the atmosphere. In contrast, ozone in the upper region of the atmosphere, called the stratosphere, has an important natural function. The ozone in the stratosphere screens Earth from harmful ultraviolet light. Ozone forms in the stratosphere through reactions of molecular oxygen that absorbs ultraviolet radiation:

$$O_2 \xrightarrow{\text{UV}} 2O \quad \text{and} \quad O_2 + O \longrightarrow O_3$$

Ozone molecules formed in the stratosphere absorb more than 90% of ultraviolet radiation, including the higher energy ultraviolet light sometimes called UV-C and UV-B radiation. In this way, ozone serves to shield the lower atmosphere and Earth's surface from a significant fraction of high-energy UV light that is part of solar radiation. Ozone concentrations stay at relatively fixed levels because ultraviolet light also decomposes ozone molecules to reform oxygen molecules:

$$2O_3 \xrightarrow{\text{UV}} 3O_2$$

The formation and decomposition of ozone produce a steady-state concentration of ozone within the stratosphere. This layer or region of ozone is sometimes called the ozonosphere, and a maximum concentration of about 10 ppm occurs in the stratosphere from 25 to 30 km (16 to 19 mi) in altitude.

Recent observations of the ozone layer using rockets, high-flying aircraft, and satellites revealed that the ozone concentration is dramatically decreasing, especially in the polar regions. This decrease in concentration is called thinning of the ozone layer, and larger decreases are called "holes" in the layer. The "holes" in the polar regions of Earth seem to vary with the seasons. The "hole" over Antarctica is most widespread from September to November, a time that corresponds to springtime in the southern hemisphere. In Fall of 1998 the ozone hole over Antarctica was the largest ever observed up to that time. "Holes" have also been observed over the north pole and some industrialized regions of the northern hemisphere.

A decrease in the ozone concentration allows more ultraviolet radiation to reach the surface of Earth. This increase in radiation causes damage to humans, other animals, plants, bacteria, and microscopic marine organisms (phytoplankton and zooplankton). For each 1% decrease in the ozone layer, there is a 2% increase in the UV radiation reaching Earth. It is estimated that a 2% increase in radiation may result in a 4% to 10% increase in basal-cell skin cancer and an 8% to 20% increase in more serious squamous-cell skin cancer in humans. In the United States, the incidence of a skin cancer called malignant melanoma has increased about 18 times its level since the 1930s. The reasons for this dramatic change are not known but may be related to changes in lifestyles in which people have more exposure to the sun and more sunburn. In most cases, there is a 10- to 20-year delay between the incidence of skin damage by sun exposure and the occurrence of melanoma. Increases in solar UV are also related to an increased incidence of human cataracts. Solar UV increases have detrimental effects on other life-forms, but the significance of these effects is not known.

The thinning of the ozone layer appears to be a result of human activity. The chemical culprits are synthetic compounds containing chlorine, fluorine, and carbon known as chlorofluorocarbons or CFCs (also known as Freons). CFCs were developed in the 1930s. Over the years, they found use as coolant gases in refrigerators and air conditioners, as propellant gases in aerosol cans, as industrial solvents, and as foaming agents in plastic products like Styrofoam and cushion materials. In the United States, they are no longer used in aerosol cans and other uses are being phased out. Halons, which are chemically similar to CFCs, find use in fire extinguishers and medical anesthetics. Halons also affect the ozone layer.

Most gaseous pollutants that enter the atmosphere have a natural sink. The term *sink* refers to a long-term repository in the environment. A sink is some place or chemical form in which a chemical ends up. Normally, a pollutant is changed by some chemical reaction and is washed from the atmosphere by the rain. Sulfur dioxide and nitrogen oxides, for example, are transformed to sulfuric acid and nitric acid, which are carried to Earth as acid rain.

One reason CFCs are used in products or as solvents is that they are very chemically inert. They are chemically stable, are nontoxic, do not support combustion, and are noncorrosive. Because CFCs are chemically inert and have no environmental sink, they are destructive to the ozone layer. CFCs released to the atmosphere by industrial processes and consumer products become semipermanent components of the atmosphere. The CFC molecules slowly diffuse to the stratosphere where high-energy UV radiation causes them to decompose to form chlorine atoms.

$$CCl_2F_2 \xrightarrow{\text{UV}} CClF_2 + Cl \text{ (chlorine atom)}$$

Chlorine atoms serve to catalyze the decomposition of ozone. A chlorine atom can react with an ozone molecule to decompose it.

$$Cl + O_3 \longrightarrow ClO + O_2$$

Then, the ClO molecule produced reacts with an oxygen atom to reform the chlorine atom.

$$ClO + O \longrightarrow Cl + O_2$$

The chlorine atom goes on to catalyze the decomposition of another ozone molecule, and so on. One chlorine atom can decompose about 100,000 ozone molecules before it reacts with some other molecule and becomes inactive. The result is that CFCs can cause a decrease in the steady-state concentration of ozone. Apparently, the ozone destruction in the polar regions is also related to the atmospheric temperature and stratospheric ice crystals, and is, thus, subject to seasonal variation.

Since the thinning of the ozone layer is caused by humanmade, synthetic gases, the phenomena can be limited and curtailed by phasing out the use of these gases. Unfortunately, most of the CFCs released in recent years will not reach the stratosphere for 10 or more years. There is no way to retrieve these gases, so the problem is likely to get worse in the future. Furthermore, even if CFC release were completely stopped, it would take many years for the ozone layer to return to normal.

CFCs are still being used as refrigerant gases, foaming agents, and industrial solvents, especially those used to clean electronics components. The United States, Canada, and several European countries have banned the use of CFCs as aerosol propellants, but they are still widely used in this way in many countries. In fact, this use still contributes to nearly one-third of the CFCs released each year. Some states and cities have passed laws to control the industrial use of CFCs and to require the recycling of CFCs from automobile air conditioners. In 1989, twelve European countries and the United States agreed to ban all CFC production by the year 2000. Together, these countries manufacture about three-fourths of the world's supply of CFCs. Industries and manufactures are currently developing alternative chemicals to substitutes for CFCs. The elimination of humanmade, ozone-destroying gases requires international cooperation. Unfortunately, the decreased production of CFCs has resulted in an increase in the smuggling of these chemicals from underdeveloped countries into the United States and Europe. A black market for CFCs has developed. Like illegal street drugs, however, the purity of black market CFCs is questionable, and selling or purchasing them is criminal behavior.

The Greenhouse Effect and Global Warming

The average temperature of Earth's atmosphere is about 15°C, and it obviously varies from region to region and season to season. It is known that carbon dioxide, water vapor, methane, and nitrous oxide (N_2O) are strong absorbers of infrared radiation. These atmospheric gases absorb energy from sunlight and from the infrared energy emitted by Earth's surface. The absorption of energy serves to keep the temperature of the atmosphere at a life-sustaining level. In a greenhouse, the glass windows transmit visible light but prevent infrared energy from leaving. This effect causes an increase in the internal temperature of the greenhouse. By analogy, atmospheric gases that trap infrared energies are called greenhouse gases. Naturally occurring greenhouse gases are vital to maintaining the temperature of the atmosphere, and the greenhouse effect is a necessary phenomenon in Earth's environment.

As shown in Table 10-2, analysis of the atmosphere reveals that concentrations of greenhouse gases have increased significantly since preindustrial times. Furthermore, the annual percentage increase of the concentrations of these gases is dramatic. Note that the table includes CFCs and lower atmosphere ozone, which are humanmade gases, that also function as greenhouse gases. Carbon dioxide is the most important of the greenhouse gases and, if its annual growth rate continues, its concentration is likely to double from current levels by 2075. Two major sources of carbon dioxide are the combustion of fossil fuels and deforestation.

Table 10-2   *Greenhouse Gases in the Atmosphere*

GAS	CONCENTRATION		ANNUAL PERCENT RATE OF INCREASE
	PREINDUSTRIAL	CURRENT	
Carbon dioxide	275 ppm	353 ppm	0.4
CFCs	0	804 ppt	5.0
Methane	0.75 ppm	1.73 ppm	1.0
Nitrous oxide	280 ppb	308 ppb	0.2
Ozone	?	35 ppb	?

Note: ppm = parts per million; ppb = parts per billion; ppt = parts per trillion.

As the concentrations of greenhouse gases increase, the average temperature of the atmosphere might increase unless some other environmental change counteracts the temperature increase. This potential increase in Earth's average temperature is called the theory of the greenhouse effect. When the concentrations of greenhouse gases reach twice the preindustrial level, climatic computer models predict a temperature rise of 1.5°C to 4.5°C. Such a doubling of concentrations could occur by 2030. A temperature increase of this magnitude might cause melting of polar ice, a rise in sea level, changes in winds and ocean currents, and the increased frequency of severe storms. Furthermore, dramatic climate changes could cause droughts and floods that disrupt agriculture or even produce a mini-ice age. Wetlands, forests, and other ecosystems could quickly change, causing impairment and extinction of some plant and animal species. Statistically, annual precipitation in the United States has increased by 6% since 1900, and worldwide the frequency of catastrophic floods, forest fires, landslides, heat waves, droughts, and windstorms has significantly increased since the 1960s.

Is the greenhouse effect increasing the temperature of the atmosphere? Variations in climate and air temperatures are being closely monitored, but generalizing about long-term effects using short-term observations is difficult. Careful analysis of Earth's temperature over the last 130 years suggests a slight increase of about 0.6°C. One-half of this increase has occurred since 1958. The 1990s was the hottest decade and 1998 one of the hottest years in the world since records have been kept. This may suggest that Earth's temperature is increasing, but it is not compelling evidence. The scientific dilemma concerning the greenhouse effect is that by the time observations show the theory to be correct it may be too late to make significant changes in the human activities that add greenhouse gases to the atmosphere. In other words, once all the scientific evidence is evaluated, it may be too late to act.

The contribution of various gases to a greenhouse effect depends on their amounts in the atmosphere and their infrared absorbing abilities. Accordingly, the rankings of the effect of greenhouse gases are carbon dioxide (50%), CFCs (20%), methane (16%), ozone (8%), and nitrous oxide (6%). Greenhouse gases are emitted by both industrial and developing countries. The greenhouse effect has global implications and is a global problem. Considering the emission of the three major greenhouse gases, the countries with the highest emissions are the United States (18%); western European countries (13%); eastern European countries, including the republics of the former USSR (13%); Brazil (11%, mainly from carbon dioxide); and China (7%).

Note that the greenhouse effect and the thinning of the ozone layer are two different and distinct global air pollution problems. They are connected because CFCs contribute to both effects and human activity produces the gases. As previously stated, the main sources of carbon dioxide emissions are combustion of fossil fuels and deforestation. When forests are destroyed and burned, carbon dioxide is formed, and when organic matter and soil humus is subjected to bacterial decay, carbon dioxide is released. Methane emission comes from the escape of natural gas, anaerobic decay of crop and animal wastes, and decay within the gut of livestock. Apparently, a significant amount of methane also comes from termite activity in woody material. Nitrous oxide is emitted as a byproduct of combustion and from bacterial activity in agriculture, especially in rice paddies. Ozone is formed by localized air pollution, and CFCs come from industrial and consumer product emission.

Some possible measures to curtail greenhouse gas emissions are as follows.

1. Increase the efficiency of energy uses: more efficient industrial processes, more efficient power plants, more fuel-efficient automobiles, and more efficient use of electricity.
2. Switch from fuels having high carbon content or lower fuel values, like coal, to fuels having low carbon content or higher fuel values, like methane or alcohols, or even to hydrogen, a noncarbon fuel.
3. Reduce the rate of deforestation and encourage reforestation.
4. Use more alternative energy sources, like solar energy, wind energy, or nuclear energy.

5. Eliminate the manufacture and use of CFCs and other ozone-destroying gases.

6. Improve the air quality of urban areas to decrease ozone levels.

The decrease greenhouse gas emission is an effort that must be done on a global scale with the cooperation of all countries of the world.

Global Warming

QUESTIONS

1. Match the following terms with the appropriate lettered definitions.

vapor *b*

    a. the volume of gas sample at a constant temperature is inversely proportional to the pressure

pressure *l*

    b. a gas that is easily changed to a liquid by altering its pressure or temperature

kinetic molecular theory *K*

    c. the volume of a gas sample at a constant pressure is directly proportional to the temperature

barometer *i*

    d. $PV = nRT$

torr *h*

    e. the volume of 1 mole of a gas at STP, or (22.4 L/mol)

atm *j*

    f. standard temperature and pressure, or 273 K and 1 atm

Boyle's law *a*

    g. the total pressure of a mixture of gases is the sum of the partial pressures of the components of the mixture

Charles' law *a*

    h. a pressure unit equal to 1 mm of Hg

ideal gas law *d*

    i. a pressure-measuring device

STP *f*

    j. a pressure unit equal to 760 torr

molar volume *e*

    k. a gas consists of a diffuse collection of particles that are in rapid, random, straight-line motion

Dalton's law *g*

    l. force per unit area

Section 10-1

2. What is vapor?

3. List five chemicals that are gases at normal Earth conditions.

Section 10-2

4. State the postulates of the kinetic molecular theory of gases. Why is this theory called the kinetic molecular theory?

5. Describe a gas in terms of the kinetic molecular theory.

6. You go for a walk and disturb a skunk. Although the skunk is a quarter of a kilometer away, you can soon detect its characteristic odor. Explain, in terms of the kinetic molecular theory, how it is possible to detect the skunk's odor, a gas, far from its source.

7. Explain why a gas can be easily compressed.

8. Explain how a gas can leak through a small hole in a container.

Sections 10-3 to 10-5

9. Define pressure. What is the difference between pressure and force?

10. Explain how gases exert pressures.

11. When a diver is under water, what is the cause of the pressure he or she experiences?

12. When the diver in Question 11 is standing on the beach, what is the cause of the pressure he or she experiences?

13. Explain how a barometer works. Describe how you would go about constructing a mercury barometer.

14. Why does a person experience "shortness of breath" when he or she is at higher altitudes?

15. Define a torr. Define an atmosphere. Give a factor that states how these two units are related.

16. What is the SI unit for pressure? What is a bar?

17. The barometer you constructed in Question 13 reads 31.6 in. of mercury. What is the pressure in torr? What is the pressure in atmospheres? (Hint: 25.4 mm = 1 in.)

18. Suppose as you construct a barometer you find that the lab is out of liquid mercury. You decide instead to use water. If the atmospheric pressure is 29.9 in. of mercury, how long a tube will you need for your water barometer? Mercury is 13.6 times as dense as water (e.g., a column of mercury 1 in. long exerts the same pressure as a column of water 13.6 in. long).

19. Change the following pressures from torr to atmospheres.

    (a) 779 torr   (b) 1147 torr

    (c) 687 torr

20. Change the following pressures from atmospheres to torr.

    (a) 2.12 atm   (b) 0.979 atm

    (c) 1.05 atm

Sections 10-6 to 10-10

21. State the following gas laws in words and in the form of an equation.

    (a) Boyle's law   (b) Charles' law

    (c) Pressure–temperature law

22. State the ideal gas law both in words and as an algebraic equation. Define each term in your equation.

23. *A sample of 1.00 g of argon gas occupies 4.17×10^2 mL at 200 K and 749 torr. Use these data to determine the value of the gas constant R having units of milliliter-torr per Kelvin-mole.

24. A sample of 1.00 mole of a gas occupies 2.45×10^4 mL at 300 K and 764 torr. Use these data to determine the value of the gas constant R having units of milliliter-torr per Kelvin-mole.

25. *How many moles of acetylene gas, C_2H_2, are contained in a 30 gal tank at 30°C and 2.25 atm? (1 gal = 3.785 L)

26. A tank of oxygen gas contains 175 L of oxygen at 27°C and 4.83 atm pressure. How many moles of oxygen are in the tank?

27. *A teakettle full of water contains a mass of 825 g of water. If all the water is boiled away and captured in a balloon, what volume would the 825 g of steam occupy at 100°C and 1.00 atm? (Hint: Find the moles from the mass of water.)

28. What volume will 1.00 g of water occupy if it is converted to steam at 100°C and 1.00 atm?

29. *A 5.00 g sample of solid carbon dioxide (dry ice) is placed in a 500 mL sealed container at 25°C. What is the pressure of the gas in the container when all the CO_2 has changed from solid to gas? (Hint: Find the moles from the mass of CO_2.) (**Danger:** Do not try this experiment.)

30. A small container of butane, C_4H_{10}, is attached to an empty 2.75 L tank at 22°C. If a 1.8 g sample of the butane is released from the container as a gas, what is the pressure of butane gas in the tank? (Hint: Find the moles from the mass of butane.)

31. *The Goodyear blimp holds 5.91×10^6 L of helium, He, at 27°C and 763 torr. What is the mass of the helium in the blimp? (Hint: This mass can be found from the number of moles.)

32. A tank contains 28.5 L of nitrogen gas at 0°C and 1.00 atm pressure. What is the mass of N_2 in the tank? (Hint: The mass can be found from the number of moles.)

33. *Nitrous oxide, N_2O, is an anesthetic gas that is very soluble in blood. During the course of an operation, a patient may have several moles of N_2O dissolved in her or his blood. After the operation, the gas is rapidly exhaled by the patient.

    (a) If the patient exhales 59.6 g of pure N_2O at 37°C, what volume will the gas occupy at 0.998 atm?

    (b) If the temperature of the gas in part (a) drops to room temperature, 20°C, what volume will the gas occupy if the pressure remains constant?

Section 10-11

34. *A gas sample in a flexible container occupies 475 mL at 32°C and 1.35 atm. If the temperature is constant, what volume does the gas occupy when

(a) the pressure is 1.70 atm?

(b) the pressure is doubled?

(c) the pressure is decreased to one-half the original value?

(d) the pressure is increased to five times the original value?

35. A gas in a flexible container occupies 250 mL at 25°C and 2.00 atm. If the temperature is constant, what volume does the gas occupy when

(a) the pressure is 1.50 atm?

(b) the pressure is tripled?

(c) the pressure is decreased to one-fourth the original value?

(d) the pressure is decreased to one-tenth the original value?

36. *Suppose an air mattress has a volume of 102 L at −13°C and 0.925 atm on a mountain pass. You then take it to the jungle where the temperature is 39°C and the pressure is 1.11 atm. What volume does the air mattress occupy in the jungle?

37. *In an airliner, a gas sample in a flexible container occupies a volume of 0.275 L at 756 torr. Suppose, while flying, the airliner loses pressure and the volume of the gas increases to 1.80 L. What is the pressure in the airliner if the temperature remains constant?

38. At sea level, a potato chip bag is found to occupy a volume of 327 mL at 762 torr. When the bag is taken to the mountains, it is found to occupy a volume of 341 mL. What is the atmospheric pressure in the mountains if the temperature is constant?

39. Suppose a helium-filled balloon occupies 6.00 L at 19.5°C and 0.989 atm. What volume will the balloon occupy at 4300 m altitude on top of Pike's Peak, Colorado, if the pressure there is 0.605 atm and the temperature is 19.5°C?

40. A gas sample in a syringe has a volume of 36.8 mL at 758 torr. The plunger of the syringe is depressed so that the gas has a volume of 24.2 mL. What is the new pressure of the gas sample if the temperature is constant?

41. An inflated balloon has a volume of 4.5 L at 1.04 atm and 25°C. What volume will the balloon occupy at a pressure of 0.916 atm if the temperature is constant?

42. While skiing in the Alps, you stop for a snack. You notice that the potato chip bags in the snack shop have expanded and are puffed up. Explain why this is so.

43. *A gas sample in a flexible container occupies 750 mL at 32°C and 2.50 atm. If the pressure remains constant, what volume does it occupy when

(a) the Celsius temperature is doubled?

(b) the Kelvin temperature is doubled?

(c) the Kelvin temperature is decreased to one-half the original value?

44. A gas in a flexible container occupies 225 mL at 25°C and 1.50 atm. If the pressure remains constant, what volume does it occupy when

(a) the temperature is 85°C?

(b) the Kelvin temperature is tripled?

(c) the Celsius temperature is decreased to one-half the original value?

45. *A gas in a rigid container occupies 300 mL at 38°C and 1.01 atm. What is the pressure when

(a) the Celsius temperature is doubled?

(b) the Kelvin temperature is doubled?

(c) the Kelvin temperature is decreased to one-half of the original value?

46. A gas in a rigid container occupies 765 mL at 41°C and 0.987 atm. What is the pressure when

(a) the Celsius temperature is tripled?

(b) the Kelvin temperature is tripled?

(c) the Kelvin temperature is decreased to one-half of the original value?

47. On the airliner in Question 37, the pressure stabilizes, but it rapidly becomes very cold. If your gas sample occupies a volume of 1.80 L at 23°C, what volume does it occupy if the temperature of the cabin drops to −37°C?

48. *Suppose a used aerosol can contains a gas at 0.998 atm at 22°C. If this can is heated in a fire to 595°C, what will the pressure of the gas be inside the can? (Danger: An aerosol can explode under these conditions. Never put an aerosol can in a fire.)

49. Your inner ear is normally open to the atmosphere through the eustacian tube, which keeps the pressure of the inner ear equal to atmospheric pressure. Suppose that the pressure is 759 torr. You get an infection, closing the eustacian tube and, thus, the inner ear from the atmosphere. When you run a fever, the temperature of your ear increases from 37°C to 42°C. What is the pressure in your inner ear if the volume remains constant?

50. Refer to Question 49. Why do your ears sometimes "pop" when you change altitudes?

51. A balloon of air occupies 7.5 L at 25°C and 0.997 atm. What volume will it occupy if it is placed in a freezer at −10°C and the pressure is constant?

52. Hydrogen stored in a metal cylinder has a pressure of 250 atm at 26°C. What will the pressure be when the cylinder is placed in liquid nitrogen at −125°C?

53. A tire is filled to a pressure of 29 psi at 22°C. After driving, the pressure reads 30 psi. Assuming the volume is constant, determine the new temperature of the tire in degrees Celsius.

54. *A gas in a flexible container occupies 1.00 L at 0°C and 760 torr. What is the volume when

    (a) the temperature is 120°C and the pressure is 0.675 atm?

    (b) the Kelvin temperature is doubled and the pressure is doubled?

    (c) the Kelvin temperature is doubled and the pressure is decreased to one-half the original value?

55. A gas in a flexible container occupies 835 mL at 19°C and 0.998 atm. What is the volume when

    (a) the temperature is 90°C and the pressure is 0.750 atm?

    (b) the Kelvin temperature is doubled and the pressure is doubled?

    (c) the Kelvin temperature is decreased to one-half the original value and the pressure is doubled?

56. *A weather balloon contains 225 L of helium gas at 29°C and 757 torr at the surface of Earth. What volume does the balloon occupy at an altitude of 30 km where the temperature is 235 K and the pressure is 10.5 torr?

57. The air in a cylinder of a diesel engine occupies 875 mL at 30°C and 1.00 atm. What is the pressure in the cylinder when the air is compressed to 62.5 mL (14 : 1 compression ratio) and heats to 475°C? When the fuel explodes in the compression chamber, the temperature increases to 2000°C for an instant. What is the pressure of the gases in the cylinder at this instant before expansion occurs?

58. A weather balloon contains 100 L of helium at 25°C and 759 torr on the ground. What volume does the balloon occupy at an altitude of 20 km where the temperature is 245 K and the pressure is 120 torr?

59. A gas sample in a flexible container occupies a volume of 6.35 L at 250 K and 0.780 atm. What volume will the gas occupy at 500 K and 2.34 atm?

60. A gas sample in a flexible container occupies a volume of 435 mL at 299 K and 1.06 atm. It is heated and the pressure is changed, which causes a change in volume. What is the new temperature if the pressure changes to 1.04 atm and the new volume is 798 mL?

Section 10-12

61. What are standard temperature and pressure (STP)?

62. What is the molar volume and how can it be used to relate the volume of a gas sample to the number of moles of gas in the sample?

63. How many moles are contained in each of the following?

    (a) 71.9 L SO_2 at STP.

    (b) 40.3 L O_2 at STP.

    (c) 528 L CO_2 at STP.

64. What volume at STP is occupied by each of the following number of moles of gas?

    (a) 1.00 mol N_2

    (b) 34.9 mol Ar

    (c) 4.32×10^3 mol CO

65. Determine the number of moles in each of the following.

    (a) How many moles of He are in a 12.5 L sample at STP?

    (b) How many moles of CO gas are in a 500 mL sample at STP?

    (c) How many moles of O_2 gas are in a 800 mL sample at STP?

66. Calculate the volume occupied by 1.00 mole of an ideal gas at 25°C and 1.00 atm pressure.

Section 10-13

67. A sample of air on Earth is primarily a mixture of N_2, O_2, and Ar. If the partial pressures of these gases are 0.780 atm, 0.209 atm, and 9.74×10^{-3} atm respectively, what is the total pressure?

68. A sample of the atmosphere of Venus is made up of CO_2, H_2O, O_2, and N_2. If the partial pressures (on Venus) of CO_2 is 18.0 atm, H_2O is 8.00×10^{-2} atm, O_2 is 8.00×10^{-2} atm, and N_2 is 1.84 atm, what is the atmospheric pressure on Venus?

69. A mixture of gases contains hydrogen at 550 torr, nitrogen at 183.3 torr, and oxygen at 225 torr. What is the total pressure of the mixture? (**Danger:** Mixtures of hydrogen and oxygen gases are very explosive.)

70. *Disparlure is a pheromone or sex attractant produced by the gypsy moth. Male gypsy moths can detect

this gas in very low concentrations. If the partial pressure of disparlure in a sample of air is 8.45×10^{-19} atm, how many moles of disparlure are there in 1.00 L of air if the temperature is 17°C? How many molecules is this?

71. A sample of oxygen gas is prepared in the lab. You collect a 628 mL sample of gas at 27°C by displacement of water, so your oxygen sample is saturated by water vapor. Water vapor is a gas and contributes to the total pressure. If the total pressure of the sample is 758 torr and the partial pressure of the water vapor is 26.7 torr, what is the partial pressure of the oxygen gas you made? How many moles of oxygen molecules are in the sample?

72. A 932 mL sample of hydrogen prepared in the laboratory is collected at 23°C by the displacement of water and becomes saturated with water vapor. Since water vapor is a gas, it contributes to the pressure of the mixture. If the total pressure of the sample 783 torr and the partial pressure of water vapor is 21.2 torr, what is the partial pressure of the hydrogen? What is the mass of H_2 in the sample? (Hint: Find the number of moles, then use the molar mass to find the number of grams.)

Section 10-14

73. *Halothane is a general anesthetic gas. If a 1.00 g sample occupies a volume of 124 mL at 25°C and 1.00 atm, what is the molar mass of halothane?

74. A 0.852 g sample of a liquid is vaporized. The vapor sample is found to occupy 153 mL at 110°C and 760 torr pressure. Calculate the molar mass of this chemical.

75. *Butane is used as a fuel. If a 100 g sample of butane occupies a volume of 42.1 L at 1.00 atm and 25°C, what is the molar mass of butane? Butane is 82.6% C and 17.4% H. Find the empirical formula of butane and use the molar mass to determine the molecular formula of butane.

76. Acetylene is a hydrocarbon used in welding. If a 6.5 g sample of acetylene occupies 5.25 L at 1.20 atm and 30°C, what is the molar mass? Acetylene is 92.3% C and 7.74% H. Find the empirical formula and use the molar mass to deduce the molecular formula of acetylene.

77. A liquid hydrocarbon is found to have the empirical formula of C_3H_6. If a 1.91 g sample of the liquid is vaporized and the vapor occupies 0.776 L at 1.03 atm and 157°C, what is the molar mass of this hydrocarbon?

Use the calculated molar mass to find the molecular formula of the hydrocarbon.

78. A 1.55 g sample of ethyl ether occupies 628 mL at 100°C and 755 torr. What is the molar mass of ethyl ether?

Section 10-15

79. *Smelling salts contain ammonium carbonate, which can decompose to form ammonia, which acts as a mild heart stimulant. The ammonium carbonate decomposes by the reaction

$$(NH_4)_2CO_3(s) \longrightarrow 2NH_3(g) + CO_2(g) + H_2O(\ell)$$

(a) How many milliliters of NH_3 at 25°C and 1.00 atm are formed from 0.500 g of $(NH_4)_2CO_3$?

(b) How many grams of ammonium carbonate are needed to form 300 mL of ammonia at 25°C and 0.950 atm?

(c) How many milliliters of CO_2 at 25°C and 1.00 atm are formed along with 30.0 mL of NH_3 at 25°C and 1.00 atm?

80. *An oxyacetylene torch burns according to the equation

$$5O_2(g) + 2C_2H_2(g) \longrightarrow 4CO_2(g) + 2H_2O(g)$$

(a) How many liters of C_2H_2 at STP are needed to produce 80.0 mL of CO_2 at 500°C and 1.10 atm?

(b) How many grams of O_2 are required to react with 10.0 L of acetylene at 0.989 atm and 20°C?

(c) How many liters of C_2H_2 at STP are needed to react with 432 L of O_2 at STP?

81. *Bicarbonate of soda is sodium hydrogen carbonate, $NaHCO_3$. If a small amount of this compound is ingested, it reacts with stomach acid by the reaction

$$NaHCO_3 + H_3O^+(aq) \longrightarrow$$
$$Na^+(aq) + 2H_2O + CO_2(g)$$

(**Caution**: Do not ingest bicarbonate of soda without medical advice.)

(a) If 2.25 g of $NaHCO_3$ react to form 253 mL of CO_2 at 37°C, what is the pressure of the CO_2? (Hint: Find the volume of CO_2 at STP, which is the volume at 1 atm. Find the new pressure when the volume is 253 mL and the temperature is 37°C.)

(b) Refer to part (a). Why does a person often burp after ingesting bicarbonate of soda?

(c) How many grams of $NaHCO_3$ are needed to form 175 mL of CO_2 at 25°C and 0.985 atm?

82. Acetylene gas can be formed by reacting calcium carbide and water:

$$CaC_2(s) + 2H_2O(\ell) \longrightarrow Ca(OH)_2(s) + C_2H_2(g)$$

(a) How many grams of CaC_2 are needed to form 5.00 L of C_2H_2 at STP?

(b) How many liters of acetylene at 30°C and 1.20 atm are formed from 100 g of CaC_2?

(c) A 20.0 g pellet of CaC_2 is placed in a sealed container having sufficient water. If the volume of the container (in excess of the water) is 10.0 L, what is the pressure of the acetylene formed at a temperature of 27°C?

(d) How many grams of $Ca(OH)_2$ are formed when 56 L of C_2H_2 at 32°C and 752 torr pressure are produced by the reaction?

83. In a catalytic muffler on an automobile, carbon monoxide is combined with oxygen:

$$2CO(g) + O_2(g) \xrightarrow{\text{Pt}} 2CO_2(g)$$

(a) How many grams of CO will react with 100 L of oxygen at 300°C and 1.00 atm?

(b) How many liters of CO at 200°C and 780 torr will react with 50.0 L of oxygen at 200°C and 780 torr?

(c) How many liters of CO_2 at 20°C and 760 torr are formed when 15.0 L of oxygen at 200°C and 780 torr react?

(d) How many grams of CO_2 are formed when 265 L of CO at 250°C and 0.300 atm react with oxygen?

Section 10-16

84. State Avogadro's law.

85. You are handed two balloons. One contains Kr(g) and one contains H_2(g). Each have identical pressure, volume, and temperature. Which balloon has the larger number of molecules of gas? Which has the greater mass? Explain your reasoning.

86. Ammonia is formed by the reaction

$$3H_2(g) + N_2(g) \longrightarrow 2NH_3(g)$$

(a) How many liters of NH_3 at STP are formed when 25 L of N_2 at STP react with hydrogen?

(b) How many liters of H_2 at 25°C and 1.00 atm are needed to react with 398 L of N_2 at 25°C and 1.00 atm?

(c) How many liters of NH_3 at STP are formed when 9.56 L of N_2 at STP reacts with 27.1 L of H_2 at STP? (Hint: This is a limiting reactant problem.)

Sections 10-17 to 10-19

87. List the five primary air pollutants.

88. What is a temperature inversion?

89. What is a secondary air pollutant? Given an example.

90. What is photochemical smog and how is a temperature inversion involved in smog formation?

91. How is ozone formed as a secondary air pollutant?

92. What is the function of ozone in the stratosphere?

93. Describe the causes and solutions to the "holes" in the ozone layer.

94. Describe the environmental causes and problems associated with the global greenhouse effect.

LIQUIDS

AND SOLIDS

11-1 CHEMICAL VISION OF SOLIDS, LIQUIDS, AND GASES

Solids, liquids, and gases are obviously different. How can we explain the differences? A sample of a solid occupies a fixed volume and is rigid. The shape of a solid can change only with some effort. A sample of a liquid has a fixed volume and is quite flexible. A liquid can be poured from one container to another or spread out on a flat surface. A sample of a gas is quite flexible and needs a container. A gas completely occupies any container in which it is placed and escapes into the atmosphere when released. Gases are typically colorless, but some have distinct colors. In any case, gases are always transparent, which means it is possible to see through them. As you know, a gas sample can be easily compressed, has a measurable pressure, and can leak through a hole in its container.

An explanation of the differences between gases, liquids, and solids relates to the dynamic behaviors of the chemical particles that compose them. Figure 11-1 illustrates the gaseous, liquid, and solid states as collections of particles. Using the kinetic molecular theory, discussed in Chapter 10, we view a gas as a diffuse collection of molecules that are in random high-speed motion. On the average, the molecules are far apart and occupy the volume by their incessant motion. In contrast, the liquid state consists of a collection of particles in a more condensed form. In a liquid, the particles are close together and are moving or wandering about within the collection. The solid state consists of particles arranged in fixed spatial positions. The particles in the solid state, however, still vibrate about their fixed positions. The solid state is a highly condensed and ordered form of matter. In a solid the particles are packed very close to one another.

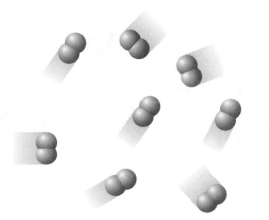

A gas is a diffuse collection of rapidly moving molecules.

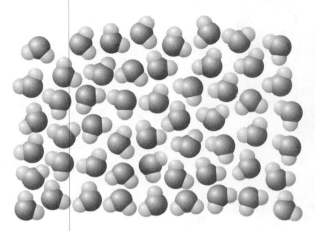

A liquid is a collection of closely packed molecules in random motion.

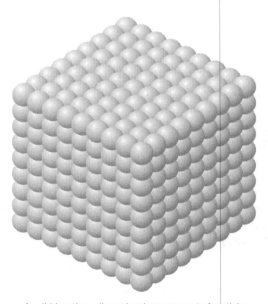

A solid is a three-dimensional arrangement of particles.

FIGURE 11-1 Idealized representations of a gas, liquid, and solid.

11-2 MOLECULAR ENERGIES

The motion of chemical particles gives them kinetic energy (KE). The kinetic energy of a moving object or particle, such as a molecule, is given by the expression $KE = \frac{1}{2}mv^2$, where m is its mass and v is its speed. A direct relation exists between kinetic energy and the speed of gas particles; higher speeds correspond to higher energies, and higher energies correspond to higher temperatures. The temperature of a sample of a material is directly related to the average kinetic energy of its particles. Heating a substance causes the chemical particles to move more rapidly, resulting in a higher temperature.

A sample of any gas, liquid, or solid compound is visualized as a collection of particles with each particle having a specific kinetic energy at a given instant. As the particles move about and collide, they exchange kinetic energy. A particle with a specific kinetic energy may have its energy changed by collision with other particles. Within a sample of a substance, not all the particles have the same kinetic energy. Because of the incessant motion of the particles and their collisions, exchanges of kinetic energy between particles occur continually. Most particles have similar kinetic energies, but some by chance may have very high or low energies.

At a given temperature, the kinetic energies of the particles in a sample of a substance fit a pattern of distribution. It is possible to predict this pattern and represent it as a Maxwell–Boltzmann distribution plot, illustrated in Figure 11-2. A Maxwell–Boltzmann distribution is shown as a plot of the percentage of particles with a particular value of kinetic energy versus possible kinetic energy values. The plot in Figure 11-2 shows two distributions corresponding to two different temperatures. In each case, the greatest percentage of particles around the peak of the plot have kinetic energies equal to or near the average. Furthermore, note that at a higher temperature, the average kinetic energy of the collection is higher.

The various states of matter represent different kinds of associations of chemical particles. In gases, the molecules are in rapid motion and attractions between molecules are not significant. The molecules are in a chaotic state and act independently. In liquids, the molecules do attract, causing them to clump into aggregations, but the particles are in continual random motion. The liquid state represents a balance between molecular attraction and kinetic motion. Attractions keep the molecules in aggregation, and the kinetic motion keeps them moving within the collection. In solids, attractive forces are dominant and the particles are fixed in a three-dimensional pattern. To understand the behaviors of liquids and solids, we need to consider some special features of molecules and how they attract and associate.

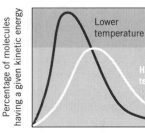

FIGURE 11-2
Maxwell–Boltzmann distributions showing two differing temperatures.

Electronegativity

11-3 ELECTRONEGATIVITY

Covalent bonds involve the sharing of electron pairs. Molecules have covalently bonded atoms. In some kinds of covalent bonds, one atom may attract the shared pair more strongly than the other atom. The result is unequal sharing of the electron pair. Therefore, the shared pair of electrons is pulled closer to one atom than the other. Electronegativity relates to the ability of atoms to attract shared electron pairs.

By definition, **electronegativity** is a measure of the tendency of an atom of an element to attract electrons in a bond. Consider a molecule of hydrogen chloride as an example. Chlorine in the molecule strongly attracts electrons. As a result, the electron cloud representing the bonding electrons shifts toward the chlorine, giving it a slight negative charge. The charge results because electrons have negative charge.

This shift does not mean that the electrons leave the covalent bond and form a negative chloride ion. The chlorine attracts the electrons in the covalent bond more strongly than the hydrogen. This behavior is related to electronegativity. We say that chlorine has a higher electronegativity than hydrogen. Measuring the ability of various elements to attract electrons in covalent bonds is done by observing their bonding behaviors and the properties of the compounds they form.

A relative scale of electronegativities was first established by Linus Pauling, an American chemist. On his **electronegativity scale**, a value of 4.0 is assigned to fluorine, the element of highest electronegativity. The electronegativities of the other elements are expressed relative to fluorine. Figure 11-3 shows a periodic table listing the electronegativities of the elements. Look at the figure and notice which elements have high electronegativities and which have low electronegativities. Also notice how the electronegativities vary in the table.

Generally, elements having small atoms with little shielding (see Sections 8-18 and 8-19) have high electronegativities. On the other hand, those elements with larger atoms and many electrons to shield the nucleus have low electronegativities. Notice that, as might be expected, the electronegativities of the elements within a given group of representative elements increase from the bottom to the top of the group. Furthermore, within a given period of representative elements the electronegativities increase from left to right. The elements in the upper right of the periodic table, excluding the noble gases, have the greatest electronegativities.

Fluorine has the highest electronegativity (4.0) of all the elements, whereas oxygen has the next highest (3.5), followed by chlorine (3.0) and

Electronegativity increases

H																	
2.1																	
Li	Be											B	C	N	O	F	
1.0	1.5											2.0	2.5	3.0	3.5	4.0	
Na	Mg											Al	Si	P	S	Cl	
0.9	1.2											1.5	1.8	2.1	2.5	3.0	
K	Ca	Sc	Ti	V	Cr	Mn	Fe	Co	Ni	Cu	Zn	Ga	Ge	As	Se	Br	
0.8	1.0	1.3	1.5	1.6	1.6	1.5	1.8	1.8	1.8	1.9	1.6	1.6	1.8	2.0	2.4	2.8	
Rb	Sr	Y	Zr	Nb	Mo	Tc	Ru	Rh	Pd	Ag	Cd	In	Sn	Sb	Te	I	
0.8	1.0	1.2	1.4	1.6	1.8	1.9	2.2	2.2	2.2	1.9	1.7	1.7	1.8	1.9	2.1	2.5	
Cs	Ba	La	Hf	Ta	W	Re	Os	Ir	Pt	Au	Hg	Tl	Pb	Bi	Po	At	
0.7	0.9	1.1	1.3	1.5	1.7	1.9	2.2	2.2	2.2	2.4	1.9	1.8	1.8	1.9	2.0	2.2	
Fr	Ra	Ac	Lanthanides: 1.1–1.2														
0.7	0.9	1.0	Actinides: 1.1–1.2														

Electronegativity increases

FIGURE 11-3
Electronegativities of the elements.

nitrogen (3.0). The greater the electronegativity of an element, the greater the tendency for atoms of that element to attract electrons when bonded. As we might expect, metals generally have the lowest electronegativity values since they are likely to lose electrons to form cations rather than gain electrons. Electronegativity differences between the two atoms involved in a covalent bond affect the distributions of bonding electrons in molecules.

ACTIVITY
11-1

Identify the element having the highest electronegativity in the following sets.
(a) S, Te, O (b) Br, Cl, F
(c) C, N, O (d) Mg, Na, K

11-4 POLAR BONDS

All molecules are electrically neutral. Unlike ions, molecules do not carry distinct positive or negative charges since within any molecule the total number of electrons equals the total number of protons. Although they are neutral, some molecules contain an electron distribution that causes them to have regions of relatively concentrated negative charge and regions of relatively concentrated positive charge. In a neutral atom, the electrons surround the nucleus, and the average center of negative charge coincides with the positive center of the nucleus. In some molecules, the average positions of electrons change because of unequal sharing of electron pairs. A diatomic hydrogen molecule is pictured in Figure 11-4. The bonding electron cloud has a greater density between the two positive nuclei. Since the two nuclei are the same, they equally share the electron pair. In such a molecule, the average centers of positive and negative charge coincide; no separation of charges exists.

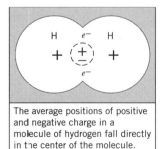

The average positions of positive and negative charge in a molecule of hydrogen fall directly in the center of the molecule.

FIGURE 11-4 The centers of positive and negative charge coincide in a hydrogen molecule.

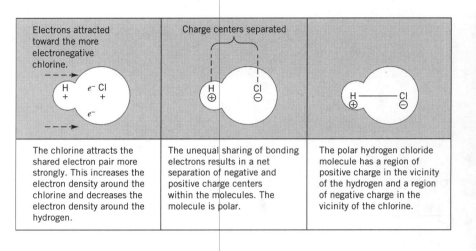

FIGURE 11-5 The centers of positive and negative charge are separated in a hydrogen chloride molecule, making it polar.

In a molecule of hydrogen chloride, HCl, the chlorine attracts the bonding electron pair more than does the hydrogen. The chlorine strongly attracts the shared electrons, giving a partial negative charge around the chlorine and a partial positive charge around the hydrogen (see Figure 11-5). In a molecule such as HCl, the average centers of positive and negative charge do not coincide. The molecule carries no overall charge, but an uneven distribution of positive and negative charges exists. The bond between the chlorine and the hydrogen is called a polar covalent bond.

A **polar covalent bond** is a covalent bond between atoms of different electronegativity. The term *polar* refers to opposite parts (e.g., the north and south poles of Earth or the opposite poles of a magnet). A polar bond has a negative region around the more electronegative atom and a positive region around the less electronegative atom. Usually, the greater the difference in electronegativity, the greater the polarity of the bond.

11-5 POLAR MOLECULES

Polar bonds in molecules affect the overall distribution of charges in the molecules. The polar bond in a simple linear molecule like hydrogen chloride results in a net separation of positive and negative charge centers within the molecule (see Figure 11-5). The unequal distribution of charge in a molecule of hydrogen chloride makes it a polar molecule. It has a negative region or negative pole and a positive region or positive pole. A **polar molecule** is a molecule that has a separation of positive and negative charge centers. Visualize a polar molecule as a molecule with a region of negative charge and a region of positive charge.

A water molecule is another good example of a polar molecule. Within the molecule, the hydrogens form polar bonds with the more electronegative oxygen. Furthermore, the water molecule happens to have an angular shape (see Section 9-18).

The polar bonds result in a slight negative charge in the region of the oxygen and in slight positive charges in the region of the hydrogens. The angular shape of the molecule results in a net separation of negative and positive charge centers.

We can easily predict whether a simple molecule is polar or nonpolar. A polar molecule must satisfy two criteria.

1. The molecule must have one or more polar bonds.
2. The molecule must have a shape that results in the net separation of positive and negative charge centers.

To make a prediction, first draw a dot structure for a molecule and predict its shape. Then decide if any of the bonds are polar by considering differences in electronegativities of the bonded atoms. All polar molecules have a **nonsymmetrical** shape and polar bonds.

To judge whether or not a molecule is polar, you can manipulate it in your imagination. Imagine what happens if you pulled outward on the polar bonds in the molecule. When you pull equally on all polar bonds involving the same kinds of atoms, does the central atom move or do the equal pulling forces cancel? In symmetrical molecules, the pulling forces cancel and the central atom would not move. A molecule of carbon dioxide has a central carbon bonded to two oxygen atoms that are more electronegative. The carbon–oxygen bonds are polar. Within the molecule, the atoms bond in a linear fashion. Imagine pulling with equal force on the oxygen atoms.

Since the molecule is linear and the two bonds are of the same type, the central carbon atom would not move. The symmetrical linear shape of the molecule results in no net separation of charge centers. The average centers of negative and positive charge happen to coincide. In other words, a symmetrical distribution of charge exists in the molecule, and it is not polar even though it has polar bonds.

If a polar molecule is not symmetrical, the central atom would move when the bonds are pulled. A water molecule has two polar hydrogen–oxygen bonds and a bent shape. If we imagine pulling outward on these bonds, the central oxygen would move. Water molecules are polar molecules.

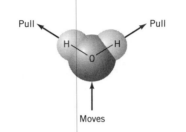

Predict whether molecules of the following compounds are polar or nonpolar.
(a) CS_2 (b) OF_2
(c) CF_4 (d) NI_3

Molecules with no net separation of charge centers are **nonpolar molecules.** Nonpolar molecules exist in two forms. One type includes those molecules in which the atoms have little or no differences in electronegativity and, thus, no polar bonds. Diatomic chlorine, Cl_2, molecules are nonpolar. (Why is this true?) The other type includes those molecules that have polar bonds but a symmetrical shape, resulting in no net separation of charge. This case is illustrated by carbon dioxide, mentioned above. Table 11-1 shows a few examples of polar and nonpolar molecules.

11-6 MOLECULAR ATTRACTIONS

Imagine two molecules touching. Atoms within the molecules are held by covalent bonds. Do the molecules attract? Yes, intermolecular attractive forces between molecules exist and are significant when molecules are in close proximity. The attractive forces between molecules are called **van der Waals, forces,** named for the Dutch scientist Johannes van der Waals, who first suggested their existence in the late 1800s. According to the kinetic molecular theory of gases, no significant attractive forces exist between gaseous molecules. Gas molecules move so rapidly and are so far apart that the attractive forces are not very important. The attractive forces become important when molecules have slower speeds and are in close proximity. In liquids and solids, particles have relatively slower speeds and are close together. In liquids and solids, attractive forces between particles are significant.

Polar molecules can attract one another through their oppositely charged regions.

Table 11-1   *Some Polar and Nonpolar Molecules*

	Compound		Polarity of Molecules
Nonpolar	Chlorine	Cl—Cl	(Nonpolar bond)
	Sulfur trioxide	O–S–O with O below	(The molecule has polar bonds but the symmetry of the molecule makes it nonpolar.)
	Carbon Tetrachloride	Cl₄C structure	(The molecule has polar bonds but the symmetry of the molecule makes it nonpolar.)
Polar	Hydrogen chloride	H—Cl	+ H — Cl − (Polar bonds with unsymmetrical molecule.)
	Ammonia	N with H's	− N, H—N—H, H + (Polar bonds with unsymmetrical molecule.)
	Water	O with H's	− O, H H + (Polar bonds with unsymmetrical molecule.)
	Ethyl alcohol	C₂H₅OH structure	+ H, O −, C structure (Polar C—O and O—H bonds with unsymmetrical molecule.)

Imagine two bar magnets. If they are far apart or move by one another quickly, the attractive forces are not noticeable. When the magnets are slowly brought closer, however, they interact. If the opposite poles of the magnets match, attractive forces pull them together. On the other hand, if like poles are aligned, repulsive forces push the magnets apart. Polar molecules interact in a similar fashion, but the attractive and repulsive forces arise from the **electric poles.** The interactions are electrostatic rather than magnetic.

When polar molecules are in close proximity, the opposite poles attract. This attraction is called **polar interaction** or **dipole interaction.** If the attractive forces are strong enough, the molecules become loosely attached.

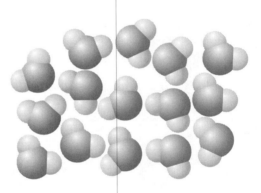

Although **intermolecular forces** of attraction are not as strong as chemical bonds, they can still cause molecules to attract and associate. Molecules aggregate because of intermolecular attractive forces.

Intermolecular attractive forces can also occur between nonpolar molecules. **Nonpolar molecular attractions** are weaker than **polar molecular attractions,** but they are important in substances consisting of nonpolar molecules. Nonpolar molecules attract one another when the negatively charged electrons within molecules are momentarily attracted to the positively charged nuclei of other molecules.

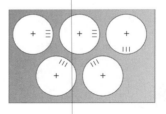

When nonpolar molecules are in close proximity, nonpolar molecular attractions cause the molecules to associate loosely or stick to each other. This is called **nonpolar interaction.**

11-7 THE LIQUID STATE

When a gas is cooled, the particles slow down and the average kinetic energy of the particles decreases. When a gas is compressed, the gaseous particles move closer. Under the proper conditions of low temperature and high pressure, a gas liquefies; that is, it converts from a gas to the liquid state. The kinetic motions of the molecules make them move apart from one another. Liquefaction occurs when the molecules move closer, attractive forces become important, and the molecules condense to the liquid state. As they condense, the molecules attract and aggregate into a group, giving the characteristic appearance of a liquid. See Figure 11-6.

The liquid state consists of molecules or, sometimes, atoms or ions packed in a relatively close manner with little space between particles. The liquid state is dynamic, however, and the molecules move and wander about the entire collection. The molecules move about randomly with short distances between collisions with other molecules and the container. These collisions result in no net loss in kinetic energy, but the molecules can exchange kinetic energy by collision. The average kinetic energy of the molecules is directly related to the temperature of the liquid. Heating a liquid increases the average speed and kinetic energy as the temperature

FIGURE 11-6
Liquefaction of a gas.

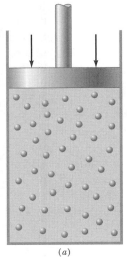
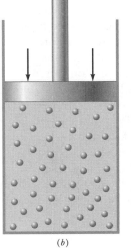
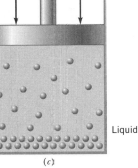

(a)

(b)

(c)

Liquid

In the gaseous state, the molecules are rapidly moving and are far apart.

As the pressure is increased and the temperature decreased, the molecules slow down and are forced closer together.

Finally, at the correct combination of temperature and pressure most of the molecules aggregate into a collection of molecules characteristic of the liquid state; liquefaction has taken place.

increases. Cooling a liquid decreases the average kinetic energy and the temperature decreases.

11-8 PROPERTIES OF LIQUIDS

The kinetic molecular view of the liquid state involving kinetic motion and intermolecular attractions is used to describe and explain some properties of liquids.

Liquids can flow. Because of the freedom of movement and the attractive forces between molecules, all liquids flow and spread out when poured on a flat surface. The attractive forces allow the collection of molecules to be transferred as a body. Hence, pouring a liquid from one container to another is possible.

Liquids have viscosity. Liquids display an internal resistance to flow. This property is called **viscosity**. Viscosity arises from the attractive forces between molecules. These forces cause the particles to stick to one another, which slows flowing. Some liquids, such as oils, are quite viscous, whereas others have low viscosities and flow readily. Consider the old saying, "slower than molasses in January." At low temperatures, the viscosity of a liquid is greater, since the attractive forces rather than the kinetic motions become more dominant.

Liquids have surface tension. Intermolecular attractive forces have a significant influence on the behavior of liquids. Have you ever floated a needle or a razor blade in a glass of water? Within a liquid, a given molecule is attracted equally by all the molecules that surround it. As illustrated in Figure 11-7, molecules on the surface boundary of a liquid are attracted only by the molecules below them and beside them. These surface molecules experience unbalanced attractive forces. Imagine you are on the surface of a swimming pool and two people are pulling on your arms on the surface and two people are pulling on your legs under the water. You would experience unbalanced tension. Unbalanced attraction on surface molecules causes a net inward pull that tends to draw these molecules into the body of the liquid. Consequently, the liquid surface is under a certain strain or tension, called

FIGURE 11-7 Surface molecules in a liquid experience unbalanced attractive forces.

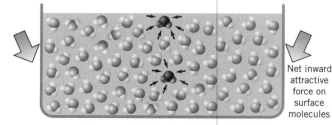

Net inward attractive force on surface molecules

the **surface tension** of the liquid. The surface of a liquid is under tension, and for liquids having relatively high surface tension, some force has to be used to penetrate the surface. A needle or razor blade can be suspended on the surface of water if the object is carefully placed so that it does not penetrate the surface. If the object is pushed below the surface, it sinks.

Liquids resist compression. When a liquid sample is compressed, a very small change in volume occurs. Liquids are essentially incompressible because the amount of free space between molecules is relatively small. The incompressibility of liquids is used in hydraulic systems such as automobile brakes. The hydraulic fluid in the brake system does not compress easily. Thus, when the brake pedal is pushed, the force is transferred by the fluid to the brake shoes.

Sketch an idealized picture to show how the insect called a water strider "walks" on water.

11-9 EVAPORATION

In a sample of a liquid, most particles have similar kinetic energies, but some have relatively high or low energies. Low-energy particles are moving more slowly, and high-energy particles have higher speeds. At a given temperature, the kinetic energies of the particles fit a Maxwell–Boltzmann distribution as mentioned in Section 11-2. Some typical Maxwell–Boltzmann distributions for a liquid sample are shown in Figure 11-8. Recall that the distributions are plots of the percentage of molecules having specific energies versus possible kinetic energies. As an analogy, imagine a plot of the percentage of cars on a highway moving at specific speeds versus the speeds. In such a plot, most cars would have speeds near or around the speed limit, relatively few would have very low speeds, and some would have very high speeds.

The distributions shown in Figure 11-8 represent two different temperatures. In each case, the greatest percentage of molecules have kinetic energies equal to or near the average. As can be seen in the figures,

FIGURE 11-8
Maxwell–Boltzmann distributions of a liquid at two differing temperatures.

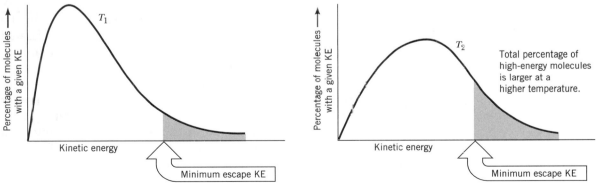

Total percentage of high-energy molecules is larger at a higher temperature.

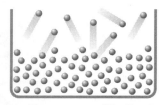

FIGURE 11-9 Idealized evaporation of a liquid.

however, a certain percentage of molecules have relatively high kinetic energies. At higher temperature, the average kinetic energy of the collection is greater and the percentage of high-energy molecules is greater. Of course, this result is expected since heating a liquid changes the speeds of the molecules and changes the kinetic energy distribution.

Note in Figure 11-8 that a certain percentage of molecules have high enough kinetic energies so that they may overcome the intermolecular attractive forces. Molecules with great enough kinetic energy on the surface of the liquid can break free of the attractive forces and escape into the vapor state. The movement of molecules from the liquid state to the vapor state is called **evaporation** (see Figure 11-9). Since the higher energy molecules are those that evaporate, the average kinetic energy of the remaining molecules decreases; thus, the temperature of the liquid state is lowered by the process of evaporation. This process is why you often feel cooler when water evaporates from your skin.

Typically, when a liquid is exposed to the atmosphere, it will evaporate completely. The rate of evaporation depends on the attractive forces between molecules, the amount of surface area of the liquid, and the distribution of kinetic energies of the molecules, which is directly related to the temperature. Therefore, if we heat a liquid or spread it out so that it has more surface area, it will evaporate faster.

11-10 BOILING

As a liquid evaporates, it forms a gaseous vapor. Imagine a liquid as it evaporates. We can picture vapor molecules rising above the liquid. As a gas, the vapor molecules above a liquid exert a measurable pressure called the **vapor pressure** of the liquid. Specific liquids have characteristic vapor pressures that depend on the temperature. The **volatility** of a liquid refers to its readiness to evaporate. A highly volatile liquid, like ethyl ether, has a relatively high vapor pressure, whereas a liquid of moderate volatility, like water, has a moderate vapor pressure. Chemicals with low vapor pressures are said to be nonvolatile. Nonvolatile chemicals do not evaporate readily.

The vapor pressure of a liquid increases as the temperature of the liquid increases. Consider what happens when a liquid sample is heated in a container open to the atmosphere. As the liquid heats, the temperature increases. The increase in temperature corresponds to an increase in the kinetic energy of the liquid particles. It follows that as the temperature increases the vapor pressure increases because a greater number of particles evaporate. As heating continues, a point is reached at which further heating does not increase the temperature of the liquid and bubbles of vapor form rapidly throughout the liquid sample. The bubbles rise to the surface and burst as the vapor escapes. When this occurs, the liquid is said to be boiling. Boiling does not occur until the vapor pressure of the liquid is equal to the prevailing atmospheric pressure to which the liquid is exposed. The tiny bubbles of gas you may see coming out of water when it is first heated are bubbles of dissolved air and are different from the large bubbles of water vapor you see when water boils vigorously.

11-11 THE BOILING POINT

When the vapor pressure of a liquid equals the pressure of its environment, there is essentially no inhibition of the evaporation process within the liquid, and very rapid evaporation process, called boiling, takes place. As boiling occurs, the temperature of the liquid remains constant. The temperature at which the vapor pressure of a liquid equals the external pressure is the **boiling point** of the liquid. Each liquid has a characteristic boiling point. The boiling depends on the prevailing external pressure. The **normal boiling point** of a liquid is the temperature at which it boils when exposed to a pressure of 1 atm. The normal boiling point of water is 100°C, but at a mile-high altitude, such as that of Denver, Colorado, water boils around 95°C. On the other hand, increased pressure, such as occurs inside a pressure cooker, raises the boiling point of water.

Adding more heat to a boiling liquid does not increase the temperature of the liquid, but it does increase the rate of evaporation. As long as a liquid is constantly heated, boiling continues until all the liquid has evaporated. The vapor formed by an evaporated liquid normally escapes into the atmosphere. If, however, the vapor is collected in some way, it can be cooled and caused to condense back to a liquid.

Special laboratory and industrial equipment is designed to allow the boiling of a liquid and the subsequent condensation of its vapors. The process of boiling a liquid and condensing the resulting vapors is called **distillation**. Distillation is used to separate mixed liquids of different volatilities and to purify liquids contaminated with nonvolatile impurities. Oil refineries include distillation columns in which crude oil is separated into various petroleum products. After separation some of the petroleum products are chemically transformed into more useful products. See Figure 11-10 on page 340. In the manufacture of distilled liquors, such as whiskey and scotch, distillation is used to boil off an alcohol–water mixture that has an increased percent of alcohol. A process somewhat similar to distillation occurs in the environment. When seawater is heated by solar energy, some water evaporates. Since the dissolved salts in seawater are not volatile, the water vapor rising from the oceans is desalinated. This vapor condenses to form rain or snow, and some of it is precipitated on land areas as "fresh" water.

Petroleum

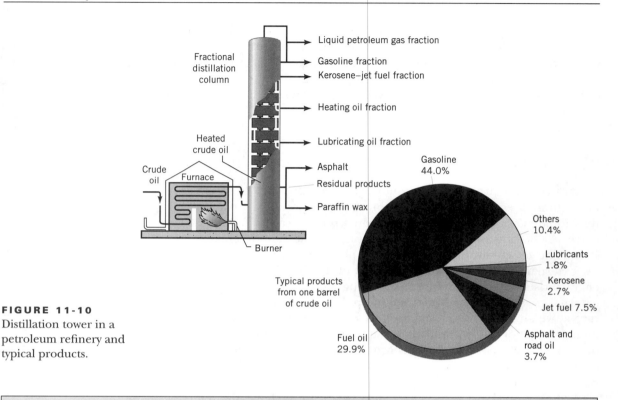

FIGURE 11-10
Distillation tower in a petroleum refinery and typical products.

Mining and Refining

In an industrial society like the United States, the most important nonfuel resources are iron ore, aluminum ore, copper ore, and minerals containing nitrogen, phosphorus, and potassium used in fertilizers. In addition, significant amounts of cement, gravel, stone, and asphalt are used. Much energy is expended in mining operations and the transportation of ore. To make the operations of any physical extraction and transportation worthwhile, the ores must contain enough of the desirable compound or element. Some low-grade ore deposits are not usable because of the economics of the mining process.

Materials extracted from the earth are processed by screening, separating, washing, and grinding, which produces wastes (some are called mine tailings) and dusts (fine solid particles suspended in air). Often, the desirable element in an ore is in an undesirable chemical form and must be converted to a desirable form. For instance, the aluminum in aluminum ore is in the form of aluminum ion (Al^{3+}) in combination with oxide ion (O^{2-}). To be useful, the aluminum ion needs to be reduced to aluminum metal. The chemical or electrical processing of an ore designed to isolate an element is called the **refining** of the ore. Refining typically involves a chemical reaction or series of chemical reactions that produce the specific element or compound sought from the ore. Table 11-2 lists some important metals obtained from ores. Also shown in the table are common uses of these metals and amounts annually used in the United States. Refining requires the use of relatively large amounts of energy, and any cooling and washing processes involve relatively large amounts of water.

Table 11-2   *Common Metals in Commercial Use*

METAL	USES	ESTIMATED ANNUAL WORLD CONSUMPTION (METRIC TONS)	ESTIMATED USE IN UNITED STATES (PERCENT)
Aluminum	Building construction, auto parts, power transmission lines, consumer containers, foils, cooking utensils, catalyst	20 million	22
Cadmium	Bearings, alloys, batteries, TV phosphors, glazes, fungicides, nuclear control rods	18 thousand	12
Copper	Electrical wires and parts, plumbing, heating, roofing, alloys (bronze, brass), electroplating, cooking utensils, catalyst, insecticides, algicides	11 million	24
Iron (steel)	Construction material, ship hulls, auto and truck bodies and parts, piping, tire belts, machines and machine parts, catalyst, magnets	740 million	13
Lead	Batteries, radiation shielding, ammunition, solder, alloys, lead tetraethyl	5 million	26
Mercury	Fungicides, bactericide, dental amalgams, catalyst, electrical switches, cells for chlorine production, paints, pulp and paper production, pharmaceuticals, fluorescent and mercury vapor lamps	7 thousand	18
Nickel	Alloys, batteries, fuel cells, ceramics, catalyst, plating	880 thousand	16
Tin	Tin plate, pewter, bronze, foils, flexible containers, solders, low-melting alloys, dental amalgam, cast type, tinned copper wire	217 thousand	15
Zinc	Alloys (brass, bronze, die-casting) galvanized iron, electroplating, auto and truck parts, dry cell batteries, fungicides, roofing	7 million	16

ACTIVITY 11-4

Look up *amalgam* in a dictionary and record its definition.

Calculate the percent by mass of gold in 18-carat gold.

Metals have various uses based on their properties and the compounds they form. In general, metals and alloys are readily cast, formed, or pressed into sheets and wires. Alloys are mixtures of two or more metals or metals and selected nonmetals. They are formulated to have useful properties different from those of the component elements. Bronze, for example, a mixture of copper and tin, typically containing 1% to 10% tin, is much harder and durable than either of its components. Silver metal is made more durable and useful in jewelry by alloying it with copper. Sterling silver contains 92.5% Ag and 7.5% Cu. Pure gold is a very soft and flexible metal. Adding copper to gold forms an alloy that is less soft and, thus, more useful. Gold is alloyed with copper and other metals, and the gold content is expressed as carats, the number of parts of gold by mass in 24 parts of the alloy. Thus, 24-carat gold is pure gold and 18-carat gold has 18 parts of gold in 24 parts of the alloy.

Iron ores are mixtures of iron oxides and silicate rocks. The principal oxides are Fe_2O_3 (hematite ore) and Fe_3O_4 (magnetite ore). These iron oxides are reduced to iron by mixing the ore with carbon, limestone, and air in a blast furnace at temperatures exceeding 1300°C. Conventional blast furnaces use mostly virgin ore along with recycled scrap iron. Newer electric arc furnaces, however, use nearly 100% recycled scrap iron. In the United States, about 50% of scraped iron and steel are recycled, but the rest is buried or placed in junkyards and may never be used again.

Pure iron is not useful since it is relatively soft and easily rusts. Carbon steels are alloys of iron containing 0.02% to 1.5% carbon. Specialty steels contain variable amounts of other metals such as nickel, chromium, vanadium, tungsten, manganese, cobalt, or zirconium in addition to carbon. Steels resist corrosion and are more useful than pure iron.

Aluminum is the most abundant metal in the lithosphere, and it is found in most clays and rocks. Relatively pure aluminum oxide, Al_2O_3, occurring as corundum (emery), ruby, and sapphire is rare. Bauxite is aluminum ore containing aluminum oxide, silicates, and some iron oxides. Bauxite is treated to isolate aluminum oxide used to make aluminum metal. Australia, Guinea, and Jamaica produce more than 60% of the world's bauxite.

Aluminum ion cannot be easily reduced chemically, so aluminum metal is made by the electrolysis of a molten solution of aluminum oxide. This process is discussed in Section 15-12. Aluminum is a low-density, nontoxic metal that is easily formed or cast into a variety of shapes, ranging from structural beams to sheets and foils. It is corrosion resistant and is a good conductor of heat and electricity. Metallic aluminum alloyed with other metals is used for such items as kitchen utensils, building decorations and structural components, aircraft construction, electrical transmission lines, mirrors, automobile parts, foils, and food and drink cans.

11-12 THE SOLID STATE

Solids

Consider what happens when a liquid is cooled rather than heated. As the liquid is cooled, the kinetic motions of the particles decrease. As cooling continues, the particles begin to occupy relatively fixed positions in space. In fact, as the liquid cools, the attractive forces become dominant and a point is reached when the particles arrange in a three-dimensional pattern in which they occupy definite spatial positions. When the particles arrange in fixed positions, the liquid has changed to a solid. The change from the liquid to the solid state is called **solidification**, **crystallization**, or **freezing**. See Figure 11-11. The temperature at which a given liquid changes to a solid is its **freezing point**.

A crystalline solid is characterized by a definite three-dimensional arrangement of particles occupying relatively fixed positions in space. Such a three-dimensional arrangement makes a **crystal lattice** or simply a **crystal**. Metals, ionic compounds, and molecular compounds can form crystalline solids. Crystals occur in a variety of shapes and sizes. Some crystals are fine powders, some are rough chunks and bits, and some have beautiful geometrical shapes. The positions that the particles occupy in the crystal lattice are **crystal lattice sites**. Molecules occupy these sites in a molecular solid. In metals, the metal atoms occupy lattice sites; in ionic compounds, cations and anions occupy lattice sites. See Figure 11-12. Some solids, such as glasses, are noncrystalline and have amorphous structures.

The particles occupying the lattice sites in a crystalline solid are held in a fixed arrangement in space. These particles, however, still have some kinetic energy that results from the particles vibrating about their positions in the crystal lattice (see Figure 11-13 on page 344). This vibratory motion allows the transfer of kinetic energy throughout the solid by collision of a particle with its neighboring particles. The vibrations allow particles in solids to exchange kinetic energy and conduct heat. The exchange of kinetic energies also allows surface particles to evaporate.

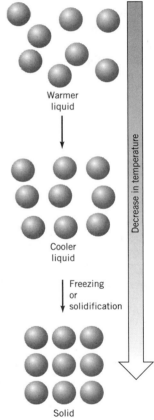

FIGURE 11-11
Solidification or freezing of a liquid. As a liquid is cooled, the particles ultimately arrange in a definite three-dimensional pattern in which they have fixed positions in space.

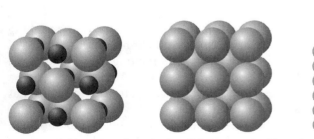

Solids as three-dimensional spatial arrangements of ions, atoms, or molecules.

FIGURE 11-12   Ionic and molecular solids.

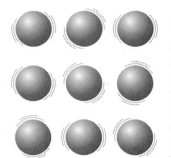

FIGURE 11-13

Particles occupying lattice sites in a crystalline solid vibrate within these positions. These vibrations and the resulting collisions between particles can transfer kinetic energy throughout the crystal.

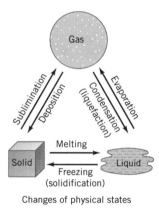

Changes of physical states

Make a drawing or sketches of molecules showing the melting of ice to form liquid water and the boiling of liquid water to give steam.

11-13 SUBLIMATION AND MELTING

As you know, some solids have distinct odors and some have no noticeable odor. At a given temperature, some particles near the surface of a solid have enough kinetic energy to break free of the solid state and enter the vapor state. A solid has a characteristic vapor pressure at a given temperature, but the vapor pressures of solids normally are much lower than those of liquids. The direct transition from the solid state to the vapor state is called **sublimation**. A common example is the sublimation of dry ice, which is solid carbon dioxide. When exposed to the atmosphere and room temperatures, dry ice sublimes and changes to carbon dioxide gas. Dry ice, however, is cold enough to cause water vapor from the air to condense on the surface as solid ice. Furthermore, the low temperatures of dry ice often cause some water vapor to condense into water droplets, forming a cloudy mist. Most solids do not sublime at room temperature, but many have enough vapor pressure so that they have detectable odors.

The average kinetic energy of the particles in a sample of a solid is directly proportional to the temperature. When a solid is heated, the particles gain enough energy so that they break free from their fixed positions in space and collect as a liquid. The change of a solid to a liquid is called **melting** or **fusion**. For a pure substance, the transition from a solid to a liquid occurs at a characteristic temperature called the **melting point**. When a solid is heated to its melting point, it begins to melt. At the melting point, additional heat causes more solid to melt, and the temperature does not change until the entire solid melts.

Because each pure chemical has a unique melting point, the melting point is used to help identify a chemical. For instance, pure aspirin, acetylsalicylic acid, has a melting point of 135°C. Suppose we measure the melting point of a substance that we think is pure aspirin and find that it does not have a melting point of 135°C. We can conclude that either the substance is impure aspirin mixed with other chemicals or that it is not aspirin at all.

11-14 CHANGES OF STATE

To melt a sample of a solid such as ice, we need to heat it to the melting point. To melt the entire sample, we continue heating until all the solid changes to liquid. Once a solid is heated to its melting point, additional heat is needed to convert the entire solid to the liquid state.

When a substance changes from one physical state to another, it has undergone a **change of state**. The term *state* refers to the solid, liquid, or gaseous state. Melting, as a change of state, requires energy to allow the particles to break free of the crystal lattice sites in the solid and enter the liquid state in which the particles are less ordered. As previously mentioned, a similar situation occurs when a liquid boils. To boil, a liquid must

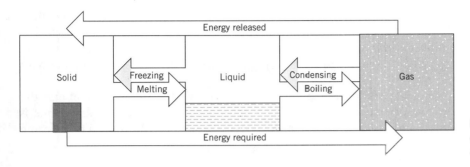

FIGURE 11-14 Energy exchanges during changes of state.

be heated to the boiling point. Heat needs to be continually supplied to continue the boiling process. The energy is needed to allow the particles to break free from the intermolecular attractions holding the particles in the liquid state and allow them to enter the vapor state. Figure 11-14 shows how energy is required to melt a solid or to boil a liquid. As shown in the figure, it follows that energy is released when a vapor condenses to a liquid and when a liquid freezes to a solid. These energy exchanges correspond to the law of conservation of energy.

11-15 ENERGY CHANGES IN MELTING AND FREEZING

The amount of energy needed to melt a specific amount of a solid at a given temperature is called the **heat of fusion**. Fusion is another word for melting. Each pure solid has a characteristic heat of fusion that depends on the nature of its crystal state and the attractive forces that exist between particles. Heats of fusion are measured by experiment and are expressed sometimes in units of kilojoules per mole (kJ/mol) and sometimes in kilocalories per mole (kcal/mol).

Since water is such a common chemical and since we are familiar with its melting and freezing behavior, it is used here as an example. When ice, which is solid water, is heated to 0°C and begins to melt, 6.008 kJ or 1.436 kcal of heat are needed to melt 1 mole of ice. The melting of ice to form liquid water is represented by equations showing the energy involved:

$$6.008 \text{ kJ} + H_2O(s) \longrightarrow H_2O(\ell)$$

$$1.436 \text{ kcal} + H_2O(s) \longrightarrow H_2O(\ell)$$

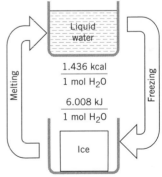

These equations show the amount of heat needed to melt 1 mole of ice. Note that energy is on the left since melting is an endothermic process.

A liquified solid normally freezes at the same temperature at which it melted. This temperature is called the freezing point. Freezing or solidification is essentially the reverse of melting. As the particles of a liquid arrange in the crystal lattice of the solid state, their excess energy is released as heat. The amount of heat released when a specific amount of liquid solidifies or freezes is called the **heat of crystallization**. Crystallization is just another name for freezing. Note that energy is needed to melt a solid, and the same amount of energy is released when the melted solid freezes or crystallizes. This equality follows the law of conservation of energy.

Since crystallization is the reverse of fusion, the heat of fusion and the heat of crystallization have the same numerical value. The difference is that fusion is endothermic and crystallization is exothermic. The freezing or crystallization of water is represented as

$$H_2O(\ell) \longrightarrow H_2O(s) + 6.008 \text{ kJ}$$

$$H_2O(\ell) \longrightarrow H_2O(s) + 1.436 \text{ kcal}$$

These equations reflect that 6.008 kJ or 1.436 kcal of heat are released when 1 mole of liquid water freezes at 0°C. Note that these equations are simply the reverse of the equations given earlier for the melting of ice.

The heat of fusion or crystallization are written as factors that relate heat to moles of water.

$$\left(\frac{6.008 \text{ kJ}}{1 \text{ mol } H_2O}\right) \quad \text{or} \quad \left(\frac{1.436 \text{ kcal}}{1 \text{ mol } H_2O}\right)$$

EXAMPLE 11-1

How many kilocalories of heat are needed to melt an 83 g sample of ice at 0°C?

This essentially stoichiometric question can be summarized as follows.

Given   Want   calculation sequence: g H_2O $\longrightarrow$ mol H_2O $\longrightarrow$? kcal
83 g ice   ? kcal

First, the mass of ice is converted to moles of water and then the heat is found using the heat of fusion factor.

$$X \text{ kcal} = 83 \text{ g} \left(\frac{1 \text{ mol } H_2O}{18.0 \text{ g}}\right)\left(\frac{1.436 \text{ kcal}}{1 \text{ mol } H_2O}\right) = 6.6 \text{ kcal}$$

EXAMPLE 11-2

How many grams of ice can be melted with 25.6 kJ of heat?

A summary of the question follows.

Given   Want   calculation sequence: kJ $\longrightarrow$ mol H_2O $\longrightarrow$? g H_2O
25.6 kJ   ? g ice

The heat of fusion is used as a factor to find the moles of water related to the kilojoules of heat. The moles of water are then converted to mass.

$$X\,g = 25.6\,\cancel{kJ}\left(\frac{1\;\cancel{mol\,H_2O}}{6.008\,\cancel{kJ}}\right)\left(\frac{18.02\;g}{1\;\cancel{mol\,H_2O}}\right) = 76.8\;g$$

How much heat is needed to melt 1.00 g of ice at 0°C?

11-16 ENERGY CHANGES IN BOILING AND CONDENSATION

As you know, energy is required to boil a liquid. When a liquid is evaporated or boiled, the particles change from a more ordered liquid state to the more random vapor state. Since energy is required for vaporization, a measurable amount of heat is required to change a specific amount of liquid to the vapor state at a given temperature. This amount of heat is called the **heat of vaporization**. Vaporization at the boiling point is, of course, known as boiling. Each liquid has a characteristic heat of vaporization. For water, the heat of vaporization at 100°C is 40.66 kJ/mole or 9.718 kcal/mole. The boiling of water is represented by the equations

$$40.66\;kJ + H_2O(\ell) \longrightarrow H_2O(g)$$

$$9.718\;kcal + H_2O(\ell) \longrightarrow H_2O(g)$$

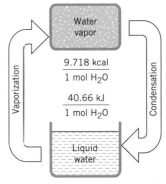

These equations show that boiling requires energy and is endothermic.

When a vapor condenses to a liquid, the particles change from the more random gaseous state to a more ordered liquid state. As expected by the law of conservation of energy, condensation releases energy. The amount of energy released when a mole of vapor condenses at a given temperature is called the **heat of condensation**. Since condensation is the reverse of vaporization, for a given liquid the heat of condensation has the same value as the heat of vaporization. The release of energy when steam condenses at 100°C is represented by the following equations written as exothermic processes:

$$H_2O(g) \longrightarrow H_2O(\ell) + 40.66\;kJ$$

$$H_2O(g) \longrightarrow H_2O(\ell) + 9.718\;kcal$$

Because the condensation of steam releases energy, steam burns are dangerous. When steam condenses on your skin, energy is released by the condensation, and the water formed is initially at 100°C.

EXAMPLE 11-3

How many kilojoules of heat are needed to completely evaporate 237 g, which is about a cup, of water, at 100°C? How many kilocalories are needed?

ACTIVITY 11-8

How much heat is needed to boil 1.00 g of liquid water at 100°C?

The heat of vaporization is used as a factor to relate moles of water to heat. A summary of the question is

Given   Want   calculation sequence: g $H_2O \longrightarrow$ mol $H_2O \longrightarrow$ heat
237 g   ? kJ or ? kcal

$$X \text{ kJ} = 237 \text{ g} \left(\frac{1 \text{ mol } H_2O}{18.02 \text{ g}} \right) \left(\frac{40.66 \text{ kJ}}{1 \text{ mol } H_2O} \right) = 535 \text{ kJ}$$

$$X \text{ kcal} = 237 \text{ g} \left(\frac{1 \text{ mol } H_2O}{18.02 \text{ g}} \right) \left(\frac{9.718 \text{ kcal}}{1 \text{ mol } H_2O} \right) = 128 \text{ kcal}$$

ACTIVITY 11-9

Why is a steam burn more dangerous than a burn from hot water?

Temperature

Mass

Heat

Specific heat
(cal/1 g 1°C) or
(J/1 g 1° C)

11-17 SPECIFIC HEAT

When you heat water on a stove or hold a piece of metal in a flame, you observe that increasing the temperature of a substance by heating it is easy. You may also observe that some substances, like metals, heat more quickly than others, like water. When an object is heated, its particles gain kinetic energy and the temperature, of course, increases. In contrast, when an object is placed in a cooler environment, its temperature decreases as its particles lose energy.

It is possible to measure the amount of heat needed to change the temperature of a sample of a given substance. The **specific heat** of a substance is defined as the number of joules or calories needed to change the temperature of 1 g by 1°C. It is observed that the specific heat of a substance varies depending on its temperature. For many substances, the specific heat does not vary greatly over small temperature ranges, so we can assume a constant value for the specific heat of a substance.

Each pure chemical has a characteristic specific heat. For instance, the specific heat of liquid water is 1.00 cal per gram-degree Celsius and that of solid aluminum is 0.212 cal per gram-degree Celsius. Hence, 1.00 cal is needed to change 1 g of water by 1°C and only 0.212 cal is needed to change 1 g of aluminum by 1°C. Thus, a sample of aluminum heats more rapidly than a sample of water of the same mass. Furthermore, the heated aluminum loses its heat more quickly than the heated water.

The specific heat of a substance is expressed as a factor relating the number of joules or calories needed to change 1 g by 1°C. For example, as a factor, the specific heat of water is

$$\left(\frac{4.18 \text{ J}}{1 \text{ g } 1°C} \right) \quad \text{or} \quad \left(\frac{1.00 \text{ cal}}{1 \text{ g } 1°C} \right)$$

The specific heat of water is 1.00 calorie per gram-degree Celsius. It has this value because the calorie was originally defined as the amount of heat required to change the temperature of 1 g of water by 1°C. Later the calorie was defined as the amount of heat equal to 4.184 joules. As a result, the specific heat of water kept its easily remembered value. Among chemicals, water has a relatively high specific heat, which has environmental consequences. A body of water heats slowly and cools slowly; therefore, it holds heat longer than other environmental objects. This property of water causes coastal regions or regions near large lakes to have less extreme temperature variations compared with inland regions.

Although the specific heat of water varies with temperature, we can assume the 1.00 cal/g°C value is valid for liquid water from 0°C to 100°C. Be aware, however, that a substance has different specific heat values in different physical states. For example, solid water (ice), liquid water, and water vapor (steam) have significantly different specific heats. When specific heats are used in calculations, appropriate values can be found in reference tables on the Web or in books.

Look closely at the units of the specific heat of water. Note that the numerator has units of energy and the denominator has two units, grams and degrees Celsius. In other words, specific heats are factors that relate three units. Specific heats are used to find the number of joules or calories needed to change the temperature of a given mass by a specified number of degrees Celsius. If the specific heat is given along with the mass of a sample of a substance, the heat needed to cause a known temperature change (ΔT) is found by multiplying the mass and the temperature change by the specific heat (sometimes represented by the symbol c_p for heat capacity at constant **pressure**):

$$\text{heat} = (\text{mass})(\Delta T)(c_p)$$

When a substance changes temperature, the temperature change or ΔT is found by subtracting the lower temperature from the higher temperature.

ACTIVITY
11-10

Would a substance with a relatively low specific heat be a good or poor conductor of heat?

EXAMPLE 11-4

How many calories of heat are needed to change the temperature of 237 g of water from 25.0°C to 35.0°C?

A summary is

GIVEN	WANT	FACTOR
237 g H_2O heated from 25.0°C to 35.0°C	? cal	(1.00 cal/g °C) specific heat of water

Note the units of specific heat. The heat needed is found by multiplying the mass and the temperature change, ΔT, by the specific heat so that the units of grams and degrees Celsius cancel. The temperature change is found by subtracting the initial temperature from the final temperature ($\Delta T = 35.0°C - 25.0°C = 10.0°C$).

$$X \text{ cal} = (237 \, \cancel{g})(10.0°\cancel{C})(1.00 \text{ cal/}\cancel{g}°\cancel{C}) = 2.37 \times 10^3 \text{ cal} \quad \text{or} \quad 2.37 \text{ kcal}$$

FIGURE 11-15
Graphical representation
of heat exchanges during
the melting, heating, and
boiling of water.

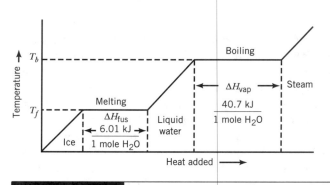

EXAMPLE 11-5

The specific heat of solid aluminum is 0.887 J/g °C. How many joules of heat are released when 365 g of aluminum cools from 80.4°C to 25.0°C?

A summary is

GIVEN	WANT	FACTOR
365 g Al cools from 80.4°C to 25.0°C	? J	(0.887 J/g °C) specific heat of aluminum

Note the units of specific heat. The number of joules is found by multiplying the mass and ΔT by the specific heat.

$$X J = (365 \text{ g})(80.4°C - 25.0°C)(0.887 \text{ J/g°C}) = 1.79 \times 10^4 \text{ J} \quad \text{or} \quad 17.9 \text{ kJ}$$

Figure 11-15 shows graphically the energies needed to melt ice, heat the resulting liquid water to its boiling point, and boil the water to vapor. The specific heat of water is used along with the heats for changes of state to find the number of joules or calories involved in the heating or cooling of water with accompanying changes of state. For instance, it takes energy to melt ice to liquid water and additional energy to heat the water from 0°C to higher temperatures.

EXAMPLE 11-6

How many calories of heat are needed to melt 50.0 g of ice at 0°C, heat the resulting water from 0°C to 100°C, and boil the water to steam at 100°C?

A summary is

GIVEN	WANT		FACTORS	
50.0 g ice melted at 0°C, heated to boiling at 100°C, and vaporized	? cal	1.436 kcal/mol heat of fusion	1.00 cal/g °C specific heat	9.718 kcal/mol heat of vaporization

Again note the units of the factors. First, the number of moles of water is found from the mass and used with the heat of fusion to find the kilocalories for melting. Next, the mass of water and the ΔT (100°C − 0°C) are multiplied by the specific heat to find the heat needed to change the temperature of the water. Then, the number of moles of water as found from the mass is multiplied by the heat of vaporization to find the kcal needed to vaporize the sample. Finally, all the results are added to find the total heat needed.

Melting $\qquad 50.0 \, g \left(\dfrac{1 \; mol \; H_2O}{18.02 \, g} \right) \left(\dfrac{1.436 \; kcal}{1 \; mol \; H_2O} \right) = 3.984 \; kcal$

Heating water $\quad (50.0 \, g)(100°C)(1 \; cal/g°C) = 5000 \; cal \; or \; 5.000 \; kcal$

Boiling $\qquad 50.0 \, g \left(\dfrac{1 \; mol \; H_2O}{18.02 \, g} \right) \left(\dfrac{9.718 \; kcal}{1 \; mol \; H_2O} \right) = 26.96 \; kcal$

Total heat $\quad$ melting $\qquad + \qquad$ heating $\qquad + \qquad$ boiling
$\qquad\qquad$ 3.984 kcal $\quad + \quad$ 5.000 kcal $\quad + \quad$ 26.96 kcal = 35.9 kcal

Note that we carried extra digits and rounded the final answer to three digits.

Silicon

Silicon and Solid-State Electronics

Silicon is the second most abundant element in the crust of Earth. Typically, it occurs in combination with oxygen in silica, SiO_2, and silicate minerals. More than 90% of Earth's crust is silica and silicate rocks. Silica or silicon dioxide occurs as quartz, ranging from beautiful large crystals to fine-grained silica sand. Silicate minerals contain silicon and oxygen in combination with various metals, the most common of which is aluminum. The basic structural unit in silica and silicates is the silicon–oxygen tetrahedra:

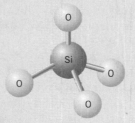

Quartz crystals consist of a three-dimensional array of these tetrahedra covalently bonded in sequence. In silicate minerals, the silicon–oxygen tetrahedra arrange in various complex networks. In such minerals, the silicon–oxygen bonds are covalent and the negatively charged open bonding positions on the oxygen atoms allow ionic bonding to various metal cations. Most silicates have complex three-dimensional structures. Some, like mica, occur in sheets and some, like asbestos, are fibrous. Clays are minerals composed of complex mixtures of aluminum silicates that form by the fracturing and weathering of rocks.

Long ago, humans discovered that clays mixed with water could be formed into desirable shapes and heated to form rigid and porous ceramic objects. Some of the oldest known ceramic vessels have been dated to 30,000 years ago. Ceramic technology is an ancient human endeavor used to make objects of art, vessels, bricks, and tiles. Various glazes and colorful minerals are added to produce nonporous and decorative objects. The ancient Sumarian civilization developed the pottery wheel, and the Chinese invented fine porcelain ceramics.

About 5000 years ago, Egyptian artisans discovered that mixtures of silica sand, sodium carbonate, and limestone heated to 1000°C fuse to form glass. Hot glass is flexible and can be manipulated and molded into various shapes. Upon cooling, it solidifies into a clear rigid, amorphous solid. As a mixture of silica and various metallic oxides, glass is viewed as a supercooled liquid rather than a crystalline solid.

Typical soft glass or soda-lime glass contains 70% by mass silicon dioxide, 15% sodium oxide (Na_2O), and 10% calcium oxide (CaO), and the rest consists of various metallic oxides. Adding some boron oxides makes heat-resistant borosilicate glass commonly called Pyrex. Various oxides lend unique colors to glass. Cobalt oxides make glass blue, manganese oxides makes it purple, chromium oxides makes it green, selenium and zinc oxides make it red, and copper oxides makes it red or blue-green. When lead oxide is added to glass, it makes a material that reflects light in a unique way, producing the brilliant glass used in leaded crystal. Adding silver chloride, AgCl, and a trace of copper compounds to glass makes photochromic glass that turns darker in sunlight and fades to clear in the absence of sunlight.

When humans discovered how to reduce iron from ores, the result was the beginnings of the Iron Age. Archeologists think that the refining of iron was first done in Africa and possibly simultaneously in China. The Iron Age reached its height in the Industrial Revolution, which began in the 1700s. Steel became abundant and was used for construction and to make locomotives (iron horses) and rails. The ultimate steel constructions are the automobile and the bicycle, which allow individual mobility. Today, the steel content of automobiles is gradually being replaced with alternative materials, but automobiles still represent the largest use of steel.

The modern technological revolution is based on silicon. Silicon computer chips are jokingly called educated grains of sand because elemental silicon is made from ordinary silica sand. Silicon dioxide is reduced to silicon by mixing it with carbon in a 3000°C furnace.

$$SiO_2 + 2C \longrightarrow Si + 2CO$$

Purer silicon is then formed by first making silicon tetrachloride that is then reduced using magnesium.

$$Si + 2Cl_2 \longrightarrow SiCl_4$$
$$SiCl_4 + 2Mg \longrightarrow 2MgCl_2 + Si$$

This product is melted and treated to remove impurities, making the ultrapure silicon needed to make electronic chips.

Solid silicon has a three-dimensional tetrahedral bonding sequence much like that found in diamond. Note that silicon is in the same family as carbon. Pure silicon is a semiconductor, but it becomes a unique semiconductor by adding low concentrations of selected impurities in a process called doping. A semiconductor is a material that conducts electricity in specific ways only under certain conditions. Silicon atoms have four valence electrons that result in a three-dimensional network of silicon atoms in solid silicon. For our discussion, a silicon crystal is shown in a simplified flat representation.

Arsenic atoms have five valence electrons. When silicon is doped with arsenic, the extra electrons are free to move under the influence of voltage. The doped silicon is called an *n*-type (*n* for negative) semiconductor.

Gallium atoms have three valence electrons. When silicon is doped with gallium, some bonding electrons are absent, leaving incomplete bonds or "holes" in the crystal.

Under the influence of voltage, electrons move about in the crystal which corresponds to the holes moving. In this case, the silicon becomes a *p*-type (*p* for positive) semiconductor.

Combining various layers of doped silicon, it is possible to fabricate electronic components such as *npn* field-effect transistors. The use of precision deposition and masking techniques allows the fabrication of tiny chips of silicon containing millions of transistors. These integrated circuit devices called microchips are used to make computer microprocessors and memory chips (RAM and ROM). Microminiaturization of computer components is the basis of the computer revolution and the information age. Integrated circuits find use in computers, calculators, digital watches and displays, alarms, sensors, telephone systems, safety devices, home appliances, and automobile parts.

FOCUS
11

Liquids and Solids

Solid: A physical state in which a sample of matter occupies a definite volume and has a definite, rigid shape. A solid is viewed as particles arranged in fixed three-dimensional spatial positions in which the particles vibrate about their fixed positions.

Liquid: A physical state in which a sample of matter occupies a definite volume and takes on the shape of the portion of the container it occupies. A liquid is viewed as a collection of particles that are closely packed and are constantly moving and wandering about within the collection.

Gas: A physical state in which a sample of matter has no definite volume, has no definite shape, and occupies the entire container. A gas is viewed as a diffuse collection of molecules that are in random high-speed motion. On the average, the molecules are far apart and occupy the volume by their incessant motion.

Electronegativity: The tendency of an atom of an element to attract electrons in a chemical bond. Except for the noble gases, the electronegativities of elements increase from bottom to top in columns and from left to right in rows of the periodic table. Fluorine is the most electronegative element, followed by oxygen and chlorine. Metals generally have low electronegativities.

Polar bond: A covalent bond between atoms of elements having different electronegativities.

Polar molecule: A molecule that has a separation of positive and negative charge centers.

To be polar, a molecule must satisfy two criteria:

1. The molecule must have one or more polar bonds.
2. The molecule must have a shape that results in a net separation of positive and negative charge centers.

Nonpolar molecules: Molecules with no net separation of charge centers.

Nonpolar molecules exist in two forms:

1. Molecules in which the atoms have little or no differences in electronegativity.
2. Molecules that have polar bonds but a symmetrical shape, resulting in no net separation of charge.

Intermolecular attractive forces are called van der Waals forces. When polar molecules are in close proximity, the opposite poles attract through polar interaction or dipole interaction. Intermolecular attractive forces also occur between nonpolar molecules. Nonpolar molecular attractions are weaker than polar molecular attractions, but they are important in substances consisting of nonpolar molecules. Nonpolar molecules attract one another when the negatively charged electrons within molecules are momentarily attracted to the positively charged nuclei of other molecules.

Vapor pressure: The pressure of a vapor associated with a liquid or a solid.

Boiling point: The temperature at which the vapor pressure of a liquid equals the prevailing pressure.

Melting point: The temperature at which a solid changes to a liquid.

Freezing point: The temperature at which a liquid changes to a solid. For a given chemical the melting point and freezing point are the same.

Heat of fusion: The amount of heat needed to melt a specific amount of a solid at a given temperature.

Heat of crystallization: The amount of heat released when a specific amount of liquid solidifies or freezes at a given temperature.

Heat of vaporization: The amount of heat required to change a specific amount of liquid to the vapor state at a given temperature.

Heat of condensation: The amount of heat released when a specific amount of vapor condenses at a given temperature.

The heat of fusion and heat of crystallization of water at 0°C expressed as factors are

$$\left(\frac{6.008 \text{ kJ}}{1 \text{ mol } H_2O}\right) \quad \text{or} \quad \left(\frac{1.436 \text{ kcal}}{1 \text{ mol } H_2O}\right)$$

The heat of vaporization and heat of condensation of water at 100°C expressed as factors are

$$\left(\frac{40.66 \text{ kJ}}{1 \text{ mol } H_2O}\right) \quad \text{or} \quad \left(\frac{9.718 \text{ kcal}}{1 \text{ mol } H_2O}\right)$$

Specific heat: The number of joules or calories needed to change the temperature of 1 g of a substance by 1°C. The specific heat is used to find the amount of heat needed to change a given mass of a substance by a certain number of degrees Celsius: heat = (mass)(ΔT)(c_p), where mass is the number of grams of a substance, ΔT is the change in number of degrees Celsius, and c_p is the specific heat.

QUESTIONS

1. Match the following terms with the appropriate lettered definitions.

gas

liquid

solid

Maxwell–
Boltzmann
distribution

electronegativity

polar molecule

Van der Waals
forces

viscosity

surface tension

evaporation

a. the internal resistance to flow

b. the movement of molecules from the liquid to the vapor state

c. intermolecular attractive forces

d. the amount of heat needed to melt a specific amount of a solid at a given temperature

e. the amount of heat released when a specific amount of liquid solidifies or freezes at a given temperature

f. the amount of heat required to change a specific amount of liquid to the vapor state at a given temperature

g. the amount of heat released when a specific amount of vapor condenses at a given temperature

h. amount of heat needed to change 1 g of a substance by 1°C

i. the rapid and uninhibited evaporation of a liquid

j. the temperature at which the vapor pressure of a liquid equals the prevailing pressure

vapor pressure	k. the pressure exerted by the vapor associated with a liquid
boiling	l. the temperature at which a solid changes to a liquid
boiling point	m. a molecule having a net separation of positive and negative charge centers
melting point	n. a measure of the tendency of the atoms of an element to attract electrons in chemical bonds
heat of fusion	o. the state of matter having a fixed volume and rigid shape
heat of crystallization	p. the state of matter having a fixed volume and a flexible shape
heat of vaporization	q. the state of matter having a flexible volume so that it completely occupies any container in which it is placed
heat of condensation	r. a plot of the fraction of molecules in a collection versus the kinetic energies of the molecules
specific heat	s. the inward attraction exerted on molecules on the surface of a liquid

Section 11-1

2. Briefly describe the nature of a solid, a liquid, and a gas and describe the differences between them.

Section 11-2

3. What is kinetic energy as it applies to chemical particles?

4. Sketch and describe a Maxwell–Boltzmann distribution of molecular kinetic energies.

Section 11-3

5. Define electronegativity. List the four most electronegative elements in descending order. Which group contains the elements of lowest electronegativity?

Sections 11-4 and 11-5

6. What is a polar covalent bond?

7. Describe a polar molecule and give an example.

8. There are two different types or categories of nonpolar molecules. Describe each type and give examples.

9. *Are molecules of the following compounds polar or nonpolar?

(a) HF (b) CS_2 (c) CH_4 (d) PCl_3 (e) CH_3OH

Section 11-6

10. What are intermolecular attractive forces?

11. Explain how polar molecules attract one another.

12. Sketch two water molecules and show what portions of the molecules can attract when they are in close proximity.

13. Explain how nonpolar molecules attract one another.

14. Sketch two nonpolar iodine, I_2, molecules and show how they attract one another.

Section 11-7

15. How do the liquid and gaseous states of matter differ in terms of the arrangement and motion of the particles that make them up?

16. Explain the process of liquefaction of a gas in terms of the temperature and pressure and the attractions between molecules.

17. Give a description of the liquid state at the molecular level.

Section 11-8

18. Define the following terms as they apply to liquids.

(a) viscosity

(b) surface tension

19. Explain the liquid properties of viscosity and surface tension in terms of the attractions that exist between the particles making up a liquid.

20. It is possible to float a needle on the surface of water. What property of water makes this possible.

21. Why does the viscosity of a liquid depend on the temperature of the liquid?

22. Explain how a hydraulic brake system functions in terms of the incompressibilty of a liquid.

Section 11-9

23. Sketch and describe a graph showing a Maxwell–Boltzmann distribution of molecular kinetic energies in a sample of a liquid. How does such a distribution change when the temperature of the sample is increased?

24. On the molecular level, explain how a liquid can evaporate.

25. Explain how the evaporation of water or other liquids from the skin can cause a cooling effect.

26. In a clothes dryer, hot air is pumped into the dryer, circulated among the clothes, and pumped out. Explain, at a molecular level, how this process facilitates the drying of wet clothes.

Sections 11-10 and 11-11

27. Define or explain the following terms.

    (a) vapor pressure of a liquid

    (b) boiling point of a liquid

    (c) volatile liquid

    (d) nonvolatile liquid

28. Why do some liquids have odors that can be detected at some distance from an open container of the liquid?

29. Describe what happens at the molecular level when a liquid boils.

30. Doctors sometimes spray ethyl chloride on skin to momentarily freeze it for minor operations. Explain how this works, considering that ethyl chloride is a very volatile liquid with a relatively low boiling point.

31. What is meant by the normal boiling point of a liquid?

32. On Mount Everest, at an elevation of 9000 m above sea level, water boils at about 68°C. Explain why.

33. A pressure cooker can cook water-based foods faster, and at a higher temperature, than a saucepan open to the atmosphere. Explain how a pressure cooker works.

34. Describe the process of distillation. Describe the refining of petroleum products.

Section 11-12

35. Describe what happens, in terms of molecular motion and attractive forces, when a liquid freezes to form a solid.

36. How do the gaseous, liquid, and solid states of matter differ in terms of the arrangement and motion of the particles that make them up?

37. Describe a crystalline solid.

38. How is it possible for the particles of a solid to exchange kinetic energy?

Section 11-13

39. What is sublimation?

40. Why do some solids have noticeable odors?

41. What is a melting point for a solid?

42. Dry ice, solid carbon dioxide, "disappears" when it is exposed to the atmosphere. Explain how this takes place at the molecular level.

43. When food is freeze-dried, it is frozen and then placed in a vacuum chamber, and the air and vapor is slowly and continually pumped out of the chamber. Explain how drying occurs under these conditions.

Sections 11-14 to 11-16

44. Explain what happens in terms of energy absorbed or released during the following:

    (a) A skier melts snow over an open fire, heats the water to its boiling point, and boils the water to form steam.

    (b) An ice cube is exposed to the room environment and allowed to melt.

    (c) Solid iodine sublimes. Sublimation is the direct change from the solid to the gaseous state.

    (d) Water vapor in air condenses to form raindrops.

    (e) Raindrops in air freeze to become hailstones.

45. Define the following terms:

    (a) heat of fusion

    (b) heat of vaporization

    (c) heat of crystallization

    (d) heat of condensation

46. A beverage can be cooled by adding ice cubes to it. Explain in terms of energy changes how ice cubes can cool a beverage.

47. Explain why ice cubes in a frost free refrigerator often decrease in size over time.

48. In regions of the country where there is an occasional frost, some people put large barrels of water in

greenhouses during frosty weather. How does this practice help prevent frost damage in the greenhouses?

49. What is meant by the melting point of a solid? What is meant by the freezing point of a solid? How are the melting and freezing point of a solid related?

50. *How many calories of heat are released when 350 g of steam condense to liquid water at 100°C?

51. How many joules of heat are needed to melt 350 g of ice to form liquid water at 0°C?

52. *At its boiling point, the heat of vaporization of a chemical having the formula of $CHCl_2F$ is 242 J/g. How many joules of heat are absorbed when 1.000 kg of this compound vaporize? How many joules of heat are released when 1.000 kg of this compound condenses at the boiling temperature?

53. *If 1188 cal of heat are released when 25.0 g of liquid stearic acid freeze, what is the heat of crystallization of stearic acid, $C_{18}H_{36}O_2$, in units of calories per mole? Give an equation representing the freezing of stearic acid and include the energy change.

54. What is the heat of fusion of stearic acid? (See Question 53.)

55. If 314.9 J of heat are required to melt 2.5 g of solid benzene at its melting point, what is the heat of fusion of benzene, C_6H_6, in units of kilojoules per mole? Give an equation representing the melting of solid benzene and include the energy change.

56. What is the heat of crystallization of benzene? (See Question 55.)

Section 11-17

57. Give a definition for specific heat.

58. Would you expect a substance with a relatively high specific heat to be a good conductor of heat or a poor conductor of heat? Why?

59. *Suppose a person drinks 0.500 gal of water with a temperature of 0°C. In the stomach, this water heats to 37°C. How many dietetic calories would the person use to warm the water? (1 dietetic calorie = 1 kcal, 1 gal = 3.785, and assume a density for water of 1.00 g/mL.)

60. How many calories are needed to heat 425 mL of water from 25.0°C to 52.5°C if the density of water is 1.00 g/mL?

61. *A water bed contains 150 gallons of water. If the heater breaks down, how many calories of heat are released as the bed cools from 35°C to 10°C? (1 gal = 3.785 L, and assume a density for water of 1.00 g/mL.)

62. A hot tub contains 580 gallons of water. How many Btus (British thermal units) of heat are needed to heat the water from 25°C to 40°C? (1 gal = 3.785 L, 1 Btu = 1.05 kJ, and assume a density for water of 1.00 g/mL.)

63. *While hiking, you notice that granite rock that has been exposed to the sun is still quite warm after sunset. You take a sample back to the laboratory and find that 229 J of heat are required to change a 32.1 g sample from 18.6°C to 27.5°C. What is the specific heat of the granite rock?

64. You do an experiment that reveals that 29.8 J of heat are required to change the temperature of a 2.33 g sample of copper metal by 33°C. Calculate the specific heat of copper.

65. *The specific heat of gold is 0.129 J/g °C. How many joules of heat are required to change the temperature of 32.1 g of gold from 27.2°C to 58.5°C?

66. Aluminum metal has a specific heat of 0.212 cal/g °C. How many calories of heat are released when a 369 g sample of aluminum cools from 165°C to 25°C?

67. *A 500 g piece of lead, sitting in the sun all day, reaches a temperature of 40°C. How many calories of heat were released when the lead cools to 10°C if the specific heat of lead is 0.031 cal/g °C? In contrast, how much heat is released when 500 g of water is cooled from 40°C to 10°C if the specific heat of water is 1 cal/g °C?

68. A 265 g sample of copper metal is sitting in the sun and heats from 25.4°C to 39.6°C. How many calories of heat are needed to heat the copper if the specific heat of copper is 0.0932 cal/g °C? In contrast, how many calories of heat are needed to heat a 265 g sample of water from 25.4°C to 39.6°C if the specific heat of water is 1.00 cal/g °C?

69. *How many calories of heat are required to heat a teacup of water, 250 mL, from 0°C to 100°C and then convert the water to vapor at 100°C? Assume water to have a density of 1.00 g/mL.

70. An ice cube tray contains 250 g of water. How many calories of heat need to be lost by the water when it is cooled from 29°C to 0°C and then frozen to ice at 0°C?

71. *Starting with 25.0 g of ice, calculate the number of joules of heat required to

(a) heat the ice from −10°C to 0°C if the specific heat of ice is 2.09 J/g °C

(b) melt the ice at 0°C

(c) heat the liquid water from 0°C to 100°C

(d) boil the water to vapor at 100°C

72. Starting with 36.5 g of ice, calculate the number of joules of heat required to

(a) melt the ice at 0°C

(b) heat the liquid water from 0°C to 100°C

(c) boil the water to vapor at 100°C

(d) heat the water vapor from 100°C to 120°C if the specific heat of water vapor is 1.97 J/g °C

Questions to Ponder

73. Using recent population data, calculate what percent the U.S. population is of the world's population. Refer to Table 11-2. How does the percent of U.S. population compare to the percent of the world's metals used by the United States? Comment on these comparisons.

74. Research the methods used to make electronic components on the silicon chips used in solid-state electronics. Summarize these methods.

WATER AND

SOLUTIONS

12-1 SOLUTIONS

When you mix sugar in water and stir the mixture, the sugar disappears. Of course, the sugar is not gone; it has become intimately mixed with the water. Sugar dissolves in the water and forms a solution. A solution is a homogeneous mixture of chemicals in which the fundamental particles of the chemicals interact and intermingle, as defined in the margin.

Chemical compounds have fixed, unchanging composition. Samples of sugar, for instance, always contain the same elements in the same fixed proportions. A solution has a known composition, but the amounts of chemicals in the mixture may vary. Sugar solutions, for instance, range from very dilute to very concentrated in sugar content. Solutions are not compounds, but unique types of mixtures. They are not gross mixtures of one substance suspended in another, such as sand in water. Instead, because of the great degree of intermingling of the particles and the forces of interaction that exist between them, the components of a solution are mixed uniformly and intimately. Filtration can separate sand suspended in water, but more involved methods are needed to separate the components of a solution. One way to get the salt out of seawater, for instance, is to evaporate the water. Seawater is boiled and the water vapor is condensed to isolate pure water. Incidentally, this process requires relatively large amounts of energy and is usually not an economical way to get pure water from seawater.

Some complex solutions, such as seawater, have many components. A simple two-component solution like sugar in water is a **binary solution.** A solution forms when one chemical dissolves another chemical. The **solvent** is the dissolver and the dissolved substance is the **solute.** When sugar is dissolved in water, the water is the solvent and the sugar is the solute. The mixing of chemicals to form a solution involves the intermixing of their constituent molecules or ions. The solute particles become

Solutions

Solution
A homogeneous mixture of two or more chemicals in which the particles intermingle on an atomic, molecular, or ionic level.

uniformly distributed among the solvent molecules to give a homogeneous mixture.

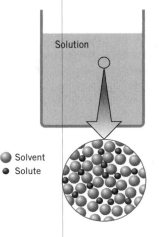

Solutions can be in the form of liquids, solids, or gases. Some metallic alloys are solid solutions in which one metal dissolves in another. All mixtures of gases are essentially solutions, but we usually do not distinguish solvents and solutes in gaseous mixtures. Liquid solutions are the most common type of solution, and water is by far the most common solvent. **Aqueous solutions** are those in which water is the solvent. Liquid solutions not involving water are simply nonaqueous solutions. Tincture of iodine, for example, is a nonaqueous solution of iodine dissolved in ethyl alcohol.

When a solute dissolves in a solvent, the various chemical particles commingle. This commingling results from the attractive forces between the particles of solute and solvent. These attractive forces involve the same kinds of intermolecular attractive forces discussed in Section 11-6. Since aqueous solutions are so common and important, let us consider how water molecules are involved in intermolecular attractions and attractions with solute particles. Recall that water molecules are polar molecules. It is the polar nature of water molecules that makes water such a useful solvent.

Latin (*aqua-*) and Greek (*hydro-*) prefixes are used with many words. Look up some words that begin with *aqu-, aqua-,* or *aqui-* and some words that begin with *hyd-, hydro-,* or *hydra-*. List two words for the Latin prefix and two for the Greek prefix.

12-2 THE HYDROGEN BOND

Water molecules are highly polar and forces of attraction between them are much stronger than most dipole-dipole attractions. The hydrogen atoms in water molecules are attracted to oxygen atoms in other water molecules (—H···O—). These attractions are called hydrogen bonds. By definition, a **hydrogen bond** is the force of attraction between a hydrogen of one molecule and an electronegative atom of another molecule. In other words, a hydrogen bond is an electrostatic force of attraction between the positive region around a hydrogen atom of one polar molecule and the negative region around a highly electronegative atom of another polar molecule. The hydrogen bond is a special type of chemical bond. It is a much weaker bond than the typical covalent or ionic bond but much stronger than simple dipole-dipole molecular attractions. Furthermore, hydrogen bonds occur only between certain polar, hydrogen-containing molecules. Hydrogen bonding commonly occurs with molecules in which hydrogen is bonded to the electronegative elements oxygen, nitrogen, or fluorine (e.g., H_2O, NH_3, HF). Hydrogen bonding is also very important in some vital biological molecules, such as proteins, deoxyribonucleic acids (DNA), and ribonucleic acids (RNA).

A sample of liquid water is pictured as a loose collection of water molecules hydrogen-bonded to one another. Typical of a liquid, the molecules of water are in constant motion within the collection. The hydrogen bonds between molecules are continually being formed and broken. The molecules move about, continually interacting with one another, breaking their hydrogen bonds with some molecules while forming new bonds with other molecules. The hydrogen bond is weak enough that it is easily broken by kinetic motion of the molecules. Consequently, liquid water consists of ever-changing groupings or clusters of polar water molecules. With this dynamic view of water in mind, let us consider the nature of water solutions.

You can observe the electrical properties of polar water molecules. To do this, you need a plastic knife, a cloth towel (dry), and a water faucet. Turn on the water faucet so that a very small stream of water is flowing. Rub the knife with the towel about 20 times (or rub it on your dry hair) and bring the edge of the knife near the stream of water. Assuming the knife has a negative electrostatic charge, draw a picture to show how the knife attracts the polar water molecules.

12-3 THE DISSOLVING PROCESS

Water is a liquid that has highly polar molecules and is an excellent solvent for many ionic and molecular compounds. Why is water such a good solvent? Polar water molecules can interact with the ions of ionic compounds. That is, the electrical charge centers of polar water molecules can attract ions since they are charged particles. The positive ends of water molecules attract negative ions and the negative ends of water molecules attract positive ions.

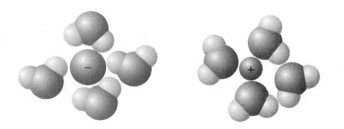

Polar water molecules can also interact with other types of polar molecules. The electric poles of water molecules attract the electric poles of other polar molecules. As a result, water can dissolve many polar molecular substances. Although water is a polar solvent, many nonpolar molecular compounds dissolve in water to some degree. For instance, ocean, river, and lake waters contain some dissolved nitrogen, oxygen, and carbon dioxide from the atmosphere. These nonpolar compounds are, however, only slightly soluble. Natural water solutions also contain a variety of dissolved ions. Seawater, for instance, has relatively high concentrations of dissolved ions, mostly sodium ions and chloride ions.

When we put some salt crystals in water and stir the mixture, the crystals become smaller and smaller and soon completely dissolve. Water can readily dissolve many ionic substances, since strong interactions occur between the ions and the polar water molecules. The interactions are strong enough to pull the ions from the solid and compel them into the solution. Figure 12-1(*b*) illustrates the dissolving of an ionic solid. The positively charged ends of the water molecules attract negative ions and surround them so that they are pulled into solution. The negatively charged ends of the water molecules attract positive ions and surround them so that they are pulled into solution. When an ionic compound, such as salt, dissolves in water, the process involves the separation of the ions composing the compound.

$$NaCl(s) \longrightarrow Na^+(aq) + Cl^-(aq)$$

The (aq) following the formulas of the ions shows they are separate ions in aqueous solution.

Many substances that have polar molecules also readily dissolve in water. Polar water molecules interact with and attract the polar solute molecules, allowing them to dissolve as illustrated in Figure 12-1(*a*). Except for a few unique cases, polar molecules do not form ions when they dissolve; instead, they remain as molecules dispersed among the water mole-

FIGURE 12-1 The dissolving of a solute in water. Water dissolves a solute by attracting the solute particles from the crystalline solid.

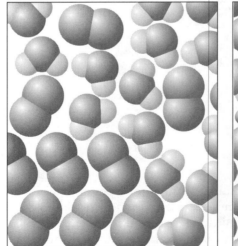

(*a*) The dissolving of a molecular solid.

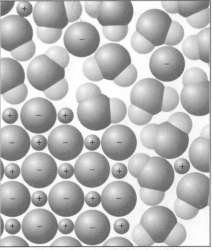

(*b*) The dissolving of an ionic solid.

cules. The dissolving of the polar molecular compound sucrose or table sugar is represented as

$$C_{12}H_{22}O_{11}(s) \longrightarrow C_{12}H_{22}O_{11}(aq)$$

The (aq) following the formula shows that the molecules are in aqueous solution.

12-4 LIKE DISSOLVES LIKE

The properties of molecular compounds relate to the polarities of their molecules. An important principle of dissolving is "like dissolves like," which means that a polar solvent, such as water, readily dissolves polar compounds. Polar compounds such as alcohols, ammonia, and sugar easily dissolve in water. In fact, since ionic compounds are essentially compounds of extreme polarity (completely separated centers of positive and negative charge), we expect many ionic compounds to dissolve in water. Those ionic compounds in which the attractive forces between ions are quite strong, however, do not readily dissolve in water. Generally, water is a good solvent for polar molecular compounds and many ionic compounds. Water can dissolve compounds composed of particles with charges or charge centers.

Waxy, oily, greasy, and fatlike compounds characteristically have nonpolar molecules. Such compounds do not dissolve in water. Nonpolar compounds do not readily dissolve in water since no polar interactions occur. Solvents or paint thinners containing hydrocarbon compounds, such as hexane, readily dissolve oily and greasy materials. The hydrocarbons are nonpolar compounds, so they interact with nonpolar solutes. To illustrate the differences, note that water is used to clean up acrylic paints, whereas paint thinners (containing nonpolar hydrocarbon solvents) are used to clean up oil-based paints. Furthermore, as expected, paint thinners and water do not dissolve in one another. **Like dissolves like** means that polar dissolves polar and that nonpolar dissolves nonpolar. Reversing this generalization, we can say that polar does not dissolve nonpolar and that nonpolar does not dissolve polar. Incidentally, latex rubber is made of nonpolar molecules. Latex rubber repels water, so water or water solutions cannot pass through. Vaseline and oil-based materials have nonpolar molecules. These materials can break down latex rubber membranes found, for example, in rubber gloves and condoms. To keep latex rubber products safe, do not let them contact oily materials.

12-5 DISSOLVING AND EQUILIBRIUM

Imagine forming a salt solution in water. When the salt mixes with water, it readily dissolves to form a solution. As more salt is added, a point is reached at which no more salt dissolves and solid salt settles to the bottom

ACTIVITY
12-3

Two stain removal procedures are given below. Tell which procedure should be used for the following stains: mustard, lipstick or grease, coffee or tea, coffee or tea with milk or cream.

Procedure A: Soak for 30 minutes in cool water. Rub with bar soap or liquid detergent and launder with a bleach safe for the fabric.

Procedure B: Place stain face down on paper towel and sponge with dry-cleaning fluid. Air dry. Rub in liquid detergent. Then launder in the hottest water safe for the fabric using a bleach safe for the fabric.

Recrystallization   Dissolving

FIGURE 12-2 Dynamic equilibrium between dissolved and undissolved solute.

To simulate dynamic equilibrium, you need a small plastic or paper cup and a water tap in a sink. Place a small hole in the bottom of the cup with a pencil or scissors. Fill the cup with water and allow it to leak out into the sink. Direct a stream of water from the tap to the cup and adjust the flow to equal the leaking of the cup. Adjust the water flow so that as much water enters the cup as leaves the cup and the water level remains constant. There is a steady-state amount of water in the cup, which is analogous to dynamic equilibrium. Water is continually entering and leaving the cup, but there is no net change in the amount of water in the cup.

of the mixture. There is a limit to how much salt we can dissolve in a given amount of water. Once a solution is saturated with dissolved salt, no additional salt dissolves unless the temperature is changed.

When a solute is placed in contact with the solvent, the solute particles enter the solution. The dissolved solute particles and solvent particles are continually moving about in the solution. Because of this motion, some solute particles migrate to a position between the solution and the undissolved solid and become reattached to the solid. This reformation of the solid solute is called crystallization. Dissolving and crystallization are opposite processes. If excess solute is present, a point is reached when the rate of dissolving equals the rate of crystallization. Figure 12-2 illustrates this situation.

The rate of dissolving refers to the number of particles entering the solution per unit time. The rate of crystallization refers to the number of particles leaving the solution per unit time. The competition between dissolving and crystallization and the eventual equalities of their rates illustrate a phenomenon called **dynamic equilibrium.** At equilibrium, crystallization is taking place just as fast as dissolving. The word *equilibrium* is a composite of the two Latin terms, *equi* (equal) and *librium* (balance). **Equilibrium** refers to a situation in which opposing forces or processes equally balance one another.

Because equilibrium is important in chemistry, getting a feeling for dynamic equilibrium is very important. Picture the situation in which many particles are dissolving at the surface of the solid. At equilibrium, just as many particles become reattached to the surface of the solid per second as leave the surface per second. A "standoff" is reached between dissolving and crystallization; the concentration of the dissolved solute is fixed and no more dissolves. At equilibrium, the concentration of the dissolved solute remains constant.

This sort of equilibrium is not static since two opposing processes are continually taking place; it is a dynamic equilibrium. Because of the equal rates, there is no net change in the amount of dissolved solute after equilibrium is reached. At equilibrium, the particles that represent the dissolved solute are not constant; the particles are continually changing between the dissolved and undissolved state. At equilibrium, however, the concentration of the dissolved solute is constant. Equilibrium is reached only if the temperature is fixed, since a change in temperature may change the rate of dissolving, and before a new state of equilibrium is reached, the concentration of dissolved solute may change.

12-6 SATURATED SOLUTIONS AND SOLUBILITY

To prepare a solution, we normally just add some solute to the solvent and stir the mixture. Typically, as we add more and more of a solid solute to a solution, we reach a point at which dynamic equilibrium exists. Once a condition of equilibrium is reached, no more solute dissolves and we have a saturated solution, as defined in the margin.

A saturated solution has a constant concentration of solute. An **unsaturated solution** simply contains less solute than a saturated solution. Most solutions we use in the laboratory are unsaturated. Preparing solutions of some solids that contain relatively large amounts of solute is possible. Often, the solubility of the solute increases with temperature. In such cases, heating a saturated solution containing some undissolved solute causes more solute to dissolve. Normally, when such a solution cools, the excess solute recrystallizes as the solid. Sometimes, however, the result is a **supersaturated solution**, a solution containing more solute than a saturated solution at the same temperature. A supersaturated solution is in an unstable state and can change to a stable saturated solution when the excess solute recrystallizes. For example, you may have noticed honey (a supersaturated solution of sugars in water) slowly crystallizing in storage. It is changing from a supersaturated solution to a saturated solution.

It is possible to dissolve 204 g of sucrose in 100 g of water at 20°C. Sucrose is quite soluble in water. In contrast, 36 g of sodium chloride and only 1.5×10^{-4} g of silver chloride dissolve in 100 g of water at 20°C. Sodium chloride is reasonably soluble, and silver chloride is essentially insoluble. For most solutes, there is a definite amount that can dissolve in a specific amount of solvent at given conditions. This amount is the solubility of the solute, as defined in the margin.

Each substance has a characteristic solubility in water at a given temperature. Solubilities normally vary with temperature. For example, the solubility of sodium chloride in water at 0°C is 35.7 g/100 g of water; at 100°C, it is 39.7 g/100 g of water.

Saturated solution
A solution in which dissolved and undissolved solute are in dynamic equilibrium.

Solubility
The maximum amount of a chemical that can be dissolved in a specific amount of solvent to make a saturated solution.

12-7 MOLARITY

A teaspoon of sugar in a cup of water gives the solution a sweet taste. The solution made from a tiny pinch of sugar in a cup of water, however, does not have a sweet taste. Both solutions contain sugar, but they differ in concentration of the solute. A common way to express **concentration** is the amount of solute per unit amount of solution. Another way to express concentration is the amount of solute per unit amount of solvent.

In chemistry, the most useful way to express solution concentrations is in terms of the number of solute particles per unit volume of solution. The number of solute particles is the number of moles. The most common expression of concentration used in chemistry is **molarity, M,** defined as:

$$\text{molarity} = \frac{\text{moles of solute}}{\text{liters of solution}} \qquad M = \frac{n}{V}$$

where n is the number of moles of solute and V is the volume of the solution in liters

Molarity
The number of moles of solute per liter of solution.

$$M = \frac{n}{V}$$

The molarity is the ratio of moles of solute to liters of solution. To calculate the molarity of a solution, we need to know both the number of moles of solute and the total volume of the solution. Note that molarity uses volume of solution, not volume of solvent. The number of moles of solute refers to the moles of formula units of an ionic compound or the moles of molecules of a molecular compound. Notice that since the number of moles of solute is related to its mass and molar mass, the molarity expression could be written as

$$M = \frac{(\text{grams solute})}{\text{liters of solution}} \left(\frac{1 \text{ mol solute}}{\text{no. of grams}} \right)$$

As seen from this expression, calculating the molarity of a solution is easy if we know the mass of solute dissolved in enough solvent to produce a specific volume of solution. Suppose a solution contains 117 g of sodium chloride dissolved in enough water to form 500 mL of solution. A summary of the question is

GIVEN	GOAL	FACTOR NEEDED
117 g NaCl	? M NaCl	molar mass of NaCl
500 mL solution		

To find the molarity of the solution, first find the number of moles of solute, then divide this by the volume of the solution in liters. The ratio of moles to volume is the molarity. The number of moles of sodium chloride is found using its molar mass ($22.99 + 35.453 = 58.44$ g/mol NaCl).

$$117 \text{ g} \left(\frac{1 \text{ mol NaCl}}{58.44 \text{ g}} \right)$$

To find the molarity, divide the number of moles of solute by the volume of the solution. Note that 500 mL is 0.500 L since 1 L is 1000 mL. That is 500 mL(1 L/1000 mL) = 0.500 L.

$$M = \left(\frac{117 \text{ g}}{0.500 \text{ L}} \right) \left(\frac{1 \text{ mol NaCl}}{58.44 \text{ g}} \right) = \left(\frac{4.00 \text{ mol NaCl}}{1 \text{ L}} \right)$$

Typically, this concentration is expressed as 4.00 M NaCl and is read 4.00 molar NaCl or 4.00 moles NaCl per liter. A concentration given in molarity can always be expressed as a factor relating moles of solute to liters of solution like the factor given above. As we will see, the molarity of a solution is used as a conversion factor in important chemical calculations.

EXAMPLE 12-1

A solution contains 225 g of glucose, $C_6H_{12}O_6$, dissolved in enough water to form 825 mL of solution. What is the molarity of the solution?

The concept involves molarity, so we need to use $M = n/V$.

GIVEN		GOAL	FACTOR NEEDED
solute	$C_6H_{12}O_6$	$? \, M \, C_6H_{12}O_6$	molar mass $C_6H_{12}O_6$
mass solute	225 g		
solution volume	825 mL or 0.825 L		

The number of moles of glucose in the solution is found using its molar mass $[6(12.01) + 12(1.0079) + 6(15.9994) = 180.2 \text{ g/mol } C_6H_{12}O_6]$.

$$225 \, g \left(\frac{1 \text{ mol } C_6H_{12}O_6}{180.2 \, g} \right)$$

Dividing by the volume in liters gives the molarity.

$$X \, M \, C_6H_{12}O_6 = \left(\frac{225 \, g}{0.825 \text{ L}} \right) \left(\frac{1 \text{ mol } C_6H_{12}O_6}{180.2 \, g} \right)$$

$$= \left(\frac{1.51 \text{ mol } C_6H_{12}O_6}{1 \text{ L}} \right) \quad \text{or} \quad 1.51 \, M \, C_6H_{12}C_6$$

ACTIVITY
12-5

A cup has a volume of 237 mL. A teaspoon of sugar has a mass of 8.0 g. What is the molarity of sucrose, $C_{12}H_{22}O_{11}$, in a solution that has 1 teaspoon of sugar dissolved in water to make 237 mL of solution?

12-8 MOLARITY AS A UNIT FACTOR

Storing chemicals in solutions is often quite convenient. When a sample of the chemical is needed, simply measure some of the solution from a storage container. If the molarity of a solution is known, it is possible to dispense a specific number of moles of the solute by measuring a definite volume of the solution. For instance, in hospitals specific amounts of chemicals are given to patients by using prescribed volumes of intravenous (IV) solutions. Furthermore, the recommended dose for a liquid over-the-counter drug is established to dispense a specific number of moles of the active ingredient in the solution.

The molarity of the solution serves as a conversion factor to find the moles of solute from the volume of solution. This situation is analogous to the use of density as a conversion factor. Density is the mass per unit volume of a substance, and molarity is the moles per unit volume of a solution. Density relates mass and volume of a substance. **Molarity** relates moles of solute and volume of a solution.

$$M = \frac{n}{V}$$

When using molarity in a calculation, include the units so that molarity becomes a conversion factor. If a solution is 4.00 M NaCl, the molarity as a factor is

$$\left(\frac{4.00 \text{ mol NaCl}}{1 \text{ L}} \right) \quad \text{or} \quad \left(\frac{4.00 \text{ mol NaCl}}{1000 \text{ mL}} \right)$$

Solution of
known molarity
M

Measured volume
of the solution
V
Number of moles
of solute
$n = VM$

Since 1 L is 1000 mL, we can use units of milliliters in molarity if we wish. That is, we can convert 1 L to milliliters by just substituting 1000 mL for 1 L. The idea of molarity was developed to give a simple way to relate amount of solute in a solution to the volume of the solution. Molarity is used in two ways.

1. To find the number of moles in a given volume of solution. This is done by multiplying the volume by the molarity.

$$n = VM$$

2. To find the volume of solution needed to contain a specific number of moles of solute. This is done by multiplying the moles by the inverted molarity or the reciprocal of the molarity

$$V = n\frac{1}{M}$$

EXAMPLE 12-2

How many moles of dissolved NaOH are in 25.0 mL of a 6.00 M NaOH solution?

Molarity is involved, so we need to use $M = n/V$. The problem is

GIVEN		GOAL	FACTOR NEEDED
solute	NaOH		molarity
molarity	6.00 M		
solution volume	25.0 mL	? mol NaOH	

We can express the molarity as a factor.

$$\left(\frac{6.00 \text{ mol NaOH}}{1 \text{ L}}\right)$$

To change 25.0 mL from milliliters to liters, we divide by 1000 or move the decimal point three places to the left: 25.0 mL = 0.0250 L. Find the number of moles by multiplying the volume in liters by the molarity.

$$X\text{ mol NaOH} = 0.0250\,\cancel{L}\left(\frac{6.00 \text{ mol NaOH}}{1\,\cancel{L}}\right) = 0.150 \text{ mol NaOH}$$

Alternatively, we can use the volume in milliliters and the molarity with 1000 mL in place of 1 L.

$$X\text{ mol NaOH} = 25.0\,\cancel{mL}\left(\frac{6.00 \text{ mol NaOH}}{1000\,\cancel{mL}}\right) = 0.150 \text{ mol NaOH}$$

ACTIVITY
12-6

A solution of acetic acid has a label that reads 0.5 M $HC_2H_3O_2$. How many moles of $HC_2H_3O_2$ are in 3 L of the solution? How many moles are in 0.1 L of the solution?

EXAMPLE 12-3

How many milliliters of a 2.00 M NaCl solution are needed to provide 0.250 mole of NaCl?

Molarity is involved, so we need to use $M = n/V$. The problem is

GIVEN			GOAL	FACTOR NEEDED
solute	NaCl			Use inverse of molarity
moles solute	0.250 mol NaCl			
molarity	2.00 M		? mL of solution	

The molarity of NaCl as a factor is

$$\left(\frac{2.00 \text{ mol NaCl}}{1 \text{ L}} \right)$$

To find the volume, multiply the given number of moles by the inverted molarity (1 L/2.00 mol NaCl) to give the volume in liters, which can then be converted to milliliters.

$$X \text{ mL} = 0.250 \text{ mol NaCl} \left(\frac{1 \text{ L}}{2.00 \text{ mol NaCl}} \right) \left(\frac{1000 \text{ mL}}{1 \text{ L}} \right) = 125 \text{ mL}$$

ACTIVITY

12-7

(a) What is the molarity of a glucose, $C_6H_{12}O_6$, solution that contains 9.01 g of glucose in 500 mL of solution? (b) What volume of a 0.100 M $C_6H_{12}O_6$ solution contains 0.0500 mole of glucose? (c) How many moles of $C_6H_{12}O_6$ are in a 10.0 mL sample of a 0.100 M $C_6H_{12}O_6$ solution?

 12-9 STANDARD SOLUTIONS

Storing substances in solutions and measuring volumes as needed is a common method of dispensing chemicals. In laboratory exercises, you are instructed to use specific volumes of solutions. Such specific volumes contain a useful number of moles of solute. Solutions of known concentration are called **standard solutions**. Often, a standard solution is prepared by weighing a sample of solute and dissolving the sample in enough solvent to give a known volume of solution. A volumetric flask, calibrated to contain a known volume of liquid, is normally used to prepare a standard solution. Figure 12-3 illustrates the use of the volumetric flask. Standard solutions have known molarities. To prepare a standard solution, we need to know three things about the solution.

1. The formula of the solute
2. The desired molarity of the solute
3. The final total volume of solution

The product of the volume and molarity of the solution gives the number of moles of solute needed to prepare the solution. The number of grams

The mass of a sample of the solute is found.	The solute is placed in a 1-liter volumetric flask.	Water is added, and the solute is brought into solution.	More water is added to make the final volume of the solution 1 liter.	Determine the molarity of the solution.
Step 1	Step 2	Step 3	Step 4	Step 5

$$\text{number of moles solute} = \text{g solute}\left(\frac{1\ \text{mol}}{\#\text{g}}\right)$$

$$M\ (\text{molarity}) = \frac{\text{No. moles solute}}{\text{Volume of solution}}$$

FIGURE 12-3 The preparation of a standard solution.

of solute needed is easily found from the number of moles using its molar mass.

EXAMPLE 12-4

How many grams of sodium chloride are needed to prepare 250 mL of a 0.500 M solution of NaCl?

Molarity is involved, so we need to use $M = n/V$. The problem is

GIVEN		GOAL	FACTORS NEEDED
solute	NaCl		molarity NaCl
molarity	0.500 M		molar mass NaCl
solution volume	250 mL or 0.250 L	? g NaCl	

We cannot find the mass of NaCl directly, but we can find the number of moles needed by multiplying the solution volume in liters by the molarity. Note that we want to cancel liters and retain moles.

$$0.250\ \cancel{L}\left(\frac{0.500\ \text{mol NaCl}}{1\ \cancel{L}}\right)$$

Next, to find the grams of NaCl, multiply the number of moles by the molar mass (58.44 g/mol NaCl). Here, we want to cancel moles and end up with grams.

$$X\ \text{g NaCl} = 0.250\ \cancel{L}\left(\frac{0.500\ \cancel{\text{mol NaCl}}}{1\ \cancel{L}}\right)\left(\frac{58.44\ \text{g}}{1\ \cancel{\text{mol NaCl}}}\right) = 7.31\ \text{g}$$

ACTIVITY
12-8

How many grams of glucose are needed to prepare 500 mL of a 0.200 M $C_6H_{12}O_6$ solution? Explain how to make such a standard solution.

The standard solution is prepared by weighing 7.31 g of sodium chloride, placing it in a 250 mL volumetric flask, and dissolving it in enough water to give 250 mL of solution.

12-10 DILUTION

Restaurants often keep cola drinks in concentrated forms and dilute them with seltzer water when needed. Aqueous solutions are made less concentrated by diluting with water. In the laboratory, it is sometimes necessary to prepare a solution of a lower molarity by dilution of a more concentrated solution.

As shown in Figure 12-4, the molarity is inversely proportional to the volume. This situation makes sense if you consider that adding water to a solution increases the volume, which decreases the molarity. As the volume gets larger, the molarity gets smaller. Molarity is the number of moles of solute per liter of solution.

$$M = \frac{n}{V}$$

When the volume of a solution is increased (the solution is diluted), the molarity decreases. Since the number of mole is constant, dividing by a larger volume gives a smaller molarity.

We can see that the number of moles of solute in a solution is the product of the volume and the molarity. If we have a known volume of a solution of specific molarity, it contains a specific number of moles.

$$n = V_c M_c$$

The subscript c refers to concentrated. Adding water to the sample dilutes it, changing the volume and, thus, the molarity. Dilution does not change the number of moles of solute. It simply makes the solution less concentrated. Thus,

$$n = V_d M_d$$

When the volume of a solution is decreased (the solvent is evaporated), the molarity increases. Since the number of moles is constant, dividing by a smaller volume gives a larger molarity.

where the subscript d refers to diluted. Dilution increases the volume, so the molarity decreases. The number of moles is unchanged by dilution.

$$n = V_c M_c \qquad n = V_d M_d$$

FIGURE 12-4 The molarity of a solution is inversely proportional to the volume.

Equating these two volume-molarity products gives the general **dilution formula:**

$$V_c M_c = V_d M_d$$

There are two common uses of this expression.

1. Given the initial molarity and volume of a concentrated solution, find the volume of the solution of lower molarity. That is, V_c, M_c, and M_d are known, from which V_d is calculated.

2. Find the volume of a more concentrated solution needed to prepare a specific volume of a more dilute solution. That is, V_d, M_d, and M_c are known, from which V_c is calculated.

Simply put, we can always calculate the fourth term in this expression if we know the three other terms.

EXAMPLE 12-5

To what volume should 10.0 mL of 15.0 M aqueous ammonia be diluted to give a 2.00 M solution?

First, summarize the data.

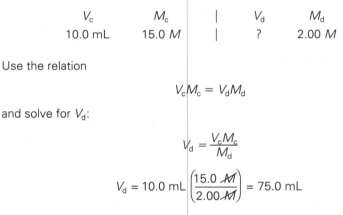

V_c	M_c		V_d	M_d
10.0 mL	15.0 M		?	2.00 M

Use the relation

$$V_c M_c = V_d M_d$$

and solve for V_d:

$$V_d = \frac{V_c M_c}{M_d}$$

$$V_d = 10.0 \text{ mL} \left(\frac{15.0 \, \cancel{M}}{2.00 \, \cancel{M}} \right) = 75.0 \text{ mL}$$

Note that using milliliters for the volume gives the calculated answer in milliliters. To prepare the less concentrated solution, 10.0 mL of the concentrated solution, is diluted with water to give a new volume of 75.0 mL.

EXAMPLE 12-6

How many milliliters of 6.00 M hydrochloric acid are needed to prepare 100 mL of 0.500 M hydrochloric acid by dilution?

First, summarize the data.

V_c	M_c		V_d	M_d
?	6.00 M		100 mL	0.500 M

Then use the relation:

$$V_c M_c = V_d M_d$$

and solve for V_c:

$$V_c = \frac{V_d M_d}{M_c}$$

$$V_c = 100 \text{ mL} \left(\frac{0.500 \cancel{M}}{6.00 \cancel{M}}\right) = 8.33 \text{ mL}$$

To prepare the less concentrated solution, 8.33 mL of the more concentrated solution is diluted with water to give a new volume of 100 mL.

ACTIVITY 12-9

(a) What volume of 5.00 *M* NaCl is needed to prepare 500 mL of 0.500 *M* NaCl by dilution? (b) To what volume should 10.0 mL of 3.00 *M* NaCl be diluted to give 1.00 *M* NaCl?

12-11 SOLUTION STOICHIOMETRY

As we know, the molarity of a solution is a conversion factor used to relate the number of moles of solute to the volume of the solution, and many laboratory chemicals are stored as solutions of known molarity. These standard solutions are sometimes called **reagent solutions** and are used to carry out chemical reactions in solution. Stoichiometric questions about solution reactions involve **solution stoichiometry.** Calculations for such questions include the usual stoichiometric conversions coupled with the molarities of reagent solutions. Stoichiometry maps are discussed in Sections 7-8 and 10-15. Molarity is also used in stoichiometric calculations and it fits into the stoichiometry map, as shown in Figure 12-5.

We start with an example using moles. A solution of sodium hydroxide and a solution of acetic acid react according to the equation

$$NaOH(aq) + HC_2H_3O_2(aq) \longrightarrow H_2O + NaC_2H_3O_2(aq)$$

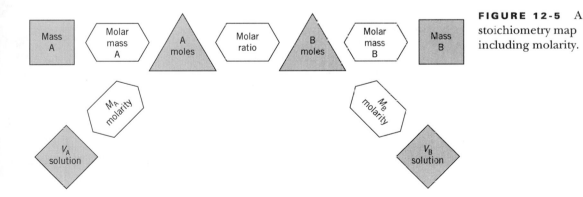

FIGURE 12-5 A stoichiometry map including molarity.

How many milliliters of 6.00 M NaOH solution are needed to react with a solution containing 0.300 mole of acetic acid, $HC_2H_3O_2$? A summary of the given information in relation to the equation is

GIVEN	WANT
0.300 mol $HC_2H_3O_2$	? mL NaOH solution
6.00 M NaOH	

We want to know what volume of 6.00 M NaOH contains enough moles to react with the 0.300 mole $HC_2H_3O_2$. Consulting the stoichiometry map, the calculation sequence is

$$\text{mol } HC_2H_3O_2 \xrightarrow{\substack{\text{molar} \\ \text{ratio}}} \text{mol NaOH} \xrightarrow{M} \text{? mL NaOH solution}$$

In the first step, find the moles of sodium hydroxide from the number of moles of $HC_2H_3O_2$ using the molar ratio from the equation

$$0.300 \text{ mol } HC_2H_3O_2 \left(\frac{1 \text{ mol NaOH}}{1 \text{ mol } HC_2H_3O_2} \right)$$

Then, find the volume of solution using the molarity so that the moles cancel and liters remain. The liters can then be converted to milliliters.

$$0.300 \text{ mol } HC_2H_3O_2 \left(\frac{1 \text{ mol NaOH}}{1 \text{ mol } HC_2H_3O_2} \right) \left(\frac{1 \text{ L}}{6.00 \text{ mol NaOH}} \right) \left(\frac{1000 \text{ mL}}{1 \text{ L}} \right)$$
$$= 50.0 \text{ mL}$$

Like any typical stoichiometric problem, no calculations are necessary until we have the complete setup of the problem.

The same general approach is used to relate the number of grams of a reactant to the volume of a reagent. The molar mass is used to find the number of moles from the mass.

EXAMPLE 12-7

How many grams of oxalic acid, $H_2C_2O_4$, react with 25.0 mL of a 0.100 M NaOH solution if the reaction is

$$2NaOH(aq) + H_2C_2O_4(aq) \longrightarrow 2H_2O + Na_2C_2O_4(aq)$$

The problem is summarized as

GIVEN	WANT	FACTOR NEEDED
25.0 mL or 0.0250 L	? g $H_2C_2O_4$	molar mass $H_2C_2O_4$
0.100 M NaOH		

Consult the stoichimetry map to follow the calculation sequence. The sequence is

$$V\,NaOH \xrightarrow{\text{molarity}} mol\ NaOH \xrightarrow{\substack{\text{molar}\\\text{ratio}}} mol\ H_2C_2O_4 \xrightarrow{\substack{\text{molar}\\\text{mass}}} ?\ g\ H_2C_2O_4$$

To put this question on a stoichiometric basis, first fnd the number of moles of NaOH from the volume and molarity of the NaOH solution.

$$0.0250\,\cancel{L}\left(\frac{0.100\ mol\ NaOH}{1\,\cancel{L}}\right)$$

Then, find the number of moles of oxalic acid using the molar ratio from the equation.

$$0.0250\,\cancel{L}\left(\frac{0.100\ \cancel{mol\ NaOH}}{1\,\cancel{L}}\right)\left(\frac{1\ mol\ H_2C_2O_4}{2\ \cancel{mol\ NaOH}}\right)$$

Finally, the number of moles of $H_2C_2O_4$ is changed to mass using its molar mass $[2(1.0079) + 2(12.01) + 4(16.00) = 90.04]$.

$$X\,g\ H_2C_2O_4 = 0.0250\,\cancel{L}\left(\frac{0.100\ \cancel{mol\ NaOH}}{1\,\cancel{L}}\right)\left(\frac{1\ \cancel{mol\ H_2C_2O_4}}{2\ \cancel{mol\ NaOH}}\right)\left(\frac{90.04\ g}{1\ \cancel{mol\ H_2C_2O_4}}\right)$$

$$= 0.113\ g$$

A C T I V I T Y
12-10

A solution of sodium hydroxide reacts with a solution cf acetic acid as represented by the equation

$$NaOH(aq) + HC_2H_3O_2(aq)$$
$$\longrightarrow NaC_2H_3O_2(aq) + H_2O$$

It is found that 20.0 mL of a 0.500 *M* NaOH solution is needed to react with the acetic acid in a sample of an acetic acid solution. How many grams of acetic acid were in the sample?

12-12 SOME OTHER SOLUTION CONCENTRATION TERMS

Percent-by-Mass (or Weight) Composition

A common way to express the amount of each component in a solution is percent. To calculate the **percent by mass** of a solution component, divide the mass of the component by the mass of the solution and then multiply by 100.

$$\text{percent component} = \frac{\text{mass component}}{\text{mass solution}} \times 100$$

$$\text{or} \quad \% = \frac{\text{no. of g component}}{\text{no. of g solution}} \times 100$$

Percent compositions are sometimes used in industry and medicine to express concentrations of solutions. For example, a salt solution called physiological saline is used in medicine; it must have a specific concentration of sodium chloride. Suppose a physiological saline solution is prepared by mixing 4.6 g of sodium chloride in enough water to make 500 g of solution. To find the percent of sodium chloride, divide the mass of NaCl by the solution mass and multiply this fraction by 100.

$$\% \text{ NaCl} = \left(\frac{4.6 \text{ g NaCl}}{500 \text{ g solution}} \right) \times 100 = 0.92\% \text{ NaCl}$$

ACTIVITY 12-11

A typical can of soda contains 40 g of sugar in 380 g of drink. Calculate the percent sugar in the soda. If a teaspoon full of sugar weighs about 8 g, how many teaspoons of sugar are in a can of typical soft drink?

EXAMPLE 12-8

A solution consists of 20.6 g of glucose dissolved in 53.5 g of water. What is the percent by mass of glucose in the solution?

The mass of glucose is 20.6 g, and the mass of solvent is 53.5 g. The total mass of the solution containing this amount of solute is found by adding the mass of the solute to the mass of the solvent.

$$\text{mass solution} = 53.5 \text{ g} + 20.6 \text{ g} = 74.1 \text{ g}$$

To find the percent by mass of glucose, divide the mass of glucose by the mass of the solution and then multiply by 100.

$$\% \text{ glucose} = \left(\frac{20.6 \text{ g glucose}}{74.1 \text{ g solution}} \right) \times 100 = 27.8\% \text{ glucose}$$

Molality

Molality, *m*, is another solution concentration term that is sometimes used in chemistry. By definition, molality is the number of moles of solute per kilogram of solvent.

$$\text{molality} = \frac{\text{no. of moles of solute}}{\text{no. of kilograms of solvent}} \quad \text{or} \quad m = \frac{n}{\text{kg solvent}}$$

where *n* is the number of moles of solute and the denominator is the number of kilograms of solvent. Note that molality expresses the amount of solute per kilogram of solvent, not per kilogram of solution. Be careful not to confuse molality with molarity since the names are very similar. A lowercase letter *m* is used as an abbreviation for molality. Molality is not as commonly used as molarity but, as we will see, it does have some important uses.

EXAMPLE 12-9

A solution contains 18.0 g of glucose dissolved in 500 g of water. What is the molality of glucose in the solution?

The concept involved is molality, so we need to use $m = n/kg$ solvent. The data are

GIVEN		GOAL	FACTOR NEEDED
solute	$C_6H_{12}O_6$	$? \, m \, C_6H_{12}O_6$	molar mass $C_6H_{12}O_6$
mass solute	18.0 g		180.2 g/mol $C_6H_{12}C_6$
solvent mass	500 g or		

0.500 kg, since there are 1000 g/kg

Since we need number of moles to find molality, we use the molar mass of the solute. Note above that the mass of solvent in grams is easily changed to kilograms by dividing by 1000 or moving the decimal point three places to the left.

First, find the number of moles of glucose in the given mass using its molar mass.

$$18.0 \, \cancel{g} \left(\frac{1 \text{ mol } C_6H_{12}O_6}{180.2 \, \cancel{g}} \right)$$

Dividing this by the number of kilograms of water gives the molality.

$$X \, m \, C_6H_{12}O_6 = 18.0 \, \cancel{g} \left(\frac{1 \text{ mol } C_6H_{12}O_6}{180.2 \, \cancel{g}} \right) \left(\frac{1}{0.500 \text{ kg } H_2O} \right)$$

$$= \left(\frac{0.200 \text{ mol } C_6H_{12}O_6}{1 \text{ kg } H_2O} \right) \quad \text{or} \quad 0.200 \, m \, C_6H_{12}O_6$$

Parts per Million

There are about 1.4×10^{21} g of dissolved aluminum in the oceans of the world. This amount of aluminum is very large, but since the total volume of the oceans is so great, the concentration of aluminum is quite low, so low, in fact, that we find it convenient to refer to the concentration in **parts per million** or **ppm**. Parts per million is analogous to percent except that percent refers to parts per hundred. For comparison, imagine yourself in a crowd of one million people; you would be one part per million. In contrast one person in 100 is 1%.

The parts per million concentration of a solute is calculated by dividing the mass of the solute in a sample of solution by the mass of the solution and multiplying by 10^6, which is a million.

$$\text{ppm} = \frac{\text{mass solute}}{\text{mass solution}} \times 10^6$$

For dilute water solutions that have densities close to 1 g/mL, the volume of the solution in milliliters may be used in place of the mass of the solution in grams. As an example of a ppm calculation, suppose a 0.500 kg sample of drinking water is found to contain 2.2 mg of fluoride ions. To calculate the ppm of F^-, we need the mass of solute in grams and the mass of solution in grams: 2.2 mg is 2.2×10^{-3} g, and 0.500 kg is 500 g. Thus, the ppm of fluoride ion is

$$\left(\frac{2.2 \times 10^{-3}\ \cancel{g}\ F^-}{500\ \cancel{g}} \right) \times 10^6 = 4.4 \text{ ppm } F^-$$

EXAMPLE 12-10

The U.S. Environmental Protection Agency (EPA) drinking water standards allow 1.00 ppm Ba^{2+}. If a 1.5 L sample of drinking water is found to contain 2.00 mg of Ba^{2+}, does the water meet or exceed the standard?

Calculate the parts per million of Ba^{2+} in the water by dividing the grams of Ba^{2+} by the number of milliliters of solution. Milliliters of solution are often used in place of the number of grams for dilute solutions.

$$\left(\frac{2.00 \times 10^{-3}\ \cancel{g}\ Ba^{2+}}{1.5 \times 10^3\ \cancel{mL}} \right) \times 10^6 = 1.3 \text{ ppm } Ba^{2+}$$

Based on this result, the drinking water sample exceeds the EPA standard for barium ion.

ACTIVITY 12-12

The population of Earth is about six billion people. What is your personal concentration in parts per billion?

Sometimes, components of mixtures or solutions occur in such low amounts that it is useful to express their concentrations in parts per billion or parts per trillion. The above expression for parts per million may be altered to calculate the **parts per billion** (ppb) or even **parts per trillion** (ppt) concentration of a solute. Such calculations are done by using 10^9 (a billion) or 10^{12} (a trillion) in place of 10^6 in the expression.

12-13 ELECTROLYTES AND NONELECTROLYTES

A rule of electrical safety is never to use an electrical appliance when you are wet or are standing in water to avoid an electrical shock caused by electricity flowing from the appliance through your body to the ground. You might wonder how water can conduct electricity. Pure water is not a good conductor of electricity. It is the presence of dissolved ions in water that make it conduct electricity.

A solution is tested for electrical conductivity by using an apparatus that has electrodes connected to a source of electrical current. If the solution conducts, it completes the circuit between the electrodes and current

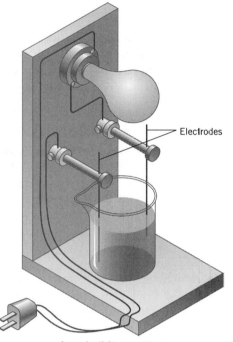

A conductivity apparatus.

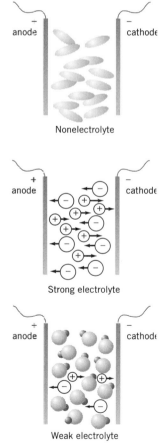

flows in the conductivity apparatus. A solution conducts electricity by the movement of ions toward the electrodes and by chemical reactions at the electrodes (see Section 15-12).

If ions are present in an aqueous solution, the solution conducts electricity. A substance that forms an aqueous solution that conducts electricity is called an **electrolyte.** Any solution that conducts is assumed to contain ions. Soluble ionic compounds are electrolytes. Molecular compounds that dissolve in water and react to form ions are also electrolytes. Incidentally, deionized water is water purified by removing ions by a special chemical process called ion exchange. Deionized water does not conduct electricity. A substance that forms an aqueous solution that does not conduct is a **nonelectrolyte.** Most soluble molecular substances (e.g., sugar and alcohol) are nonelectrolytes. Characterizing a substance as a nonelectrolyte means it does not form ions when it dissolves.

Conductivity measurements reveal that solutions of some electrolytes are strong conductors, whereas comparable solutions of others are quite weak conductors. According to these tests, some electrolytes are classified as strong electrolytes and others as weak electrolytes. The difference comes from the extent to which substances form ions when they dissolve in water. Substances that dissolve in water and separate or react completely to form no products other than positive and negative ions are **strong electrolytes.** Strong electrolytes are ionic compounds and a few special molecular compounds. All soluble ionic compounds are strong electrolytes since they separate into ions when they dissolve. **Weak electrolytes** are substances that dissolve in water and react with water to only a slight extent, producing relatively low concentrations of ions. Low concentrations of ions account for the slight conduction of electricity.

12-14 STRONG AND WEAK ELECTROLYTES

Suppose we had a water solution of 1 molar hydrogen chloride (1 M HCl) and a water solution of 1 molar acetic acid (1 M $HC_2H_3O_2$). We can assume they have the same amount of dissolved solute per liter of solution since they have the same molarity. Both solutions are found to conduct electricity, so we classify hydrogen chloride and acetic acid as electrolytes. The hydrogen chloride solution, however, is a much stronger conductor of electricity than the acetic acid solution. Hydrogen chloride is a strong electrolyte, and acetic acid is a weak electrolyte.

This difference in conductivity suggests that the acetic acid solution has a lower concentration of ions than does the hydrogen chloride solution. Since they are both electrolytes and since we are comparing solutions of the same concentration, we could expect the solutions to contain the same concentrations of ions. They do not, however. The difference in the concentration of ions is a result of the extent to which these two substances react chemically with water to form ions.

When some molecular substances, such as hydrogen chloride, dissolve in water, they react completely to form ions. The reaction between hydrogen chloride and water is

$$HCl(g) + H_2O \longrightarrow H_3O^+(aq) + Cl^-(aq)$$

Note that hydrogen chloride and water react to give chloride ions and H_3O^+ ions, called hydronium ions. We'll learn more about this important ion in Chapter 13.

Some molecular electrolytes dissolve in water and react slightly with the water, forming relatively low concentrations of ions. In other words, these compounds readily dissolve in water but relatively few of the molecules react with water. For example, when acetic acid dissolves in water, the reaction is

$$HC_2H_3O_2(aq) + H_2O \rightleftharpoons H_3O^+(aq) + C_2H_3O_2^-(aq)$$

This equation has a double arrow between reactants and products to convey that it is a reversible chemical reaction involving a dynamic **chemical equilibrium.** The reactants react to form the products and the products in turn react to reform the reactants. The result is a mixture of reactants and products. With acetic acid, a weak electrolyte, the equilibrium favors the reactant side, which means there is only a slight reaction with water. The

double arrow showing a smaller arrow to the right and a larger arrow to the left ($\longleftarrow$) suggests a slight reaction toward the product side. The reaction favors the reactant side, and the equilibrium mixture has relatively high concentrations of reactants and relatively low concentrations of products.

In previous discussions, we used chemical equations showing reactants and products separated by a single arrow. When conveying that a reaction involves chemical equilibrium is important, we write a double arrow. Actually, all chemical reactions are theoretically reversible and involve chemical equilibrium. When a reaction favors the formation of the product and the reaction mixture contains very little of the original reactants, a single arrow is used in an equation. Furthermore, if showing that a reaction involves chemical equilibrium is not important use a single arrow. You will see some equations written with double arrows and some with single arrows. See Chapter 14 for more discussion of chemical equilibrium.

12-15 SUBSTANCES IN SOLUTION

When working with solutions, a question arises concerning what formulas should be used to represent the dissolved solutes. This point is important in developing a chemical vision. An aqueous solution is always clear; we can see through it. If the dissolved solute is colorless, the solution is colorless. If the dissolved solute is colored, the solution is colored. As part of our chemical vision, we need a way to picture the particles found in solutions. The picture depends on whether the solute is a strong electrolyte, weak electrolyte, or nonelectrolyte. We know that a weak electrolyte reacts somewhat with water to produce some ions. Ammonia is a weak electrolyte and reacts with water.

$$NH_3(aq) + H_2O \longleftrightarrow NH_4^+(aq) + OH^-(aq)$$

Notice that the sizes of the arrows in the equation reveal that ammonia reacts only slightly with water. Writing the formulas of all the species when we want to represent a solution of ammonia is not convenient nor is it necessary. Since NH_3 is the species present in the greatest concentration and since ions are present in relatively low concentrations, a solution of ammonia is represented most conveniently by the formula $NH_3(aq)$.

It is standard practice to represent a **solution of a weak electrolyte** by the formula of the most abundant species in solution. If wanted, this formula is followed by (aq) to show a water solution. Keep in mind that water is always present in an aqueous solution, and its presence is either implied or suggested by (aq). With a weak electrolyte, the most abundant or major species in solution is the molecular form of the dissolved substance.

EXAMPLE 12-11

An aqueous solution of acetic acid, $HC_2H_3O_2$, is a weak conductor of electricity. How should an aqueous solution of acetic acid be represented? That is, how do we picture the major particles that populate the solution?

Since acetic acid is a weak electrolyte, representing a solution of acetic acid by its molecular formula is best:

$$HC_2H_3O_2(aq)$$

Nonelectrolytes do not react with water when they dissolve; their molecules just commingle with water molecules. **Solutions of nonelectrolytes** are represented by the molecular formula followed by (aq) if wanted. For instance, an aqueous solution of the nonelectrolyte ethyl alcohol is represented as $C_2H_5OH(aq)$.

A chemical that is a strong electrolyte is completely ionized in solution. The major species in solution are the cations and anions produced when the substance dissolves. For most ionic substances, it is a simple task to decide which ions are involved. A **solution of an ionic compound** is represented by the formulas of the cation and the anion in the compound. Picture the ions as separate rather than combined by ionic bonds. Write their formulas as separate ions followed by (aq) to show an aqueous solution. That is, decide which ions make up a compound and write the formulas of the cation and anion separately. For instance, a solution of NaCl is shown as $Na^+(aq) + Cl^-(aq)$ since the cation and anion separate when salt dissolves.

EXAMPLE 12-12

How is a solution of potassium chloride, KCl, represented? How is a solution of sodium carbonate, Na_2CO_3, represented?

We represent solutions of a soluble ionic compound by the formulas of its constituent cation and anion. Therefore, we show a solution of potassium chloride as $K^+(aq) + Cl^-(aq)$.

Sodium carbonate contains the sodium ion and the carbonate ion, so we show a solution as $2Na^+(aq) + CO_3^{2-}(aq)$. Notice that there are twice as many Na^+ as CO_3^{2-} corresponding to Na_2CO_3.

For strong electrolytes that are molecular rather than ionic, we must know how the substance reacts with water. A **solution of a molecular strong electrolyte** is represented by the formulas of the ions produced in the reaction between water and the strong electrolyte. Hydrogen chloride gas is a strong electrolyte. When it dissolves in water, it reacts according to the equation

$$HCl(g) + H_2O \longrightarrow H_3O^+(aq) + Cl^-(aq)$$

Since hydronium ions and chloride ions are the species formed when HCl dissolves in water, we should represent an aqueous solution of HCl, called hydrochloric acid, as

$$H_3O^+(aq) + Cl^-(aq)$$

Table 12-1 lists some common electrolytes and nonelectrolytes. There are four important ideas to keep in mind when representing substances in aqueous solution.

1. Ionic compounds are shown as the formulas of the separate cation and anion.

Table 12-1 *Some Strong Electrolytes, Weak Electrolytes, and Nonelectrolytes*

COMPOUND NAME	FORMULA	MAJOR SPECIES IN SOLUTION
Molecular Strong Electrolytes		
Sulfuric acid	H_2SO_4	$H_3O^+(aq) + HSO_4^-(aq)$
Hydrogen chloride	HCl	$H_3O^+(aq) + Cl^-(aq)$
Nitric acid	HNO_3	$H_3O^+(aq) + NO_3^-(aq)$
Soluble Ionic Compounds		
Sodium chloride	NaCl	$Na^+(aq) + Cl^-(aq)$
Potassium sulfate	K_2SO_4	$2K^+(aq) + SO_4^{2-}(aq)$
Ammonium nitrate	NH_4NO_3	$NH_4^+(aq) + NO_3^-(aq)$
Sodium hydroxide	NaOH	$Na^+(aq) + OH^-(aq)$
Potassium hydroxide	KOH	$K^+(aq) + OH^-(aq)$
Barium hydroxide	$Ba(OH)_2$	$Ba^{2+}(aq) - 2OH^-(aq)$
Sodium acetate	$NaC_2H_3O_2$	$Na^+(aq) + C_2H_3O_2^-(aq)$
Silver nitrate	$AgNO_3$	$Ag^+(aq) + NO_3^-(aq)$
Weak Electrolytes		
Ammonia	NH_3	$NH_3(aq)$
Hydrogen fluoride	HF	$HF(aq)$
Hydrogen sulfide	H_2S	$H_2S(aq)$
Phosphoric acid	H_3PO_4	$H_3PO_4(aq)$
Acetic acid	$HC_2H_3O_2$	$HC_2H_3O_2(aq)$
Nonelectrolytes		
Methyl alcohol	CH_3OH	$CH_3OH(aq)$
Ethyl alcohol	C_2H_5OH	$C_2H_5OH(aq)$
Sucrose	$C_{12}H_{22}O_{11}$	$C_{12}H_{22}O_{11}(aq)$
Hydrogen peroxide	H_2O_2	$H_2O_2(aq)$

Write formulas that represent the chemical particles
in the following solutions.

(a) hydrogen peroxide,
H_2O_2, a nonelectrolyte

(b) nitric acid, HNO_3, a
strong molecular electrolyte
like HCl

(c) ammonium sulfate,
$(NH_4)_2SO_4$, an ionic
compound

2. Molecular compounds that are strong electrolytes are shown as the ions they form in water. (As shown in Table 12-1, the three common molecular strong electrolytes are HCl, HNO_3, and H_2SO_4.)

3. Molecular compounds that are weak electrolytes are shown by the formulas of the molecules.

4. Nonelectrolytes are shown by the formulas of the molecules.

12-16 NET-IONIC EQUATIONS

Water solutions are convenient for storing some chemicals and useful for carrying out some reactions. Particles of a chemical dissolved in water can intermix, move about the solution, and collide with any other dissolved particles. In solution, **ionic compounds** and **strong molecular electrolytes** form ions, which are sometimes involved in chemical reactions. To write an equation for a reaction in solution, we have to know the formulas of the reactants and products. The formulas of products can come from experimental observations and, sometimes, by prediction based on information about the solution. To know the nature of the solute particles in a solution, we need to know whether a solute is a strong electrolyte, a weak electrolyte, or a nonelectrolyte.

Let's look at an example of an equation for a reaction that occurs in solution. When a solution of the ionic compound calcium chloride, $CaCl_2$, mixes with a solution of the ionic compound sodium carbonate, Na_2CO_3, the insoluble solid $CaCO_3$ forms as a product. The solutions of the initial ionic compounds are represented by the separate ions they form upon dissolving.

calcium chloride	$Ca^{2+}(aq) + 2Cl^-(aq)$
sodium carbonate	$2Na^+(aq) + CO_3^{2-}(aq)$

The product of the reaction is $CaCO_3$, so the equation for the reaction is

$$Ca^{2+}(aq) + 2Cl^-(aq) + 2Na^+(aq) + CO_3^{2-}(aq) \longrightarrow CaCO_3(s) + 2Cl^-(aq) + Na^+(aq)$$

Using your chemical vision of solutions, imagine what happens when the two solutions mix. The ions in one solution mix with the ions of the other solution. The ions move about and interact with one another and with water molecules. The Ca^{2+} and the CO_3^{2-} form ionic bonds that are stronger than the attraction of water molecules, resulting in the formation of solid $CaCO_3$. Notice that the Na^+ and Cl^- do not react in any way. They remain dissolved and distributed among the water molecules of the solution. In the above equation, these ions appear unchanged on both sides of the arrow.

When reactions occur in water, some particles react and others do not react. Ions that do not react are known as **spectator ions** because they just "watch" the reaction. You might wonder why they are present in the first place. All solutions are electrically neutral, having as much positive as negative charge. A solution of a pure cation or a pure anion is not possible. Spectator ions are ions that do not take part in a reaction but are present to maintain equal amounts of negative and positive charge in the solution.

Often equations representing reactions in solution do not include the spectator ions. Such equations for solution reactions are called net-ionic equations. A **net-ionic equation** shows the particles that react and the particles formed. It does not include any spectator ions. Deleting the spectator ions gives the net-ionic equation for the reaction of solutions of calcium chloride and sodium carbonate:

$$Ca^{2+}(aq) + CO_3^{2-}(aq) \longrightarrow CaCO_3(s)$$

This equation is both chemically balanced and charge balanced. **Charge balance** means that an equation involving ions must have the same net amount of positive and negative charge on both sides of the arrow. A reaction cannot produce a net change in electrical charge; it cannot produce excess positive charge or negative charge.

In net-ionic equations, the (aq) notation is used to emphasize that the reaction is taking place in a water solution. For convenience, the (aq) may be omitted. To write a net-ionic equation for a reaction, use the following pattern:

1. Write the formulas of all major particles present in the solutions.
2. Write the formulas of the product or products of the reaction.
3. Delete any particles that are not involved in the reaction; normally, these are spectator ions.
4. Balance the equation by making sure it is chemically balanced and charge balanced.

EXAMPLE 12-13

A solution of the ionic compound NaOH is added to a solution of the weak electrolyte acetic acid, $HC_2H_3O_2$. Write a net-ionic equation for the reaction that produces acetate ion, $C_2H_3O_2^-$, and water.

First, write the formulas of the particles that are present in the solutions that mix. Show ionic compounds as separate ions and weak electrolytes by their molecular formulas.

$$Na^+ + OH^- + HC_2H_3O_2$$

Second, write the formulas of the products.

$$Na^+ + OH^- + HC_2H_3O_2 \longrightarrow C_2H_3O_2^- + H_2O$$

Third, delete any particles that do not react and are not changed. The sodium ion does not react in this case.

$$OH^- + HC_2H_3O_2 \longrightarrow C_2H_3O_2^- + H_2O$$

Fourth, balance the net-ionic equation, both chemically and by charge. In this case, the equation balances chemically and has the same net charge ($1-$) on both sides of the arrow. It is balanced.

EXAMPLE 12-14

A water solution of hydrochloric acid (remember that this strong electrolyte is represented as H_3O^+ and Cl^-, as shown in Table 12-1) is mixed with a solution of potassium carbonate, K_2CO_3. Write a net-ionic equation for the reaction that forms CO_2 gas and water.

First, write the formulas of the particles that are present in the solutions. Hydrochloric acid is a strong electrolyte, and potassium carbonate is an ionic compound.

$$H_3O^+ + Cl^- + 2K^+ + CO_3^{2-}$$

Second, write the formulas of the products.

$$H_3O^+ + Cl^- + 2K^+ + CO_3^{2-} \longrightarrow CO_2 + H_2O$$

Third, delete any particles that do not react and are not changed. In this case, delete Cl^- and K^+.

$$H_3O^+ + CO_3^{2-} \longrightarrow CO_2 + H_2O$$

Fourth, balance the equation both chemically and by charge. To balance the $2-$ charge of CO_3^{2-}, we need to place a coefficient of 2 in front of H_3O^+, which gives $2+$ and $2-$ charges on the left, resulting in net zero charge on the left to balance zero charge on the right. To balance the equation chemically, we need a coefficient of 3 for water to give 6 hydrogens and 5 oxygens on each side.

$$2H_3O^+ + CO_3^{2-} \longrightarrow CO_2 + 3H_2O$$

ACTIVITY 12-14

Hard-water scale may be removed from a teakettle by pouring vinegar in the kettle and allowing it to sit overnight. In the morning, rinse and scrub the kettle. Write an equation for the reaction of acetic acid with solid calcium carbonate to give aqueous calcium acetate, carbon dioxide gas, and water.

12-17 PRECIPITATION REACTIONS

With your chemical vision, you can imagine an ionic compound in solution as separate cations and anions. The ions are in continuous motion and migrate about the solution, momentarily interacting with one another. If the compound corresponding to the cation and anion is soluble and if enough water is present, the ions remain in solution. In such cases, the attractions between oppositely charged ions are not great

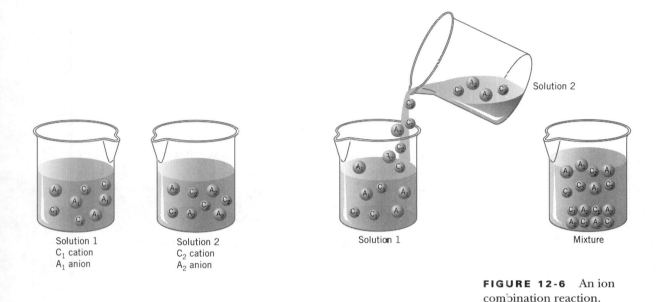

FIGURE 12-6 An ion combination reaction.

enough to cause the ions to bond to form a crystalline ionic solid. If any of the mixed ions are the ions of an **insoluble ionic compound,** the ions bond to form crystals and the insoluble solid separates from the solution; this process is illustrated in Figure 12-6. A reaction of ions in solution to form a solid compound is called an **ion combination** or a **precipitation reaction. Precipitate** is the term used to name the insoluble solid that separates from the solution.

A good example of a precipitation reaction is the reaction of calcium ion and carbonate ion described in Section 12-16.

$$Ca(aq)^{2+} + CO_3^{2-}(aq) \longrightarrow CaCO_3(s)$$

When any solutions containing the calcium ion and carbonate ion mix and the concentrations of the ions are high enough, solid calcium carbonate precipitates. **Hard water** is a term used for natural waters that contain relatively high concentrations of calcium ions and other ions. The use of hard water sometimes results in the precipitation of calcium carbonate to form scale in hot-water pipes and teakettles. Some natural limestone deposits, such as the White Cliffs of Dover in England, are precipitated calcium carbonate. The stalactites and stalagmites in limestone caves are calcium carbonate deposits.

Some mineral deposits were formed by natural precipitation of ionic compounds. In nature, precipitation may occur quite slowly, forming large crystals. In the laboratory, however, where the concentrations of ions are relatively high, precipitation usually occurs very quickly, producing very small finely divided crystals. Precipitates form as soon as the solutions mix and then slowly settle from the mixture. Precipitates have a variety of col-

Calcium

ors and appearances. They may be flocculent (fluffy), gelatinous (jelly-like), coagulated (curdy or lumpy), or crystalline (powdery or silky) in appearance.

Calcium in the Body

Calcium is an important nutrient mineral in our diets, and it forms compounds in bones and teeth. Calcium ions serve as body electrolytes and play a part in controlling the heartbeat. About 1.5% to 2.0% of body weight is calcium, most of which occurs in bones and teeth. Usually, a normal diet supplies sufficient calcium. Dairy products are a particularly rich source of calcium. Some postmenopausal women take calcium supplements to help prevent deterioration of bones.

Bones and tooth enamel are made rigid and strong by calcium carbonate, $CaCO_3$, and hydroxyapatite, $Ca_5(PO_4)_3OH$. Tooth enamel is made more resistant to decay when a fluoride ion replaces the hydroxide ion in hydroxyapatite.

$$Ca_5(PO_4)_3OH + F^- \longrightarrow Ca_5(PO_4)_3F + OH^-$$

Fluoride ions form a stronger bond and harden the enamel, making decay less likely. Compounds containing fluoride ions are added to some toothpastes to "fight" cavities. Fluoride ions are also present in most public water supplies, either as a natural mineral or added in low concentrations as a dental prophylactic. Too much fluoride ion is undesirable because it can cause tooth enamel to become mottled (blotched with brown spots).

A calcium imbalance in the body can result in the formation of kidney stones. Often, kidney stones are mixed precipitates of insoluble calcium compounds such as calcium phosphate and calcium oxalate. When small crystals of insoluble calcium compounds form stones in the kidneys, they are very painful. Most of them pass from the body naturally or after medication. Sometimes they are treated by lithotripsy, which uses sound waves to break up the stones for easier passage.

ACTIVITY

12-15

(a) Using the information that 1.5% to 2.0% of body weight is calcium, calculate the number of kilograms of calcium in your body. (b) Read the label of a toothpaste container and give the name or the formula of the active ingredient containing fluoride ions.

12-18 SOLUBILITY RULES FOR IONIC COMPOUNDS

Predicting a precipitation reaction when solutions of ionic compounds are mixed is possible. To make a prediction, you need to know which ionic compounds are soluble and which are insoluble. Ionic compounds vary in solubility, ranging from **very soluble** through **soluble** and **slightly soluble** to **insoluble.**

very soluble   soluble | slightly soluble   insoluble

⟵―――――――――――――――――――――――――⟶

An ionic compound is called **soluble** if it is possible to make a solution of the compound that is at least 0.1 M. We can consider an ionic compound that is not soluble enough to make a 0.1 *M* solution as slightly soluble or **insoluble.** Every chemical compound dissolves to some extent. Many, however, are quite insoluble. An example of a very insoluble compound is barium sulfate, $BaSO_4$. Barium ions are toxic, so it is dangerous to ingest them. Before being X-rayed in intestinal areas, it is standard practice to have patients drink a barium sulfate slurry (a mixture of the compound with water that is not a solution). The compound is so insoluble it passes through the intestinal tract without dissolving in any significant amount. Barium is a strong absorber of X rays, so it helps in the visualization of the intestinal passages.

The solubilities of many common ionic compounds in water are summarized in solubility rules based on experimental observations. If it is possible to prepare at least a 0.1 *M* solution of a compound, it is soluble. Typical ionic compounds contain metallic cations in combination with nonmetal anions or polyatomic anions. For our purposes, the solubility rules given here are useful. Keep in mind that the rules do not include all ionic compounds and that there may be some exceptions to the general rules. Refer to the rules or the summary given in Table 12-2 when you want to check the solubility of an ionic compound.

1. Nearly all ionic compounds containing sodium ions, Na^+, potassium ions, K^+, or ammonium ions, NH_4^+, are **soluble.**

2. Nearly all compounds containing nitrate ions, NO_3^-, or acetate ions, $C_2H_3O_2^-$, are **soluble.**

3. All compounds containing sulfate ions, SO_4^{2-}, are **soluble** except

calcium sulfate	$CaSO_4$
strontium sulfate	$SrSO_4$
barium sulfate	$BaSO_4$
lead(II) sulfate	$PbSO_4$

Table 12-2   *Solubility List of Common Ionic Compounds*

ANION	CATIONS THAT FORM PRECIPITATE	CATIONS THAT DO NOT FORM PRECIPITATE
NO_3^-	None	All other common
$C_2H_3O_2^-$	None	All other common
SO_4^{2-}	Ca^{2+}, Sr^{2+}, Ba^{2+}, Pb^{2+}	Most other common
Cl^-, Br^-, I^-	Ag^+, Hg_2^{2+}, Pb^{2+}	Most other common
OH^-	Most	Na^+, K^+, Ba^{2+}
F^-	Mg^{2+}, Ca^{2+}, Sr^{2+}, Ba^{2+}, Pb^{2+}	Most other common
$C_2O_4^{2-}$, CO_3^{2-}, CrO_4^{2-}, PO_4^{3-}	Most	NH_4^+, group IA

Almost all ionic compounds of Na^+, K^+, and NH_4^+ are soluble.

4. All ionic compounds containing chloride ions, Cl^-, bromide ions, Br^-, or iodide ions, I^-, are **soluble** except silver, mercury(I), and lead(II) compounds of these ions.

$$
\begin{array}{lll}
AgCl & Hg_2Cl_2 & PbCl_2 \\
AgBr & Hg_2Br_2 & PbBr_2 \\
AgI & Hg_2I_2 & PbI_2
\end{array}
$$

5. All ionic compounds containing fluoride ions, F^-, are **soluble** except

calcium fluoride	CaF_2
strontium fluoride	SrF_2
barium fluoride	BaF_2
lead(II) fluoride	PbF_2

6. All common ionic compounds containing hydroxide ions, OH^-, are **insoluble** except (Note the change to insoluble.)

sodium hydroxide	NaOH
potassium hydroxide	KOH
barium hydroxide	$Ba(OH)_2$

7. Most ionic compounds containing carbonate ions, CO_3^{2-}, chromate ions, CrO_4^{2-}, oxalate ion, $C_2O_4^{2-}$, or phosphate ions, PO_4^{3-}, are **insoluble** except group IA and ammonium ion compounds.

12-19 PREDICTING PRECIPITATION REACTIONS

It is possible to predict a precipitation reaction that occurs when solutions of ionic compounds are mixed. A prediction is done by considering all the ions mixed and deciding whether any set of cations and anions forms an insoluble compound. The solubility rules are used to decide which compounds are insoluble. When a solution of one chemical containing cations and anions is mixed with a solution of another chemical containing cations and anions, there are two possible ion combinations (see Figure 12-7).

$$cation_1 + anion_1 + cation_2 + anion_2$$

If either combination gives an insoluble compound, a precipitation reaction is predicted and a net-ionic equation may be written for the reaction. Note that sometimes it is possible to form two different precipitates.

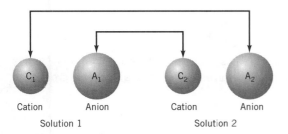

Cation   Anion   Cation   Anion

Solution 1   Solution 2

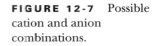

FIGURE 12-7 Possible cation and anion combinations.

EXAMPLE 12-15

Write a net-ionic equation for any reaction that occurs when a solution of silver nitrate, $AgNO_3$, is mixed with a solution of sodium chloride, NaCl.

Both solutes are ionic compounds, so we can show them as separate cations and anions.

$$Ag^+ + NO_3^- + Na^+ + C^-$$

Consider the two possible combinations. By solubility rule 2, all nitrates are soluble, so sodium nitrate is not expected to precipitate. By solubility rule 4, silver chloride is insoluble. Thus, we predict that silver ions react with chloride ions to give silver chloride.

$$Ag^+ + Cl^- \longrightarrow AgCl$$

Check to see if the net-ionic equation is balanced chemically and by charge.

EXAMPLE 12-16

Write a net-ionic equation for any reaction that occurs when a solution of aluminum nitrate, $Al(NO_3)_3$, is mixed with a solution of NaOH.

The ions in the solutions of these ionic compounds are

$$Al^{3+} + NO_3^- + Na^+ + OH^-$$

All nitrates are soluble so sodium nitrate does not precipitate. Aluminum hydroxide is insoluble, so the equation for the precipitation reaction is

$$Al^{3+} + 3OH^- \longrightarrow Al(OH)_3$$

To balance the charge, we use three OH^- and one Al^{3+}.

EXAMPLE 12-17

Write a net-ionic equation for any reaction that occurs when a solution of potassium acetate, $KC_2H_3O_2$, is mixed with a solution of sodium chloride.

When a solution of calcium chloride, $CaCl_2$, and a solution of Na_3PO_4 are mixed, a reaction occurs that forms a solid. Write a balanced net-ionic equation for the reaction.

Write the net-ionic equation for any reaction that occurs when a solution of lead(II) nitrate, $Pb(NO_3)_2$, is mixed with a solution of potassium chloride, KCl.

The ions involved are

$$K^+ + C_2H_3O_2^- + Na^+ + Cl^-$$

By solubility rule 2, all acetates are soluble, and by solubility rule 4, KCl is soluble. Thus, neither combination gives a precipitate and we predict no reaction.

EXAMPLE 12-18

Write a net-ionic equation for any reaction that occurs when a solution of barium hydroxide, $Ba(OH)_2$, is mixed with a solution of zinc sulfate, $ZnSO_4$.

The ions in the solutions are

$$Ba^{2+} + OH^- + Zn^{2+} + SO_4^{2-}$$

By solubility rule 3, barium sulfate is insoluble, and by solubility rule 5, zinc hydroxide is insoluble. Here, two ion combinations occur. The equation showing both reactions is

$$Ba^{2+} + 2OH^- + Zn^{2+} + SO_4^{2-} \longrightarrow Zn(OH)_2 + BaSO_4$$

Or, we could show each precipitation in a separate equation.

$$Zn^{2+} + 2OH^- \longrightarrow Zn(OH)_2$$
$$Ba^{2+} + SO_4^{2-} \longrightarrow BaSO_4$$

 **COLLIGATIVE PROPERTIES**

When we mix commercial antifreeze with water, the resulting solution has a freezing point that is lower than pure water. In addition, its boiling point is higher than pure water. Salt mixed with ice and water in an ice-cream maker lowers the freezing point so that the ice-cream mixture freezes. Salt in water also causes the boiling point of the solution to become higher than that of pure water. The presence of a solute in a solvent makes the properties of the solution different from those of the pure solvent.

Freezing point depression and boiling point elevation are properties of solutions known as colligative properties. **Colligative properties** relate to the concentration of solute particles, not to the chemical identity of these particles. The term *colligative* refers to the collective effect of the solute particles. These properties are generally the same no matter what solute is involved; they depend on the concentration, not on the identity of the solute particles. This situation is true for the boiling point when the solute is nonvolatile. A nonvolatile solute is one that does not evaporate or boil from the solution. Being nonvolatile is not a prerequisite for freezing point depression.

Pure water freezes at 0°C. A water solution containing 15% by mass of sodium chloride freezes at around −10°C. The solute depresses the freezing point of the solvent. When a solute dissolves in a solvent, the resulting solution typically boils at a higher temperature than the pure solvent and freezes at a lower temperature than the pure solvent. These phenomena are called **boiling point elevation** and **freezing point depression** (see Figure 12-8). The main component of commercial antifreeze is ethylene glycol. The ethylene glycol lowers the freezing point of the water in the radiator so that it does not freeze in cold weather. The antifreeze also elevates the boiling point so that it does not boil as readily as pure water. Thus, commercial antifreeze products are sold as coolants. Some plants and animals have special solutes in the solutions of their cells that protect them from freezing in cold weather.

The change in the freezing point of a solvent is directly proportional to the amount of solute in the solution. If the amount of solute is expressed in terms of molality, *m*, the change in freezing point is given by

$$\Delta T = K_f m$$

Why does the fluid added to a car radiator act both as a coolant and antifreeze?

Solute
Solvent

(*a*)

(*b*)

Boiling point elevation: The presence of solute molecules interferes with the ability of the solvent molecules to evaporate. Consequently, a higher temperature is needed to make the solution boil. The increase in the boiling point of a solvent caused by the addition of a nonvolatile solute is called boiling point elevation.

Freezing point depression: The presence of solute particles interferes with the ability of the solvent molecules to freeze to form crystals. Consequently, a lower temperature is needed to reach the freezing point. The decrease in the freezing point of a solvent caused by the addition of a solute is called freezing point depression. Freezing point depression is used in the making of homemade ice cream. Salt is added to an ice–water mixture to lower the freezing temperature enough to freeze the ice-cream mixture.

FIGURE 12-8   Boiling point elevation and freezing point depression.

In this expression, ΔT is the change in freezing point, m is the molality of the solute, and K_f is a proportionality constant called the **molal freezing point depression constant.** Each solvent has a unique value for this constant. For water, the value of K_f is (1.86°C/1 m), which means that 1 mole of solute per kilogram of water depresses the freezing point by 1.86°C. Note that the units of the constant are °C/m. To find the freezing point depression for a water solution, we simply multiply the constant by the molality of the solute. See Section 12-12 to review molality. If the solute is a strong electrolyte, however, we need to know the total ionic concentration since the ions act as independent particles. Consequently, the equation works best for solutes that are not strong electrolytes.

Calculate the freezing point and boiling point of a 2.00 m glucose solution. For water, K_f is (1.86°C/1 m) and K_b is (0.52°C/1 m).

EXAMPLE 12-19

Calculate the freezing point of a 2.00 m sucrose solution. Since the question involves the freezing point of a water solution, we use $\Delta T = K_f m$.

$$\Delta T = (1.86°C/1 \ m)(2.00 \ m) = 3.72°C$$

Here, ΔT is the freezing point depression or decrease in freezing point. Thus, to find the value of the freezing point, we need to subtract ΔT from the freezing point of pure water.

$$\text{solution freezing point} = 0°C - 3.72°C = -3.72°C$$

The change in the boiling point of a solvent is directly proportional to the amount of solute in the solution. If the amount of solute is expressed in terms of molality, m, the change in boiling point is given by

$$\Delta T = K_b m$$

In this expression, ΔT is the change in boiling point, m is the molality of the solute, and K_b is a proportionality constant called the **molal boiling point elevation constant.** Each solvent has a unique value for this constant. For water, the value of K_b is (0.52°C/1 m), which means that 1 mole of solute per kilogram of water elevates the boiling point by 0.52°C. Like freezing point depression, the boiling point equation works best with solutes that are not strong electrolytes, but it can be modified for use with strong electrolyte solutions.

 **OSMOSIS**

When red blood cells are placed in pure water, water enters the cells, causing them to swell and eventually burst. On the other hand, when red blood cells are placed in a concentrated salt solution, water migrates out and they shrivel and shrink. These two processes result from the same phenomenon, known as osmosis. **Osmosis** is a phenomenon involving solutions separated by a membrane. The membrane (e.g., blood cell membrane) acts as a barrier between two solutions. It has the property of allowing certain types of molecules to pass through while preventing the passage of other species in solution. Such a membrane is a **semipermeable membrane,** since it is only permeable to selected species.

Semipermeable membranes that allow the passage of water but not solutes are **osmotic membranes.** Imagine a solution separated from a sample of pure water (or another solution of lower concentration of solute) by an osmotic membrane. An osmotic membrane allows water but not solute to migrate through the membrane. In such a situation, the water molecules spontaneously penetrate the membrane from both directions but not at equal rates. More water molecules move from the pure solvent side of the membrane to the solution side than move from the solution side to the pure solvent side. Figure 12-9 illustrates this type of osmosis.

The migration of water through a semipermeable membrane occurs because of a **concentration gradient.** A gradient exists when the concentrations are different on either side of the membrane. The natural tendency is for the solvent to move from a region where it is present in higher concentration to a region where it is present in lower concentration. The net result of osmosis is the transfer of water across the membrane to the side having the lower concentration of water. Note here we are referring to the concentration of water not solute. A dilute solution has a relatively low concentration of solute and a relatively high concentration of water. A

FIGURE 12-9 Osmosis and osmotic pressure.

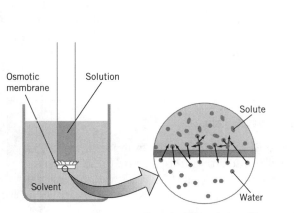

Osmosis: Enlarged view of two solutions separated by a semipermeable membrane.

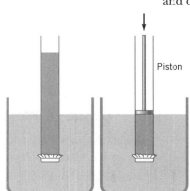

Osmotic pressure: In the osmosis process, solvent molecules pass from the less concentrated solution to the more concentrated solution. The process of osmosis can be reversed by applying pressure to the more concentrated side. The amount of pressure needed just to counteract the osmosis process is called the osmotic pressure. Pressures in excess of the osmotic pressure can cause reverse osmosis.

concentrated solution has a relatively high concentration of solute and a lower concentration of water. Thus, water moves from the dilute solution of solute into the concentrated solution of solute.

The mechanism of osmosis is not completely understood, but it appears to occur between aqueous solutions separated by an osmotic membrane no matter what solute is involved. Osmosis is stopped by exerting an opposing pressure or force per unit area on the solution side of the membrane, as illustrated in Figure 12-9. The applied pressure counteracts the natural flow of solvent through the membrane. Another way to view this situation is that osmosis results in a pressure being exerted on the solution of higher concentration of solute. This pressure is called **osmotic pressure.** Osmotic pressure is counteracted by applying a pressure to the side of the membrane with the higher concentration of solute. The osmotic pressure of a solution is measured by finding the opposing pressure needed to balance the pressure of osmosis. In some solutions, the osmotic pressure is quite great and may reach hundreds of atmospheres. Osmotic pressure is one factor involved in the uptake of water by trees and other plants.

Red blood cells are surrounded by osmotic membranes, and the blood cells contain various dissolved solutes. When red blood cells are placed in a concentrated salt solution, water moves from the cells to the salt water by osmosis. The cells shrivel and shrink; this process is called crenation. On the other hand, blood cells swell and burst in freshwater because water enters the cells by osmosis. Inward osmosis that swells and bursts the cells is called hemolysis. Typically, water passes through an osmotic membrane from the side having the lower concentration of solute to the side having the higher concentration of solute. To prevent crenation or hemolysis of blood cells, drugs and medicines administered intravenously must contain concentrations of solutes similar to those found in body fluids. Such medicinal solutions are known as **isotonic solutions.**

Drinking Water

Drinking Water Standards

According to the Federal Safe Drinking Water Act, the EPA requires public water systems to test for a variety of dissolved chemicals and bacteria in the water they supply. For these potential water pollutants, the EPA has established maximum contaminant levels (MCL), which are legal standards. Any public system with 15 or more connections must notify its customers of any violation of MCLs in supplied water. Nearly 15% of Americans use private water sources. The EPA recommends that these water sources be tested periodically for contaminants, especially nitrate ions and coliform (*E. Coli*) bacteria.

The EPA standards include a long list of organic chemicals that could show up in drinking water from industrial sources and the agricultural use of pesticides. The standards also include testing for radioactivity and coliform bacteria. The presence of excess coliform bacteria is an indication that the water could be contaminated with sewage or other harmful bacteria. Here is a list of the MCLs for some common chemicals that occur dissolved in water in ionic forms.

Chemical	MCL (ppm)	Chemical	MCL (ppm)
Aluminum	1	Arsenic	0.05
Barium	1	Cadmium	0.01
Chloride	500	Chromium	0.05
Copper	1	Fluoride	2
Iron	0.3	Lead	0.015
Manganese	0.05	Mercury	0.002
Nitrate	45	Selenium	0.01
Silver	0.05	Sulfate	500
Zinc	5	Total dissolved solids	1000

FOCUS 12

Solutions

Solution: A mixture of two or more chemicals in which the particles intermingle on an atomic, molecular, or ionic level.

Like dissolves like means that a polar solvent dissolves polar solutes. A polar solvent like water readily dissolves polar molecular compounds and many ionic compounds. A nonpolar solvent dissolves nonpolar solutes. Nonpolar hydrocarbon solvents readily dissolve waxy, oily, greasy, and fatlike compounds that characteristically have nonpolar molecules.

Saturated solution: A solution in which dissolved and undissolved solute are in dynamic equilibrium.

Solubility: The maximum amount of a chemical that dissolves in a specific amount of solvent to make a saturated solution.

Molarity: The number of moles of solute per liter of solution.

When a solution is diluted with solvent, the molarity of the solute decreases. The dilution equation is $M_c V_c = M_d V_d$, where M and V are molarities and volumes, c is concentrated, d is diluted.

The concentration of a solute may be expressed in terms of percent by mass

$$\text{percent component} = \frac{\text{mass component}}{\text{mass solution}} \times 100$$

$$\text{or} \quad \% = \frac{\text{no. of g component}}{\text{no. of g solution}} \times 100$$

The concentration of a solute or impurity may be expressed in terms of parts per million or ppm

$$\text{ppm} = \frac{\text{mass solute}}{\text{mass solution}} \times 10^6$$

Molality: The number of moles of solute per kilogram of solvent.

$$\text{molality} = \frac{\textbf{number of moles of solute}}{\textbf{number of kilograms of solvent}} \quad \text{or} \quad m = \frac{n}{\textbf{kg solvent}}$$

Electrolytes: Compounds that form solutions that conduct electricity.

Strong electrolytes: Soluble ionic compounds and a few molecular compounds (HCl, HNO_3, H_2SO_4) that dissolve in water to form ions.

Weak electrolytes: Soluble molecular compounds that react slightly with water upon dissolving to form relatively low concentrations of ions.

Nonelectrolytes: Compounds that form solutions that do not conduct electricity. Soluble molecular compounds that do not form ions when dissolved in water.

Precipitation reaction: An ion combination reaction in which cations and anions combine to form an insoluble ionic compound.

Freezing point depression: When a solute is dissolved in a solvent, the freezing point of the solution is lower than the freezing point of the pure solvent. The freezing point depression equation is $\Delta T = K_f m$, where ΔT is the change in freezing point, m is the molality of the solute, and K_f is the molal freezing point depression constant. For water, K_f is ($1.86°C/1$ m).

Boiling point elevation: When a (nonvolatile) solute is dissolved in a solvent, the boiling point of the solution is higher than the boiling point of the pure solvent. The boiling point elevation equation is $\Delta T = K_b m$, where ΔT is the change in boiling point, m is the molality of the solute and K_b is the molal boiling point elevation constant. For water, K_b is ($0.52°C/1$ m).

Osmosis: The phenomenon in which water passes from a solution of lower concentration of solute to a solution of higher concentration of solute through a semipermeable membrane.

QUESTIONS

1. Match the following terms with the appropriate lettered definitions.

solution

solvent

solute

hydrogen bond

like dissolves like

dynamic equilibrium

saturated solution

solubility

a. the number of moles of solute per liter of solution

b. the number of moles of solute per kilogram of solvent

c. a solution in which dissolved and undissolved solute are in dynamic equilibrium

d. the amount of solute needed to form a specific amount of a saturated solution

e. a solution containing more solute than a saturated solution at a given temperature

f. a situation in which two opposing processes balance one another but continually occur

g. the force of attraction between a hydrogen in one molecule and an electronegative atom in another molecule

h. a solute that forms a solution that conducts electricity

supersaturated solution

i. a solute that forms a solution that does not conduct electricity

molarity

j. an equation for a reaction occurring in solution showing only those molecules and ions that are reactants and products

standard solution

k. the component of a solution that dissolves the other components

percent by mass

l. the component of a solution that is dissolved by the solvent

molality

m. properties of solutions that depend only on the number of solute particles and not the nature of the particles

electrolyte

n. the phenomenon in which water passes through a membrane from a solution of lower concentration of solute to a solution of higher concentration of solute

nonelectrolyte

o. a solution of known concentration

chemical equilibrium

p. the ratio of the mass of a component to the mass of the mixture multiplied by 100

net-ionic equation

q. a reversible chemical reaction in which two opposing reactions are in dynamic equilibrium

precipitation reaction

r. an ion combination reaction in which cations and anions combine to form an insoluble ionic compound

colligative properties

s. a homogeneous mixture of two or more components in which the particles intermingle on an atomic, molecular, or ionic level

osmosis

t. polar dissolves polar and nonpolar dissolves nonpolar

Section 12-1

2. Define the following terms.
   (a) solution
   (b) solvent
   (c) solute

Section 12-2

3. Explain how polar molecules attract one another. What is a hydrogen bond?

4. Sketch four water molecules and show what portions of the molecules can attract when they are in close proximity.

Section 12-3

5. Describe what happens, in terms of the interaction between water molecules and ions, when the ionic compound potassium chloride, KCl, dissolves in water.

6. Why does a solid solute usually dissolve faster when it is as tiny crystals rather than larger crystals? Why do most solid solutes dissolve faster when the solute and solvent are stirred together?

7. Sodium dichromate, $Na_2Cr_2O_7$, is an orange-yellow solid. The color is due to the presence of the dichromate ion, which has an orange-yellow color. When sodium dichromate is dissolved in water, the entire solution takes on an orange-yellow color. Explain this in terms of the dissolving process.

8. Most gases are less soluble in hot water and more soluble in cold water. Natural waters contain dissolved oxygen and nitrogen from the atmosphere.
   (a) When tap water is heated, small gas bubbles form. Explain this observation.
   (b) When natural waters become overly heated by power plant or industrial wastewaters, fish and other aquatic life cannot survive. Explain this.

9. Most gases are more soluble at higher pressures and less soluble at lower pressures. Carbonated water is water saturated with carbon dioxide under pressure. Why does part of the carbon dioxide gas bubble off when a bottle of a carbonated beverage is opened?

Section 12-4

10. What does "like dissolves like" mean?

11. Explain why water is such a good solvent for ionic compounds.

12. Suppose you want a good solvent for a compound that has nonpolar molecules. Do you think water is the best solvent? Why?

Sections 12-5 and 12-6

13. What is meant by the term *dynamic equilibrium*?

14. Define the following terms.
   (a) saturated solution
   (b) solubility
   (c) supersaturated solution
   (d) unsaturated solution

15. A bottle on the laboratory shelf is marked "saturated NaCl." The bottle contains a clear liquid on top of some white crystals. Describe the nature of this solution in terms of dissolving, recrystallization, and dynamic equilibrium.

Sections 12-7 and 12-8

16. Give a definition for molarity. How is the symbol M interpreted?

17. A solution in the laboratory has a label marked 2.00 M KNO_3. Explain its meaning.

18. *A sample of wine vinegar is found to contain 18.9 g of acetic acid, $HC_2H_3O_2$, in 341 mL of the solution. Calculate the molarity of acetic acid in the vinegar.

19. A 325 mL sample of beer is found to contain 9.75 g of ethyl alcohol, C_2H_6O. Calculate the molarity of ethyl alcohol in the beer.

20. *You analyze a cup of coffee and find that it contains 88 mg of caffeine, $C_8H_{10}N_4O_2$. What is the molarity of caffeine in your coffee if 1 cup contains 237 mL of coffee?

21. A 49.5 mL sample of medicinal hydrogen peroxide solution is found to contain 1.49 g of H_2O_2. What is the molarity of H_2O_2 in the solution?

22. Following a dose of an antibiotic, 1.18×10^6 units of penicillin G ($C_{16}H_{18}O_4N_2S$) are recovered from 165 mL of the patient's urine. What is the molarity of penicillin G in the urine if 10^6 units of penicillin equal 0.600 g?

23. A sodium chloride solution is prepared by adding 1.50 moles of NaCl to 1.00 L of water. Explain why this solution is not 1.50 M in NaCl.

24. *A solution is prepared by dissolving 20.5 mL of glacial acetic acid, $HC_2H_3O_2$, in enough water to produce 115 mL of solution. If the density of glacial acetic acid is 1.049 g/mL, determine the molarity of the acetic acid solution. Glacial acetic acid is a common name for pure acetic acid.

25. A solution is prepared by dissolving 28.4 mL of ethyl alcohol, C_2H_6O, in enough water to produce 0.275 L of solution. If the density of pure ethyl alcohol is 0.790 g/mL, determine the molarity of the ethyl alcohol solution.

26. *A solution is prepared by dissolving 19.5 g of Na_2SO_4 in enough water to form 785 mL of solution. When the solute dissolves, it forms ions according to the equation

$$Na_2SO_4(s) \longrightarrow 2Na^+(aq) + SO_4{}^{2-}(aq)$$

   (a) What is the molarity of the $SO_4{}^{2-}$ in the solution?
   (b) What is the molarity of the Na^+ in the solution?

27. Iron(III) sulfate, $Fe_2(SO_4)_3$, dissolves in water and forms ions according to the equation

$$Fe_2(SO_4)_3(s) \longrightarrow 2Fe^{3+}(aq) + 3SO_4{}^{2-}(aq)$$

If 36.2 g of iron(III) sulfate are dissolved in 350 mL of solution, find

   (a) the molarity of the sulfate ion, $SO_4{}^{2-}$, in the solution
   (b) the molarity of the iron(III) ion, Fe^{3+}, in the solution

28. *How many milliliters of a 0.215 M potassium nitrate solution are needed to provide 0.0323 mole of dissolved KNO_3?

29. How many milliliters of a 0.789 M acetic acid solution are needed to provide 0.475 mole of $HC_2H_3O_2$?

30. *A solution is 0.363 M in H_2O_2. How many milliliters of this solution contain 1.75 g of H_2O_2?

31. A $MgCl_2$ solution is 1.50 M. How many milliliters of this solution contain 16.3 g of $MgCl_2$?

Section 12-9

32. What is a standard solution? Describe how to prepare a standard solution.

33. *Determine the number of grams of solute needed to prepare the following solutions.
   (a) 300 mL of a 0.925 M KOH solution
   (b) 520 mL of a 0.487 M $NaNO_3$ solution

(c) 1.17 L of a 0.125 M C_2H_5OH (ethyl alcohol) solution

(d) 60 mL of a 0.180 M $C_6H_{12}O_6$ (glucose) solution

(e) 725.0 mL of a 2.12 M Li_2CO_3 solution

34. Determine the number of grams of solute needed to prepare the following solutions.

(a) 235 mL of a 4.95×10^{-2} M $HC_2H_3O_2$ (acetic acid) solution

(b) 8.30 L of a 0.396 M $FeCl_3$ solution

(c) 740 mL of a 0.200 M NaOH solution

(d) 3.15 L of a 0.500 M C_2H_5OH (ethanol) solution

(e) 910 mL of a 6.75×10^{-3} M K_2CO_3 solution

35. *Plasma is the fluid portion of blood. The concentration of acetylsalicylic acid (aspirin), $C_9H_8O_4$, in your plasma is 2.99×10^{-4} M after you take two aspirin. If the volume of your plasma is 5.85 L, how many grams of aspirin are in your blood?

36. A person can just taste the sweetness in 2.0×10^{-2} M sucrose, or table sugar, $C_{12}H_{22}O_{11}$. How many grams of sucrose are contained in a cup of tea having a volume of 250 mL that is 2.0×10^{-2} M in sucrose?

37. *A cleaning solution is 9.75 M in ammonia, NH_3. How many grams of ammonia are contained in 1.00 qt of the cleaning solution? (1 qt = 0.946 L)

Section 12-10

38. *To what volume should 25.0 mL of a 1.37 M KI solution be diluted so that the solution is 0.375 M?

39. To what volume should 150 mL of 5.0 M acetic acid be diluted to give a 0.100 M solution?

40. *You need 0.250 M hydrochloric acid for an experiment, but all you can find on the shelf is 6.00 M hydrochloric acid. If you need 100 mL of 0.250 M acid, how many milliliters of the 6.00 M acid are needed to prepare it?

41. How many milliliters of 12 M NH_3 are needed to prepare 75 mL of 1.0 M NH_3 by dilution?

42. *Potassium chloride, KCl, is used to treat people with hypokalemia (low blood potassium levels). Potassium Chloride Injection, USP, is a 3.00 M solution of KCl. You must dilute this to 0.075 M before giving it to a patient. To prepare 50 mL of the dilute solution, what volume of the more concentrated KCl solution do you have to use?

43. How many milliliters of 6.00 M hydrochloric acid are needed to prepare 10.0 L of 0.100 M hydrochloric acid by dilution?

44. *How many milliliters of a 6.00 M sulfuric acid solution are needed to prepare 1.2 L of a 0.050 M sulfuric acid solution by dilution?

45. How many milliliters of 12.00 M nitric acid are needed to prepare 220 mL of 0.150 M nitric acid solution by dilution?

46. *An $FeCl_3$ solution is 2.50 M. Find the molarity of the following.

(a) a 10.0 mL sample diluted with water to a volume of 40.0 mL

(b) a 75.0 mL sample allowed to evaporate to a volume of 40.0 mL

(c) A 1.25 L sample in which an additional 25.0 g of $FeCl_3$ is dissolved (assume no change in volume)

47. An NaCl solution is 2.00 M. Find the molarity of the following.

(a) a 10.0 mL sample diluted with water to a volume of 30.0 mL

(b) a 100 mL sample allowed to evaporate to a volume of 60.0 mL

(c) A 1.10 L sample in which an additional 10.0 g of NaCl is dissolved (assume no change in volume)

Section 12-11

48. *How many milliliters of 1.00 M hydrochloric acid are needed to dissolve 15.00 g of aluminum metal in acid, if the reaction is

$$2Al(s) + 6HCl(aq) \longrightarrow 2AlCl_3(aq) + 3H_2(g)$$

49. How many grams of calcium metal can react with 118 mL of 0.375 M hydrochloric acid if the reaction is

$$Ca(s) + 2HCl(aq) \longrightarrow CaCl_2(aq) + H_2(g)$$

50. *How many liters of carbon dioxide gas measured at STP can form when 40.0 mL of 12.00 M HCl reacts with calcium carbonate by the reaction

$$CaCO_3(s) + 2HCl(aq) \longrightarrow$$
$$CaCl_2(aq) + CO_2(g) + H_2O(\ell)$$

51. How many liters of carbon dioxide gas measured at 21°C and 1.00 atm can form when 225 mL of 1.00 M nitric acid react with calcium carbonate by the reaction

$$CaCO_3(s) + 2HNO_3(aq) \longrightarrow$$
$$Ca(NO_3)_2(aq) + CO_2(g) + H_2O(\ell)$$

Section 12-12

52. *A cup of tea is 12.5% by mass sugar, $C_{12}H_{22}O_{11}$. If a cup of tea contains 300 g of solution, how many grams of sugar did you add?

53. A solution is prepared by dissolving 0.727 g of potassium chloride in 90.0 g of water. What is the percent by mass of KCl in the solution?

54. *A 3.0% by mass solution of hydrogen peroxide, H_2O_2, is sold in drugstores as a mild antiseptic. If a bottle of hydrogen peroxide contains 448 g of solution, how many grams of H_2O_2 does it contain?

55. A physiological saline solution contains 0.92% NaCl by mass. How many grams of NaCl are in 320 g of this solution?

56. *Ocean water contains 1.31 g of magnesium ions, Mg^{2+}, per kilogram of ocean water. Calculate the parts per million concentration of Mg^{2+} ions in ocean water.

57. Ocean water contains 0.401 g of calcium ions, Ca^{2+}, per kilogram of ocean water. Calculate the parts per million concentration of Ca^{2+} ions in ocean water.

58. *Ocean water contains 1.39×10^{-4} ppm silver ions, Ag^+. If the volume of the ocean is 1.5×10^9 km^3 and the average density is 1.005 g/cm^3, calculate how many grams of Ag^+ are contained in the ocean.

59. Bromine is extracted from ocean water by reacting bromide ions, Br^-, with chlorine.

$$Cl_2(aq) + 2Br^-(aq) \longrightarrow Br_2(\ell) + 2Cl^-(aq)$$

If ocean water contains 6.8×10^{-2} g of bromide ions, Br^-, per kilogram, how many kilograms of water must be processed to obtain 1 metric ton, 1000 kg, of bromine, Br_2?

60. *A concentration of 200 ppm CO_2 dissolved in water is lethal to fish. If a 2.0 L sample of Mississippi River water contains 0.68 g of dissolved CO_2, is the water likely to be lethal to fish?

61. The EPA standards for drinking water allow an MCL of 0.01 ppm Cd^{2+}. If a 1.25 L sample of drinking water contains 0.014 mg of Cd^{2+}, does the water meet or exceed the maximum allowable concentration?

62. *Medicinal hydrogen peroxide solution (3.0%) contains 3.0 g of H_2O_2 in 97 g of water. What is the molality of hydrogen peroxide in the solution?

63. A solution contains 35.0 g of glucose in 100 g of water. What is the molality of glucose in the solution?

Sections 12-13 and 12-14

64. What are electrolytes and nonelectrolytes?

65. What is meant by the statement that a solution contains ions?

66. Is it possible for a solution to contain only positive ions and no negative ions? Why?

67. What simple test is used to find out whether a solution contains ions?

68. Why is it dangerous to handle electrical equipment while you are immersed in water or standing in water?

69. What is the difference between strong electrolytes and weak electrolytes?

70. Why would you expect a solution of a soluble ionic compound to conduct electricity?

Section 12-15

71. *Give the formulas of the ions that are present in aqueous solutions of the following ionic compounds.

    (a) sodium sulfate, Na_2SO_4

    (b) potassium dichromate, $K_2Cr_2O_7$

    (c) sodium fluoride, NaF

    (d) lithium nitrate, $LiNO_3$

    (e) ammonium chloride, NH_4Cl

    (f) silver nitrate, $AgNO_3$

72. Give the formulas of the ions that are present in aqueous solutions of the following ionic compounds.

    (a) ammonium sulfate, $(NH_4)_2SO_4$

    (b) potassium permanganate, $KMnO_4$

    (c) lithium carbonate, Li_2CO_3

    (d) calcium nitrate, $Ca(NO_3)_2$

    (e) sodium chromate, Na_2CrO_4

    (f) sodium phosphate, Na_3PO_4

73. *Give the formulas of the major species that are found in solutions of the following compounds.

    (a) nitric acid, HNO_3 (strong electrolyte like HCl)

    (b) ethyl alcohol, C_2H_5OH (nonelectrolyte)

    (c) hydrogen sulfide, H_2S (weak electrolyte)

    (d) sucrose, $C_{12}H_{22}O_{11}$ (nonelectrolyte)

    (e) hydrogen fluoride, HF (weak electrolyte)

    (f) copper(II) sulfate, $CuSO_4$ (strong electrolyte)

    (g) hydrogen peroxide, H_2O_2 (nonelectrolyte)

    (h) calcium chloride, $CaCl_2$ (strong electrolyte)

    (i) ammonia, NH_3 (weak electrolyte)

74. Give the formulas of the major species that are found in solutions of the following compounds.

    (a) glucose, $C_6H_{12}O_6$ (nonelectrolyte)

    (b) potassium chloride, KCl (strong electrolyte)

    (c) methyl alcohol, CH_3OH (nonelectrolyte)

    (d) magnesium sulfate, $MgSO_4$ (strong electrolyte)

    (e) phosphoric acid, H_3PO_4 (weak electrolyte)

    (f) ethylene glycol, $C_2H_6O_2$ (nonelectrolyte)

    (g) sodium hydrogen carbonate, $NaHCO_3$ (strong electrolyte)

    (h) silver nitrate, $AgNO_3$ (strong electrolyte)

    (i) oxalic acid, $H_2C_2O_4$ (weak electrolyte)

Section 12-16

75. What is a net-ionic equation? Give an example.

76. What are spectator ions?

77. *Write balanced net-ionic equations for the following reactions.

    (a) Some zinc metal, Zn, is added to a solution containing dissolved copper(II) chloride and produces copper metal, Cu, and $Zn^{2+}(aq)$.

    (b) $Cl_2(aq)$ reacts with a solution containing $Br^-(aq)$ and produces $Br_2(aq)$ and $Cl^-(aq)$.

    (c) A solid piece of zinc is added to a hydrochloric acid solution and produces hydrogen gas, water, and zinc ion, $Zn^{2+}(aq)$.

    (d) A solution of K_2CrO_4 is added to a solution of $BaCl_2$ and produces solid $BaCrO_4$.

    (e) A solution of NaOH is added to a solution of Na_2HPO_4 and produces H_2O and $PO_4^{3-}(aq)$.

    (f) Solid $CaCO_3$ is added to a solution of hydrochloric acid and produces CO_2, H_2O, and $Ca^{2+}(aq)$. [Hint: $CaCO_3(s)$ is one reactant, and hydrochloric acid, the other reactant, is $H_3O^+(aq)$ and $Cl^-(aq)$.]

78. Write balanced net-ionic equations for the following reactions.

    (a) $Cl_2(aq)$ reacts with a solution containing $I^-(aq)$ to give $Cl^-(aq)$ and I_2.

    (b) Some copper, Cu, metal is added to a solution containing dissolved $AgNO_3$; a reaction occurs and produces Ag metal and $Cu^{2+}(aq)$.

    (c) A solution of KI is added to a solution of $Pb(NO_3)_2$, and solid PbI_2 is formed.

    (d) A solution of KOH is added to a solution of $NaHCO_3$, producing H_2O and $CO_3^{2-}(aq)$.

    (e) A piece of solid Na metal reacts with water to form hydrogen gas, hydroxide ion, OH^-, and sodium ion, Na^+.

    (f) A solution of the weak electrolyte oxalic acid, $H_2C_2O_4$, is mixed with a solution of NaOH; in the reaction, water and oxalate ion, $C_2O_4^{2-}$, are formed.

Sections 12-17 to 12-19

79. What is a precipitation reaction? Give an example.

80. What are the differences between soluble, slightly soluble, and insoluble ionic compounds?

81. Give an example of an insoluble ionic compound.

82. *Write balanced net-ionic equations for any precipitation reactions that occur when the following solutions are mixed. Use the general approach to predicting precipitation reactions and refer to the solubility rules in Section 12-18.

    (a) mercury(I) nitrate solution and sodium bromide solution

    (b) potassium fluoride solution and barium nitrate solution

    (c) lithium chloride solution and magnesium chloride solution

    (d) ammonium nitrate solution and sodium nitrate solution

    (e) potassium hydroxide solution and calcium chloride solution

    (f) barium nitrate solution and sodium phosphate solution

    (g) lead(II) acetate solution and sodium iodide solution

83. Write balanced net-ionic equations for any precipitation reactions that occur when the following solutions are mixed. Use the general approach to predicting precipitation reactions, and refer to the solubility list given in Section 12-18.

    (a) lead(II) nitrate solution and potassium chloride solution

    (b) sodium fluoride solution and calcium nitrate solution

    (c) ammonium chloride solution and sodium chloride solution

(d) sodium hydroxide solution and magnesium chloride solution

(e) barium nitrate solution and sodium sulfate solution

(f) sodium chloride solution and zinc chloride solution

(g) silver nitrate solution and potassium bromide solution

84. *Predict and give balanced net-ionic equations for any precipitation reactions that occur when the following solutions are mixed. Base the predictions on the solubility list given in Section 12-18.

(a) $MgCl_2$ solution and $Ba(OH)_2$ solution

(b) $Ca(NO_3)_2$ solution and NaF solution

(c) $Pb(NO_3)_2$ solution and Na_3PO_4 solution

(d) $FeCl_3$ solution and $Pb(NO_3)_2$ solution

(e) KOH solution and NaCl solution

(f) $AgNO_3$ solution and NaI solution

(g) $FeCl_3$ solution and $Ba(OH)_2$ solution

(h) $Ba(OH)_2$ solution and $CdSO_4$ solution

(i) NaCl solution and $(NH_4)_2SO_4$ solution

(j) $Pb(NO_3)_2$ solution and K_2CrO_4 solution

(k) $Mg(C_2H_3O_2)_2$ solution and NaF solution

85. What is the function of calcium in the human body?

Sections 12-20 and 12-21

86. What are colligative properties of solutions?

87. How are the freezing and boiling points of a solvent affected by the presence of a solute?

88. Ethylene glycol is added to the radiator system of an automobile as an antifreeze and coolant. How does ethylene glycol function as an antifreeze and coolant?

89. To make homemade ice cream, it is necessary to cool the mixture below 0°C. How does an ice and salt mixture provide the required cooling?

90. *Calculate the freezing point of a 3.0% hydrogen peroxide solution. See Question 62.

91. Calculate the freezing point of a solution that contains 50.0 g of sucrose dissolved in 200 g of water.

92. *Calculate the boiling point of a solution that contains 50.0 g of sucrose dissolved in 200 g of water.

93. The freezing point of a glucose solution is −3.2°C. In this solution, how many grams of glucose are dissolved in each 100 g of water?

94. Describe the phenomenon of osmosis.

95. Cellular membranes in plants and animals are osmotic membranes. Explain the following in terms of osmosis.

(a) Sliced fruit forms its own juice when sugar is added.

(b) Drinking salt water can cause dehydration of the body and sometimes death.

(c) Solutions injected into the body intravenously must have the same ionic concentrations as blood; they are called isotonic solutions.

(d) Prunes swell when they are placed in water.

(e) Cucumbers shrink when they are placed in a concentrated salt solution to make pickles.

ACIDS AND

BASES

13-1 THE NATURE OF ACIDS AND BASES

Vinegar has a distinct and characteristic odor and sour taste. Vinegar is a water solution of acetic acid, and it serves as a good example of an acid in aqueous solution. Taste buds on the sides of our tongues respond to sour tastes. Citric acid in lemon juice tastes sour, as does ascorbic acid or vitamin C. You've undoubtedly tasted other sour foods. The chemical structures of the molecules of acetic acid, citric acid, and vitamin C are quite different, but they do share certain chemical properties. In fact, a unique chemical property of these acids, and other acids, accounts for the sour taste of acid solutions. In times past, chemists often tasted chemicals to test their properties. This practice resulted in many cases of sickness and death. You should not taste or smell laboratory chemicals indiscriminately since many are toxic. It is interesting that the properties of some chemicals have been discovered in modern times by accidental tasting. The hallucinogenic drug LSD was discovered by a chemist who accidentally ingested some of it and experienced the startling effects. Similarly, the artificial sweetener aspartame (NutraSweet) was discovered when a chemist accidentally tasted it.

It is possible to classify compounds according to similarities in chemical properties, that is, similarities in the kinds of chemical reactions they undergo. Chemists noted long ago that certain substances, now known as acids, are characterized by a set of common properties. The word *acid* comes from the Latin *acidus* (sour). Here are some experimental observations on solutions of **acids.**

1. Acids taste sour in diluted solutions and, in concentrated solutions, cause skin burns by dissolving tissue.
2. Acids change the color of litmus, a vegetable dye, from blue to red.

407

3. Acids chemically react with and neutralize the effect of chemical bases (neutralization reactions).
4. Acids react with certain metals to dissolve them and give metal ions and hydrogen gas.

Another class of compounds, called bases, includes compounds that are in a sense the chemical opposites of acids. Here are some experimental observations on solutions of **bases.**

1. Bases taste bitter and feel slippery in diluted solutions and, in concentrated solutions, cause skin burns by dissolving tissue.
2. Bases change the color of litmus from red to blue.
3. Bases chemically react with and neutralize the effect of acids (neutralization reactions).

Chemists made some first known bases by strongly heating substances isolated from water extracts of wood ashes. Chemists viewed compounds made in this way as the basis, or base, from which other compounds could be made. Hence, these compounds came to be called bases.

Vinegar and Acetic Acid

Vinegar (from the old French for sour wine) has been used as a food condiment for thousands of years. It is used in salad dressings and as a pickling agent. Typical vinegar is a water solution containing 4% to 8% acetic acid by mass. According to the U.S. Food and Drug Administration (FDA), the minimum legal limit for vinegar is 4% acetic acid. Commercial vinegar is made by the fermentation of malt, barely, apple cider, wine, or other fruit juices. The fermenting agent is a mold (e.g., *Mycoderma aceti*) known as "mother of vinegar." Some vinegar is distilled to remove coloring matter and other impurities; it is then called white vinegar. Acetic acid is also present in sourdough bread. In making this bread, the dough is leavened with a species of yeast that cannot metabolize the carbohydrate maltose. The presence of maltose encourages the growth of the bacteria *Lactobacillus sanfrancisco*, which produces acetic acid and lactic acid from maltose. In years passed, prospectors in Alaska and the Yukon came to be called sourdoughs since they often carried some sourdough yeast starter for bread making.

Acetic acid is manufactured industrially by the reaction of methyl alcohol and carbon monoxide.

$$CH_3OH(g) + CO(g) \xrightarrow[175°C \quad 35 \text{ atm}]{Rh} CH_3CO_2H(g)$$

Pure acetic acid is known as glacial acetic acid since it freezes to form ice-like crystals at 17°C. The major uses of industrial acetic acid are as solvent and a reactant in the production of the plastics polyvinyl acetate and cellulose acetate.

Acids and Bases

13-2 DEFINITIONS FOR ACIDS AND BASES

As a deeper understanding of the nature of chemicals developed, it became apparent that many substances could be classified as acids or bases. Several chemical definitions for acids and bases are possible, and we consider the most useful definitions for our discussion.

In 1884, a Swedish chemist, Svante Arrhenius, proposed the first significant definitions of acid and base. He defined an acid as a chemical that forms hydrogen ions (H^+) in water solution and a base as a chemical that forms hydroxide ions (OH^-) in water solution. In the Arrhenius view, the behavior of hydrogen chloride as an acid is represented as

$$HCl \longrightarrow H^+ + Cl^-$$

and the behavior of sodium hydroxide as a base is represented as

$$NaOH \longrightarrow Na^+ + OH^-$$

The Arrhenius theory established that hydrogen ions and hydroxide ions are related to acid and base behavior in water. The theory, however, was not general enough to include all chemicals that behave as acids and bases. More general definitions come from the **Brønsted–Lowry** theory of acids and bases. Definitions are given in the margin.

A **proton, (H^+),** is a positively charged hydrogen ion. It is a hydrogen atom without an electron. A proton is a unique chemical particle; it is a hydrogen ion, but unlike other ions, it has no electrons and is a bare nucleus. A lone proton is formed by the breaking of a covalent bond involving hydrogen and another element so that the shared electron pair remains with the other element. In this discussion, hydrogen ions are called protons. The loss of a proton by an acid is represented generally as

$$H\text{-}A \longrightarrow \quad H^+ + :A^-$$

where H^+ is the proton, A is an element that covalently bonds to hydrogen, and : represents the electrons of the covalent bond.

When an acid loses a proton, a base gains it. Thus, a base is called a proton acceptor. Free protons are not produced by acids; instead, they are passed from acids to bases. Consequently, a **base** is a species that can form a covalent bond with a proton. A base gains or accepts a proton by forming a covalent bond with it. Before continuing your study of acids and bases, you should review the nomenclature of acids given in Sections 6-9 and 6-10.

13-3 ACID–BASE REACTIONS

A chemical reaction between an acid and a base is simply called an acid–base reaction. Let's look at an example of an acid–base reaction in detail. A solution of acetic acid has the typical properties of an acid. A solu-

Acid
A chemical species that can lose a hydrogen ion or a proton in a chemical reaction; a proton donor.

Base
A chemical species that can gain a hydrogen ion or a proton in a chemical reaction; a proton acceptor.

tion of sodium hydroxide has the typical properties of a base. When these two solutions are mixed, the acid and base neutralize each other's properties in a chemical reaction. An acid reacts with a base in a proton transfer or acid–base reaction (see Figure 13-1). An acid–base reaction occurs when a solution of NaOH is mixed with a solution of $HC_2H_3O_2$. The reaction involves molecular acetic acid and the hydroxide ion present in a sodium hydroxide solution. The net-ionic equation (see Section 12-16) for the reaction is

$$HC_2H_3O_2 + OH^- \longrightarrow H_2O + C_2H_3O_2^-$$

Look closely at the formulas and note that an H^+ transfers from $HC_2H_3O_2$ to OH^- to give H_2O and $C_2H_3O_2^-$.

Any acid–base reaction is represented generally as

$$\underset{\text{acid}}{H-A} + \underset{\text{base}}{B} \longrightarrow \underset{\text{acid}}{H-B^+} + \underset{\text{base}}{A^-}$$

In such a chemical reaction, an acid loses a proton to a species that can bond to the proton more readily than the original acid. Thus, we say that an acid, H—A, loses a proton to form a base, A^-, and a base, B, gains a proton to form an acid, $H-B^+$.

Notice that in an acid–base reaction, there is an acid as a reactant and an acid as a product. An acid donates a proton to a base to give a proton containing species that is a potential acid. The acid that is a reactant is considered a stronger acid than the acid formed as a product. The **stronger acid** is the acid that has a greater tendency to lose a proton. Furthermore, in any acid–base reaction, there is a base as a reactant and a base as a product. The base that is a reactant is considered a stronger base than the base formed as a product. The **stronger base** is the base that has a greater tendency to gain a proton.

$$\underset{\substack{\text{stronger} \\ \text{acid}}}{H-A} + \underset{\substack{\text{stronger} \\ \text{base}}}{B} \longrightarrow \underset{\substack{\text{weaker} \\ \text{acid}}}{H-B^+} + \underset{\substack{\text{weaker} \\ \text{base}}}{A^-}$$

In simple terms, when an acid–base reaction occurs, the stronger acid and stronger base form a weaker acid and a weaker base. In other words, when an acid–base reaction occurs, the weaker acid and base are products.

Typically, acid–base reactions are reversible and involve chemical equilibrium, so double arrows are often, but not always, shown in the equations. When the larger arrow is shown to the right ($\rightleftharpoons$), the acid–base reaction is extensive in the direction of the products. In such a case, the weaker acid and base are on the product side. When the larger arrow is directed to the left ($\rightleftharpoons$), the reaction is not extensive toward the products but only occurs to a slight extent. (See Section 12-14.) In the case of a slight reaction, the weaker acid and base are on the reactant side. In general, when you see a double arrow in an equation for any acid–base reaction, the larger arrow points toward the weaker acid and base.

13-4 IDENTIFYING ACIDS AND BASES

Every Brønsted–Lowry acid has a corresponding base formed by the loss of a proton, and each base has a corresponding acid formed by the gain of a proton. An acid and the base it forms by the loss of a proton are called a **conjugate acid–base pair.** Likewise, a base and the acid it forms by gaining a proton are considered a **conjugate acid–base pair.** In the acid–base reaction between acetic acid and hydroxide ions, referred to in the previous section, $HC_2H_3O_2$ and $C_2H_3O_2^-$ are a conjugate acid–base pair and H_2O and OH^- are a conjugate acid–base pair.

$$\overbrace{HC_2H_3O_2 + OH^-}^{\text{conjugate pair}} \underset{}{\rightleftharpoons} \underbrace{H_2O + C_2H_3O_2^-}_{\text{conjugate pair}}$$

The double arrow suggests that the reaction is extensive toward the product side; H_2O and $C_2H_3O_2^-$ are the weaker acid and base. In any acid–base reaction, there are two sets of conjugate acid–base pairs. Each pair is related by the loss or gain of a proton.

When an acid loses a proton, it is converted to its conjugate base. The base that gains the proton is, in turn, converted to its conjugate acid (see Figure 13-1). Acid–base reactions are common types of reactions in water solutions. In any Brønsted–Lowry acid–base reaction, it is possible to label each acid and base. The acetic acid–hydroxide ion reaction is labeled as

$$
\begin{array}{cccc}
\text{acid 1} & \text{base 2} & \text{acid 2} & \text{base 1} \\
HC_2H_3O_2 \ + & OH^- & \rightleftharpoons \quad H_2O \ + & C_2H_3O_2^- \\
\text{stronger} & \text{stronger} & \text{weaker} & \text{weaker} \\
\text{acid} & \text{base} & \text{acid} & \text{base}
\end{array}
$$

The labels mean that acid 1 pairs with base 1 and that base 2 pairs with acid 2. In the reaction, the stronger acid, acetic acid, reacts with the stronger base, the hydroxide ion, to form the weaker acid, water, and the weaker base, the acetate ion. Be careful with this terminology. The stronger an acid, the greater its ability to lose a proton; the stronger a base, the greater its ability to gain a proton. The term *stronger* means more chemically reactive; stronger acids and stronger bases react to form weaker acids and weaker bases.

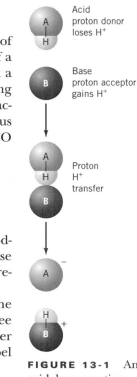

FIGURE 13-1 An acid–base reaction.

EXAMPLE 13-1

Label the conjugate acid–base pairs in the following equation and indicate which acid is the stronger acid and which base is the stronger base.

$$HC_2H_3O_2(aq) + NH_3(aq) \rightleftharpoons NH_4^+(aq) + C_2H_3O_2^-(aq)$$

On the left, we pick the species that loses an H^+ as the acid and the species that gains an H^+ as the base. On the right, the acid is the conjugate acid of the reactant base and the base is the conjugate base of the reactant acid. The stronger acid and base are those that more readily lose and gain protons. Thus, according to the size of the double arrow, the stronger acid and base are on the left.

acid 1		base 2			acid 2		base 1
$HC_2H_3O_2(aq)$	$+$	$NH_3(aq)$	$\rightleftharpoons$		$NH_4^+(aq)$	$+$	$C_2H_3O_2^-(aq)$
stronger		stronger			weaker		weaker
acid		base			acid		base

EXAMPLE 13-2

Indicate which species are conjugate acid–base pairs in the equation

$$H_3O^+(aq) + F^-(aq) \rightleftharpoons HF(aq) + H_2O$$

CAUTION: Hydrofluoric acid is very dangerous and can cause serious skin burns.

The conjugate acid–base pairs are related by the loss and gain of an H^+. So, we look for the acid as the reactant that loses an H^+ to form its conjugate base as a product. The base is the reactant that gains an H^+ to form its conjugate acid as a product.

	ACIDS	BASES
conjugate pair	H_3O^+	H_2O
conjugate pair	HF	F^-

ACTIVITY 13-1

For the following acid–base reaction, label the conjugate acid–base pairs (see Example 13-1) and make a table of the acids and bases (see Example 13-2).
$$H_2SO_4 + H_2O \longrightarrow$$
$$H_3O^+(aq) + HSO_4^-(aq)$$

13-5 ACIDIC AND BASIC SOLUTIONS

Many common acids are stored as solutions made by dissolving them in water. When the acid hydrogen chloride is dissolved in water, the following acid–base reaction occurs:

HCl	$+$	H_2O	$\longrightarrow$	$H_3O^+(aq)$	$+$	$Cl^-(aq)$
acid		base		hydronium ion		chloride ion

Hydrogen chloride is an example of an acid that is a strong electrolyte since it reacts almost completely with water to form hydronium ions and chloride ions. Note that since the reaction is essentially complete, no double arrow is used in the equation. In other words, HCl in water is 100% ionized.

Many bases are also stored as water solutions. When ammonia gas is mixed with water, the following equilibrium reaction occurs:

NH_3	$+$	H_2O	$\rightleftharpoons$	$NH_4^+(aq)$	$+$	$OH^-(aq)$
base		acid		ammonium ion		hydroxide ion

Ammonia is quite soluble in water but reacts to only a slight extent, which means that ammonia is the weaker base and H_2O is the weaker acid in the reaction. The relative sizes of the arrows in the equation show that the reaction does not occur extensively to the right. Note that water behaves as an acid in this reaction but as a base in the reaction with hydrogen chloride. Water may act as an acid in some reactions and a base in other reactions; this behavior is compatible with its structure.

gain of a proton forms ··· hydronium ion

loss of a proton forms :Ö—H$^-$ hydroxide ion

Make sketches of a water molecule, a hydronium ion, and a hydroxide ion.

Whether water acts as an acid or a base depends on the chemical species that reacts with it.

An animal that can exist both on land and in water is an amphibian. A chemical species that can act as an acid or a base is said to be amphiprotic. An amphiprotic species can play two roles. An **amphiprotic** species is one that can gain a proton or lose a proton, depending on the other chemicals with which it is mixed. Since water is amphiprotic, some proton transfer occurs between individual water molecules in any sample of pure water or any water solution. That is, water molecules react with one another to a slight extent as shown by the equilibrium equation

$$H_2O + H_2O \rightleftharpoons H_3O^+ + OH^-$$

This reversible reaction goes only slightly to the right, producing, in pure water, equal but relatively low concentrations of hydronium and hydroxide ions. Hence, a sample of water contains mostly water molecules along with very minor concentrations of H_3O^+ and OH^-. Pure water or water solutions that contain neither acids nor bases have equal concentrations of hydronium and hydroxide ions. Such solutions are called **neutral solutions.**

When an acid is mixed with water, it reacts with water to form hydronium ions. If we represent an acid by the general formula HA, the equation for the reaction of an acid with water is

$$HA + H_2O \rightleftharpoons H_3O^+ + A^-$$

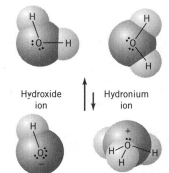

Hydroxide ion   Hydronium ion

A water solution that has a concentration of hydronium ions that is greater than that in pure water is called an **acidic solution.** When an acid dissolves in water, an acidic solution results. In highly acidic solutions, the concentrations of hydronium ions are relatively high, whereas in slightly acidic solutions, the concentrations of hydronium ions are relatively low but still higher than pure water.

A base mixed with water reacts to form some hydroxide ions. If we use B as a general formula for a base, the equation for the reaction of a base with water is

$$B + H_2O \rightleftharpoons BH^+ + OH^-$$

A water solution that has a concentration of hydroxide ions greater than in pure water is called a **basic solution.** Dissolved in water, bases form basic solutions also known as **alkaline solutions.** In highly basic solutions, the concentrations of hydroxide ions are relatively high, whereas in slightly basic solutions, the concentrations of hydroxide ions are relatively low but still higher than pure water.

13-6 AN ACID–BASE TABLE

We know that water solutions of hydrogen chloride and acetic acid are acidic, but solutions of these acids differ in one very important way. As discussed in Section 12-14, hydrogen chloride is a strong electrolyte and acetic acid is a weak electrolyte. The reaction of each of these acids with water reveals how they differ.

$$HCl + H_2O \longrightarrow H_3O^+(aq) + Cl^-(aq)$$

$$HC_2H_3O_2 + H_2O \rightleftharpoons H_3O^+(aq) + C_2H_3O_2^-(aq)$$

Hydrogen chloride reacts completely with water to form hydronium ions and chloride ions; that is, it is 100% ionized in solution. Hydrogen chloride is an example of a strong acid. Other common strong acids are nitric acid, HNO_3, and sulfuric acid, H_2SO_4. Recall that these are also the three common strong molecular electrolytes. The term *strong acid* is used because an acid of this type mixed with water reacts completely with water to form hydronium ions and the anions corresponding to its conjugate base.

In contrast to the strong acid hydrogen chloride, acetic acid has only a slight tendency to react with water to form hydronium ions and acetate ions. Acids that react with water only to a slight extent to form relatively low concentrations of hydronium ions and the corresponding conjugate base are called **weak acids.** Many acids are weak acids. When solutions of various weak acids are compared, it is found that weak acids vary in their ability to react with water; that is, some have a greater tendency to lose protons to water than others. The strength of a weak acid relative to other weak acids is determined by experiment. In fact, it is possible to rank or list acids according to their strengths.

Some common acids ranked in order of decreasing strength are shown in Table 13-1. The stronger the acid, the higher it ranks in the table. The three common strong acids are listed at the top, but even among the weak acids there is variation in their strengths. Be careful to distinguish between

Table 13-1   *Acid–Base Table of Common Brønsted–Lowry Acids and Bases*

	ACIDS		BASES		
Strong Acids React completely with water to form H_3O^+ and conjugate base	Sulfuric acid	H_2SO_4	HSO_4^-	Hydrogen sulfate ion	*Very Weak Bases* Will not react with H_3O^+ to form conjugate acid
	Hydrogen chloride	HCl	Cl^-	Chloride ion	
	Nitric acid	HNO_3	NO_3^-	Nitrate ion	
Weak Acids Do not react extensively with water	Hydronium ion	H_3O^+	H_2O	Water	*Weak Bases*
	Oxalic acid	$H_2C_2O_4$	$HC_2O_4^-$	Hydrogen oxalate ion	
	Hydrogen sulfate ion	HSO_4^-	SO_4^{2-}	Sulfate ion	
	Phosphoric acid	H_3PO_4	$H_2PO_4^-$	Dihydrogen phosphate ion	
	Hydrogen fluoride	HF	F^-	Fluoride ion	
	Hydrogen oxalate ion	$HC_2O_4^-$	$C_2O_4^{2-}$	Oxalate ion	
	Acetic acid	$HC_2H_3O_2$	$C_2H_3O_2^-$	Acetate ion	
	Carbon dioxide (aq)	$(CO_2 + H_2O)$	HCO_3^-	Hydrogen carbonate ion	
	Hydrogen sulfide	H_2S	HS^-	Hydrogen sulfide ion	
	Dihydrogen phosphate ion	$H_2PO_4^-$	HPO_4^{2-}	Hydrogen phosphate ion	
	Hydrogen sulfite ion	HSO_3^-	SO_3^{2-}	Sulfite ion	
	Ammonium ion	NH_4^+	NH_3	Ammonia	
	Hydrogen cyanide	HCN	CN^-	Cyanide ion	
	Hydrogen carbonate ion	HCO_3^-	CO_3^{2-}	Carbonate ion	
	Hydrogen phosphate ion	HPO_4^{2-}	PO_4^{3-}	Phosphate ion	
	Hydrogen sulfide ion	HS^-	S^{2-}	Sulfide ion	
	Water	H_2O	OH^-	Hydroxide ion	*Strong Base*

Decreasing strength ↓ (left margin)   Increasing base strength ↑ (right margin)

"strong acid" and "stronger acid." Strong acid has a definite meaning. All strong acids are 100% ionized in water.

Any acid has a corresponding conjugate base which it can form by the loss of a proton. The conjugate base of each acid is included in Table 13-1. The bases are listed in the right-hand column. The conjugate bases are also listed according to relative strength, but they are listed with the weakest at the top and the strongest at the bottom. This arrangement makes sense because the stronger an acid, the weaker its conjugate base, and the weaker an acid, the stronger its conjugate base.

Look closely at Table 13-1. Note that water is listed as a very weak acid at the bottom of the list of acids and as a weak base near the top of the list of bases. Any of the strong acids, when mixed with water, reacts to give hydronium ions, H_3O^+, and the conjugate base. In reference to Table 13-1, using HX for a strong acid this reaction is represented as

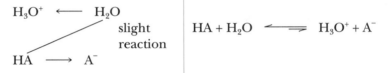

Thus, the **hydronium ion** is the strongest acid that can exist in water.

Weak acids include ionic species and molecular substances. For example, H_2S is a molecular weak acid and the hydrogen sulfide ion, HS^-, is an even weaker acid. All weak acids react only slightly with water, so the major species in a weak acid solution is the acid itself. Note the location of weak acids in Table 13-1 relative to H_3O^+. Any weak acid, represented as HA, reacts with water only to a slight extent.

$$H_3O^+ \longleftarrow H_2O$$
slight reaction
$$HA \longrightarrow A^-$$

$$HA + H_2O \rightleftharpoons H_3O^+ + A^-$$

As an example, consider that a solution of hydrogen sulfide contains H_2S as the major species since it only reacts with water to a slight extent.

$$H_2S + H_2O \rightleftharpoons H_3O^+ + HS^-$$

When carbon dioxide is dissolved in water to give carbonated water, the solution is acidic. Since CO_2 has no proton, we can explain its acidic nature as a close association and reaction with water. The equation for the reaction of carbon dioxide with water to give an acidic solution is

$$CO_2 + 2H_2O \rightleftharpoons H_3O^+ + HCO_3^-$$

Because of this behavior, carbon dioxide is included in the table as $CO_2 + H_2O$. You may see this written as H_2CO_3, called carbonic acid, but there is little evidence for the existence of such a molecule. If it does form, it quickly reacts with water or separates into carbon dioxide and water.

Returning to Table 13-1, note that most common bases are negatively charged ionic species. The only exceptions are water and ammonia, NH_3. Water and ammonia are molecular bases. Other molecular bases are known but are not included in the table. In a water solution of any base, the base is the major species. When a base, as represented as B, that is weaker than OH^- is in water, it reacts only slightly with water

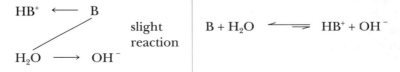

For example, in a solution of ammonia, $NH_3(aq)$ is the major species.

$$NH_3 + H_2O \rightleftharpoons \underset{\text{minor}}{NH_4^+(aq) + OH^-(aq)}$$
$$\underset{\text{major}}{}$$

For a solution of an ionic compound, the major species are the ions. In a solution of NaOH, Na^+ and OH^- are the major species.

$$NaOH(s) \longrightarrow \underset{\text{major}}{Na^+(aq) + OH^-(aq)}$$

A solution of NaOH is viewed as a strong base since it contains relatively high concentrations of hydroxide ion. Hydroxide ion is the strongest base that can exist in water.

13-7 COMMON ACID AND BASE SOLUTIONS

In the laboratory, you may encounter water solutions of strong acids such as hydrochloric acid, sulfuric acid, or nitric acid. These acids are also widely used in industry for various chemical processes. Strong acid solutions contain hydronium ions and the conjugate base (see Table 13-1). To represent the major species in these solutions, we write the formulas of both ions.

hydrochloric acid solution $H_3O^+(aq) + Cl^-(aq)$
sulfuric acid solution $H_3O^+(aq) + HSO_4^-(aq)$
nitric acid solution $H_3O^+(aq) + NO_3^-(aq)$

A solution of a strong acid always contains the hydronium ion as a major ion. **Solutions of strong acids** are good sources of hydronium ions.

A weak acid does not react extensively with water. Thus, a solution of a molecular weak acid contains the unreacted acid as the major species. A solution of a **molecular weak acid** is represented by the formula of the acid. For example, a water solution of hydrogen fluoride, HF, is represented as $HF(aq)$, and a water solution of phosphoric acid, H_3PO_4, is represented as $H_3PO_4(aq)$. Likewise, weak acids that are ionic species are represented by the formula of the ion. How do we prepare solutions containing these ions? Any soluble ionic compound is represented in solution by the formulas of the ions comprising the compound. If we want a solution containing hydrogen sulfate ions, for example, we dissolve $NaHSO_4$ in water. The solution is represented as

$$Na^+(aq) + HSO_4^-(aq)$$

Incidentally, this compound is the main ingredient in some toilet bowl cleaners. On the label of such a product, it may be listed as sodium bisulfate, a common name for sodium hydrogen sulfate. In a solution of this compound, the sodium ion is present to balance the charge, but the acid in solution is the hydrogen sulfate ion.

ACTIVITY
13-3

How would you prepare a water solution containing the weak acid ammonium ion, NH_4^+?

How would you prepare a solution containing the base phosphate ion, PO_4^{3-}?

Make a table with columns for strong acids, weak acids, weak bases, and strong bases. Give the formula for one example of each of these. Also, give the formulas of the major species found in a solution of each of your examples.

Solutions of bases are similarly represented by the major species in solution. A solution of a **molecular base** is represented by the formula of the base. Thus, a water solution of ammonia is represented as $NH_3(aq)$.

Note in Table 13-1 that many ionic bases are ions that you learned about in Chapters 6 and 9. As ions, these bases occur in many ionic compounds. Solutions of these bases are formed by dissolving soluble ionic compounds. For example, a solution of potassium carbonate, K_2CO_3, is represented as

$$2K^+(aq) + CO_3^{2-}(aq)$$

The carbonate ion is the base in the solution.

As Table 13-1 reveals, the hydroxide ion, OH^-, is the strongest base that can exist in water. Note its location in the table. A solution containing hydroxide ions is prepared by dissolving a soluble ionic compound containing the hydroxide ion. The common soluble hydroxide ion–containing compounds are the strong bases sodium hydroxide, $NaOH$, potassium hydroxide, KOH, and barium hydroxide, $Ba(OH)_2$. In aqueous solutions, these **strong bases** are represented as the hydroxide ion and the cation.

sodium hydroxide solution	$Na^+(aq) + OH^-(aq)$
potassium hydroxide solution	$K^+(aq) + OH^-(aq)$
barium hydroxide solution	$Ba^{2+}(aq) + 2OH^-(aq)$

Sodium hydroxide solutions are the most common sources of hydroxide ions for use in laboratory work, and sodium hydroxide is by far the most important industrial hydroxide. It is used in many chemical manufacturing processes, and in paper making, water treatment, aluminum ore processing and petroleum refining.

13-8 NEUTRALIZATION REACTIONS

Some acid–base reactions are truly amazing. A concentrated solution of hydrochloric acid is quite dangerous; it can dissolve metals and cause skin burns. A concentrated solution of sodium hydroxide is also very dangerous. When these solutions are mixed in the correct proportions, however, the result is a neutral salt solution, and the dangerous properties are no longer present. **DANGER: Never mix concentrated acid and base solutions, since they react with vigor and heat.** Most acid–base reactions are not this dramatic, but in all acid–base reactions, the original acid and base are altered chemically.

When we mix a solution containing an acid with a solution containing a base, an acid–base reaction may occur. A solution of a strong acid contains hydronium ions, potent proton donors. A solution of a strong base contains hydroxide ions, potent proton acceptors. When a solution of hy-

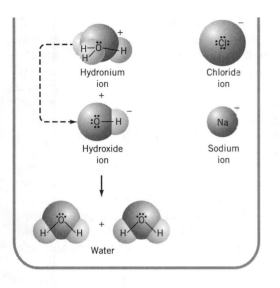

FIGURE 13-2 The neutralization reaction between hydrochloric acid and sodium hydroxide. The hydronium ions and hydroxide ions react and the sodium ions and chloride ions are spectator ions.

drochloric acid is mixed with a solution of sodium hydroxide, the ions in the solutions are

$$H_3O^+(aq) + Cl^-(aq) \quad \text{and} \quad Na^+(aq) + OH^-(aq)$$

Mixing the solutions brings all the ions into contact.

$$H_3O^+(aq) + Cl^-(aq) + Na^+(aq) + OH^-(aq)$$

The hydronium ion reacts with the hydroxide ion as shown by the net ionic equation

$$H_3O^+(aq) + OH^-(aq) \rightleftharpoons 2H_2O$$

The sodium ion and the chloride ion are spectator ions and do not react. Thus, the acid–base reaction results in a solution of sodium ions and chloride ions, as shown in Figure 13-2.

The reaction between a solution of a soluble hydroxide and a solution of a strong acid always produces water. When the solutions are added in the correct proportions so that enough H_3O^+ is present to react with all the OH^-, the resulting solution is neither acidic nor basic, but it is neutral. For this reason, these reactions are called **neutralization reactions**.

13-9 SOME FACTS ABOUT ACIDS AND BASES

One reason acid solutions must be handled with care is that they can cause burns to the skin. Acids can dissolve and destroy tissue. Wart removers, for instance, are typically solutions of weak acids. Solutions having relatively high concentrations of hydronium ions are especially dangerous. Strong acid solutions always have higher concentrations of hydronium ions than

WEAR YOUR
SAFETY GOGGLES

"Do as you ought'er always add acid to water"

Anonymous

weak acid solutions of comparable molarities. For instance, a 6 M hydrochloric acid solution contains 6 moles of hydronium ions per liter, whereas a 6 M acetic acid solution contains only about 1×10^{-2} mole of hydronium ions per liter. Although all acid solutions should be handled carefully, great care should be used with strong acid solutions. **Always wear safety goggles when working with laboratory acids** since your eyes are much more sensitive than your skin.

In the laboratory, you may encounter solutions of the common strong acids: sulfuric, nitric, and hydrochloric. Always read the labels of bottles containing these acids very carefully. Sometimes, acid solutions are designated as concentrated or dilute. The molarities of typical **concentrated and dilute acid solutions** are listed in Table 13-2. Whatever its concentration, an acid solution should be handled with respect. Especially avoid adding water to concentrated acids, since the heat released upon mixing may cause the water to boil and splatter the acid. Instead slowly and carefully add acids to water.

The only strong acid that you may find in a hardware or a pool supply store is hydrochloric acid. It is sold as **muriatic acid** and is used to adjust the acidity of swimming pools and as a cleaning agent. It should be used with gloves and safety goggles. Sulfuric acid is used in lead storage batteries in automobiles. Be careful not to spill battery acid on your skin or clothing. Concentrated sulfuric acid is especially dangerous since it not only causes skin burns but can absorb water from skin tissue and undergo other reactions with body chemicals. If you happen to spill acid on yourself or your clothing, rinse it off with large quantities of water. You can treat acid on the skin or clothing with a paste made from baking soda (sodium bicarbonate). Never try to neutralize an acid burn with a solution of a strong base because the mixture generates heat and makes the burn worse. Since sulfuric acid is dangerous, never watch a battery that is being charged or jump started. Batteries sometimes explode while being charged, so just stand clear and do not look.

Basic solutions with high concentrations of hydroxide ion are also potentially dangerous. These solutions can react with and dissolve skin, so handle them with care. As Table 13-1 shows, the hydroxide ion is the strongest base in water; solutions of sodium hydroxide, NaOH, and potassium hydroxide, KOH, are the most common sources of hydroxide ions.

Table 13-2 *Molarities of Concentrated and Dilute Laboratory Acids*

Formula	Acid	MOLARITY (M)	
		Concentrated	Dilute
$HCl(aq)$	Hydrochloric	12	6
$HNO_3(aq)$	Nitric	16	6
$H_2SO_4(aq)$	Sulfuric	18	3

The common name for sodium hydroxide is lye or caustic soda, and potassium hydroxide is sometimes called caustic potash. The term *caustic* is well chosen, since solutions of these compounds are quite caustic and dissolve proteins, vegetable oils, and animal fats. In fact, solutions of these bases are the main components of commercial drain cleaners; they dissolve hair, fatty grease deposits, and other organic matter. Naturally, these solutions should be handled with great care. Strong base solutions can cause severe eye damage so, just as was emphasized with strong acids, **always wear eye covering or goggles when working with solutions of strong bases.**

Some basic solutions are commonly used as household cleaners; since the hydroxide ion helps to dissolve fats and vegetable oils, it "cuts grease." Note the position of ammonia in Table 13-1. Ammonia is used in a variety of household cleaning solutions. Sometimes, Na_3PO_4 (commonly called trisodium phosphate or TSP) and Na_2CO_3 (commonly called washing soda) solutions are used for cleaning purposes. Note the positions of the phosphate ion and the carbonate ion in Table 13-1. TSP is strong enough that it can injure your skin and eyes. Use rubber gloves and goggles when working with TSP.

As you know, common baking soda or bicarbonate of soda is sodium hydrogen carbonate, $NaHCO_3$. The hydrogen carbonate ion, HCO_3^- (sometimes called the bicarbonate ion), is amphiprotic. Look for HCO_3^- on the acid and base sides of Table 13-1. The hydrogen carbonate ion reacts with the hydronium ion or the hydroxide ion. Thus, sodium hydrogen carbonate solid, or as a solution, is used for treatment of acid and base spills. Solutions of $NaHCO_3$, or powdered $NaHCO_3$, are usually available in the laboratory for emergencies.

Occasionally, the product formed when a solution of a strong acid is added to a solution of a weak base is only slightly soluble. One example is the reaction that occurs between a strong acid solution and a solution containing the carbonate ion.

$$2H_3O^+ + CO_3^{2-} \rightleftharpoons 3H_2O + CO_2(g)$$

During this reaction, slightly soluble carbon dioxide gas bubbles out of solution and provides a test for the presence of carbonate ions. If a strong acid solution is added to a test solution and carbon dioxide gas is formed, carbonate ions or hydrogen carbonate ions are in the solution.

Use Table 13-1 to write equations that show why the carbonate ion test works for either CO_3^{2-} or HCO_3^-.

Three other bases in Table 13-1 form insoluble gases when mixed with strong acids. **CAUTION: All these gases are very dangerous and toxic.** The bases are the fluoride ion, the sulfide ion, and the cyanide ion. They react with the hydronium ion as follows.

$H_3O^+(aq) + F^-(aq) \rightleftharpoons H_2O + HF(g)$ a very dangerous poisonous gas

$2H_3O^+(aq) + S^{2-}(aq) \rightleftharpoons 2H_2O + H_2S(g)$ a poisonous gas with a rotten-egg odor

$H_3O^+(aq) + CN^-(aq) \rightleftharpoons H_2O + HCN(g)$ a dangerous poisonous gas with an almondlike odor

Never add acid solutions to solutions containing fluoride ions, sulfide ions, or cyanide ions since poison gases are formed. **Never** unnecessarily smell a solution.

13-10 STOICHIOMETRY AND ACID–BASE REACTIONS

By federal law, a commercial vinegar solution must have at least a set minimum concentration of acetic acid. The label on commercial vinegar may read 5% acidity, which corresponds to an acetic acid molarity of about 0.8 M $HC_2H_3O_2$. If a product does not have the minimum concentration of acetic acid, it cannot be called vinegar. Suppose we had a sample of commercial vinegar and wanted to know the concentration of acetic acid in the sample. A solution of a base is added to the acid solution to supply a base to react with the acid. If the molarity of the base solution is known and we measure the volume of base solution needed to supply enough base to react with the acid completely, the amount of acid is determined by stoichiometric calculations. To accomplish such an analysis, we start with a carefully measured volume of the acetic acid solution. Then we add a sodium hydroxide solution of known concentration to react with the acetic acid. Since we know the equation for the reaction taking place and the volume and concentration of the sodium hydroxide solution used, we can calculate the amount of acetic acid in the vinegar. The reaction between acetic acid and the hydroxide ion is given by the equation

$$HC_2H_3O_2 + OH^- \rightleftharpoons C_2H_3O_2^- + H_2O$$

Let us assume, for example, that 33.3 mL of a 0.500 M solution of OH^- is required to react with the acetic acid in a 20.0 mL sample of the commercial vinegar. We want to calculate the molarity of the acid solution using these data. This calculation is simply a solution stoichiometry question of the type discussed in Section 12-15.

	$HC_2H_3O_2$ +	$OH^- \rightleftharpoons C_2H_3O_2^- + H_2O$
moles	mol $HC_2H_3O_2$	$\Longleftarrow$ mol OH^-
molarity	? M	0.500 M
solution volume	20.0 mL or 0.0200 L	33.3 mL or 0.0333 L

The calculation sequence for this question is

$$V\ OH^- \xrightarrow{M\ OH^-} mol\ OH^- \xrightarrow{molar\ ratio} mol\ HC_2H_3O_2 \xrightarrow{1/V} M\ HC_2H_3O_2$$

We find the number of moles of hydroxide ion involved in the reaction by using $n = VM$.

$$0.0333\ \cancel{L} \left(\frac{0.500\ mol\ OH^-}{1\ \cancel{L}}\right)$$

Next, from the balanced equation, we see that the molar ratio relating the hydroxide ion and acetic acid is

$$\left(\frac{1 \text{ mol HC}_2\text{H}_3\text{O}_2}{1 \text{ mol OH}^-} \right)$$

We multiply the number of moles of hydroxide ion by this ratio to give the number of moles of acetic acid involved in the reaction.

$$0.0333 \,\cancel{L} \left(\frac{0.500 \,\cancel{\text{mol OH}^-}}{1 \,\cancel{L}} \right) \left(\frac{1 \text{ mol HC}_2\text{H}_3\text{O}_2}{1 \,\cancel{\text{mol OH}^-}} \right)$$

Finally, to find the concentration of the acetic acid in moles per liter, divide the number of moles by the initial volume of the acid sample in liters.

$$0.0333 \,\cancel{L} \left(\frac{0.500 \,\cancel{\text{mol OH}^-}}{1 \,\cancel{L}} \right) \left(\frac{1 \text{ mol HC}_2\text{H}_3\text{O}_2}{1 \,\cancel{\text{mol OH}^-}} \right) \left(\frac{1}{0.0200 \text{ L}} \right)$$

$$= \left(\frac{0.833 \text{ mol HC}_2\text{H}_3\text{O}_2}{1 \text{ L}} \right)$$

This vinegar solution meets the official vinegar standard.

13-11 TITRATIONS

Let's consider the actual laboratory technique used to find the molarity of an acetic acid solution. Such a determination is accomplished by using a solution sample of carefully measured volume. A solution of known concentration of sodium hydroxide is added to the sample of acid in the proper proportion so that the number of moles of acetic acid can be calculated using stoichiometric methods. Mixing measured amounts of solutions for purposes of determining the concentration of one solution or the number of grams of some species in solution is known as a **titration.**

Titration is a very important experimental method used for the quantitative analysis of solutions. In practice, the addition and measurement of the volume of the solution of known concentration is performed using a **buret,** pictured in Figure 13-3. A titration is done by filling a buret with the known solution, called the **titrant,** and placing a sample of the solution to be titrated in a flask. The titrant is then slowly delivered to the flask until the necessary amount has been added. The point at which the necessary volume of titrant has been added is termed the **end point** of the titration.

The end point of a titration is usually detected by placing a small amount of a substance called an **indicator** in the reaction flask. The indicator is chosen so that it reacts with the titrant when the end point is reached. The idea is that when all the titrated species has reacted with the titrant, any additional amount of titrant reacts with the indicator. Usually, the reaction of the indicator and titrant produces a colored product that is

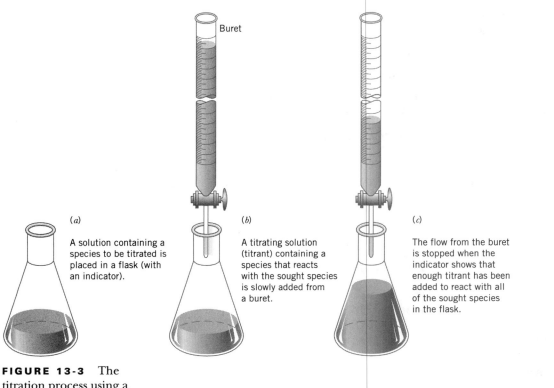

(a)

A solution containing a species to be titrated is placed in a flask (with an indicator).

Buret

(b)

A titrating solution (titrant) containing a species that reacts with the sought species is slowly added from a buret.

(c)

The flow from the buret is stopped when the indicator shows that enough titrant has been added to react with all of the sought species in the flask.

FIGURE 13-3 The titration process using a buret.

visible. **Phenolphthalein** is an indicator that is pink in basic solutions and colorless in acidic solutions. Some indicators are colored to begin with and produce a different colored product, so the color change shows the end point. Once the end point is found, the volume of titrant used to reach the end point is read from the buret. See Figure 12-5 for a stoichiometry map to find the mass of a titrated species.

13-12 TITRATION CALCULATIONS

When a titration is done using a buret, the main piece of experimental data is the volume of titrant used to reach the end point in the titration. The product of the volume and concentration of the titrant gives the number of moles of the titrant. In turn, the number of moles of the titrated species is found using the appropriate molar ratio from the balanced equation for the titration reaction.

If we want to find the number of grams of some species in a solution by titration, the experimental data needed from the titration are

volume of titrant used in titration $= V_t$ molarity of titrant $= M_t$

We use these data and the molar ratio from the equation to find the number of moles of the titrated species. Then, find the mass of this species from the number of moles using its molar mass. The general calculation sequence is

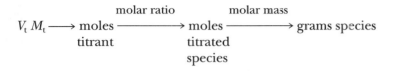

EXAMPLE 13-3

A solution of ammonia requires 32.5 mL of a 0.227 M hydrochloric acid solution to titrate it to the end point. Calculate the number of grams of ammonia in the original solution if the reaction is

$$H_3O^+ + NH_3 \rightleftharpoons NH_4^+ + H_2O$$

A summary is

	titrant		unknown		
	H_3O^+	+	NH_3	$\rightleftharpoons$	$NH_4^+ + H_2O$
mass			? g		
molar mass			17.03 g/mol		
moles	mol H_3O^+	$\Longrightarrow$	mol NH_3		
molarity	0.227 M				
volume	32.5 mL or				
	0.0325 L				

We find the number of moles of hydronium ion used in the titration from the volume and the molarity of the hydrochloric acid t trant. Then, find the moles of ammonia in the titrated solution from the moles of hydronium ion. Finally, find the grams of ammonia from the number of moles and its molar mass.

$$X\text{g } NH_3 = 0.0325\,\cancel{L}\left(\frac{0.227\,\cancel{\text{mol } H_3O^+}}{1\cancel{L}}\right)\left(\frac{1\,\cancel{\text{mol } NH_3}}{1\,\cancel{\text{mol } H_3O^+}}\right)\frac{17.03\text{ g}}{1\,\cancel{\text{mol } NH_3}} = 1.26 \times 10^{-1}\text{ g}$$

EXAMPLE 13-4

It is found that 37.2 mL of a 0.575 M sodium hydroxide solution is needed to titrate a 25.0 mL sample of a solution of oxalic acid, $H_2C_2O_4$, to the end point. What is the molarity of the oxalic acid solution?

The equation for the reaction and a summary of the question are

	unknown		titrant		
	$H_2C_2O_4$	+	$2OH^-$	$\rightleftharpoons$	$2H_2O + C_2O_4^{2-}$
moles	mol $H_2C_2O_4$	$\Longleftarrow$	mol OH^-		
molarity	? M_a		0.575 M_b		
volumes	$V_a = 25.0$ mL		$V_b = 37.2$ mL		
	or 0.0250 L		or 0.0372 L		

The data include the volume of titrant, the molarity of the titrant, the volume of the solution titrated, and the molar ratio from the balanced equation. Since we want the molarity of the acid, the calculation sequence is

$$V_b M_b \xrightarrow{\text{molar ratio}} \text{moles titrant} \xrightarrow{} \text{moles titrated species} \xrightarrow{1/V_a} M_a$$

Using the data in the calculation gives

$$0.0372 \, L \left(\frac{0.575 \, \text{mol OH}^-}{1 \, L} \right) \left(\frac{1 \, \text{mol H}_2\text{C}_2\text{O}_4}{2 \, \text{mol OH}^-} \right) \left(\frac{1}{0.0250 \, L} \right) = \left(\frac{0.428 \, \text{mol H}_2\text{C}_2\text{O}_4}{1 \, L} \right)$$

$$= 0.428 \, M \, \text{H}_2\text{C}_2\text{O}_4$$

13-13 SIMPLIFIED TITRATION CALCULATIONS

The molarity of the acid, M_a, is found by titrating a solution of an acid to the end point with a solution of a base. In such a titration, measuring the volume of the original acid sample is necessary before it is titrated. The experimental data needed from the titration to find M_a are

volume of base needed in the titration $\quad\quad\quad\quad V_b$

molarity of base $\quad\quad\quad\quad\quad\quad\quad\quad\quad\quad\quad\quad M_b$

initial volume of the acid solution of unknown molarity $\quad V_a$

Figure 13-4 shows that in general the calculation setup to find the molarity of the acid in a titration is

$$M_a = M_b \frac{V_b}{V_a} \left(\begin{array}{c} \text{molar} \\ \text{ratio} \end{array} \right)$$

In an acid–base titration in which there is a 1:1 molar ratio of acid to base, the general calculation setup is simply

$$M_a = M_b \frac{V_b}{V_a}$$

which is rearranged to give an expression to find any of the terms:

$$M_a V_a = M_b V_b$$

$$M_a = M_b \frac{V_b}{V_a} \left(\begin{array}{c} \text{molar} \\ \text{ratio} \end{array} \right)$$

When the molar ratio is one to one, the relation simplifies to

$$M_a = M_b \frac{V_b}{V_a}$$

The relation can be rearranged to give

$$M_a V_a = M_b V_b$$

FIGURE 13-4
Simplified calculations for an acid–base titration.

EXAMPLE 13-5

A sodium hydroxide solution is used to titrate a hydrochloric acid solution. The equation for the neutralization reaction in this titration is

$$\text{H}_3\text{O}^+ + \text{OH}^- \rightleftharpoons 2\text{H}_2\text{O}$$

If 23.54 mL of a 0.1025 M solution of NaOH is required to titrate a 20.00 mL sample of hydrochloric acid to the end point, what is the molarity of the hydrochloric acid?

The data are

$$\begin{aligned}
\text{volume of base used} &= V_b = 23.54 \text{ mL} \\
\text{molarity of base} &= M_b = 0.1025 \; M \\
\text{volume of acid titrated} &= V_a = 20.00 \text{ mL} \\
\text{molarity of acid} &= M_a = ?
\end{aligned}$$

Note that the molarity of the NaOH is also the molarity of the CH^-. The molar ratio is 1:1, so the molarity of the unknown acid is simply

$$M_a = M_b \frac{V_b}{V_a} = 0.1025 \; M \left(\frac{23.54 \text{ mL}}{20.00 \text{ mL}} \right) = 0.1206 \; M \; H_3O^+$$

Notice how the calculation fits the general setup shown above. Incidentally, solution volumes in liters could just as easily be used to give the same results. (Explain why.)

13-14 K_w, the Water Constant

Acidic solutions are characterized by the presence of the hydronium ion. In fact, it is the hydronium ion that is responsible for the similar acidic properties of such solutions. Some solutions have high concentrations of hydronium ions and some have low concentrations. As mentioned in Section 13-5, in any sample of water or a water solution, the hydronium ion and the hydroxide ion are in equilibrium with water.

$$H_2O + H_2O \rightleftharpoons H_3O^+ + OH^-$$

In pure water or a neutral aqueous solution, the concentrations of hydronium and hydroxide ions are equal. In a neutral solution at 25°C, the concentrations of each of these two ions as measured by experiment are both $1.0 \times 10^{-7} \; M$. These facts are expressed as

$$[H_3O^+] = [OH^-] = 1.0 \times 10^{-7} \; M$$

where the square brackets represent concentration in moles per liter of the species enclosed in the bracket. In pure water or any water solution, the concentrations of hydronium and hydroxide ions are fixed. If the concentration of one of these ions increases, however, the other decreases in proportion. That is, if a solution becomes more acidic because of an increase in the hydronium ion concentration, it becomes proportionally less basic since the hydroxide ion concentration decreases. On the other hand, if a solution becomes more basic because of an increase in the hydroxide ion concentration, it becomes proportionally less acidic since the hydronium ion concentration decreases. The relation between the concentrations of these ions shown as an algebraic proportion is

$$[H_3O^+] = \frac{K_w}{[OH^-]}$$

This inverse proportion between the hydronium and hydroxide ion concentrations means that higher hydronium ion concentrations correspond to lower hydroxide ion concentrations and vice versa. The hydronium ion–hydroxide ion expression is rearranged to show the product of the two concentrations is equal to the proportionality constant.

$$K_w = [H_3O^+][OH^-]$$

This proportion suggests that if $[OH^-]$ increases, $[H_3O^+]$ decreases and vice versa. The proportionality constant represented by K_w (read K–W) is called the ion product of water. Its value is determined by experiment. The typical value for K_w is 1.0×10^{-14}. Thus, in pure water or water solutions, the product of the concentrations of hydronium and hydroxide ions has a constant value of 1.0×10^{-14} at 25°C.

When an acid or a base is dissolved in water, the hydronium ion–hydroxide ion equilibrium is affected. An acid in solution reacts with water to increase the concentration of hydronium ions. As the hydronium ion concentration increases, the hydroxide ion concentration decreases proportionally. An acidic solution is characterized by a concentration of hydronium ions that is greater than 1.0×10^{-7} M and a concentration of hydroxide ions that is less than 1.0×10^{-7} M. Remember that the product of the two concentrations equals 1.0×10^{-14}.

A base in solution increases the concentration of hydroxide ions. The increased hydroxide ion concentration results in a lower concentration of hydronium ions. They vary in inverse proportion. A basic solution is characterized by a concentration of hydroxide ions that is greater than 1.0×10^{-7} M and a concentration of hydronium ions that is less than 1.0×10^{-7} M.

1.0 *M*	10^0 *M*
0.1 *M*	10^{-1} *M*
0.01 *M*	10^{-2} *M*
0.001 *M*	10^{-3} *M*
0.000 1 *M*	10^{-4} *M*
0.000 01 *M*	10^{-5} *M*
0.000 001 *M*	10^{-6} *M*
0.000 000 1 *M*	10^{-7} *M*
0.000 000 01 *M*	10^{-8} *M*
0.000 000 001 *M*	10^{-9} *M*
0.000 000 000 1 *M*	10^{-10} *M*
0.000 000 000 01 *M*	10^{-11} *M*
0.000 000 000 001 *M*	10^{-12} *M*
0.000 000 000 000 1 *M*	10^{-13} *M*
0.000 000 000 000 01 *M*	10^{-14} *M*

13-15 pH AND pOH

The concentration of hydronium ion in solutions can vary greatly. Some solutions are highly acidic; they have relatively high concentrations of hydronium ion. A 1 *M* hydrochloric acid solution, for instance, has a hydronium ion concentration of 1 *M*. Some solutions are highly basic; they have relatively high concentrations of hydroxide ions and correspondingly low concentrations of hydronium ions. A 1 *M* sodium hydroxide solution, for instance, has a hydroxide ion concentration of 1 *M* and a hydronium ion concentration of 1×10^{-14} *M*. In pure water and neutral solutions, the concentration of hydronium ions is 1.0×10^{-7} *M*. Common solutions of acids and bases have hydronium ion concentrations ranging from around 1 *M* to 1×10^{-14} *M* and corresponding hydroxide ion concentrations ranging from around 1×10^{-14} *M* to 1 *M*. Let's see how the concentrations of these ions vary in inverse proportion. For instance, if a solution is 1.0 *M* H_3O^+, the OH^- concentration is 1.0×10^{-14} *M*. The reason is provided by the ion product of water.

$$K_w = [H_3O^+][OH^-] = 1.0 \times 10^{-14}$$

Therefore,

$$[OH^-] = \frac{K_w}{[H_3O^+]}$$

If $[H_3O^+] = 1.0$ *M* and $K_w = 1.0 \times 10^{-14}$, then

$$[OH^-] = \frac{1.0 \times 10^{-14}}{1.0} = 1.0 \times 10^{-14} \ M$$

The hydroxide concentration is 1.0×10^{-14} *M* when the hydronium ion concentration is 1.0 *M*. Similarly, when the hydroxide concentration is 1.0 *M*, the hydronium ion concentration is 1.0×10^{-14} *M*.

The concentration of hydronium ion in a solution is expressed as molarity. Since the possible concentrations cover a very wide range, however, a special scale, called the pH (read P–H) scale, has been devised to express hydronium ion concentrations in aqueous solutions. Similarly, various hydroxide ion concentrations in solutions are expressed on a pOH (read P–O–H) scale. Definitions are given in the margin.

pH

pH
The negative logarithm of the hydronium ion concentration. pH = $-log[H_3O^+]$, where $[H_3O^+]$ represents the molarity of the hydronium ion.

pOH
The negative logarithm of the hydroxide ion concentration. pOH = $-log[OH^-]$, where $[OH^-]$ represents the molarity of the hydroxide ion.

Logarithms

Let us pause to review the nature of logarithms. It is possible to express any number in terms of 10 raised to a power. Simple examples are 100 expressed as 10^2 and 0.001 expressed as 10^{-3}. Any number may be expressed as a power of 10, for example, 2 as a power of 10 is $10^{0.3}$ and 9.5 as a power of 10 is $10^{0.98}$. The base 10 logarithm of a number is the power to which 10 is raised to give the number. In general, any number, *y*, is expressed as $y = 10^x$. The log of the number *y* is *x*. Thus, the log of 10^2 is 2, the log of 10^{-3} is -3, the log of 2 is 0.3, and the log of 9.5 is 0.98. The numerical value of the logarithm of a number is found using a scientific calculator. To find the logarithm of a number using a typical calculator with a log key, just key in the number and press the log key. You may have a different kind of calculator, so you need to know how to find a logarithm on it. See Appendix 2. Using a calculator with a log key, practice by finding the logarithms of 100, 0.001, 2, and 9.5. You'll need a calculator with a log key to calculate pH or pOH values.

The definitions of pH and pOH allow for the expression of ion concentrations as powers of 10 rather than normal numbers. In fact, the term *pH* is viewed as power (p) of the hydronium ion concentration and pOH as the power of the hydroxide ion concentration. pH and pOH allow the expression of concentrations on a logarithmic scale, which makes them simple numbers. The logarithms of hydronium and hydroxide ion concentrations

$pH = -log[H_3O^+]$

are typically negative (i.e., $\log 10^{-7} = -7$), however, the definitions of pH and pOH include the negative logarithm. Hence, the logarithm has its sign changed or multiplied by -1 during the calculation of pH or pOH, for example, $-\log 10^{-7} = -(-7) = 7$. By this convention, the pH or pOH of a solution normally has a positive value. pH is a convenient way to express the acidity of a solution, and pOH is a convenient way to express the basicity of a solution. The pH scale reduces the range of potential hydronium ion concentrations to a scale ranging from about 0 to 14. It is possible, however, to have pH values lower than 0 and higher than 14.

$[H_3O^+]$	10^0	10^{-1}	10^{-2}	10^{-3}	10^{-4}	10^{-5}	10^{-6}	10^{-7}	10^{-8}	10^{-9}	10^{-10}	10^{-11}	10^{-12}	10^{-13}	10^{-14}
pH	0	1	2	3	4	5	6	7	8	9	10	11	12	13	14

Note that $1\,M = 10^0 M$ since $10^0 = 1$. The pOH scale reduces the range of potential hydroxide ion concentrations to a scale ranging from about 14 to 0.

$[OH^-]$	10^{-14}	10^{-13}	10^{-12}	10^{-11}	10^{-10}	10^{-9}	10^{-8}	10^{-7}	10^{-6}	10^{-5}	10^{-4}	10^{-3}	10^{-2}	10^{-1}	10^0
pOH	14	13	12	11	10	9	8	7	6	5	4	3	2	1	0

For a given solution, the sum of its pH and pOH is 14.00. This useful fact is expressed as

$$pH + pOH = 14.00$$

This expression is used to calculate the pOH from the pH of a solution or the pH from the pOH. For example, if we know that a solution has a pH of 9.0, the pOH is $14.00 - 9.0 = 5.0$. If the pOH of a solution is 4.0, the pH is 10.0.

13-16 PH AND POH CALCULATIONS

If we know the hydronium ion concentration in a solution, the pH is easily calculated using the relation $pH = -\log[H_3O^+]$. If we know the hydroxide ion concentration in a solution, the pOH is easily calculated using $pOH = -\log[OH^-]$. If we know the pH, we can find the pOH by subtracting from 14.00 or vice versa to get the pH from the pOH.

Note, on the pH scale, that there is a 10-fold difference in acidity between subsequent pH units. Therefore, a given pH value corresponds to a hydronium ion concentration that is 10 times greater than the next higher pH value on the scale. At pH 4, the $[H_3O^+]$ is 10 times greater than at pH 5, 100 times greater than at pH 6, 1000 times greater than at pH 7, and so on. A solution of pH 0 has a hydronium concentration that is 100 trillion times more than a solution of pH 14. Keep in mind that higher values of pH correspond to lower concentrations of hydronium ions, which means that higher pH values correspond to solutions of lower acidity. This fact is a consequence of the negative logarithm in the definition of pH. It is interesting that an acidic solution having a hydronium ion concentration greater than 1 M has a negative pH. For instance, the pH of an extremely acidic solution having a hydronium ion concentration of 10 M is -1.00; pH $= -\log[H_3O^+]$ and $-\log(10) = -1.00$. Such highly concentrated solutions are not common, however. (Why are such solutions very dangerous to handle?)

For solutions with $[H_3O^+]$ ranging from 1 M to 10^{-14} M, the pH ranges from 0 to 14. Solutions of lower pH are more acidic than solutions of higher pH. Solutions with pH less than 7 are **acidic,** whereas solutions with pH greater than 7 are **basic.** Basic solutions are sometimes called alkaline solutions. A neutral solution at 25°C has a pH of 7 and pOH of 7. Notice that solutions with lower pOH are more basic than solutions of higher pOH.

	highly acidic				slightly acidic		neutral	slightly basic					highly basic	
pH	0	1	2	3	4	5	6	7	8	9	10	11	12	13	14
pOH	14	13	12	11	10	9	8	7	6	5	4	3	2	1	0

A pH is used as a convenient means of expressing the hydronium ion concentration in a solution. If $[H_3O^+]$ is a whole-number power of 10, the pH is

$$[H_3O^+] = 10^{-pH}$$

which means that the pH is the numerical value of the power of 10 with its sign changed. Thus, a solution with $[H_3O^+] = 10^{-14}$ M has a pH of 14, and a solution with $[H_3O^+] = 10^{-9}$ M has a pH of 9. Similarly, if $[OH^-]$ is an integral power of 10, the pOH is easily determined as

$$[OH^-] = 10^{-pOH}$$

If $[OH^-] = 10^{-8}$, the pOH is 8, and if $[OH^-] = 10^{-2}$, the pOH is 2.

When the hydronium ion concentration is not an integral power of 10, the concentration is used in the definition of pH and the logarithm must be determined. For example, a solution with $[H_3O^+] = 2 \times 10^{-3}$ M has a pH of

$$pH = -\log(2 \times 10^{-3})$$

A hand calculator that allows the input of exponential numbers and has a base 10 logarithm key (log) is used to calculate pH. To do this on a typical calculator, first key in the hydronium ion concentration. Then, find the log using the logarithm key and change the sign of the result to give the negative log. See Appendix 2 for more on the use of a calculator for pH calculations using various calculators. Here, using a typical calculator, the calculation sequence is 2 EE 3 $+/-$ log $+/-$ = 2.69 . . . = 2.7. Round to one digit past the decimal since the data had only one digit. When reporting a pH, the digits on the left of the decimal correspond to the exponent and do not count as significant digits. Only those digits to the right of the decimal point are significant digits.

EXAMPLE 13-6

What is the pH of a solution that has a hydronium ion concentration of 9.5×10^{-9} M? What is the pOH of the solution? Is the solution acidic or basic?

To find the pH using a typical hand calculator, just key in the hydronium ion concentration, push the log key, and change the sign (for the negative log).

$$9.5 \times 10^{-9} \log +/- 8.0222 \ldots$$

or 8.02 to two significant digits. The pOH of this solution is $14.00 - 8.02 = 5.98$. The pH is greater than 7 and the pOH is less than 7, so the solution is basic.

ACTIVITY 13-7

What are the pOH and pH of a solution that has $[OH^-] = 3.3 \times 10^{-3}$ M? Is the solution acidic or basic?

The formulas $[H_3O^+] = 10^{-pH}$ and $[OH^-] = 10^{-pOH}$ are also used to find the $[H_3O^+]$ or the $[OH^-]$ corresponding to given values of pH or pOH. Calculations of this type are discussed in Appendix 2.

Hydronium ion is present in all water solutions. Gastric juices, citrus fruit juices, soft drinks, vinegar, urine, milk, and some natural freshwaters are acidic. Pure water is neutral, whereas blood is slightly basic. Table 13-3 lists the pH of some common solutions.

Table 13-3   *pH Values of Some Common Solutions*

pH		
0	1 M hydrochloric acid	0
1	Gastric juices	1.0–3.0
2	Lemon juice	2.2–2.4
	Vinegar	2.4–3.4
3	Wines	2.8–3.8
4	Carbonated water (saturated)	3.8
5	Black coffee	4.8–5.2
6	Milk	6.3–6.6
	Saliva	6.5–7.5
7	Pure water	7.0
	Blood plasma	7.3–7.5
8	0.1 M sodium hydrogen carbonate	8.4
9	Soap solutions	8.5–10.0
10	Saturated magnesium hydroxide	10.5
11	1 M aqueous ammonia	11.6
12	Saturated calcium hydroxide (limewater)	12.4
13	1 M sodium phosphate (TSP)	13.2
14	1 M sodium hydroxide	14

13-17 ESTIMATING THE pH OR pOH OF SOLUTIONS

An estimate of the pH of a solution may be determined without using a calculator. Suppose we had a sample of lemon juice having a hydronium ion concentration of 3.0×10^{-3} M. The pH is found using the concentration in the pH definition.

$$pH = -\log(3.0 \times 10^{-3})$$

To estimate, we first round the concentration down to 1 times the power and then up to 10 times the power. Round 3.0×10^{-3} down to 1×10^{-3} and up to 10×10^{-3} (which equals 1×10^{-2}). In this way we know that the pH is between 2 and 3, so it must be 2 point something or 2.?. The ? represents the something. The simple pattern for obtaining an estimate of the pH or pOH of a solution is as follows.

1. If the ion concentration is 1 with a whole number power such as 1×10^{-n} or 1.0×10^{-n}, the pH or pOH value is $n.0$ or $n.00$. For instance, if $[H_3O^+] = 1.0 \times 10^{-5}$, then the pH = 5.00.
2. Otherwise, use the power in the $\times 10$ part of the concentration, change its sign, and subtract 1, and
3. the approximate pH or pOH value is this number point something.

As an example, let's estimate the pH of a soap solution having a hydronium ion concentration of 8.5×10^{-10}. The power is -10, so changing the sign and subtracting 1 from 10 suggests that the pH is 9 point something. Since the sum of pH and pOH is 14, the pOH must be 4 point something. In other words, 9 point something plus 4 point something equals 14. You can calculate the precise pH with a calculator. You can use the quick-estimation method, however, to decide whether a solution is acidic or basic and to double check your calculation of the precise pH or pOH.

ACTIVITY
13-8

(a) Estimate the pH of a sample of wine that has a hydronium ion concentration of 1.8×10^{-4} M. Is it acidic or basic?

(b) Estimate the pOH of a solution of KOH that has a hydroxide ion concentration of 5.0×10^{-3} M. What is the approximate pH of the KOH solution?

FOCUS
13

Acids, Bases, pH, and pOH

BRØNSTED–LOWRY ACIDS AND BASES

Acid: A chemical species that can lose a hydrogen ion or a proton in a chemical reaction. A proton donor.

Base: A chemical species that can gain a hydrogen ion or a proton in a chemical reaction. A proton acceptor.

A proton transfer or acid–base reaction is represented as

acid 1		base 2		acid 2		base 1
H—A	+	B	$\rightleftharpoons$	H—B$^+$	+	A$^-$
stronger		stronger		weaker		weaker
acid		base		acid		base

Conjugate acid-base pair: An acid and the base that it forms by the loss of a proton or a base and the acid it forms by gaining a proton.

Acidic solution: A water solution that has a concentration of hydronium ion that is greater than that in pure water. When an acid dissolves in water, an acidic solution results.

Basic solution: A water solution that has a concentration of hydroxide ion that is greater than that in pure water. When a base dissolves in water, a basic solution results.

A solution of a strong acid always contains hydronium ion as a major ion. Three common strong acids are

hydrochloric acid solution	$H_3O^+(aq) + Cl^-(aq)$
sulfuric acid solution	$H_3O^+(aq) + HSO_4{}^-(aq)$
nitric acid solution	$H_3O^+(aq) + NO_3{}^-(aq)$

A weak acid does not react extensively with water, and a solution of a molecular weak acid contains the unreacted acid as the major species. A solution of a molecular weak acid is represented by the formula of the acid, and an ionic weak acid is represented by its formula.

A solution of a weak base is represented by the formula of the base.

The common strong bases are sodium hydroxide, NaOH, potassium hydroxide, KOH, and barium hydroxide, Ba(OH)$_2$. In aqueous solutions, these compounds are represented as hydroxide ion and the cation.

Titration: A method of chemical analysis that involves mixing measured amounts of solutions for purposes of determining the concentration of one solution or number of grams of some species in solution.

End point: The point at which the necessary volume of titrant has been added in a titration.

The molarity of an acid solution is found by titrating a known volume with a base solution of known molarity. Calculations for titrations follow typical solution stoichiometry calculation sequences. When the molar ratio in an acid–base titration is one to one, a titration calculation is simply

$$M_a = M_b \frac{V_b}{V_a}$$

The product of the concentrations of hydronium ion and hydroxide ion in water solutions is a constant: $K_w = [H_3O^+][OH^-]$.

pH is the negative logarithm of the hydronium ion concentration.

$pH = -\log[H_3O^+]$, where $[H_3O^+]$ represents the molarity of hydronium ion.

pOH is the negative logarithm of the hydroxide ion concentration.

---

pOH = −log[OH⁻], where [OH⁻] represents the molarity of hydroxide ion.

					neutral										
	highly				slightly	slightly						highly			
	acidic				acidic	basic						basic			
pH	0	1	2	3	4	5	6	7	8	9	10	11	12	13	14
pOH	14	13	12	11	10	9	8	7	6	5	4	3	2	1	0

The sum of the pH and pOH is 14.00. **pH + pOH = 14.00.**

To find the pH from the hydronium ion concentration, use **pH = −log[H₃O⁺]**.

To find the pOH from the hydroxide ion concentration, use **pOH = −log[OH⁻]**.

To estimate the pH or pOH of a solution using the hydronium ion or hydroxide ion concentration:

1. If the ion concentration is 1×10^{-n} or 1.0×10^{-n}, the value is n.0 or n.00.
2. Otherwise, use the power in the $\times 10$ part of the concentration, change its sign, and subtract 1, and
3. the estimated value is this number point something.

To find the hydronium ion concentration from the pH, use **[H₃O⁺] = 10⁻ᵖᴴ**. (See Appendix 2.)

To find the hydroxide ion concentration from the pOH, use **[OH⁻] = 10⁻ᵖᴼᴴ**. (See Appendix 2.)

QUESTIONS

1. Match the following terms with the appropriate lettered definitions.

acid	a. a solution in which the concentration of hydronium ion is greater than the hydronium ion concentration in pure water
base	b. a solution in which the concentration of hydroxide ion is greater than the hydroxide ion concentration in pure water
strong acid	c. a solution having equal concentrations of hydronium ion and hydroxide ion
weak acid	d. capable of acting as an acid or a base
conjugate acid–base pair	e. −log[OH⁻]
amphiprotic	f. −log[H₃O⁺]

neutral solution	g. a species that can lose a proton in a chemical reaction
acidic solution	h. a species that can gain a proton in a chemical reaction
basic solution	i. an acid that dissolves in water to form nothing but hydronium ion and the conjugate base of the acid
titration	j. an acid that dissolves in water and reacts slightly to form relatively low concentrations of hydronium ion and the conjugate base of the acid
K_w	k. [H₃O⁺][OH⁻] = 1.0×10^{-14}
pH	l. an acid and a base related by the loss and gain of a proton

pOH m. adding a measured amount of a solution of known concentration to a solution of unknown concentration to determine the unknown concentration

Section 13-1

2. List the common properties of water solutions of acids. List the common properties of water solutions of bases.

Section 13-2

3. Give the Brønsted–Lowry definitions for an acid and a base.

4. What is a proton in reference to acid–base reactions?

Section 13-3

5. Describe an acid–base reaction and give an example.

6. What is the meaning of a double arrow in an equation for an acid–base reaction?

Section 13-4

7. What is a conjugate acid–base pair? Give examples.

8. *Complete the following table by supplying the conjugate acid or base as needed.

ACID	BASE
HNO_2	——
——	H_2O
HSO_4^-	——
——	NH_3
H_2O	——

9. For each of the following acid–base reactions, pick out the acids and the bases and indicate the conjugate acid–base pairs.

 (a) $HNO_3 + H_2O \longrightarrow H_3O^+(aq) + NO_3^-(aq)$
 (b) $OH^-(aq) + NH_4^+(aq) \rightleftharpoons NH_3(aq) + H_2O$
 (c) $H_3O^+(aq) + F^-(aq) \rightleftharpoons HF(aq) + H_2O$

10. For each of the following acid–base reactions, pick out the acids and the bases and indicate the conjugate acid–base pairs.

 (a) $H_2SO_4 + H_2O \longrightarrow H_3O^+(aq) + HSO_4^-(aq)$

 (b) $OH^-(aq) + H_3PO_4(aq) \rightleftharpoons$
       $H_2O + H_2PO_4^-(aq)$
 (c) $H_3O^+(aq) + OH^-(aq) \rightleftharpoons H_2O + H_2O$

Section 13-5

11. What is an acidic solution compared with a neutral solution? What is a basic solution compared with a neutral solution?

12. What does it mean when a species is amphiprotic?

13. Water is amphiprotic. Write the electron dot structure for water and explain how it can behave as an acid or a base.

Section 13-6

14. Explain the structure of Table 13-1.

15. Explain the difference between a weak acid and a strong acid.

Section 13-7

16. What is a strong acid and what are the major species in a solution of a strong acid? Give an example.

17. What is a weak acid and what are the major species in a solution of a weak acid? Give an example.

18. What are the major species in a solution of a strong base? What are the major species in a solution of a weak base?

19. According to Table 13-1, what is the strongest acid that can exist in water?

20. According to Table 13-1, what is the strongest base that can exist in water?

Section 13-8

21. What is a neutralization reaction?

22. Give the net-ionic equation for the neutralization reaction that occurs when a solution of nitric acid is added to a solution of sodium hydroxide.

23. Give the net-ionic equation for the neutralization reaction that occurs when a solution of hydrochloric acid is mixed with a solution of potassium hydroxide.

Section 13-9

24. Why are some acid and base solutions potentially dangerous?

25. Which bases in Table 13-1 produce poison gases when mixed with a solution of a strong acid? **CAUTION: Never mix compounds or solutions containing these ions with acids.**

26. Why do 1.0 M solutions of hydrochloric acid and acetic acid differ in their concentrations of the hydronium ion?

27. Which acid is used as an electrolyte in lead storage batteries?

28. Why should you not watch a car battery as it is being jump started?

29. Why is baking soda, $NaHCO_3$, useful for the treatment of acid spills?

Sections 13-10 to 13-13

30. What is a titration?

31. Define the following terms with respect to a titration.

   (a) buret
   (b) titrant
   (c) indicator
   (d) end point

32. *A vinegar sample is titrated with a sodium hydroxide solution to the end point. Using the equation and the experimental data given below, calculate the molarity of the acetic acid in the vinegar.

$$HC_2H_3O_2(aq) + OH^-(aq) \rightleftharpoons H_2O + C_2H_3O_2^-(aq)$$

M NaOH	Volume NaOH	Volume vinegar
0.515 M	21.15 mL	10.0 mL

33. *Calculate the molarity of a hydrochloric acid solution if 27.5 mL of a 0.675 M KOH solution are needed to titrate a 20.0 mL sample of the acid solution to the end point. The reaction involved is

$$H_3O^+(aq) + OH^-(aq) \rightleftharpoons H_2O + H_2O$$

34. *A sample of sodium hydrogen phosphate, Na_2HPO_4, is dissolved in water and titrated with a sodium hydroxide solution to the end point. If 19.4 mL of a 0.789 M NaOH solution were required to titrate the sample, calculate the mass in grams of Na_2HPO_4 in the sample. The reaction is

$$HPO_4^{2-}(aq) + OH^-(aq) \rightleftharpoons H_2O + PO_4^{3-}(aq)$$

35. A sample of a cleaning solution containing the base ammonia was titrated with a hydrochloric acid solution. Using the equation and titration data given below, calculate the molarity of the ammonia in the cleaning solution.

$$H_3O^+(aq) + NH_3(aq) \rightleftharpoons H_2O + NH_4^+(aq)$$

Molarity M hydrochloric acid	Volume acid used	Volume NH_3 solution titrated
1.10 M	9.28 mL	10.50 mL

36. A vinegar sample is titrated with a sodium hydroxide solution. Using the reaction and experimental data given below, calculate the molarity of acetic acid in the vinegar.

$$HC_2H_3O_2(aq) + OH^-(aq) \rightleftharpoons H_2O + C_2H_3O_2^-(aq)$$

Molarity M NaOH	Volume NaOH used	Volume vinegar solution titrated
0.768 M	21.2 mL	25.0 mL

37. A sample of an ammonia solution is titrated with 0.100 M hydrochloric acid. It requires 38.5 mL of the acid to titrate the ammonia. If the reaction is

$$H_3O^+(aq) + NH_3(aq) \rightleftharpoons NH_4^+(aq) + H_2O$$

calculate the number of grams of NH_3 in the sample.

38. Calculate the molarity of a KOH solution if 48.52 mL of a 0.165 M hydrochloric acid solution are required to titrate 20.0 mL of the base solution. The reaction is

$$H_3O^+(aq) + OH^-(aq) \longrightarrow 2H_2O$$

Section 13-14

39. What is K_w and what does it represent?

40. How is the ion product of water useful?

Section 13-15

41. Give a definition of pH. Give a definition of pOH.

42. Describe acidic and basic solutions in terms of the hydronium ion concentration and the hydroxide ion concentration. What are they in terms of pH and pOH?

43. What are the concentrations of the hydronium ion and the hydroxide ion in pure water or a neutral solution at 25°C? Explain why they are equal.

44. Draw a number line from 0 to 14 to represent pH values. Indicate on the line the regions corresponding to acidic solutions and basic solutions. Draw an arrow pointing to the pH of neutral solutions. List pOH values corresponding to pH values on your line.

Sections 13-16 and 13-17

45. *Estimate and calculate the pH of the following solutions and indicate whether a solution is acidic or basic.

   (a) an ammonia solution that has
   8.9×10^{-9} M H_3O^+

(b) a sample of cerebrospinal fluid that has
$2.7 \times 10^{-8} \, M \, H_3O^+$

(c) a sample of urine that has $6.9 \times 10^{-7} \, M \, H_3O^+$

(d) a citric acid solution that has
$9.9 \times 10^{-4} \, M \, H_3O^+$

(e) a 1.0 M NaOH solution that has
$1.0 \times 10^{-14} \, M \, H_3O^+$

(f) a 0.50 M hydrochloric acid solution, a strong acid

(g) a sample of gastric juice that has
$3.2 \times 10^{-2} \, M \, H_3O^+$

(h) a sample of rainwater that has
$1.1 \times 10^{-7} \, M \, H_3O^+$

(i) a sample of acid rainwater that has
$2.3 \times 10^{-4} \, M \, H_3O^+$

(j) a sample of carbonated water that has
$2.8 \times 10^{-4} \, M \, H_3O^+$

(k) a sample of blood plasma that has
$4.0 \times 10^{-8} \, M \, H_3O^+$

46. Estimate and calculate the pH of the following solutions and indicate whether a solution is acidic or basic.

(a) an ammonia solution that has
$8.4 \times 10^{-9} \, M \, H_3O^+$

(b) a sample of saliva that has $3.5 \times 10^{-7} \, M \, H_3O^+$

(c) a sample of rainwater that has
$1.8 \times 10^{-7} \, M \, H_3O^+$

(d) apple juice that has $9.3 \times 10^{-4} \, M \, H_3O^+$

(e) a sample of milk that has $2.3 \times 10^{-7} \, M \, H_3O^+$

(f) a 0.10 M NaOH solution that has
$1.0 \times 10^{-13} \, M \, H_3O^+$

(g) a sample of wine that has $3.8 \times 10^{-3} \, M \, H_3O^+$

(h) a sample of coffee that has $6.2 \times 10^{-5} \, M \, H_3O^+$

(i) a solution of soap that has $9.7 \times 10^{-8} \, M \, H_3O^+$

(j) a sample of lemon juice that has
$2.4 \times 10^{-2} \, M \, H_3O^+$

(k) a sample of hydrochloric acid that has
$5.5 \, M \, H_3O^+$

47. *Estimate and calculate the pOH of the following solutions and indicate whether a solution is acidic or basic.

(a) an ammonia solution that has
$2.3 \times 10^{-5} \, M \, OH^-$

(b) a sample of cerebrospinal fluid that has
$6.2 \times 10^{-6} \, M \, OH^-$

(c) a sample of soap solution that has
$6.9 \times 10^{-6} \, M \, OH^-$

(d) a citric acid solution that has
$1.0 \times 10^{-11} \, M \, OH^-$

(e) a 1.0 M NaOH solution

(f) a 0.50 M hydrochloric acid solution, a strong acid

(g) a sample of bile juice that has
$3.2 \times 10^{-6} \, M \, OH^-$

(h) a sample of alkaline lake water that has
$9.2 \times 10^{-5} \, M \, OH^-$

(i) a sample of acid rainwater that has
$7.7 \times 10^{-11} \, M \, OH^-$

(j) a sample of white wine that has
$2.1 \times 10^{-10} \, M \, OH^-$

(k) a sample of blood plasma that has
$6.0 \times 10^{-7} \, M \, OH^-$

48. Estimate and calculate the pOH of the following solutions and indicate whether the solution is acidic or basic.

(a) an ammonia solution that has
$2.0 \times 10^{-6} \, M \, OH^-$

(b) a sample of saliva that has $3.5 \times 10^{-7} \, M \, OH^-$

(c) a sample of rainwater that has
$1.9 \times 10^{-8} \, M \, OH^-$

(d) apple juice that has $7.0 \times 10^{-12} \, M \, OH^-$

(e) a sample of milk that has $8.1 \times 10^{-8} \, M \, OH^-$

(f) a 0.10 M KOH solution

(g) a sample of beer that has $3.6 \times 10^{-9} \, M \, OH^-$

(h) a sample of coffee that has $4.0 \times 10^{-10} \, M \, OH^-$

(i) a solution of soap that has $8.5 \times 10^{-6} \, M \, OH^-$

(j) a sample of lemon juice that has
$7.6 \times 10^{-13} \, M \, OH^-$

(k) a 0.5 M solution of nitric acid, a strong acid

49. Estimate and calculate the pOH of the solutions in parts (a) to (e) of Question 46.

50. Estimate and calculate the pH of the solutions in parts (a) to (e) of Question 47.

51. Use various resources to find information about acid rain. Describe its sources and environmental effects.

CHAPTER 14

CHEMICAL

EQUILIBRIUM

 14-1 **RATES OF REACTIONS**

When the reactants involved in a reaction come into contact, a reaction may occur spontaneously. Typically, however, a reaction needs initiation with some source of energy. We need to heat the reactants to encourage the reaction. Charcoal, for example, needs heating before it reacts spontaneously with oxygen in air. Some reactions are fast and some are slow. In air bags, used as safety devices in automobiles, a fast chemical reaction is required but one that is only initiated by an impact. An air bag contains the unique chemical called sodium azide, NaN_3, as a solid. Upon impact, an electrical source initiates its decomposition to form sodium metal and nitrogen gas.

$$2NaN_3 \longrightarrow 2Na + 3N_2$$

This reaction occurs very rapidly and the nitrogen gas expands to fill the bag within 0.04 s. The process only works because the reaction occurs at a very fast rate.

The speed at which a reaction takes place is expressed as the **rate of the reaction.** The rate of a reaction depends on several factors, especially on the exact way in which the reactants interact to give products. Since a reaction involves the breaking and forming of chemical bonds, contact between the species making up the reactants is quite important. That is, for the reaction to continue, the species making up the reactants must come in contact through collisions. This idea is called the **collision theory** of chemical reactions. According to the theory, the chemical particles of the reactants must collide with sufficient energy to break chemical bonds. Collisions can produce intermediate transition species in which old bonds break and new bonds form, ultimately giving the products of the reaction.

The equation for the explosion of nitroglycerine is

$$4C_3H_5(NO_3)_3\,(\ell) \longrightarrow 6N_2(g)$$
$$+ 10H_2O(g) + 12CO_2(g) + O_2(g)$$

The instant a 100 g sample of nitroglycerine explodes, assume the gases formed have a temperature of 500°C and occupy a volume of 0.250 L. What is the pressure of this mixture of gases under these conditions?

439

Imagine a chemical reaction starting as the reactants mix. The atoms, molecules, or ions that compose the reactants intermingle and collide with one another. Collisions are very important in chemical reactions, since the contact of colliding species provides a way in which bonds are broken and new bonds formed. Collisions alone do not cause a reaction. A specific set of species must collide with sufficient energy and proper spatial orientation for a reaction to occur. Nevertheless, with many collisions, enough are successful to give the products of the reaction. How fast the process occurs is expressed as the rate of the reaction. For most reactions, the rate is measured by observing how fast a reactant is used or how fast a product is formed. Rates are expressed as grams per unit time or moles per unit time. Time units are normally seconds or minutes, but in some reactions they are hours or days.

14-2 FACTORS AFFECTING REACTION RATES

We can illustrate factors that affect the rate of a reaction using a familiar example. The burning of wood includes the reaction of cellulose, the major carbohydrate in wood, with oxygen in the air. As it begins to burn and release heat, the wood becomes hotter, and it burns at a faster rate. The rate of the reaction increases with temperature. Another way to get the wood to burn faster is to break it up into smaller pieces. The state of subdivision of the wood increases the reaction rate. If we limit the air supply, the wood does not burn as quickly and may just smolder. Lowering the concentration of oxygen decreases the reaction rate. On the other hand, if we fan the fire, increasing the air supply, it burns more rapidly. Increasing the concentration of oxygen increases the reaction rate. Using Figure 14-1 for reference, we will consider the effects of temperature, subdivision, and concentration in more detail. The effect of catalysts on reactions is discussed in Section 14-3.

Temperature

An increase in the temperature of a reaction system always causes the chemical particles to move more rapidly. Thus, heating the reactants allows more frequent collisions and an increased reaction rate. More important, heating produces more high-energy particles, which react more readily. In other words, collisions of these high-energy particles are more likely to form products, and the rate of the reaction increases accordingly. See Figure 14-1(*a*).

As an example, consider the reaction of hydrogen gas and nitrogen gas to form ammonia gas. For a room temperature mixture of hydrogen gas and nitrogen gas, the rate of the reaction is so slow that no detectable reaction occurs. Even with an iron catalyst, the reaction at room temperature is

Chemical Reactions

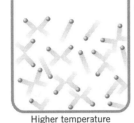

Lower temperature   Higher temperature

(a) An increase in temperature increases the speed of the particles. This causes more frequent and more successful collisions, thereby increasing the rate of the reaction.

Larger pieces of a reactant   Many smaller pieces of a reactant

(b) An increase in the state of subdivision provides for more contact between reactants and thus more frequent collisions. This increases the rate of the reaction.

Lower concentration   Higher concentration

(c) An increase in the concentration increases the number of particles in a given volume. This increases the frequency of collisions, and thus the rate of the reaction increases.

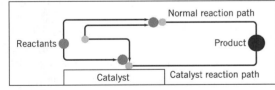
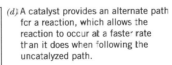

Normal reaction path

Reactants   Product

Catalyst   Catalyst reaction path

(d) A catalyst provides an alternate path for a reaction, which allows the reaction to occur at a faster rate than it does when following the uncatalyzed path.

FIGURE 14-1   Factors affecting reaction rates.

very slow. We need to heat the reaction system to about 425°C so the rate is high enough to form significant amounts of ammonia in a reasonable time.

$$N_2(g) + 3H_2(g) \xrightarrow[425°C]{Fe} 2NH_3(g)$$

Subdivision of Reactants

Starch, another carbohydrate, is the main component of wheat grain. Oxygen in the air reacts spontaneously with starch, but the rate of the reaction of oxygen with the individual grains of wheat is so low that it is not noticed. On the other hand, some empty grain elevators filled with tiny pieces of finely divided starch have exploded violently, destroying the buildings. The rate of the reaction of starch with oxygen increases dramatically when the starch is present as very tiny pieces called grain dust.

When solids and liquids are involved in a reaction, the subdivision of the materials influences the rate. Subdivision makes sense if we realize that tinier pieces have greater **surface area** so they provide much greater contact of the reactants and more collisions. To increase the rate of a reaction, divide the reactants into smaller parts and mix them to increase contact. Solids may be broken into pieces or powdered, and liquids can be sprayed to form tiny droplets called aerosols. See Figure 14-1(*b*).

Concentration

Acetylene, C_2H_2, is a hydrocarbon fuel used in oxyacetylene welding and steel-cutting torches. Acetylene gas burns readily in air, giving off heat.

$$2C_2H_2(g) + 5O_2(g) \longrightarrow 4CO_2(g) + 2H_2O(g) + \text{energy}$$

Mixing acetylene gas and pure oxygen gas increases the rate of the reaction in an oxyacetylene welding torch. The rate increases so much that the flame of the torch produces a temperature high enough to melt metals. The rate of reaction in the torch increases because the concentration of oxygen is increased. Concentration refers to the number of particles in a given volume. Air is only about 21% oxygen, so pure oxygen is more concentrated in oxygen than air. When the concentration increases, the frequency of collisions is increased and the reaction speeds up. In other words, higher concentrations of reactants result in more collisions.

Generally, the rate of a reaction involving gases or chemicals dissolved in water is changed by changing the number of reactant particles in a given volume, that is, by changing the concentration of one or more of the reactants. Increasing the concentration increases the reaction rate, and decreasing the concentration decreases the reaction rate. See Figure 14-1(*c*). For species dissolved in aqueous solution, the concentration or molarity increases by dissolving more of the species. For gaseous reactants, the concentration is increased by using a purer form of the gas or by compressing the gas. Compressing a gas results in more molecules per unit volume. For instance, the industrial production of ethyl alcohol, C_2H_5OH, involves the reaction of ethylene, C_2H_4, and water. To raise the rate of the reaction to a productive level, the reaction is carried out in a closed system at 400°C and very high pressures.

$$C_2H_4(g) + H_2O(g) \xrightarrow[\text{pressure}]{400°C} C_2H_5OH(g)$$

14-3 CATALYSTS

Catalysts are very important in controlling the rates of reactions. A **catalyst** is a chemical that increases the speed or rate of a chemical reaction without undergoing permanent chemical change. Many industrial and biological processes need catalysts for efficiency; the reactions are too slow without them. Catalysts are fascinating chemicals. The reaction between hydrogen and nitrogen gas to form ammonia does not occur at a significant rate, even at high pressures and high temperatures. If some finely divided iron is added to the reaction vessel, however, the rate of the reaction increases dramatically and the reaction becomes productive. The iron is not a reactant and does not chemically change in the reaction. The iron functions as a catalyst.

The H_2O_2 in an aqueous solution of hydrogen peroxide does not decompose at room temperature. A small amount of manganese(IV) oxide, MnO_2, catalyst causes the H_2O_2 to decompose rapidly at room temperature. Since the catalyst is not a reactant or product, we show it above the arrow in the equation.

$$H_2O_2(aq) \xrightarrow{\text{MnO}_2} 2H_2O(\ell) + O_2(g)$$

If you place some medicinal hydrogen peroxide disinfectant on a cut, a different catalyst, called an enzyme, present in blood and tissue, causes the rapid decomposition of the H_2O_2. The oxygen gas that is formed causes the characteristic bubbling and fizzing. This natural enzyme is called hydrogenperoxidase. It is present in the cells of plants and animals.

Specific catalysts are found to catalyze specific reactions, but some reactions have no known catalyst. A catalyst for the synthesis of ammonia was sought for many years. Chemists tested many chemicals as possible catalysts. In time, finely divided iron was found to work as a catalyst, and it is now used in the industrial synthesis of ammonia. Generally, as described below, there are three types of catalysis.

Heterogeneous Catalysts

Heterogeneous catalysts are surface or contact catalysts; they are typically metals, metal oxides, or other solids immersed in a gaseous or liquid mixture of the reactants. They function by catalyzing the reaction on the surface of the solid. A common application of a heterogeneous catalyst is the catalytic converter on automobiles. One air pollutant in the exhaust of the internal combustion engine is carbon monoxide, CO. Most of the carbon monoxide is converted to carbon dioxide by mixing the exhaust gases with air and passing this mixture over a bed of finely divided platinum. The equation for the reaction is

$$2CO(g) + O_2(g) \xrightarrow{\text{Pt}} 2CO_2(g)$$

Unburned hexane, C_6H_{14}, in automobile exhaust reacts with oxygen in the catalytic converter. Write a balanced equation for the burning of hexane and show the Pt as a catalyst.

The platinum catalyst allows the reaction to occur rapidly at the exhaust temperatures. The catalytic converter also catalyzes the reaction of unburned hydrocarbon from gasoline with oxygen to give carbon dioxide and water. The reactions occur on the surface of the platinum.

Another example of a heterogeneous catalyst is rhodium. Gaseous methyl alcohol or methanol, CH_3OH, and carbon monoxide, CO, do not noticeably react without a catalyst. When solid rhodium is placed in the reaction vessel, a rapid reaction occurs, forming acetic acid.

$$CH_3OH(g) + CO(g) \xrightarrow{\text{Rh}} CH_3CO_2H(g)$$

The rhodium serves as a contact catalyst. It makes the reaction occur quickly enough for the commercial manufacture of acetic acid. Without the catalyst, the reaction is too slow.

Homogeneous Catalysts

Homogeneous catalysts function by being intimately mixed with the reactants in gaseous mixtures or liquid solutions. In other words, the catalyst is in the same physical state as the reactants and, unlike heterogeneous catalysis, such catalysis does not involve surface reactions. For example, when a compound containing iron(III) ions dissolves in a water solution of hydrogen peroxide, the H_2O_2 decomposes quickly.

$$H_2O_2(aq) \xrightarrow{\text{Fe}^{3+}} 2H_2O(\ell) + O_2(g)$$

The iron(III) ion catalyst is not a solid but is a dissolved ion, homogeneous with the components of the solution. Of course, a negative ion is also in the solution to balance the iron(III) ion, but the iron ion is the catalyst.

Another example of a homogeneous catalyst is the catalyst in commercial epoxy glue. When a solution containing the catalyst is mixed with the glue reactants, the catalyst encourages a reaction that forms the hardened products. Without the catalyst, the reaction is very slow and the glue is not useful.

Enzyme Catalysts

The numerous chemical reactions that occur in the body are referred to collectively as metabolism. Metabolic reactions include digestion and the reactions in which certain food molecules are used by the body for energy. Other metabolic reactions involve the breakdown of body cells and the formation of new cells. Almost all the thousands of metabolic reactions occurring in our bodies require specific biological catalysts. These catalysts are synthesized by the body. Biological catalysts are known as **enzymes.** Literally thousands of enzymes are involved in life processes. These enzymes allow a variety of specific reactions to occur at useful rates. In our bodies, enzymes catalyze reactions that without the enzymes occur very slowly at body temperatures. For example, the human body produces many specialized digestive enzymes to aid the digestion of food molecules. Without these digestive enzymes, foods do not break down fast enough for absorption into the body.

14-4 How Catalysts Function

How a catalyst works may seem a bit mysterious. In some reactions, it is not known in detail how the catalyst functions. It is known, however, that the catalyst provides an alternate path or sequence of intermediate reactions that is faster than the uncatalyzed reaction. That is, the catalyst provides another way for the reaction to occur but gives the same products. See Figure 14-1(*d*). A catalyzed reaction can occur up to 10 times faster than an uncatalyzed reaction. Under specific conditions, the reaction follows the faster catalyzed route rather than the uncatalyzed route. With rhodium used in the manufacture of acetic acid, it appears that methanol molecules are attracted to the surface of the solid rhodium. A given methanol molecule arranges on the surface so that when a carbon monoxide molecule collides with it, acetic acid forms. Thus, rhodium is a heterogeneous surface catalyst.

A homogeneous catalyst is of the same physical state as the reactants. Normally, such catalysts are mixed with solutions of the reactants and some are used in reactions involving gases. Homogeneous catalysts typically react with a reactant to form an intermediate product. The intermediate product then reacts further to give the ultimate product and to regenerate the catalyst. As an example, reread the part of Section 10-19 that describes how chlorine catalyzes the decomposition of ozone in the stratosphere.

Enzymes are composed of macromolecules of varying complexity. All enzymes contain protein. Some enzymes are composed entirely of protein. The chemical reactions involved in life processes need specific enzymes that must be present at the correct time and place. It is thought that an enzyme functions by interacting with one of the reactants involved in a reaction. Such a reactant is called the substrate. A given enzyme will act as a catalyst for only one kind of reaction. An obvious question is, How can an enzyme distinguish one kind of reactant from another?

The commonly accepted view of enzyme behavior is the lock-and-key theory, illustrated in Figure 14-2. According to this theory, the enzyme has a definite three-dimensional structure arranged in a way that allows the substrate molecule to fit into the structure. Only a specific kind of substrate can fit into a given enzyme. Once the enzyme and substrate form an aggregate, the substrate is held in place and exposed for the reaction. Enzymes allow metabolic reactions to readily occur at body temperatures. Without enzymes, these reactions would not occur fast enough to maintain

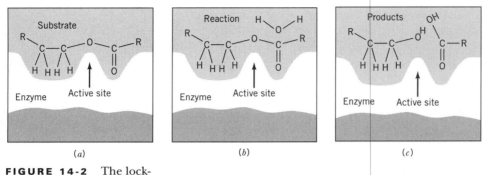

(a)   (b)   (c)

FIGURE 14-2   The lock-and-key theory of enzyme function. (*a*) The three-dimensional shape of the macromolecular enzyme in region of the active site is arranged to accommodate the molecule with which the enzyme works. The active site is that part of the enzyme that has the major catalytic function. (*b*) The molecule is held in position by the enzyme while the reaction takes place. (*c*) The product or products of the reaction leave the enzyme, and the active site is available to catalyze the reaction of another molecule.

life processes. After an enzyme-catalyzed reaction occurs, the product moves away from the enzyme, leaving it unchanged and available to catalyze the reaction of another substrate molecule.

Certain metabolic processes involve a long sequence of enzyme-catalyzed reactions. Each enzyme must be present at the correct time for the process to occur. If any of the required enzymes is missing, the process is disrupted. Since such metabolic processes are vital to life, the absence of an enzyme can lead to illness and even death.

The absence of certain vitamins and minerals in the diet can lead to enzyme deficiencies. A normal diet typically supplies sufficient vitamins and minerals. Most vitamins and minerals are used by the body to form important enzymes. Some hereditary conditions and diseases result from the inability of the body to produce specific enzymes. For instance, many people have lactose intolerance caused by the body not making the digestive enzyme called lactase. For them, the lactose in dairy products cannot be digested, which causes nausea and discomfort.

An **inhibitor** is a chemical that slows or decreases the rate of a chemical reaction. Very often an inhibitor slows a reaction by preventing the function of a catalyst. Sometimes inhibitors are desirable in reactions and sometimes they are not. For example, leaded gasoline, in an automobile with a catalytic converter, can poison the catalyst in the converter. The lead inhibits the catalyst by coating it with lead and lead compounds. Some antibacterial drugs function by inhibiting specific enzymes in bacteria, preventing the growth and multiplication of the bacteria. Furthermore, some anticancer drugs inhibit enzymes to prevent the rapid growth and multiplication of cells. On the other hand, some inhibitors act as general poisons. Mercury and lead compounds in the body can inhibit a variety of enzymes, causing the body to malfunction. The result is heavy-metal poisoning that can cause permanent damage or death. Several potent poisons, such as cyanide ion, bind to and inhibit the functioning of critical enzymes. A sufficient dose of cyanide ion results in death in a matter of seconds by interfering with an enzyme involved in certain metabolic reactions. The absence of the enzyme causes death as a result of the inability to use oxygen.

14-5 CHEMICAL EQUILIBRIUM

When reactants are mixed, they react to give products at a certain rate. In many reactions, as the reactants form, products react in a reverse reaction to reform the reactants. When this occurs, the reaction is called **reversible** and is viewed as a system involving **chemical equilibrium.** To get a picture or feeling for chemical equilibrium, consider the apparatus shown in Figure 14-3. The figure shows two containers of water, each equipped with a pump that works to pump the water over the barrier. Imagine that we start with the reactant side full and the product side empty. As the water is pumped to the product side, it is pumped back to the reactant side. If both pumps are working, where does the water go? If both pumps are working, the system soon reaches an equilibrium-like steady state in which there is some water on the reactant side and some on the product side. The pumping of the water continues, but a "standoff" is reached in which the rate of the forward process equals the rate of the reverse process. A double arrow could be used to represent the balanced steady state. If the pumps are of equal strength, the amounts of water on the reactant and product side are the same. This case could be represented by double arrows of equal size ($\rightleftharpoons$). If the forward pump is stronger than the reverse pump, a steady

Equilibrium

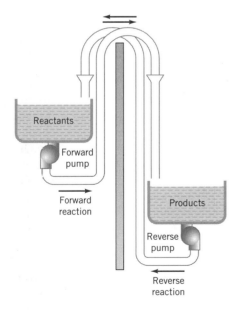

FIGURE 14-3 Two containers are separated by a barrier. Each container has a pump that pumps water over the barrier. The pumps operate in opposite directions. When the rate of forward pumping equals the rate of reverse pumping, the system is in "equilibrium."

state is still reached. In such a case, however, the amount of water on the product side is greater than the amount on the reactant side. This case could be represented by a double arrow suggesting that the system favors the product side ($\rightleftharpoons$). In contrast, if the reverse pump is stronger than the forward pump, a steady state is still reached. In this case, however, the amount of water on the reactant side is greater than the amount on the product side. This case could be represented by a double arrow suggesting that the system favors the reactant side ($\rightleftharpoons$). Chemical reactions, of course, do not involve pumps, but they can involve a forward reaction and a reverse reaction. Forward and reverse reactions are opposing processes that are occurring simultaneously but in opposite directions.

14-6 EQUILIBRIUM SYSTEMS

Dynamic Chemical Equilibrium
The state that exists when the rate of a forward reaction equals the rate of a reverse reaction.

To illustrate the idea of a reversible chemical reaction involving chemical equilibrium, consider the following reaction:

$$N_2O_4(g) \rightleftharpoons 2NO_2(g)$$
$$\text{colorless} \qquad \text{reddish brown}$$

This reversible reaction includes two opposing reactions: colorless dinitrogen tetroxide, N_2O_4, decomposes to form reddish brown nitrogen dioxide, NO_2, and nitrogen dioxide combines to form dinitrogen tetroxide.

The two reactions in this example occur in opposition. Dynamic chemical equilibrium results when two opposing reactions occur, producing a mixture of reactants and products in which the relative amounts are fixed. Equilibrium is reached when the rates of the two opposing reactions become equal. Both reactions occur, but since equilibrium exists, the net result is a continuous cyclic situation in which the reactant gives the products and the products react to give the original reactant. At equilibrium, no further change in the system is observed.

A reversible reaction has forward and reverse directions. Typically, the **forward reaction** is written from left to right in the equation and the **reverse reaction** is written from right to left. Of course, in the equation for an equilibrium reaction, we can write either side on the left or right. Therefore, the designation of the forward and reverse reaction is arbitrary and is used only as an aid in discussing the reaction. In the reversible reaction used in this example, the forward reaction, the decomposition of dinitrogen tetroxide, competes with the reverse reaction, the combination of nitrogen dioxide. Since both reactions are taking place, the net result is a standoff that produces a mixture of reactants and products.

Normally, it is not possible to see a chemical reaction come to equilibrium as it occurs, but in this equilibrium, the NO_2 has a reddish brown color and N_2O_4 is colorless. Suppose we place some pure dinitrogen tetroxide in a sealed glass vessel as illustrated in Figure 14-4. The dinitrogen tetroxide reacts to give nitrogen dioxide. As the nitrogen dioxide forms, it begins to react to give dinitrogen tetroxide. As equilibrium is approached, the mixture changes from colorless to a slight reddish brown color, corresponding to the equilibrium mixture of the two gases. How soon equilibrium is reached will depend on the rates of the reactions involved. At equilibrium, the mixture takes on a uniform color because it contains a specific concentration of each gas.

Imagine starting with some pure nitrogen dioxide. It reacts to form dinitrogen tetroxide, and as the dinitrogen tetroxide is formed, it reacts to give nitrogen dioxide. This process is illustrated in Figure 14-5. As equilibrium is approached, the original reddish brown color of the nitrogen dioxide fades to a lighter shade, corresponding to the equilibrium mixture of the two gases. Again at equilibrium, the mixture takes on a uniform color since it contains specific concentrations of each gas. Recall from Section 12-6 that dynamic equilibrium occurring in a saturated solution causes a fixed concentration of dissolved solute. At equilibrium, the relative concentrations of reactants and products is fixed. As shown in Figures 14-4 and 14-5, the nitrogen dioxide–dinitrogen oxide equilibrium reaction produces fixed concentrations of both gases. These fixed concentrations are called **equilibrium concentrations.** At equilibrium, the ratio of the concentrations of the products and reactant has a constant numerical value, corresponding to the constant equilibrium concentrations.

Decomposition of
dinitrogen tetroxide to
form nitrogen dioxide

Combination of
nitrogen dioxide to
form dinitrogen
tetroxide

FIGURE 14-4   Idealized equilibrium between dinitrogen tetroxide and nitrogen dioxide. A pure sample of dinitrogen tetroxide produces, in time, an equilibrium mixture of N_2O_4 and NO_2 in which the concentrations of the reactant and products remain constant. The ratio of the square of the NO_2 concentration to the N_2O_4 concentration has a constant value at a fixed temperature. See Figure 14-5.

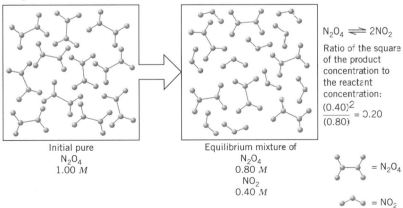

Initial pure
N_2O_4
1.00 M

Equilibrium mixture of
N_2O_4
0.80 M
NO_2
0.40 M

$N_2O_4 \rightleftharpoons 2NO_2$

Ratio of the square
of the product
concentration to
the reactant
concentration:
$$\frac{(0.40)^2}{(0.80)} = 0.20$$

= N_2O_4

= NO_2

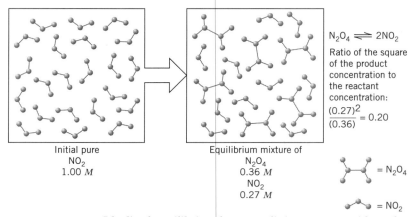

FIGURE 14-5 Idealized equilibrium between dinitrogen tetroxide and nitrogen dioxide. A pure sample of nitrogen dioxide produces, in time, an equilibrium mixture of N_2O_4 and NO_2 in which the concentrations of the reactants and product remain constant. The ratio of the square of the NO_2 concentration to the N_2O_4 concentration has a constant value at a fixed temperature. See Figure 14-4.

14-7 Equilibrium Constants

When the dinitrogen tetroxide–nitrogen dioxide reaction reaches equilibrium, the result is a steady-state mixture of reactants and products. The forward and reverse reactions continue to occur at equilibrium, but the reactions balance out to give constant concentrations of both nitrogen dioxide and dinitrogen tetroxide.

At equilibrium, the ratio of the concentrations of the products to the reactants is equal to a constant numerical value, characterizing any reaction system involving chemical equilibrium. This ratio is called the equilibrium constant. An equilibrium system represented by the general equation

$$a\mathrm{A} + b\mathrm{B} \rightleftharpoons c\mathrm{C} + d\mathrm{D}$$

is described by an **equilibrium constant expression** of the general form

$$K_{eq} = \frac{[\mathrm{C}]^d[\mathrm{D}]^d}{[\mathrm{A}]^a[\mathrm{B}]^b}$$

where K_{eq} is the equilibrium constant and the square brackets refer to the concentrations of the products and reactants in moles per liter. Each concentration is raised to a power corresponding to its coefficient in the equilibrium equation. To find the numerical value of K_{eq} for an equilibrium system, it is necessary to measure the equilibrium concentrations of the products and reactants at a constant temperature. The ratio of the molar concentrations raised to the appropriate powers gives the numerical value of K_{eq} for the equilibrium system at the specified temperature.

As an example, let us write an equilibrium constant expression for the reaction

$$N_2O_4(g) \rightleftharpoons 2NO_2(g)$$

The equation has a coefficient of 2 for the nitrogen dioxide. In a sense, there are two products.

$$N_2O_4(g) \rightleftharpoons NO_2(g) + NO_2(g)$$

We include the concentration of both products and the reactant in the equilibrium constant expression

$$K_{eq} = \frac{[NO_2][NO_2]}{[N_2O_4]}$$

Note that $[NO_2][NO_2]$ is the same as $[NO_2]^2$. So, the algebraically equivalent form is

$$K_{eq} = \frac{[NO_2]^2}{[N_2O_4]}$$

The above expression reveals that at equilibrium, the ratio of the square of the concentration of nitrogen dioxide to the concentration of dinitrogen tetroxide is a constant. See Figures 14-4 and 14-5.

The reactants and products in an equilibrium system may be liquids, solids, gases, or species in solution. Note that only gases and substances in solution can have variable concentrations in equilibrium reactions. Any solid or liquid in a reaction has a constant concentration, a fixed number of moles per liter, because the concentration of a solid or a liquid is directly related to its density. Since the density is constant, the concentration is constant. Thus, **gases and solutes in solution are included in equilibrium constant expressions**, and liquids and solids are not included. Recall from Section 12-13 that liquid water reacts with itself.

$$2H_2O \rightleftharpoons H_3O^+(aq) + OH^-(aq)$$

The equilibrium constant expression for this reaction is

$$K_w = [H_3O^+][OH^-]$$

Note that the equilibrium constant expression for this reaction does not include water since water is a liquid. Also, the constant is given the special symbol K_w but it is a K_{eq}.

The value of the equilibrium constant for a reaction is calculated using the concentrations of the species involved in the reaction as measured using special experimental methods. In other words, the equilibrium constant is found experimentally by measuring the molar concentration of each species at equilibrium. At 25°C, the pH of pure water is 7.00, corresponding to a hydronium ion concentration of 1.0×10^{-7} M. In pure water, the concentrations of hydronium and hydroxide ions are equal, so the value of the equilibrium constant for water is

$$K_w = [H_3O^+][OH^-] = (1.0 \times 10^{-7})(1.0 \times 10^{-7}) = 1.0 \times 10^{-14}$$

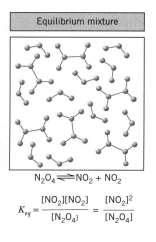

Equilibrium mixture

$N_2O_4 \rightleftharpoons NO_2 + NO_2$

$$K_{eq} = \frac{[NO_2][NO_2]}{[N_2O_4]} = \frac{[NO_2]^2}{[N_2O_4]}$$

EXAMPLE 14-1

The dinitrogen tetroxide-nitrogen dioxide reaction may be carried out at a variety of temperatures. At any of these temperatures, the equilibrium equation for the reaction is

$$N_2O_4(g) \rightleftharpoons 2NO_2(g)$$

Analysis of an equilibrium mixture for the reaction at 100°C gives the following concentration data in moles per liter.

$$[N_2O_4] = 0.36 \qquad [NO_2] = 0.27$$

The equilibrium constant expression for the reaction is

$$K_{eq} = \frac{[NO_2]^2}{[N_2O_4]}$$

The value of the equilibrium constant is found simply by substituting the equilibrium concentrations into the expression.

$$K_{eq} = \frac{[NO_2]^2}{[N_2O_4]} = \frac{(0.27)^2}{(0.36)} = 0.20$$

Note that the value of the equilibrium constant in Example 14-1 is close to 1. It is neither a very large nor a very small value. An equilibrium constant value around 1, say between 0.1 and 10, shows that the equilibrium mix-

Table 14-1   *Some Equilibrium Systems*

Formation of nitrogen oxide gas from nitrogen gas and oxygen gas at 25°C	$N_2 + O_2 \rightleftharpoons 2NO$
Formation of nitrogen dioxide gas from nitrogen oxide gas and oxygen gas at 25°C	$2NO + O_2 \rightleftharpoons 2NO_2$
Formation of ammonia gas from nitrogen and hydrogen gas at 425°C	$N_2 + 3H_2 \rightleftharpoons 2NH_3$
Reaction of ammonia and water at 25°C	$NH_3(aq) + H_2O \rightleftharpoons NH_4^+(aq) + OH^-(aq)$
Reaction of hydrogen fluoride and water at 25°C	$HF(aq) + H_2O \rightleftharpoons H_3O^+(aq) + F^-(aq)$
Dissolving of silver chloride in water at 25°C	$AgCl(s) \rightleftharpoons Ag^+(aq) + Cl^-(aq)$
Formation of calcium carbonate from its aqueous ions at 25°C	$Ca^{2+}(aq) + CO_3^{2-}(aq) \rightleftharpoons CaCO_3(s)$

ture contains similar concentrations of reactants and products. Such a value reveals that the equilibrium reaction does not significantly favor one side of the reaction or the other. On the other hand, an equilibrium constant value that is much smaller than 0.1 suggests that the equilibrium largely favors the reactants, and a value much greater than 10 suggests that the equilibrium largely favors the products. For instance, $K_w = 1 \times 10^{-14}$ for the reaction of water molecules, suggests the reaction only occurs to an extremely slight extent. Further examples of equilibrium reactions are given in Table 14-1.

A change in temperature upsets an equilibrium reaction, producing a new ratio of product-to-reactant concentrations. At a new temperature, the new equilibrium is reflected by a different value of the equilibrium constant. Consequently, a measured value for an equilibrium constant applies at a specific temperature.

14-8 THE EXTENT OF A REACTION

As discussed in Section 14-7, calculating the value of an equilibrium constant using experimentally measured concentrations of the reactants and products is possible. The numerical value of K_{eq} reflects the extent of the reaction. A relatively large value of K_{eq} suggests that the reaction produces a higher concentration of products than reactants. An equation for an equilibrium favoring the products shows a larger arrow toward the product side.

$$a\text{A} + b\text{B} \rightleftharpoons c\text{C} + d\text{D}$$

A relatively small value of K_{eq} suggests that the concentration of reactants is higher than that of the products. The equilibrium favors the reactant side. An equation for this equilibrium state gives the larger arrow toward the reactants.

$$a\text{A} + b\text{B} \rightleftharpoons c\text{C} + d\text{D}$$

When the equilibrium concentrations of reactants and products are close to one another, K_{eq} has a value close to 1. In such a case, the two arrows are shown about the same size.

$$a\text{A} + b\text{B} \rightleftharpoons c\text{C} + d\text{D}$$

Note that the equations given in Table 14-1 have arrows that reflect the extent of the reactions. When showing the extent of a reaction is not necessary or the favored direction is not known, equal arrows are used ($\rightleftharpoons$).

14-9 EQUILIBRIUM CONSTANT EXPRESSIONS

By convention, an equilibrium constant expression is written with the concentrations of the products in the numerator and the concentrations of the reactants in the denominator. To deduce an **equilibrium constant expression** for an equilibrium reaction, we use the following pattern. This pattern applies to the equilibrium reactions discussed in this book. There are some reactions for which these guidelines do not apply.

1. Write an equilibrium equation and note whether the species involved are gases, solids, liquids, or in aqueous solution.
2. Enclose the formulas of all products that are gases or aqueous species in square brackets for the numerator of the expression. If a species has a coefficient greater than 1, raise its concentration to the power given by the coefficient.
3. Enclose the formulas of all reactants that are gases or aqueous species in square brackets for the denominator of the expression using the coefficients as powers.
4. This ratio of the product concentrations to the reactant concentrations, each raised to a power corresponding to its coefficient in the balanced equation, is equal to the constant, K_{eq}.

EXAMPLE 14-2

The equilibrium equation for the reaction of nitrogen gas and hydrogen gas to give ammonia gas is

$$N_2(g) + 3H_2(g) \rightleftharpoons 2NH_3(g)$$

What is the equilibrium constant expression?

Since all the components are gases, their concentrations are included in the expression and each is raised to the power corresponding to its coefficient.

$$K_{eq} = \frac{[NH_3]^2}{[N_2][H_2]^3}$$

EXAMPLE 14-3

The equilibrium equation for the reaction of nitrogen and oxygen gas to produce nitrogen oxide gas is

$$N_2(g) + O_2(g) \rightleftharpoons 2NO(g)$$

What is the equilibrium constant expression?

Since all the components are gases, their concentrations are included in the expression and each is raised to the power corresponding to its coefficient.

$$K_{eq} = \frac{[NO]^2}{[N_2][O_2]}$$

ACTIVITY 14-3

Write the equilibrium constant expressions for the following equilibrium reactions:
(a) $PbCl_2(s) \rightleftharpoons Pb^{2+}(aq) + 2Cl^-(aq)$
(b) $CaCO_3(s) \rightleftharpoons CaO(s) + CO_2(g)$

EXAMPLE 14-4

The equilibrium equation for the dissolving of calcium carbonate is

$$CaCO_3(s) \rightleftharpoons Ca^{2+}(aq) + CO_3^{2-}(aq)$$

What is the equilibrium constant expression?

$CaCO_3$ is a solid and its concentration cannot change, so it does not appear in the expression. The aqueous ionic species, however, do appear.

$$K_{eq} = [Ca^{2+}][CO_3^{2-}]$$

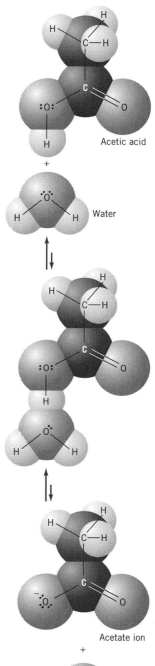

Acetic acid

+

Water

14-10 EQUILIBRIUM OF ACETIC ACID

An example of an equilibrium reaction in water is the reversible reaction between acetic acid and water.

$$HC_2H_3O_2(aq) + H_2O \rightleftharpoons H_3O^+(aq) + C_2H_3O_2^-(aq)$$

When pure acetic acid is added to water, the $HC_2H_3O_2$ reacts with H_2O to form H_3O^+ (hydronium ion) and $C_2H_3O_2^-$ (acetate ion). As these two ions begin to accumulate, they react with one another to reform the acetic acid and water. In time, the rates of the two reactions become equal and chemical equilibrium is reached. At equilibrium acetic acid and water are continually reacting to give hydronium ions and acetate ions, and the ions are continually reacting to give the original reactants. The equilibrium constant expression for the reaction is

$$K_{eq} = \frac{[H_3O^+][C_2H_3O_2^-]}{[H_2O][HC_2H_3O_2]}$$

Note that since water is a liquid and since very little water is used in the reaction, the concentration of water is constant, which is shown as

$$K_{eq}[H_2O] = \frac{[H_3O^+][C_2H_3O_2^-]}{[HC_2H_3O_2]} = K_a$$

This kind of equilibrium constant for a weak acid like acetic acid is represented by the symbol K_a, where a stands for acid. The equilibrium reaction involving acetic acid produces relatively low concentrations of ions as shown by a K_a value at 25°C of 1.8×10^{-5}.

$$K_a = \frac{[H_3O^+][C_2H_3O_2^-]}{[HC_2H_3O_2]} = 1.8 \times 10^{-5}$$

The numerical value of the constant is a relatively small number, indicating that the numerator is small and the denominator is large. Thus, a solution of acetic acid is a mixture of acetic acid molecules and water molecules with relatively low concentrations of hydronium and acetate

Acetate ion

+

Hydronium
ion

ions. The ion concentrations are low and the molecular acetic acid has by far the highest concentration of all the species in the solution.

A state of dynamic equilibrium in a reaction is denoted by the use of a double arrow in the equation for the reaction. Keep in mind that the relative sizes of the arrows show which species exist at greater concentrations at equilibrium. The larger arrow points in the direction of the species present at greater concentrations. The equilibrium is said to favor the direction of the larger arrow. The reversible reaction between acetic acid and water (see the equation given above) favors the water and acetic acid side of the equilibrium. In other words, water and acetic acid slightly react and produce relatively low concentrations of hydronium and acetate ions. Hence, a solution of acetic acid is a weak electrolyte and acetic acid is called a weak acid.

14-11 EQUILIBRIUM CONSTANTS FOR WEAK ACIDS

When a weak acid is in water, it reacts with water in a reversible chemical reaction (see Section 13-6). The **equilibrium reaction of any weak acid, HA,** in water is represented by

$$HA + H_2O \rightleftharpoons H_3O^+ + A^-$$

where HA is the general formula for a weak acid and A^- is the conjugate base. Equilibrium is reached when the rates of the forward and reverse reactions are equal. Freshly prepared solutions of acids quickly reach equilibrium. At equilibrium, the solution contains constant concentrations of the various species. The concentration of water is constant since very little is used in the reaction. The actual values of the concentrations of the other species depend on the acidic properties of the particular acid involved. Remember that weak acids vary in strength which means that they vary in the extent with which they react with water.

For a weak acid equilibrium system, the ratio of the concentrations of the products to the concentrations of the reactants is constant. This situation is represented by the general **equilibrium constant expression for any weak acid.**

$$K_a = \frac{[H_3O^+][A^-]}{[HA]}$$

where K_a is called the **acid ionization constant** or the **acid constant** for a weak acid.

An acid constant expression states that the product of the molar concentrations of the hydronium ion and the conjugate base of the acid divided by the molar concentration of the weak acid is constant. This constant has the same numerical value for any solution of a given acid at a specific temperature. Each weak acid has a unique K_a value.

Acid ionization constants are determined experimentally by preparing a solution of a weak acid and measuring the equilibrium concentrations of

ACTIVITY
14-4

Use the formulas for formic acid, HCO_2H, and its conjugate base, HCO_2^-, and show how they fit the general weak acid equilibrium equation given in Section 14-11.

the hydronium ion, the weak acid, and its conjugate base. The concentration values are used in the equilibrium expression to calculate the numerical value of K_a.

EXAMPLE 14-5

Calculate the value of the acid ionization constant, K_a, for acetic acid if a solution of acetic acid is experimentally found to have the following equilibrium concentrations:

$[HC_2H_3O_2]$	$[H_3O^+]$	$[C_2H_3O_2^-]$
0.10 M	$1.34 \times 10^{-3} M$	$1.34 \times 10^{-3} M$

K_a is found by use of the equilibrium expression for the reaction as written from the equilibrium equation. The equilibrium equation for acetic acid in water is

$$HC_2H_3O_2 + H_2O \rightleftharpoons H_3O^+ + C_2H_3O_2^-$$

The acid constant expression is the ratio of the product concentrations to reactant concentration.

$$K_a = \frac{[H_3O^+][C_2H_3O_2^-]}{[HC_2H_3O_2]}$$

Calculate the K_a value by substituting the known concentrations into this expression.

$$K_a = \frac{[H_3O^+][C_2H_3O_2^-]}{[HC_2H_3O_2]} = \frac{(1.34 \times 10^{-3})(1.34 \times 10^{-3})}{(0.10)} = 1.3 \times 10^{-5}$$

ACTIVITY 14-5

When lactic acid, $HC_3H_5O_3$ dissolves in water, it reacts with water to give hydronium ions and lactate ions, $C_3H_5O_3^-$. Write an equation showing this reaction as an equilibrium and write an acid constant expression from the equation. A solution of the acid is found to contain 0.025 M $HC_3H_5O_3$, 1.8×10^{-3} M H_3O^+, and 1.8×10^{-3} M $C_3H_5O_3^-$. Calculate the value of the K_a for lactic acid.

The numerical value of K_a reflects the extent of the reaction of the acid with water, which reflects the strength of the acid. The stronger a weak acid, the larger its K_a value, of course, because higher concentrations of hydronium ion and conjugate base relative to the concentration of unreacted weak acid correspond to a larger value of K_a. The K_a values for weak acids are always less than 1, so the double arrows used for the reactions of weak acids in water are typically written as $\rightleftharpoons$. The K_a values for some common acids are given in Table 14-2.

Sometimes it is convenient to express the relative strengths of weak acids as pK_a values defined as $pK_a = -\log K_a$. These pK_a values are calculated using the same methods as pH and pOH. Using pK_a values is a way to express the relative strengths of weak acids without using exponential numbers. The negative logarithm in the definition of pK_a means that relatively stronger weak acids have lower numerical values of pK_a and weaker acids have correspondingly higher pK_a values. Table 14-2 also lists the pK_a values for the acids in the table. Notice that when weak acids are listed ac-

Table 14-2 K_a *and* pK_a *Values for Some Weak Acids*

ACID	FORMULA	K_a	pK_a
Oxalic acid	$H_2C_2O_4$	5.9×10^{-2}	1.23
Hydrogen sulfate ion	HSO_4^-	1.0×10^{-2}	2.00
Phosphoric acid	H_3PO_4	7.1×10^{-3}	2.15
Hydrogen fluoride	HF	6.9×10^{-4}	3.16
Hydrogen oxalate ion	$HC_2O_4^-$	5.2×10^{-5}	4.28
Acetic acid	$HC_2H_3O_2$	1.8×10^{-5}	4.74
Carbon dioxide + water	$CO_2 + H_2O$	4.4×10^{-7}	6.36
Dihydrogen phosphate ion	$H_2PO_4^-$	6.2×10^{-8}	7.21
Hydrogen cyanide	HCN	5.8×10^{-10}	9.24
Hydrogen carbonate ion	HCO_3^-	4.7×10^{-11}	10.33
Hydrogen phosphate ion	HPO_4^{2-}	4.5×10^{-13}	12.35

cording to decreasing K_a or increasing pK_a values, the result is an acid–base table like Table 13-1. For practice calculate the pK_a value for each acid in Table 13-1.

14-12 LE CHÂTELIER'S PRINCIPLE

Before exploring some practical applications of equilibrium, we need to consider the behavior of equilibrium systems. When a chemical system is at equilibrium, it remains in this state indefinitely unless the equilibrium is disturbed in some manner. An equilibrium system shifts in one direction or another when some factor upsets the equilibrium. A general principle regarding this situation is called Le Châtelier's principle in honor of the French chemist who first stated it in 1888. See the margin.

An equilibrium system is in a state of balance. When something is done to upset the balance, the equilibrium shifts in a way that allows the attainment of a new state of balance. To illustrate Le Châtelier's principle, we will use the equilibrium reaction involved in the formation of ammonia from hydrogen and nitrogen (see Figure 14-6).

$$N_2(g) + 3H_2(g) \rightleftharpoons 2NH_3(g) + energy$$

At equilibrium, the equilibrium concentrations of the reactants and products are fixed unless something is done to change them.

What factors affect an equilibrium system?

1. A change in the concentration of a species that is a reactant or product
2. A change in temperature of the equilibrium mixture
3. A change in the pressure of an equilibrium mixture involving one or more gases

Le Châtelier's Principle

When a factor affecting an equilibrium system is changed, the equilibrium shifts in a direction that tends to counteract the change.

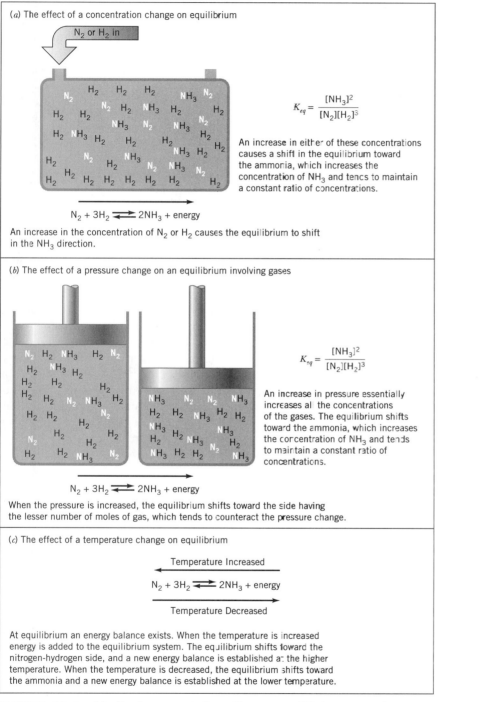

(a) The effect of a concentration change on equilibrium

N₂ or H₂ in

$$K_{eq} = \frac{[NH_3]^2}{[N_2][H_2]^3}$$

An increase in either of these concentrations causes a shift in the equilibrium toward the ammonia, which increases the concentration of NH₃ and tends to maintain a constant ratio of concentrations.

$N_2 + 3H_2 \rightleftharpoons 2NH_3 + energy$

An increase in the concentration of N₂ or H₂ causes the equilibrium to shift in the NH₃ direction.

(b) The effect of a pressure change on an equilibrium involving gases

$$K_{eq} = \frac{[NH_3]^2}{[N_2][H_2]^3}$$

An increase in pressure essentially increases all the concentrations of the gases. The equilibrium shifts toward the ammonia, which increases the concentration of NH₃ and tends to maintain a constant ratio of concentrations.

$N_2 + 3H_2 \rightleftharpoons 2NH_3 + energy$

When the pressure is increased, the equilibrium shifts toward the side having the lesser number of moles of gas, which tends to counteract the pressure change.

(c) The effect of a temperature change on equilibrium

Temperature Increased

$N_2 + 3H_2 \rightleftharpoons 2NH_3 + energy$

Temperature Decreased

At equilibrium an energy balance exists. When the temperature is increased energy is added to the equilibrium system. The equilibrium shifts toward the nitrogen-hydrogen side, and a new energy balance is established at the higher temperature. When the temperature is decreased, the equilibrium shifts toward the ammonia and a new energy balance is established at the lower temperature.

FIGURE 14-6   Idealized representation of how an equilibrium system is affected by concentration, pressure, and temperature changes.

A change in the amount of a solid or liquid that is a reactant or product in an equilibrium system does not affect the equilibrium because the concentration of a solid or liquid is not changed by changing the amount. A simple example to illustrate this point is a saturated solution of sodium chloride. In a saturated solution, a dynamic equilibrium exists between dissolved and undissolved solute. If more solid sodium chloride is added to a saturated solution, no more salt dissolves. In other words, the equilibrium is not affected by adding the solid, and dissolving more salt by adding solid salt is not possible.

Concentration

When a system is at equilibrium, the concentration of each species is constant. The **concentration** of a gaseous or dissolved species is changed by adding more of it to the system or by removing some of it from the system. When this is done, the equilibrium shifts in the direction that tends to diminish an increase in concentration or replenish a decrease in concentration. In other words, the equilibrium shifts away from an increase and toward a decrease.

As illustrated in Figure 14-6(*a*), if more hydrogen or nitrogen is added to the equilibrium system, the equilibrium shifts toward the ammonia side. This shift tends to diminish the increased concentration of nitrogen or hydrogen and maintain the equilibrium balance of the relative concentrations of the species. If the concentration of ammonia is decreased by removal of ammonia, the equilibrium shifts toward the ammonia side, which tends to replenish the decrease and maintain the equilibrium balance of concentrations. The equilibrium ratio of products to reactants must remain constant if the system is in equilibrium.

Pressure

What happens to an equilibrium system involving gases when the total pressure of the system is changed? Recall from Chapter 10 that the pressure of a gas sample, even one containing a mixture of gases, is given by $P = nRT/V$. One way to increase the pressure is to decrease the total volume of the system by compressing it. The ideal gas law shows that the pressure of a gas mixture depends on the total number of molecules, that is, the number of moles of gas. An equilibrium system can counteract an increase in pressure by shifting in the direction that gives the lesser number of moles which corresponds to fewer molecules of gas. This shift allows the concentrations of the gases to change so that the ratio of products to reactants remains constant.

As illustrated in Figure 14-6(*b*), the ammonia reaction involves 1 mole of nitrogen and 3 moles of hydrogen, forming 2 moles of ammonia. Thus, four moles of gas react to give two moles of gas. An increase in pressure causes the equilibrium to shift toward the ammonia side since this produces fewer moles of gas and reduces the pressure. On the other hand, a decrease in the total pressure of the equilibrium system causes the equilibrium to shift in the direction that forms the greater number of moles of gas. A decrease in the total pressure of the ammonia equilibrium causes a shift toward the nitrogen–hydrogen side. A pressure change does not change the equilibrium constant but can change the relative amounts of reactants and products. In general, a decrease in pressure causes a shift in equilibrium to make more moles of gas, and an increase in pressure causes a shift to make fewer moles of gas.

Temperature

An equilibrium system involves the production and consumption of energy. The ammonia equilibrium is exothermic in the forward direction and endothermic in the reverse direction.

$$\xrightarrow{\text{\hspace{1em}exothermic direction\hspace{1em}}}$$

$$N_2(g) + 3H_2(g) \rightleftharpoons 2NH_3(g) - \text{energy}$$

$$\xleftarrow{\text{\hspace{1em}endothermic direction\hspace{1em}}}$$

At equilibrium, the same amount of energy produced in the forward reaction is used in the reverse reaction. An energy balance exists at equilibrium, and no heat is lost or gained by the system.

As shown in Figure 14-6(*c*), a change in temperature of the reaction system causes a shift in equilibrium. An increase in temperature increases the energy of the system. The equilibrium shifts in the direction that diminishes the increase in energy and tends to maintain the energy balance. In other words, an **increase in temperature** causes the equilibrium to shift in the direction of the endothermic reaction. When the ammonia equilibrium system is heated, the equilibrium shifts in the nitrogen–hydrogen direction.

A **decrease in temperature** of an equilibrium system causes the equilibrium to shift in the direction of the exothermic reaction, which tends to replenish the decrease in energy. When the ammonia equilibrium system is cooled, the equilibrium shifts in the direction that increases the yield of ammonia. It is interesting that this process is not used in the industrial manufacture of ammonia because it does not produce ammonia fast enough. Recall from Section 14-2 that reaction rates are increased by increasing the temperature. The nitrogen–hydrogen reaction is very slow at room temperature, even when a catalyst is present. Cooling the reaction system to shift the equilibrium makes the reaction occur even more slowly. Consequently, to make the reaction occur fast enough, it is run at elevated temperatures. This example shows where increasing the rate of a reaction by heating is more important than shifting the equilibrium by cooling. In other words, it does not do any good to try to shift the equilibrium if the reaction is so slow that equilibrium cannot be readily attained. In many

ACTIVITY 14-6

Carbonated beverages contain dissolved carbon dioxide gas. The equilibrium for the carbon dioxide is represented as

$$CO_2(g) \rightleftharpoons CO_2(aq)$$

Use this equation and Le Châtelier's principle to explain why carbon dioxide gas bubbles out of solution when a can or bottle of a beverage is opened.

ACTIVITY 14-7

Carbonated beverages contain dissolved carbon dioxide gas. The equilibrium for the carbon dioxide is represented as

$$CO_2(g) \rightleftharpoons CO_2(aq)$$

When a beverage is heated, carbon dioxide gas bubbles out. Which side of the above equilibrium does energy belong? Explain your choice.

Write an equilibrium equation and an equilibrium constant expression for the dissolving of nitrogen gas in water and use it, along with Le Châtelier's principle, to explain why nitrogen is more soluble at higher pressures and less soluble at lower pressures.

chemical reactions, however, lower temperatures or elevated temperatures are used to shift the equilibrium in the direction of desirable products.

Catalyst

Recall that a catalyst is a species that increases the rate of a reaction and is not chemically changed. A catalyst does not cause equilibrium to shift because it has the same effect on the rates of the forward and reverse reactions. Putting a catalyst in a system already in equilibrium has no effect. A catalyst does, however, cause a system to reach equilibrium more quickly since it speeds both the forward and reverse reactions. In fact, in many industrial reactions a catalyst is used to allow the reaction to reach equilibrium quickly. Biological catalysts or enzymes allow many reactions in the body to reach equilibrium quickly.

The Bends

The bends or decompression illness can affect deep-sea divers and scuba divers. Air contains oxygen and nitrogen. The oxygen has a physiological function, but nitrogen is physiologically inert. Air that you breathe mixes with blood in the lungs and the blood becomes saturated with nitrogen gas. The equilibrium reaction is

$$N_2(g) \rightleftharpoons N_2(aq)$$

Higher air pressures cause increased concentrations of dissolved nitrogen. At higher pressures, nitrogen is more soluble.

As divers descend in water, they are subjected to increased pressures. Diving to 34 ft is equivalent to about 1 atmosphere of pressure above the prevailing atmospheric pressure. The increase in pressure causes more nitrogen than normal to dissolve in the blood. Incidentally, if too much nitrogen dissolves in the blood, a diver may experience nitrogen narcosis, a dangerous effect very similar to excess alcohol in the blood. When a diver rises too quickly, the decrease in pressure causes the release of nitrogen gas into the blood. The released gas forms bubbles in the tissue and blood vessels. The result is extreme pain in the joints and muscles and loss of bladder and rectal muscle control. If not treated, the bends can result in coma and death. Hyperbaric pressure chambers are used to treat bends victims. The pressure inside the chamber is increased causing the nitrogen bubbles to dissolve in the blood. The pressure is then slowly decreased to allow time for the nitrogen to be exhaled from the lungs. Proper diving procedures to avoid the bends include slowly ascending to provide a gradual decrease in pressure. It is interesting that astronauts who space walk in soft space suits also risk the bends. (Why?) To prevent the bends resulting from a space walk, an astronaut prebreathes a mixture of oxygen and noble gases to rid the blood of nitrogen. Noble gases are not as soluble as nitrogen.

14-13 HYDRONIUM IONS IN WEAK ACID SOLUTIONS

As discussed in Chapter 13, solutions of weak acids have typical acidic properties that reflect the presence of hydronium ions. The hydronium ion concentration in a solution of a weak acid is found from the acid ionization constant expression for the acid. A monoprotic weak acid in solution exists in equilibrium with water.

$$HA + H_2O \rightleftharpoons H_3O^+ + A^-$$

In solutions of typical weak acids, we can assume that the concentration of H_3O^+ and the concentration of A^- are equal since they are formed in equal amounts by the reaction.

$$[H_3O^+] = [A^-]$$

Thus, we can substitute $[H_3O^+]$ for $[A^-]$ in the acid ionization constant expression.

$$K_a = \frac{[H_3O^+][A^-]}{[HA]} = \frac{[H_3O^+][H_3O^+]}{[HA]} = \frac{[H_3O^+]^2}{[HA]} = K_a$$

Solving this expression for $[H_3O^+]^2$ gives

$$[H_3O^+]^2 = K_a[HA]$$

and taking the square root of both sides gives a general equation for finding the **hydronium ion concentration in a solution of any weak acid.**

$$\mathbf{[H_3O^+] = \sqrt{K_a[HA]}}$$

To find the hydronium ion concentration, we need to know the concentration of the weak acid and its K_a value. Incidentally, this general equation does not apply to very low concentrations of weak acids that have relatively large K_a values.

EXAMPLE 14-6

Determine the concentration of hydronium ions in a 0.83 *M* solution of acetic acid. The K_a for acetic acid is 1.8×10^{-5}.

Given: 0.83 M $HC_2H_3O_2$ and $K_a = 1.8 \times 10^{-5}$
Goal: find the $[H_3O^+]$

The square root of the product of the K_a and the concentration of the weak acid gives the hydronium ion concentration.

$$[H_3O^+] = \sqrt{(1.8 \times 10^{-5})(0.83)} = \sqrt{1.5 \times 10^{-5}} = 3.9 \times 10^{-3}$$

A calculator is used to find square roots. Using a typical calculator, just key in the number and push the square root key.

ACTIVITY 14-9

Estimate and calculate the pH of the acetic acid solution discussed in Example 14-6.

The pH of a solution is the negative logarithm of the hydronium ion concentration. If we want to find the pH of a weak acid solution of known concentration, we can find the hydronium ion concentration and then use it to calculate the pH.

EXAMPLE 14-7

What is the estimated and precise pH of a 1.0 M formic acid, HCO_2H, solution if the K_a of formic acid is 1.9×10^{-4}?

First, find the $[H_3O^+]$ from the square root of the product of the K_a and the concentration of the acid.

$$[H_3O^+] = \sqrt{(1.9 \times 10^{-4})(1.0)} = \sqrt{1.9 \times 10^{-4}} = 1.38 \times 10^{-2}$$

Next, use the hydronium ion concentration to find the pH.

$$pH = -\log[H_3O^+] = -\log(1.38 \times 10^{-2})$$

We estimate that the pH is 1 point something. The pH found with a calculator is 1.86.

Buffers

14-14 BUFFER SOLUTIONS

Pure water has a pH of 7.00. When a small amount of acid is added to water, the pH decreases; the solution becomes acidic. When a small amount of base is added to pure water, the pH increases; the solution becomes basic. Blood serum, the fluid base of blood, behaves differently than water. Blood serum has a pH of 7.4. When a small amount of a strong acid or strong base is added to serum, the pH changes only a slight amount, if at all. The difference between serum and pure water is that serum is a buffer solution. A **buffer solution** is a solution that resists a change in pH on addition of small amounts of an acid or base. Serum is a naturally occurring buffer system, and it contains more than one buffering agent to protect the blood from unexpected changes in acidity. Changes in acidity cause enzymes to malfunction and may lead to coma or death.

It is possible to prepare a simple buffer solution in the laboratory by mixing the appropriate chemicals. A **simple buffer solution** contains a mixture of a weak acid and the conjugate base of that acid at approximately equal concentrations. Some complex buffer solutions are mixtures of more than one acid and base. Blood contains several buffer systems. One buffer system in blood involves equilibrium between the dihydrogen phosphate ion, $H_2PO_4^-$, and its conjugate base, the hydrogen phosphate ion, HPO_4^{2-}.

$$H_2PO_4^- + H_2O \rightleftharpoons HPO_4^{2-} + H_3O^+$$
conjugate acid   conjugate base

This equilibrium, along with other buffer systems in blood, serves to maintain a fairly constant pH. The relatively constant pH results from a shift in the equilibrium as predicted by Le Châtelier's principle. That is, when some acid or base is added to a buffer, the equilibrium shifts in the direction that tends to absorb the change in acidity. In a buffer system, a balance exists between the weak acid and its conjugate base. If the balance is upset by changing the hydronium ion concentration, the system shifts to counteract the change. The term *buffer* comes from the idea that the shift in equilibrium tends to counteract or cushion a change in the hydronium ion concentration. A sponge is a good analogy for a buffer system in the sense that the system absorbs an increase in hydronium ion concentration or generates hydronium ion to counteract a decrease.

The function of the dihydrogen phosphate ion–hydrogen phosphate ion buffer is illustrated in Figure 14-7. If some OH^- is added to such a solution, it reacts with some H_3O^+ and is neutralized. The temporary decrease in H_3O^+ concentration causes a shift in the equilibrium toward the right. Thus, the $H_2PO_4^-$ reacts with water to replenish most of the hydronium ions used in the reaction with hydroxide ions.

If some H_3O^+ ions are added to the solution, a right to left reaction occurs between the H_3O^+ ions and the HPO_4^{2-} ions to produce additional

FIGURE 14-7   A pictorial representation of a buffer system.

(*a*) A buffer system has concentrations of weak acid and its conjugate base that are high when compared to the concentration of H_3O^+

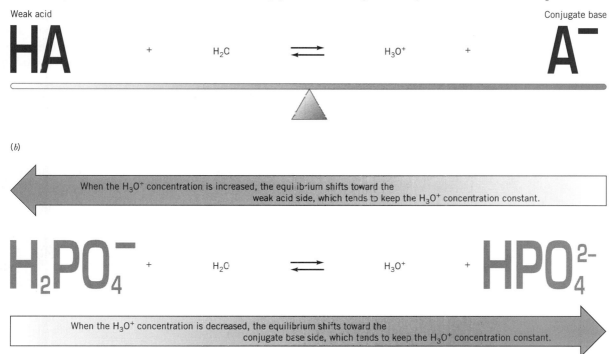

H_2O and $H_2PO_4^-$. This shift removes most of the added H_3O^+ ions from the system. Thus, because of the equilibrium between $H_2PO_4^-$ and HPO_4^{2-}, the H_3O^+ concentration and pH remain at nearly a constant level when small amounts of acid or base are added to the system. A buffer system has a limited capacity to resist changes in H_3O^+ concentration, however, and if too much acid or base is added, the capacity of the buffer will be exceeded. Nevertheless, the dihydrogen phosphate–hydrogen phosphate ion buffer system in blood helps to maintain the hydronium ion concentration of blood within critical limits.

Buffers are used in chemistry when we want a solution to have a specific pH and to resist a change in pH on the addition of small amounts of acid or base. Buffer solutions that maintain a specific pH are prepared by adding the appropriate acid and its conjugate base in specific proportions. Buffer preparation is often done by mixing an ionic compound containing the conjugate base with a solution of the weak acid. For example, a specific buffer solution is prepared by adding sodium acetate to an acetic acid solution. The sodium acetate dissolves to form acetate ions, the conjugate base of acetic acid. Various weak acids and their conjugate bases are used to make buffer systems of varying pH.

14-15 FINDING THE pH OF BUFFER SOLUTIONS

The acid ionization constant expression for an acid is used to find the hydronium ion concentration of a buffer solution that contains a known concentration of a weak acid and its conjugate base. A K_a value for the acid is also needed in the calculation. Tabulations of K_a values for various acids are available in chemical reference books. As an example, the determination of the hydronium ion concentration in a buffer solution that is 2.0 M in acetic acid and 1.5 M in acetate ion is shown here. The equilibrium equation is

$$HC_2H_3O_2(aq) + H_2O \rightleftharpoons H_3O^+(aq) + C_2H_3O_2^-(aq)$$

and the acid ionization constant expression is

$$K_a = \frac{[H_3O^+][C_2H_3O_2^-]}{[HC_2H_3O_2]} = 1.8 \times 10^{-5}$$

We are given that $[C_2H_3O_2^-] = 1.5$ and $[HC_2H_3O_2] = 2.0$ and we want $[H_3O^+]$. The expression is first solved for $[H_3O^+]$, and the given values are used to find the value of $[H_3O^+]$.

$$[H_3O^+] = 1.8 \times 10^{-5} \frac{[HC_2H_3O_2]}{[C_2H_3O_2^-]}$$

$$[H_3O^+] = 1.8 \times 10^{-5} \frac{(2.0)}{(1.5)} = 2.4 \times 10^{-5}$$

Using this $[H_3O^+]$ value, we find that the buffer has an estimated pH of 4 point something and precise pH of 4.62. You can confirm this answer by calculating the pH using the hydronium ion concentration.

In general, the **hydronium ion concentration for any buffer solution** of a weak acid and its conjugate base

$$HA + H_2O \rightleftharpoons H_3O^+ + A^-$$

is given by

$$[H_3O^+] = K_a \frac{[HA]}{[A^-]}$$

$$[H_3O^+] = K_a \frac{[HA]}{[A^-]}$$

where K_a is the known acid constant, $[HA]$ is the concentration of the weak acid, and $[A^-]$ is the concentration of its conjugate base. If wanted, the pH of a buffer can be calculated from the value of the hydronium ion concentration.

EXAMPLE 14-8

What is the pH of a buffer solution consisting of a 1.0 *M* solution of ammonium ions in equilibrium with 2.0 *M* ammonia? The K_a of NH_4^+ is 5.6×10^{-10}.

First, we consider the equilibrium reaction of the acid by showing the reaction of the acid and water to give the hydronium ion and the conjugate base of the acid. In this case, the ammonium ion is the acid and ammonia is its conjugate base.

$$NH_4^+ + H_2O \rightleftharpoons NH_3 + H_3O^+$$
$$\text{weak acid} \qquad \text{conjugate base}$$

The hydronium ion concentration of the buffer is found using the following data in the general expression for a buffer:

$[NH_4^+]$	$[NH_3]$	K_a
1.0 *M*	2.0 *M*	5.6×10^{-10}

$$[H_3O^+] = K_a \frac{[NH_4^+]}{[NH_3]} = 5.6 \times 10^{-10} \frac{(1.0)}{(2.0)} = 2.8 \times 10^{-10}$$

Finally, the pH is found using $pH = -\log[H_3O^+]$. The pH is estimated as 9 point something. The pH as found using a calculator is 9.55.

A buffer solution contains 0.25 *M* $H_2PO_4^-$ and 0.16 *M* HPO_4^{2-}. The K_a of $H_2PO_4^-$ is 6.23×10^{-8}. What is the pH of the buffer solution?

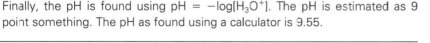

Chemical Equilibrium

Chemical equilibrium: The state of a reversible chemical reaction when the rate of the forward reaction equals the rate of the reverse reaction.

At equilibrium, the ratio of the concentrations of the products to the reactants is equal to a constant value, characterizing a reaction system involving chemical equilibrium. This ratio is called the equilibrium constant.

An equilibrium system represented by the general equation

$$aA + bB \rightleftharpoons cC + dD$$

is described by an equilibrium constant expression of the general form

$$K_{eq} = \frac{[C]^c [D]^d}{[A]^a [B]^b}$$

where K_{eq} is the equilibrium constant and the square brackets refer to the concentrations of the products and reactants in moles per liter. Each concentration is raised to a power corresponding to its coefficient in the equilibrium equation. Steps used to derive the equilibrium constant expression for an equilibrium system are as follows.

1. Write an equilibrium equation and note whether the species involved are gases, solids, liquids, or in aqueous solution.
2. Enclose the formulas of all products that are gases or aqueous species in square brackets for the numerator of the expression. If any of these species has a coefficient greater than 1, show its concentration raised to a power given by the coefficient.
3. Enclose the formulas of all reactants that are gases or aqueous species in square brackets for the denominator of the expression using the coefficients as powers.
4. The ratio of the product concentrations to the reactant concentrations, each raised to a power corresponding to its coefficient in the balanced equation, is equal to the constant K_{eq}.

The relative sizes of double arrows in equilibrium equations relate to the relative numerical values of K_{eq}.

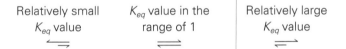

| Relatively small K_{eq} value | K_{eq} value in the range of 1 | Relatively large K_{eq} value |

Le Châtelier's principle: When a factor affecting an equilibrium system is changed, the equilibrium shifts in a direction that tends to counteract the change. Factors that affect an equilibrium system are as follows.

1. A change in the concentration of some gaseous or dissolved species that is a reactant or product. Equilibrium shifts away from an increase in concentration or toward a decrease in concentration.
2. A change in temperature of the equilibrium mixture. When a mixture is heated, the equilibrium shifts in the endothermic direction, and when a mixture is cooled, the equilibrium shifts in the exothermic direction.
3. A change in the pressure of an equilibrium mixture involving one or more gases. When the pressure is increased, the equilibrium shifts toward the side having fewer moles of gas, and when the pressure is decreased, the equilibrium shifts toward the side having more moles of gas.

The equilibrium reaction of any weak acid, HA, in water is represented by the general equation

$$HA + H_2O \rightleftharpoons H_3O^+ + A^-$$

where HA is the acid and A^- is its conjugate base. For a weak acid equilibrium system, the ratio of the concentrations of the products to the concentrations of the reactants is constant.

$$K_a = \frac{[H_3O^+][A^-]}{[HA]}$$

where K_a is called the acid ionization constant or the acid constant for a weak acid.

The general equation for finding the hydronium ion concentration in a solution of any weak acid is

$$[H_3O^+] = \sqrt{K_a[HA]}$$

To find the hydronium ion concentration, multiply the concentration of the weak acid by its K_a value and take the square root of the product.

Relative strengths of weak acids are expressed by their K_a or pK_a values. The definition of pK_a is

$$pK_a = -\log K_a$$

Buffer solution: A solution that resists a change in pH on addition of an acid or base. A simple buffer solution contains a mixture of a weak acid and the conjugate base of that acid at approximately equal concentrations. The hydronium ion concentration for a buffer solution of a weak acid and its conjugate base is found using

$$[H_3O^+] = K_a \frac{[HA]}{[A^-]}$$

where K_a is the known acid constant, [HA] is the concentration of the weak acid, and $[A^-]$ is the concentration of its conjugate base.

QUESTIONS

1. Match the following terms with the appropriate lettered definitions.

rate of reaction

a. when a factor affecting an equilibrium system is changed, the equilibrium shifts in a direction that tends to counteract the change

catalyst

b. biological catalyst

enzyme

c. speed at which a chemical reaction occurs

chemical equilibrium

d. a chemical that increases the rate of a reaction but is not changed chemically

equilibrium constant expression

e. the ratio of the concentrations of the products to the concentration of the reactants in an equilibrium system

K_a

f. a solution that resists a change in pH upon addition of small amounts of acid or base

Le Châtelier's principle

g. $[H_3O^+][A^-]/[HA]$

buffer solution

h. a reversible chemical reaction in which the rate of the forward reaction equals the rate of the reverse reaction

pK_a

i. $-\log K_a$

Section 14-1

2. What is meant by the rate of a chemical reaction?
3. What is a chemical explosion?
4. How does an airbag work?

Section 14-2

5. Tell how each of the following can alter the rate of a chemical reaction.

(a) temperature

(b) state of subdivision

(c) concentration

6. Suppose you want to burn some coal in air quickly. Explain how each of the following factors can be changed to increase the rate of the combustion of coal.

(a) temperature

(b) state of subdivision

(c) concentration

7. Explain how the burning of wood is encouraged by each of the following. Indicate the factor that affects the rate in each case.

(a) A match is used to start the fire.

(b) Kindling is used to get the fire started.

(c) Air is blown on a freshly lit fire.

(d) The firewood is loosely stacked with numerous air spaces.

8. What is a catalyst and how does a catalyst alter the rate of a chemical reaction?

9. *In each of the following cases, explain what happens in terms of the factor or factors that affect reaction rates.

(a) Flour dust in a flour mill can cause an explosion.

(b) Nitrogen and oxygen in the air combine in the piston chamber during high-temperature operation of the internal combustion engine:

$$N_2 + O_2 \longrightarrow 2NO$$

(c) Increasing pressure in the synthesis of ammonia speeds up the reaction

$$N_2 + 3H_2 \longrightarrow 2NH_3$$

(d) Microorganisms can make ammonia from N_2 at low temperature and pressure.

(e) Fish become very active and sometimes die in heated water.

(f) Smoking close to a hospital oxygen tent can cause a fire.

(g) Wood shavings are easier to ignite than larger pieces of wood.

(h) A lighted candle will extinguish when burned in a closed container.

10. In each of the following cases, explain what happens in terms of the factor or factors that affect reaction rates.

(a) Swimmers become very lethargic and sometimes die in extremely cold water.

(b) Decreasing the air supply in a Bunsen burner gives a "cooler" flame.

(c) A person can become weak and lose energy at high altitudes.

(d) A safety match ignites when it is rubbed on the striking surface.

(e) A piece of aluminum metal, when coated with mercury, will quickly react with oxygen in the air to form aluminum oxide.

(f) Coal dust in a coal mine can cause an explosion.

(g) A wad of cotton will burn more rapidly than a piece of wood even though they both are composed of cellulose.

(h) Methyl alcohol can be made from carbon monoxide and hydrogen by carrying out the following reaction at high pressures in the presence of a catalyst.

$$CO + 2H_2 \longrightarrow CH_3OH$$

11. Describe the chemical reaction used in an oxyacetylene torch.

Sections 14-3 and 14-4

12. What is a catalyst?

13. Describe or define the following.

(a) heterogeneous catalyst

(b) homogeneous catalyst

(c) enzyme

14. Explain the function of a catalytic converter on an automobile.

15. How does a catalyst work?

16. What is an inhibitor?

Sections 14-5 to 14-9

17. What are reversible chemical reactions and chemical equilibrium.

18. How are double arrows used to express the extent of an equilibrium reaction?

19. What is an equilibrium constant, how is it determined, and how does the value of a constant indicate the direction and extent of an equilibrium reaction?

20. *Experimental observation of the following equilibrium system gives the equilibrium concentrations shown. Give the acid ionization constant expression and calculate the value of K_a.

$$HC_2H_3O_2 + H_2O \rightleftharpoons H_3O^+ + C_2H_3O_2^-$$

equilibrium $[HC_2H_3O_2]$ $[H_3O^+]$ $[C_2H_3O_2^-]$

concentrations

in moles per liter 2.12 6.2×10^{-3} 6.2×10^{-3}

21. Experimental observation of the following equilibrium system gives the equilibrium concentrations shown. Give the equilibrium constant expression and calculate the value of K_{eq}.

$$2HI(g) \rightleftharpoons H_2(g) + I_2(g)$$

equilibrium $[HI]$ $[H_2]$ $[I_2]$

concentrations

in moles per liter 0.20 0.64 0.64

22. *Write equilibrium constant expressions for the following equilibrium reactions.

(a) $2HCl(g) \rightleftharpoons H_2(g) + Cl_2(g)$

(b) $2SO_2(g) + O_2(g) \rightleftharpoons 2SO_3(g)$

(c) $3O_2(g) \rightleftharpoons 2O_3(g)$

(d) $HCN(aq) + H_2O \rightleftharpoons H_3O^+(aq) + CN^-(aq)$

(e) $NH_3(aq) + H_2O \rightleftharpoons NH_4^+(aq) + OH^-(aq)$

23. Write equilibrium constant expressions for the following equilibrium reactions.

(a) $C(s) + CO_2(g) \rightleftharpoons 2CO(g)$

(b) $HNO_2(aq) + H_2O \rightleftharpoons H_3O^+(aq) + NO_2^-(aq)$

(c) $CH_4(g) + H_2O(g) \rightleftharpoons CO(g) + 3H_2(g)$

(d) $CO_2(aq) + 2H_2O \rightleftharpoons$
$$H_3O^+(aq) + HCO_3^-(aq)$$

(e) $2NO(g) + Br_2(g) \rightleftharpoons 2NOBr(g)$

24. Explain the meaning of a double arrow in an equation and how arrows are used to convey information about the extent of a reaction.

25. *For each of the following equilibrium reactions, give an appropriate equilibrium equation. Use the value of the equilibrium constant as a guide for writing a larger arrow and a smaller arrow in the appropriate directions.

(a) Ethylene, C_2H_4, gas and steam in equilibrium with gaseous ethyl alcohol. Note that water is a gas in the reaction and is included in the equilibrium constant expression.

$$K_{eq} = \frac{[C_2H_5OH]}{[C_2H_4][H_2O]} = 8.2 \times 10^3$$

(b) Carbon monoxide gas and steam in equilibrium with hydrogen gas and carbon dioxide gas.

$$K_{eq} = \frac{[H_2][CO_2]}{[CO][H_2O]} = 3.4 \times 10^3$$

(c) Dissolved carbon dioxide and water in equilibrium with bicarbonate ion and hydronium ion in your kidneys.

$$K_a = \frac{[H_3O^+][HCO_3^-]}{[CO_2]} = 4.3 \times 10^{-7}$$

(d) Hydrogen fluoride in solution and water are in equilibrium with hydronium ion and fluoride ion.

$$K_a = \frac{[H_3O^+][F^-]}{[HF]} = 7.0 \times 10^{-4}$$

26. For each of the following equilibrium reactions, give an appropriate equilibrium equation. Use the value of the equilibrium constant as a guide for writing a larger arrow and a smaller arrow in the appropriate directions.

(a) Phosphorus pentachloride gas in equilibrium with phosphorus trichloride gas and chlorine gas.

$$K_{eq} = \frac{[PCl_3[Cl_2]}{[PCl_5]} = 1.8$$

(b) Hydrogen gas and chlorine gas in equilibrium with hydrogen chloride gas.

$$K_{eq} = \frac{[\text{HCl}]^2}{[\text{H}_2][\text{Cl}_2]} = 2.5 \times 10^4$$

(c) Solid NH_4HS in equilibrium with ammonia gas and hydrogen sulfide gas.

$$K_{eq} = [\text{NH}_3][\text{H}_2\text{S}] = 1.8 \times 10^{-4}$$

(d) The weak base ammonia and water in equilibrium with ammonium ion and hydroxide ion.

$$K_{eq} = \frac{[\text{NH}_4^+][\text{OH}^-]}{[\text{NH}_3]} = 1.8 \times 10^{-5}$$

Sections 14-10 and 14-11

27. Write the equation for the equilibrium reaction of propionic acid, $C_2H_5CO_2H$, in water and give the equilibrium constant expression for the reaction.

28. An acetic acid solution contains 2.0 M $HC_2H_3O_2$, 3.6×10^{-5} M H_3O^+, and 3.6×10^{-5} M $C_2H_3O_2^-$. Calculate the value of the acetic acid in water equilibrium constant using these data. Based on this value, why is acetic acid classified as a weak acid?

Section 14-7

29. *Write acid constant expressions for the following weak acids in water.
(a) $H_2PO_4^-(aq) + H_2O \rightleftharpoons H_3O^+(aq) + HPO_4^{2-}(aq)$
(b) $HCO_3^-(aq) + H_2O \rightleftharpoons H_3O^+(aq) + CO_3^{2-}(aq)$
(c) $NH_4^+(aq) + H_2O \rightleftharpoons H_3O^+(aq) + NH_3(aq)$

30. *Calculate the value of the acid constant, K_a, for each of the following solutions having the equilibrium concentrations shown.
(a) $HSO_4^-(aq) + H_2O \rightleftharpoons H_3O^+(aq) + SO_4^{2-}(aq)$
    0.25 M 5.5×10^{-2} M 5.5×10^{-2} M
(b) $HCO_3^-(aq) + H_2O \rightleftharpoons H_3O^+(aq) + CO_3^{2-}(aq)$
    1.15 M 8.1×10^{-6} M 8.1×10^{-6} M
(c) $NH_4^+(aq) + H_2O \rightleftharpoons H_3O^+(aq) + NH_3(aq)$
    3.75 M 4.6×10^{-5} M 4.6×10^{-5} M

31. Using the values of the acid constants calculated in Question 30, rank the three weak acids from strongest to weakest. How does this ranking correspond to the positions of these acids in Table 13-1?

32. Write equilibrium constant expressions for the following weak acids in water.
(a) $HF(aq) + H_2O \rightleftharpoons H_3O^+(aq) + F^-(aq)$

(b) $HC_2O_4^-(aq) + H_2O \rightleftharpoons H_3O^+(aq) + C_2O_4^{2-}(aq)$
(c) $H_3PO_4(aq) + H_2O \rightleftharpoons H_3O^+(aq) + H_2PO_4^-(aq)$

33. Calculate the value of the acid constant, K_a, for each of the following solutions having the equilibrium concentrations shown.
(a) $HF(aq) + H_2O \rightleftharpoons H_3O^+(aq) + F^-(aq)$
    0.50 M 1.32×10^{-2} M 1.32×10^{-2} M
(b) $HC_2O_4^-(aq) + H_2O \rightleftharpoons H_3O^+(aq) + C_2O_4^{2-}(aq)$
    2.00 M 1.01×10^{-2} M 1.01×10^{-2} M
(c) $H_3PO_4(aq) + H_2O \rightleftharpoons H_3O^+(aq) + H_2PO_4^-(aq)$
    0.876 M 2.5×10^{-2} M 2.5×10^{-2} M

34. Using the values of the acid constants calculated in Question 33, rank the three acids from the strongest to the weakest. How does this ranking correspond to the positions of these acids in Table 13-1?

Section 14-12

35. Give a statement of Le Châtelier's principle.

36. *Using Le Châtelier's principle, predict any shift in equilibrium in the reaction

$$\text{energy} + \text{H}_2(g) + \text{I}_2(g) \rightleftharpoons 2\text{HI}(g)$$

when the following occur.
    (a) The equilibrium system is cooled.
    (b) Extra H_2 is added to the system.
    (c) A catalyst is added to the system.
    (d) The pressure of the equilibrium system is decreased.
    (e) The concentration of HI is increased.

37. Using Le Châtelier's principle, predict any shift in equilibrium in the reaction

$$\text{CO}(g) + \text{H}_2\text{O}(g) \rightleftharpoons \text{CO}_2(g) + \text{H}_2(g) + \text{energy}$$

when the following occur.
    (a) The temperature of the equilibrium system is increased.
    (b) The concentration of $H_2O(g)$ is increased.
    (c) A catalyst is added to the system.
    (d) The equilibrium system is heated.
    (e) The pressure of the equilibrium system is decreased.

38. *Using Le Châtelier's principle, predict any shift in equilibrium in the reaction

$$\text{N}_2(g) + 3\text{H}_2(g) \rightleftharpoons 2\text{NH}_3(g) + \text{energy}$$

when the following occur.

(a) The temperature of the equilibrium system is increased.

(b) The concentration of H_2 is increased.

(c) The concentration of N_2 is decreased.

(d) The temperature of the equilibrium system is decreased.

(e) The pressure on the equilibrium system is increased.

39. Using Le Châtelier's principle, predict any shift in equilibrium in the reaction

$$energy + C(s) + H_2O(g) \rightleftharpoons CO(g) + H_2(g)$$

when the following occur.

(a) The temperature of the equilibrium system is increased.

(b) The concentration of H_2 is increased.

(c) A catalyst is added to the system.

(d) The concentration of CO is decreased.

(e) The pressure of the equilibrium system is increased.

40. *Using Le Châtelier's principle, predict any shift in equilibrium in each of the following systems.

(a) $energy + H_2O(s) \rightleftharpoons H_2O(\ell)$, when the system is cooled.

(b) $2CO(g) + O_2(g) \rightleftharpoons 2\,CO_2(g)$, when a catalyst is added.

(c) $Mg(OH)_2(s) \rightleftharpoons Mg^{2+}(aq) + 2OH^-(aq)$, when extra OH^- from sodium hydroxide is added.

(d) $CO_2(aq) + 2H_2O \rightleftharpoons$
$H_3O^+(aq) + HCO_3^-(aq)$,
when H_3O^+ from hydrochloric acid is added.

(e) $CaCO_3(s) \rightleftharpoons CaO(s) + O_2(g)$, when the pressure is decreased. (Pressure does not affect solids.)

(f) $energy + 2H_2O(g) \rightleftharpoons 2H_2(g) + O_2(g)$, when the temperature is decreased and the pressure is increased.

(g) $Ag^+(aq) + Cl^-(aq) \rightleftharpoons AgCl(s)$, when Cl^- from sodium chloride is added.

41. Using Le Châtelier's principle, predict any shift in equilibrium in each of the following systems.

(a) $energy + 2N_2O(aq) \rightleftharpoons 2N_2(g) + O_2(g)$, when the system is cooled.

(b) $2SO_2(g) + O_2(g) \rightleftharpoons 2SO_3(g)$, when a catalyst is added.

(c) $CaCO_3(s) \rightleftharpoons Ca^{2+}(aq) + CO_3^{2-}(aq)$, when extra Ca^{2+} from calcium chloride is added.

(d) $NH_3(aq) + H_2O \rightleftharpoons NH_4^+(aq) + OH^-(aq)$, when OH^- from sodium hydroxide is added.

(e) $H_2(g) + CO_2(g) \rightleftharpoons H_2O(g) + CO(g)$, when the pressure is decreased.

(f) $2NO(g) + Br_2(g) \rightleftharpoons 2NOBr(g) + energy$, when the temperature is decreased and the pressure is increased.

(g) $Pb^{2+}(aq) + 2Cl^-(aq) \rightleftharpoons PbCl_2(s)$, when Cl^- from sodium chloride is added.

42. Carbon dioxide, CO_2, and water are in equilibrium with hydrogen carbonate ion, HCO_3^-, and hydronium ion, H_3O^-, in your blood. Write an equilibrium equation. Using Le Châtelier's principle, explain what happens to the concentration of hydronium ion in your blood when you become anxious and start to hyperventilate, or breathe rapidly, which removes carbon dioxide from your body.

43. In the internal combustion engine, a gasoline–air mixture is compressed in the piston chamber and burned producing a high temperature. Nitrogen oxide, NO, an air pollutant, is formed in the internal combustion engine by the reaction

$$energy + N_2(g) + O_2(g) \rightleftharpoons 2NO(g)$$

According to Le Châtelier's principle, what condition in the internal combustion engine causes a shift in this equilibrium toward the formation of NO?

44. Hemoglobin (represented symbolically as Hb) is a complex protein found in red blood cells. Hemoglobin is involved in oxygen transport in the body according to the symbolic reaction

$$Hb + O_2 \rightleftharpoons HbO_2$$

Using Le Châtelier's principle, explain how hemoglobin transports oxygen from the lungs, where there is a high concentration of oxygen, to the cells fed by blood capillaries, where there is a low concentration of oxygen.

45. Explain the bends or decompression illness using Le Châtelier's principle.

Section 14-13

46. *Find the hydronium ion concentration of a 0.10 M solution of $NaHSO_4$. The K_a for the weak acid HSO_4^- is 1.2×10^{-2}. (See Question 30.) What is the pH of this solution?

47. Find the hydronium ion concentration of a 2.25 M solution of HF. The K_a for this weak acid is 3.5×10^{-4}. What is the pH of this solution?

48. *Formic acid is a weak acid having a K_a value of 1.9×10^{-4}. What is the hydronium ion concentration in a 0.315 M solution of formic acid? What is the pH of this solution?

49. Vinegar, an acetic acid solution, typically has a concentration of 0.83 M $HC_2H_3O_2$. If the K_a of acetic acid is 1.8×10^{-5}, what is the hydronium ion concentration in vinegar? What is the pH of vinegar?

Sections 14-14 and 14-15

50. Describe a simple buffer solution and give an example of a buffer system.

51. Use the following general equation for an acid in equilibrium with its conjugate base to explain how a buffer solution functions upon the addition of small amounts of acid or base.

$$HA + H_2O \rightleftharpoons H_3O^+ + A^-$$

52. *Calculate the hydronium ion concentration of a buffer solution containing 1.25 M acetic acid, $HC_2H_3O_2$, and 0.75 M sodium acetate, $NaC_2H_3O_2$. ($K_a = 1.8 \times 10^{-5}$ for $HC_2H_3O_2$)

53. *Calculate the hydronium ion concentration and pH of a buffer solution containing 2.5 M formic acid, HCO_2H, and 1.8 M potassium formate, $KHCO_2$. ($K_a = 1.9 \times 10^{-4}$ for HCO_2H)

54. *Calculate the pH of a buffer solution containing 1.5 M sodium dihydrogen phosphate, NaH_2PO_4, and 3.0 M sodium hydrogen phosphate, Na_2HPO_4. The equilibrium is

$$H_2PO_4^-(aq) + H_2O \rightleftharpoons H_3O^+(aq) + HPO_4^{2-}(aq)$$

and $K_a = 6.23 \times 10^{-8}$ for $H_2PO_4^-$, which is the acid species in the buffer.

55. *Calculate the pH of a buffer solution containing 1.0 M HCO_2H and 0.92 M HCO_2^-. ($K_a = 1.9 \times 10^{-4}$ for HCO_2H)

56. Calculate the hydronium ion concentration of a buffer solution containing 0.75 M acetic acid, $HC_2H_3O_2$, and 0.50 M sodium acetate, $NaC_2H_3O_2$. ($K_a = 1.8 \times 10^{-5}$ for $HC_2H_3O_2$)

57. Calculate the hydronium ion concentration of a buffer solution containing 2.3 M HCO_2H and 1.4 M HCO_2^-. ($K_a = 1.9 \times 10^{-4}$ for HCO_2H)

58. Calculate the pH of a buffer solution containing 1.6 M sodium dihydrogen phosphate, NaH_2PO_4, and 1.20 M sodium hydrogen phosphate, Na_2HPO_4. The equilibrium is

$$H_2PO_4^-(aq) + H_2O \rightleftharpoons H_3O^+(aq) + HPO_4^{2-}(aq)$$

and $K_a = 6.23 \times 10^{-8}$ for $H_2PO_4^-$, which is the acid in the buffer.

59. Calculate the pH of a buffer solution containing 1.80 M sodium oxalate, $Na_2C_2O_4$, and 3.10 M sodium hydrogen oxalate, $NaHC_2O_4$. The equilibrium is

$$HC_2O_4^-(aq) + H_2O \rightleftharpoons H_3O^+(aq) + C_2O_4^{2-}(aq)$$

and $K_a = 5.2 \times 10^{-5}$ for $HC_2O_4^-$, which is the acid species in the buffer.

CHAPTER

15

OXIDATION AND

REDUCTION

 ## 15-1 OXIDATION NUMBERS

In chemical reactions, elements change the ways in which they are chemically combined. Various types of reactions are known. Precipitation or ion combination reactions were discussed in Chapter 12, and acid–base or proton transfer reactions were discussed in Chapter 13. This chapter discusses another category of reactions known as oxidation–reduction or electron transfer reactions.

Oxidation Numbers

The way in which an element is chemically combined is described by an oxidation number. To illustrate this idea, consider that when in ionic compounds fluorine occurs as the fluoride ion, F^-, and when in molecular compounds fluorine occurs as part of a single covalent bond. An oxidation number of -1 is assigned to fluorine to describe its chemical behavior. An **oxidation number** or the **oxidation state** of an element shows the kind of monatomic ion it forms or the kind of covalent bonds it forms.

From another view, an oxidation number is the charge an element has in a monatomic ion or the hypothetical charge it would have in a covalent bond if the shared electrons are assigned to the more electronegative element (see Section 11-3 for a discussion of electronegativity). For example, the oxidation number of oxygen in the oxide ion, O^{2-}, is -2. In the compound water, if we assign the shared electrons to the more electronegative oxygen, the oxygen would have a hypothetical charge of $2-$, since it would have two more electrons than a neutral oxygen atom. Each hydrogen would have a hypothetical charge of $1+$, since each would have one electron less than a neutral hydrogen atom.

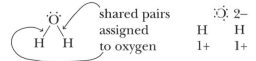

shared pairs
assigned
to oxygen

$\overset{..}{\underset{..}{O}}$ 2–

H   H
1+   1+

In water, oxygen has an oxidation number of -2 and each hydrogen has an oxidation number of $+1$.

In molecular hydrogen, H_2, the shared electrons are assigned equally to the two hydrogen atoms. Neither hydrogen is viewed as having an excess nor deficiency of electrons, so the oxidation number of each hydrogen is zero. Likewise, in molecular oxygen, O_2, each oxygen has an oxidation number of 0.

An element may have different oxidation numbers in different compounds. For instance, sulfur forms the sulfide ion, S^{2-}, and can form two covalent bonds with less electronegative elements (e.g., H_2S) and, thus, is assigned the -2 oxidation number. Sulfur also forms two, four, and sometimes six covalent bonds with elements of higher electronegativity, for which sulfur is assigned oxidation numbers of $+2$, $+4$, and $+6$, respectively. Note that oxidation numbers are numbers preceded by signs. The sign used with an oxidation number relates to the charge of a monatomic ion formed by an element or to its electronegativity compared with elements with which it covalently bonds. An oxidation number for an element consists of a number with a positive or negative sign. If an element is more electronegative than the element with which it bonds, it is given a negative oxidation number. The less electronegative element is given a positive oxidation number.

As we will see, oxidation numbers are useful in describing certain kinds of reactions. To find an oxidation number or oxidation state of an element in an ion or compound, we follow oxidation number rules. These rules are based on patterns that elements follow when they enter chemical combinations.

1. The oxidation number or oxidation state of an element in the natural form uncombined with other elements is zero, 0. For example, the oxidation state of oxygen in O_2 is 0, the oxidation state of chlorine in Cl_2 is 0, and the oxidation state of sodium in metallic sodium, Na, is 0.

2. The oxidation number of an element in a monatomic ion is given by the charge of the ion. For example, the oxidation number of sodium in Na^+ is $+1$, and the oxidation number of sulfur in S^{2-} is -2.

3. Hydrogen has an oxidation number of $+1$ when it is covalently bonded to nonmetals. When it combines with metals, in binary ionic compounds called hydrides, it has an oxidation number of -1.

4. Oxygen has an oxidation number of -2 in the vast majority of its compounds. Oxygen has a -1 oxidation number in compounds called peroxides, such as hydrogen peroxide, H_2O_2.

5. The algebraic sum of the oxidation numbers of the elements in a compound equals zero. This means that if we add all the positive oxidation numbers and all the negative oxidation numbers, the sum must be zero. For example, in water, H_2O, each of the two hydrogen atoms is in the $+1$ oxidation state, and the oxygen is in the -2 oxidation state. The sum of the oxidation numbers ($+1$, $+1$, and -2) equals zero.

$$H_2 \quad | \quad O$$
$$2(+1) + (-2) = +2 - 2 = 0$$

6. The sum of the oxidation numbers of the elements in a polyatomic ion equals the charge of the ion. For example, in the hydroxide ion, OH^-, the hydrogen is in the +1 oxidation state and the oxygen is in the −2 oxidation state. The sum of the oxidation numbers corresponds to the charge of the ion, which is 1−.

$$O \qquad H^-$$
$$(-2) + (+1) = -1$$

7. Metals normally have positive oxidation numbers in compounds. Nonmetals have negative oxidation numbers when combined with metals. Nonmetals, however, can have positive oxidation numbers when combined with more electronegative nonmetals. The more electronegative element always takes on the negative oxidation number.

Table 15-1   *Common Oxidation Numbers of Elements*

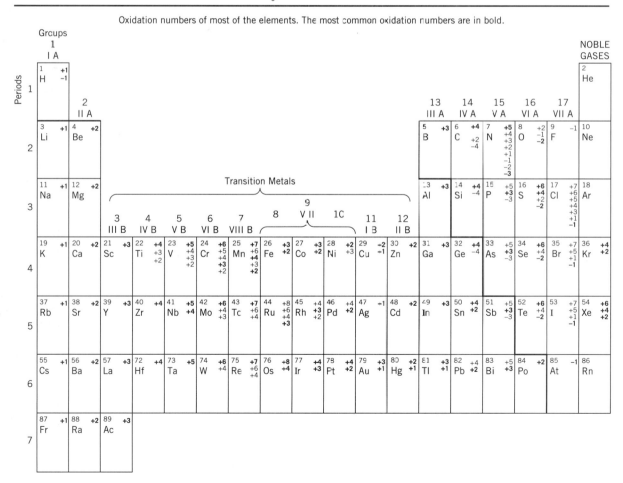

Table 15-1 is a periodic table that shows the common oxidation numbers of most elements. Notice that the most likely negative oxidation number for group 16 or VIA elements is -2 and for group 17 or VIIA elements is -1.

We can figure out the oxidation number elements in a compound or a polyatomic ion by assigning the expected oxidation numbers to the other elements then applying rule 5 or rule 6. Hydrogen is typically $+1$ and oxygen is typically -2. For example, in the compound SO_3, the expected oxidation number of oxygen is -2. Therefore, the oxidation number of sulfur must be $+6$ to match the total of the negative oxidation numbers $[3(-2) = -6]$.

EXAMPLE 15-1

What is the oxidation number of nitrogen in ammonia, NH_3?

Assigning an oxidation number of $+1$ to hydrogen and using rule 5 gives

$$N \quad H_3$$
$$x + 3(+1) = 0$$
$$x + (+3) = 0$$

Therefore, $x = -3$, so nitrogen has an oxidation number of -3.

EXAMPLE 15-2

What is the oxidation number of sulfur in the hydrogen sulfate ion, HSO_4^-?

The oxidation number of sulfur in the hydrogen sulfate ion, HSO_4^-, is found by assigning a $+1$ oxidation number to hydrogen and a -2 oxidation number to oxygen. The oxidation number of sulfur must have a value that results in the overall ionic charge of -1. So, applying rule 6, we have

$$H \quad S \quad O_4^- \; \text{Charge}$$
$$(+1) + x + 4(-2) = -1$$
$$(+1) + x + (-8) = -1$$
$$x = -1 + 8 - 1 = +6$$

The oxidation number of S is $+6$.

EXAMPLE 15-3

What is the oxidation number of chromium in the dichromate ion, $Cr_2O_7^{2-}$?

Assigning an oxidation number of -2 to oxygen and using rule 6 gives

$$Cr_2 \quad O_7^{2-} \quad \text{Charge}$$
$$2x + 7(-2) \quad = (-2)$$

ACTIVITY 15-1

What is the oxidation number of carbon in the hydrogen carbonate ion, HCO_3^-? What are oxidation numbers of sulfur in S, SO_2, H_2SO_3, and SO_3?

$$2x + (-14) \quad = (-2)$$
$$2x = -2 + 14 = +12$$
$$x = +6$$

Each chromium atom has an oxidation number of -6.

15-2 ELECTRON TRANSFER REACTIONS

Chemical bonds involve electrons and chemical reactions involve the breaking and forming of chemical bonds. A very common type of chemical reaction involves the transfer of electrons between atoms. It is important to realize, however, that not all reactions involve electron transfer. A simple example of an electron transfer is the reaction between iron and sulfur to form iron(II) sulfide (see Figure 15-1).

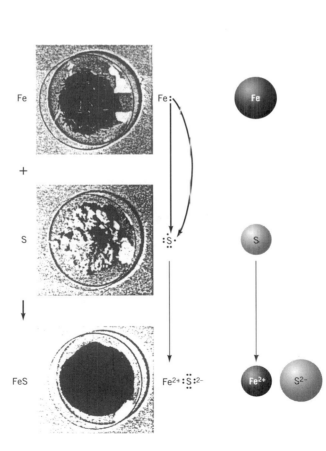

FIGURE 15-1 The reaction between iron and sulfur.

Note that electrons transfer or move from the iron to the sulfur giving the iron(II) cation and the sulfide anion. The resulting cations and anions form ionic bonds. An important fact about electron transfer reactions is that the oxidation numbers of the elements change during the reaction. In the above reaction, the oxidation number of iron changes from 0 to +2 and the oxidation number of sulfur changes from 0 to −2. Electron transfer reactions are also called **oxidation–reduction** reactions. By definition, **oxidation** is the loss of electrons or an increase in oxidation number. By definition, **reduction** is the gain of electrons or a decrease in oxidation number. Since one element loses electrons and another gains them, oxidation and reduction happen simultaneously. In the above reaction, iron undergoes oxidation and sulfur undergoes reduction. The term *oxidation* was used originally because many elements react with oxygen to form oxides. Reduction came from reactions in which pure metals were refined from metallic compounds found in ores. Oxidation and reduction now have more general meanings related to electron transfer.

Metals commonly form cations and nonmetals form anions. Reactions between metals and nonmetals are examples of simple electron transfer reactions. When sodium reacts with chlorine, electron transfer occurs.

$$Cl_2 \longrightarrow 2 \cdot \ddot{\underset{\cdot\cdot}{Cl}}\colon \qquad \begin{matrix} Na \cdot\!\!\!\curvearrowright\!\! \cdot \ddot{\underset{\cdot\cdot}{Cl}}\colon \\ Na \cdot\!\!\!\curvearrowright\!\! \cdot \ddot{\underset{\cdot\cdot}{Cl}}\colon \end{matrix} \longrightarrow \begin{matrix} Na^{+} \colon\!\ddot{\underset{\cdot\cdot}{Cl}}\colon^{-} \\ Na^{+} \colon\!\ddot{\underset{\cdot\cdot}{Cl}}\colon^{-} \end{matrix}$$

Show the reaction of Mg and F$_2$ as an electron transfer process.

Electron transfer does not always involve the formation of ions. In some reactions, an element changes its state of combination by bonding to an element with a higher electronegativity. For example, consider the reaction of carbon with oxygen.

$$C + O_2 \longrightarrow CO_2$$

The oxidation number of carbon changes from 0 to +4 and the oxidation number of oxygen changes from 0 to −2. Carbon and oxygen become covalently bonded. Carbon has been oxidized and oxygen has been reduced although the electrons have not completely transferred from one atom to another. Nevertheless, we consider such reactions to be electron transfer reactions. If a reaction involves electron transfer, the oxidation numbers of some elements change. Thus, if you have an equation for a reaction, recognizing the reaction as an electron transfer is easy. Just look for oxidation number changes. Keep in mind, however, that not all the elements in a reaction need to change oxidation number.

15-3 SOME TYPES OF ELECTRON TRANSFER REACTIONS

We can classify simple electron transfer reactions into specific categories or types, and we can confirm that they are electron transfer reactions by noting the oxidation numbers of the elements involved.

Synthesis or Combination Reactions

Synthesis or combination reactions involve the synthesis or formation of a compound from simpler chemicals.

1. Forming compounds from elements:

$$3H_2 + N_2 \longrightarrow 2NH_3$$ ammonia synthesis

$$S + O_2 \longrightarrow SO_2$$ the first step in the synthesis of sulfuric acid

2. Forming a compound from an element and a compound:

$$2SO_2(g) + O_2(g) \xrightarrow{V_2O_5} 2SO_3(g)$$ the second step in the synthesis of sulfuric acid

$$2CO(g) + O_2(g) \xrightarrow{Pt} 2CO_2(g)$$ reaction in the catalytic converter on cars

Decomposition Reactions

Decomposition reactions involve a compound decomposing into simpler chemicals.

1. Forming elements from compounds:

$$2H_2O \xrightarrow{electrolysis} 2H_2 + O_2$$ electrical decomposition of water

$$2HgO \xrightarrow{heat} 2Hg + O_2$$ thermal decompostion

2. Forming an element and compound from a compound:

$$2H_2O_2 \xrightarrow{enzyme} 2H_2O + O_2$$ enzyme catalyzed decomposition of hydrogen peroxide

Displacement Reactions

Displacement reactions involve one element exchanging for another.

$$HgS + O_2 \xrightarrow{heat} Hg + SO_2$$ prepares mercury from mercury ore

$$SiO_2 + 2C \xrightarrow{heat} Si + 2CO$$ prepares silicon from silica sand

$$6Na - Fe_2O_3 \longrightarrow 3Na_2O + 2Fe$$ used in air bags to consume sodium metal

$$C + H_2O \xrightarrow{\text{heat}} CO + H_2 \qquad \text{the first step in the synthesis of methanol}$$

Combustion

Combustion reactions in air are also called burning. A common type of combustion is the reaction of organic (carbon-containing) compounds with oxygen. The complete combustion of a carbon–hydrogen or a carbon–hydrogen–oxygen compound always produces carbon dioxide and water.

$$C_2H_5OH + 3O_2 \longrightarrow 2CO_2 + 3H_2O \qquad \text{combustion of ethanol}$$

$$2C_8H_{18} + 25O_2 \longrightarrow 16CO_2 + 18H_2O \qquad \text{combustion of octane, a major component of gasoline}$$

Sometimes combustion occurs with insufficient oxygen, which results in the formation of carbon monoxide instead of carbon dioxide. The following equation shows the incomplete combustion of octane:

$$2C_8H_{18} + 17O_2 \longrightarrow 16CO + 18H_2O$$

In the internal combustion engine, most of the gasoline undergoes complete combustion to give carbon dioxide and water, but a small amount undergoes incomplete combustion, which results in some carbon monoxide in the exhaust. Carbon monoxide gas is a deadly poison.

A C T I V I T Y
15-3

Deduce the oxidation numbers of C, H, and O in the reactants and products of the reaction

$$C + H_2O \longrightarrow CO + H_2$$

15-4 ELECTRICAL CELLS

When you turn on a flashlight or a hand calculator, electrical energy is instantly available to operate the device. Batteries supply electrical energy through electron transfer reactions. Electron transfer reactions in batteries are exothermic reactions in that energy is released but most of the energy takes the form of electrical energy.

When a piece of zinc metal is added to a solution containing copper(II) ions, the following reaction occurs spontaneously:

$$Zn(s) + Cu^{2+}(aq) \longrightarrow Zn^{2+}(aq) + Cu(s)$$

In this reaction, zinc metal changes to zinc ions and copper(II) ions change to copper metal. Note how the oxidation number of zinc changes and how the oxidation number of copper changes.

The zinc–copper reaction is done in another interesting way. A piece of zinc (the zinc electrode) is placed in a solution containing zinc ions and a piece of copper (the copper electrode) is placed in a solution containing copper(II) ions. Of course, negative ions are also present in these solutions, but the metal ions are involved in the chemical reaction. A conducting bridge, called a salt bridge, connects the two solutions, as shown in Figure

15-2. A salt bridge is a tube that contains a solution of an ionic compound called the electrolyte. The bridge provides a way to have electrical contact between the two solutions without actually mixing them. The conducting bridge contains electrolyte ions that move within the bridge from one solution to the other providing electrical contact. Since the two solutions are not in direct contact, no reaction occurs. When the electrodes are connected to one another by a metal wire, however, an electron transfer reaction occurs. The electrons flow through the wire from one side of the system to the other. The reaction at the zinc electrode (the anode) is

$$Zn(s) \longrightarrow Zn^{2+}(aq) + 2e^-$$

The lost electrons flow through the wire to the copper electrode (the cathode). The copper(II) ions gain electrons at the copper electrode.

$$Cu^{2+}(aq) + 2e^- \longrightarrow Cu(s)$$

When the electrodes are connected by a wire, the reactions occur spontaneously and electrons flow through the wire as electrical current. The current flow is a source of useful electrical energy. The zinc–copper setup described above is an example of a **voltaic cell** or **galvanic cell.** A voltaic cell is an apparatus in which electrode reactions are separated by a salt bridge. When the electrodes are connected by an external conductor, electrical work is produced as electrons flow. **Batteries** of all sorts are voltaic cells specially packaged to provide electrical energy from electron transfer reactions (see Section 15-11). When the electrodes in a battery are connected by a metallic conductor, electricity flows as the reaction takes place.

In the copper–zinc cell described above, zinc metal changes to zinc ions and copper(II) ions change to copper metal. To get an equation for the overall reaction, add the two electrode reactions.

$$Zn(s) \longrightarrow Zn^{2+}(aq) + 2e^-$$
$$\underline{Cu^{2+}(aq) + 2e^- \longrightarrow Cu(s)}$$
$$Zn(s) + Cu^{2+}(aq) + 2e^- \longrightarrow Cu(s) + Zn^{2+}(aq) + 2e^-$$

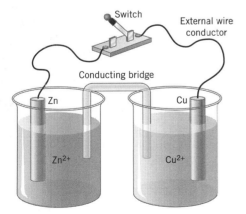

FIGURE 15-2   Galvanic or voltaic cell.

The electrons lost by the zinc are gained by the copper ions; after cancel-ing the electrons on both sides, the equation becomes

$$Zn(s) + Cu^{2+}(aq) \longrightarrow Cu(s) + Zn^{2+}(aq)$$

Notice that this is the same equation for the spontaneous reaction that oc-curs when zinc metal is placed directly in a solution containing copper(II) ions. Electron transfer reactions typically occur when reactants are in di-rect contact. It is possible, however, to carry out some electron transfer re-actions by separating the reaction into two halves in a voltaic cell.

15-5 OXIDATION AND REDUCTION DEFINED

Consider the electron transfer reaction between sodium metal and chlo-rine gas.

$$2Na(s) + Cl_2(g) \longrightarrow 2NaCl(s)$$

In an electron transfer reaction, elements change their chemical combina-tions. As a result, they change oxidation number. Sodium has an oxidation number of 0 in metallic sodium and an oxidation number of +1 in NaCl. Chlorine has an oxidation number of 0 in chlorine gas and oxidation number of −1 in NaCl. In the reaction, sodium has undergone oxidation and chlorine has undergone reduction. See the definitions in the margin and Figure 15-3.

Electron transfer reactions are oxidation–reduction reactions or redox reactions. You can recognize them by noting whether changes occur in the oxidation numbers of any elements involved in the reaction. Since oxida-tion and reduction occur simultaneously, a redox reaction must involve an increase in the oxidation number of one element and a decrease for an-other element.

Another example of an electron transfer reaction is the reaction that occurs when sodium metal is placed in water.

$$2Na(s) + 2H_2O \longrightarrow 2Na^+(aq) + 2OH^-(aq) + H_2(g)$$

In this reaction, sodium changes from elemental sodium to a monatomic sodium ion and hydrogen combined with oxygen in water changes to ele-mental hydrogen gas. Sodium is not a spectator ion because it changes from a solid metal to an ion in the reaction. The hydrogen covalently bonded to oxygen in water becomes covalently bonded to itself in hydro-gen gas. Look closely at the oxidation numbers of the elements involved in the reaction. The oxidation number of each sodium has changed from 0 in sodium metal to +1 in the sodium ion. Sodium must have lost electrons.

$$\overset{0}{2Na} + 2H_2O \longrightarrow \overset{+1}{2Na^+} + 2OH^- + H_2$$

Oxidation

Loss of electrons; the in-crease in oxidation num-ber of an element.

Reduction

Gain of electrons; the de-crease in oxidation num-ber of an element.

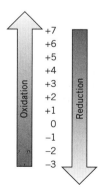

FIGURE 15-3
Oxidation is an increase in oxidation number, and reduction is a decrease in oxidation number.

The oxidation number of hydrogen has changed from $+1$ in water to 0 in hydrogen gas. Hydrogen must gain electrons in the reaction.

$$\overset{+1}{2Na + 2H_2O} \longrightarrow 2Na^+ + 2OH^- + \overset{0}{H_2}$$

The oxidation number of oxygen is -2 in water, on the left, and -2 in the hydroxide ion, on the right. Oxygen has not changed oxidation number, so it has not lost or gained electrons, and two of the hydrogen atoms have not changed oxidation number. In the reaction, sodium has undergone oxidation and hydrogen has undergone reduction.

15-6 HALF-REACTIONS

For convenience and clarification, an equation for a redox reaction is often separated into two parts. One part involves the loss of electrons (oxidation), and the other involves the gain of electrons (reduction). For example, the reaction between sodium metal and water is represented as

$$Na(s) \longrightarrow Na^+(aq) + e^- \qquad \text{oxidation; loss of electrons}$$

$$2e^- + 2H_2O \longrightarrow 2OH^-(aq) + H_2(g) \qquad \text{reduction; gain of electrons}$$

Such representations of a redox reaction are called **half-reactions** or **half-equations.** One half-reaction represents the oxidation half of the overall reaction, and the other represents the reduction half of the overall reaction. Of course, one half-reaction does not occur without the other half-reaction taking place simultaneously. The equation for a complete redox reaction is a combination of the two half-equations.

In a redox reaction, the number of electrons lost must balance the electrons gained, since no electrons are created or destroyed in a reaction. Electrons are lost by one element and gained by another. To incorporate this into the complete reaction, the half-reactions often must be adjusted by multiplying one or both by the appropriate numbers to balance the electrons lost with the electrons gained. The sodium or oxidation half of the sodium–water reaction given above is multiplied by 2; two electrons lost balance the two electrons gained in the reduction half.

$$2Na \longrightarrow 2Na^+ + 2e^-$$

$$2e^- + 2H_2O \longrightarrow 2OH^- + H_2$$

After multiplying to balance the electrons, the two half-equations are added to give the complete redox equation.

$$2Na + 2e^- + 2H_2O \longrightarrow 2Na^+ + 2e^- + 2OH^- + H_2$$

The electrons appearing on both sides are canceled to give

$$2Na + 2H_2O \longrightarrow 2Na^+ + 2OH^- + H_2$$

Many redox reactions are not easily balanced using the typical trial-and-error methods. Systematic methods used to balance half-equations and redox equations for more complicated reactions are discussed in Appendix 4.

15-7 Oxidizing and Reducing Agents

In a redox reaction, one species acts as an electron donor. It loses electrons and is oxidized since oxidation is loss of electrons. Sodium metal in the sodium–water reaction is an electron donor. The electrons lost by the electron donor reduce another species. An **electron donor** that reduces another species is called a reducing agent. A **reducing agent** is a species that reduces another species and is itself oxidized in the process.

The species that gains electrons in a redox reaction is an electron acceptor. By accepting electrons, it is reduced since gain of electrons is reduction. Water in the sodium–water reaction accepts or gains electrons from sodium metal. The **electron acceptor** is called an oxidizing agent because it oxidizes another species. An **oxidizing agent** oxidizes another species and is itself reduced in the process.

Identifying the reducing agent in a redox reaction is usually easy. It is that species having an element that undergoes an increase in oxidation number. The element changes from a lower oxidation number to a higher oxidation number. In the sodium–water reaction, sodium is the reducing agent and its oxidation number changes from 0 to +1.

$$
\begin{array}{ccc}
0 & & +1 \\
2Na \; + \; 2H_2O & \longrightarrow & 2Na^+ + 2OH^- + H_2
\end{array}
$$

reducing agent   reduction
reduces hydrogen
in water

Also, identifying the oxidizing agent is usually just as easy since it is the species having an element that undergoes a decrease in oxidation number. The element changes from a higher oxidation number to a lower oxidation number. In the sodium–water reaction, water is the oxidizing agent since the oxidation number of hydrogen changes from +1 to 0.

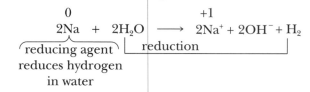

$$
\begin{array}{ccc}
\text{oxidation} & & \\
+1 & & 0 \\
2Na \; + \; 2H_2O & \longrightarrow & 2Na^+ + 2OH^- + H_2
\end{array}
$$

oxidizing agent

Each redox reaction must have an oxidizing agent and a reducing agent. We identify the oxidizing and reducing agents by noting the oxidation number changes of elements in the reaction.

EXAMPLE 15-4

Label the oxidizing and reducing agent in the redox reaction

$$S + O_2 \longrightarrow SO_2$$

The sulfur undergoes an increase in oxidation number from 0 to +4. It is oxidized, so oxygen is the oxidizing agent. The oxygen changes oxidation number from 0 to −2. It is reduced, so sulfur is the reducing agent.

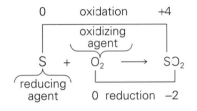

EXAMPLE 15-5

Label the oxidizing and reducing agents in the redox reaction

$$5Fe^{2+} + MnO_4^- + 8H^+ \longrightarrow 5Fe^{3+} + Mn^{2+} + 4H_2O$$

Look for any change in ion charge or change in combination. The iron changes its oxidation number from +2 to +3; it is oxidized. The manganese changes its oxidation number from +7 in MnO_4^- to +2 in Mn^{2+}; it is reduced. Hydrogen and oxygen do not change their oxidation numbers in the reaction. Hydrogen is +1 and oxygen is −2 on both sides of the equation. The iron is oxidized by the permanganate ion, so MnO_4^- is the oxidizing agent. Note that the manganese in the permanganate ion changes oxidation number but that the entire ion is viewed as the oxidizing agent. The permanganate ion is reduced by the iron(II) ion, so Fe^{2+} is the reducing agent.

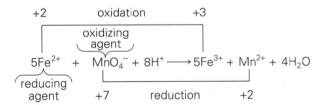

The hydrogen ion, H^+, in this reaction is neither an oxidizing nor a reducing agent. It is involved in the reaction as a carrier of charge and appears in the balanced equation. As you learned in Chapter 13, the H^+ or proton exists in water as hydronium ion H_3O^+. When dealing with redox reactions, it is more convenient to use H^+ instead of H_3O^+. In any case, hydrogen ion participates in many redox reactions as a charge carrier. It is involved in the reaction as a carrier of positive charge to help balance the charge.

Write the oxidation and reduction half-reactions and label the oxidizing and reducing agents in the redox reaction

$$2H^+ + Zn \longrightarrow Zn^{2+} + H_2$$

15-8 MEASURING STRENGTHS OF OXIDIZING AND REDUCING AGENTS

When an oxidizing and a reducing agent are mixed, a redox reaction may occur. Oxidizing and reducing agents vary in their ability to react with each other. How extensively oxidizing and reducing agents react depends on the relative strength of the agents with which they are mixed.

When an oxidizing agent reacts, it is reduced and it forms a potential reducing agent. For example, chlorine, Cl_2, can act as an oxidizing agent, and the chloride ion, Cl^-, can act as a reducing agent.

$$\text{oxidizing agent} \quad Cl_2 + 2e^- \longrightarrow 2Cl^-$$

$$\text{reducing agent} \quad 2Cl^- \longrightarrow Cl_2 + 2e^-$$

When a reducing agent reacts, it is oxidized and it forms a potential oxidizing agent. For example, zinc metal is a reducing agent and zinc ion formed by the oxidation of zinc is a potential oxidizing agent.

$$\text{reducing agent} \quad Zn \longrightarrow Zn^{2+} + 2e^-$$

$$\text{oxidizing agent} \quad Zn^{2+} + 2e^- \longrightarrow Zn$$

Oxidation and reduction processes are potentially reversible. It is possible to couple or associate each oxidizing agent with its reduced form that is a reducing agent. Such pairs are called **redox couples.** Chlorine, Cl_2, and the chloride ion, Cl^-, are a redox couple, and the zinc ion, Zn^{2+}, and zinc metal, Zn, are a redox couple.

Oxidizing agents and reducing agents vary in strength. The stronger the oxidizing agent, the greater its tendency to oxidize reducing agents. In comparison, the reducing agent couple of a stronger oxidizing agent is a weaker reducing agent. Fluorine, F_2, is a very strong oxidizing agent, and its reduced form, the fluoride ion, F^-, is a very weak reducing agent. The stronger a reducing agent, the greater its tendency to reduce oxidizing agents. In comparison, the oxidizing agent couple of a stronger reducing agent is a weaker oxidizing agent. Sodium metal is a very strong reducing agent, and its oxidized form, sodium ion, Na^+, is a very weak oxidizing agent.

How is it possible to measure the strength of an oxidizing agent or a reducing agent? One way is to incorporate the agent into a galvanic cell. In a cell, two half-reactions are physically separated and compared. Recall from Section 15-4 that when the two electrodes in contact with the half-reactions are connected by an external circuit, a spontaneous electron transfer can occur. As the reaction occurs, the oxidizing agent gains electrons and the reducing agent loses electrons. A measure of the tendency for two half-reactions to occur is found by placing a voltmeter in the circuit. A voltmeter is an electronic device that measures the electromotive force between two locations in an electrical circuit. Electromotive force is a force that acts when there is a difference in electrical charge between two points. A voltage is measured by attaching a voltmeter to the terminals of a battery or the electrodes of a galvanic cell. The greater the voltage reading, the greater the tendency for spontaneous electron transfer to occur. The strength of an oxidizing agent is measured by comparing it, in a galvanic cell, to the hydrogen ion–hydrogen half-reaction.

$$2H^+(aq) + 2e^- \longrightarrow H_2(g)$$

The electrode for this half-reaction is called the **hydrogen electrode** when it is part of a galvanic cell (see Figure 15-4). To make measurements of cell voltages, a specific half-reaction is set up as part of the galvanic cell. The voltage between this half of the cell and the hydrogen electrode is measured using a voltmeter under the conditions that each dissolved species has a concentration of 1 mole per liter and gases are at 1 atm pressure.

By convention, if the oxidizing agent gains electrons and the hydrogen electrode loses electrons, the half-reaction being tested has a positive cell voltage. For example, a fluorine–fluoride half-reaction gives a voltage of +2.87 volts when coupled with the hydrogen electrode.

$$F_2(g) + 2e^- \longrightarrow 2F^-(aq) \qquad (+2.87 \text{ volts})$$

A half-reaction involving permanganate ion mixed with hydrogen ion (called an acidified solution) gives a voltage of +1.49 volts.

$$MnO_4^-(aq) + 8H^+(aq) + 5e^- \longrightarrow Mn^{2+}(aq) + 4H_2O \qquad (+1.49 \text{ volts})$$

This result means fluorine is a stronger oxidizing agent than acidified permanganate ion because the fluorine cell has a higher voltage.

Oxidizing agents are placed in cells, and the cell voltages are measured against the hydrogen electrode. These measurements are used to rank oxidizing agents according to their strengths. When a half-cell involving zinc

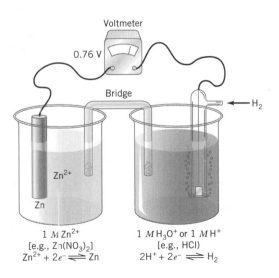

1 M Zn^{2+}
[e.g., $Zn(NO_3)_2$]
$Zn^{2+} + 2e^- \rightleftharpoons Zn$

1 M H_3O^+ or 1 M H^+
[e.g., HCl]
$2H^+ + 2e^- \rightleftharpoons H_2$

FIGURE 15-4 A galvanic cell, including a hydrogen electrode and a voltmeter.

ions and zinc metal is tested against hydrogen, the zinc ions do not gain electrons; instead, the zinc metal loses electrons and the hydrogen ions, in the hydrogen electrode, gain electrons. (See Figure 15-4.) When this happens, the measured voltage associated with the half-reaction is given a negative sign.

$$Zn^{2+}(aq) + 2e^- \longrightarrow Zn(s) \qquad (-0.76 \text{ volt})$$

Oxidizing the negative voltage shows that the zinc ion is a weaker oxidizing agent than the hydrogen ion. Furthermore, it also means the zinc ion is a weaker oxidizing agent than any oxidizing agent having a positive voltage.

Table 15-2 is a brief **redox table** in which some common oxidizing agents are ranked according to strength. They are ranked in order of decreasing cell voltages. Since each oxidizing agent has a corresponding reducing agent, the reducing agents are also listed according to strength. Stronger oxidizing agents have weaker reducing agents in the couple, so the reducing agents in the table are listed according to increasing strength. That is, the table starts with weaker reducing agents at the top, and the strengths increase down the table. Note that the completely balanced half-reaction is given for each oxidizing agent in the table. Also note that some oxidizing agents have the hydrogen ion included in their half-reactions. The hydrogen ion is not acting as an acid but is acting to balance the charge in the reaction. If the solution is not acidified using an acid like sulfuric acid, the reaction does not occur.

Table 15-2   *A Redox Table Including Some Common Oxidizing and Reducing Agents*

	OXIDIZING AGENT			REDUCING AGENT	CELL VOLTAGE
Fluorine	$F_2(g)$	$+ 2e^- \rightleftharpoons$	$2F^-(aq)$	Fluoride ion	$+2.87$
Hydrogen peroxide	$H_2O_2(aq) + 2H^+(aq) + 2e^- \rightleftharpoons$		$2H_2O$	Water	$+1.78$
Permanganate ion	$MnO_4^-(aq) + 8H^+(aq) + 5e^- \rightleftharpoons 4H_2O +$		$2Mn^{2+}(aq)$	Manganese(II) ion	$+1.49$
Chlorine	$Cl_2(g)$	$+ 2e^- \rightleftharpoons$	$2Cl^-(aq)$	Chloride ion	$+1.36$
Dichromate ion	$Cr_2O_7^{2-}(aq) + 14H^+(aq) + 6e^- \rightleftharpoons 7H_2O +$		$2Cr^{3+}(aq)$	Chromium(III) ion	$+1.33$
Bromine	$Br_2(\ell)$	$+ 2e^- \rightleftharpoons$	$2Br^-(aq)$	Bromide ion	$+1.07$
Nitrate ion	$NO_3^-(aq) + 4H^+(aq) + 3e^- \rightleftharpoons 2H_2O +$		$NO(g)$	Nitrogen oxide	$+0.96$
Iron(III) ion	$Fe^{3+}(aq)$	$+ 1e^- \rightleftharpoons$	$Fe^{2+}(aq)$	Iron(II) ion	$+0.77$
Iodine	$I_2(s)$	$+ 2e^- \rightleftharpoons$	$2I^-(aq)$	Iodide ion	$+0.54$
Carbon dioxide	$2CO_2(g) + 2H^+(aq) + 2e^- \rightleftharpoons$		$H_2C_2O_4(aq)$	Oxalic acid	$+0.49$

15-9 USING A REDOX TABLE

Redox reactions are predicted using Table 15-2. A prediction is based on the following: When an oxidizing agent and reducing agent mix, a reaction occurs if a weaker oxidizing agent and weaker reducing agent form as products. The agents in the table are listed according to strength. Generally, any oxidizing agent reacts with any reducing agent below it in the table. In contrast, no oxidizing agent reacts with a reducing agent lying above it in the table.

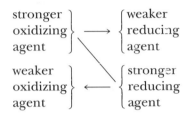

In any reaction predicted from the table, the oxidizing agent reacts with the reducing agent to form the products given in the table. For example, suppose a solution containing iron(III) ions, Fe^{3+}, mixes with a solution containing iodide ions, I^-. In Table 15-2, the iron(III) ion, Fe^{3+}, is found as an oxidizing agent, and the iodide ion, I^-, is found as a reducing agent.

$$Fe^{3+} + e^- \longrightarrow Fe^{2+}$$
$$I_2 + 2e^- \longleftarrow 2I^-$$

Since the oxidizing agent is above the reducing agent, the predicted reaction is

$$Fe^{3+} + e^- \longrightarrow Fe^{2+} \qquad \text{reduction}$$
$$2I^- \longrightarrow I_2 + 2e^- \qquad \text{oxidation}$$

Note that the oxidation part is the reverse of the reaction in the table. To write a complete net-ionic equation for the redox reaction, copy the half-reaction of the oxidizing agent from the table along with the reversed half-reaction of the reducing agent. Then, multiply each half-reaction by the appropriate factors so that the number of electrons gained equals the number of electrons lost. Start with the two half-reactions reversing the reducing agent reaction.

$$Fe^{3+} + e^- \longrightarrow Fe^{2+}$$
$$2I^- \longrightarrow I_2 + 2e^-$$

Multiplying the iron half-reaction by 2 and adding the half-reactions gives the final equation.

$$2Fe^{3+} + 2e^- \longrightarrow 2Fe^{2+}$$
$$\underline{\qquad 2I^- \longrightarrow I_2 + 2e^-}$$
$$2Fe^{3+} + 2I^- \longrightarrow 2Fe^{2+} + I_2$$

EXAMPLE 15-6

What reaction occurs when an acidified solution containing dichromate ions, $Cr_2O_7^{2-}$, is mixed with a solution containing bromide ions, Br^-? The acidified solution supplies hydrogen ions.

The possible reactants are $Cr_2O_7^{2-}$, H^+, and Br^-. By using Table 15-2, we note that the position of the acidified dichromate ion is above the bromide ion.

$$14H^+ + Cr_2O_7^{2-} + 6e^- \longrightarrow 2Cr^{3+} + 7H_2O$$
$$Br_2 + 2e^- \longleftarrow 2Br^-$$

Therefore, we write the reverse of the half-reaction of the reducing agent, Br^-.

$$2Br^- \longrightarrow Br_2 + 2e^-$$

Then, we multiply the bromine half by 3 to balance the electrons. The final equation is obtained by adding the two half-reactions.

$$14H^+ + Cr_2O_7^{2-} + 6e^- \longrightarrow 2Cr^{3+} + 7H_2O$$
$$6Br^- \longrightarrow 3Br_2 + 6e^-$$
$$\overline{14H^+ + Cr_2O_7^{2-} + 6Br^- \longrightarrow 3Br_2 + 2Cr^{3+} + 7H_2O}$$

EXAMPLE 15-7

What reaction occurs when an acidified solution containing dichromate ions, $Cr_2O_7^{2-}$, mixes with a solution containing chloride ions, Cl^-?

The possible reactants are $Cr_2O_7^{2-}$, H^+, and Cl^-. Noting the positions of the acidified dichromate ion and the chloride ion in Table 15-2, we predict no reaction, since the couple with Cl^- is above the couple with $Cr_2O_7^{2-}$ in the table.

$$Cl_2 + 2e^- \longrightarrow 2Cl^- \qquad \text{no reaction}$$
$$14H^+ + Cr_2O_7^{2-} + 6e^- \longrightarrow 2Cr^{3+} + 7H_2O$$

ACTIVITY

15-5

Give a balanced equation for any reaction that occurs when nitric acid is mixed with a solution of sodium iodide.

 15-10 **THE ACTIVITY SERIES OR ELECTROMOTIVE SERIES OF METALS**

Table 15-3 shows another redox table that includes some common metals, water, and the hydrogen ion. This table is often called the **electromotive series** of metals or the **activity series** of metals because it lists common metals according to their strength as reducing agents. Note that metals, as reducing agents, are listed on the right in the table. Stronger reducing agents occur at the bottom of the list, and they decrease in strength from the bottom to the top of the table. You can use this table to predict reactions in the same manner as Table 15-2. That is, any oxidizing agent reacts with any reducing agent below it in the table and does not react with a reducing agent above it.

Metal ions are listed in the table as oxidizing agents; metals are listed as reducing agents. The table shows which metals and metal ions react with one another. A specific metal ion reacts with any metal that appears below it in the table. When zinc metal is placed in a solution containing copper(II) ions, Cu^{2+}, a reaction occurs. In the table, the oxidizing agent Cu^{2+} is above the couple involving the reducing agent, Zn.

$$Cu^{2+} + 2e^- \longrightarrow Cu \quad\quad \text{reduction}$$
$$Zn^{2+} + 2e^- \longleftarrow Zn \quad\quad \text{oxidation}$$

Thus, the predicted reaction is zinc metal reacting with copper(II) ion to give copper metal and zinc ion. The copper half-equation is taken from the table and added to the reverse of the zinc half-equation:

$$Cu^{2+} + 2e^- \longrightarrow Cu$$
$$\underline{\phantom{Cu^{2+} + 2e^-}\,Zn \longrightarrow Zn^{2+} + 2e^-}$$
$$Cu^{2+} + Zn \longrightarrow Cu + Zn^{2+}$$

Recall that this was the spontaneous electron transfer reaction mentioned in Section 15-4.

Table 15-3 lists metals according to their chemical activity as reducing agents. The more active metals are at the bottom of the table, and the less

Table 15-3   *The Electromotive Series of Metals or the Activity Series of Metals*

	OXIDIZING AGENT		REDUCING AGENT		CELL VOLTAGE	
Silver ion	$Ag^+(aq)$	$+ 1e^- \rightleftharpoons$	$Ag(s)$	Silver	$+0.80$	
Mercury(II) ion	$Hg^{2+}(aq)$	$+ 2e^- \rightleftharpoons$	$Hg(\ell)$	Mercury	$+0.79$	
Iron(III) ion	$Fe^{3+}(aq)$	$+ 1e^- \rightleftharpoons$	$Fe^{2+}(aq)$	Iron(II) ion	$+0.77$	
Copper(II) ion	$Cu^{2+}(aq)$	$+ 2e^- \rightleftharpoons$	$Cu(s)$	Copper	$+0.34$	
Hydrogen ion	$2H^+(aq)$	$+ 2e^- \rightleftharpoons$	$H_2(g)$	Hydrogen	0.00	
Lead(II) ion	$Pb^{2+}(aq)$	$+ 2e^- \rightleftharpoons$	$Pb(s)$	Lead	-0.13	
Nickel(II) ion	$Ni^{2+}(aq)$	$+ 2e^- \rightleftharpoons$	$Ni(s)$	Nickel	-0.25	These metals dissolve in acid solutions.
Cobalt(II) ion	$Co^{2+}(aq)$	$+ 2e^- \rightleftharpoons$	$Co(s)$	Cobalt	-0.28	
Cadmium ion	$Cd^{2+}(aq)$	$+ 2e^- \rightleftharpoons$	$Cd(s)$	Cadmium	-0.40	
Iron(II) ion	$Fe^{2+}(aq)$	$+ 2e^- \rightleftharpoons$	$Fe(s)$	Iron	-0.41	
Zinc ion	$Zn^{2+}(aq)$	$+ 2e^- \rightleftharpoons$	$Zn(s)$	Zinc	-0.76	
Water	$2H_2O$	$+ 2e^- \rightleftharpoons$	$H_2(g) + 2OH^-(aq)$	Hydrogen	-0.83	
Aluminum ion	$Al^{3+}(aq)$	$+ 3e^- \rightleftharpoons$	$Al(s)$	Aluminum	-1.67	
Magnesium ion	$Mg^{2+}(aq)$	$+ 2e^- \rightleftharpoons$	$Mg(s)$	Magnesium	-2.38	
Sodium ion	$Na^+(aq)$	$+ 1e^- \rightleftharpoons$	$Na(s)$	Sodium	-2.71	These metals react spontaneously in water.
Calcium ion	$Ca^{2+}(aq)$	$+ 2e^- \rightleftharpoons$	$Ca(s)$	Calcium	-2.76	
Potassium ion	$K^+(aq)$	$+ 1e^- \rightleftharpoons$	$K(s)$	Potassium	-2.92	

active metals are at the top of the table. An active metal is a strong reducing agent, so it is easily oxidized. Less active metals are weaker reducing agents and are less easily oxidized. Sometimes copper occurs in nature as the metal since it is a metal of relatively low activity. Active metals, such as sodium and potassium, are never found in nature. Furthermore, samples of these metals must be carefully stored since they are very reactive. They are usually stored in containers of kerosene since they do not readily react with hydrocarbons. Samples of metals of intermediate activity, such as iron, oxidize slowly when exposed to the environment. Over time, pure iron, for example, readily rusts and corrodes in the environment. See the end of this section for more about rust.

Use Table 15-3 as an activity series of metals to predict reactions that may occur between various metals and metal ions in solution. Generally, any metal ion reacts with any metal below it to oxidize the metal to its ion and reduce the original ion to its metallic form. More active metals are those that are strong reducing agents at the bottom of the table. These active metals readily form ions. The activity series reveals some practical information. The Statue of Liberty is made of copper metal. It has a green color due to the oxidation of copper to form various copper compounds. These compounds form a coating on the copper that prevents it from further oxidation. Unfortunately, the statue was built on an iron frame. According to the activity series, any solutions of copper ions that contact iron react with the iron to give iron(II) ions. This reaction caused the iron frame to corrode and lose its strength. Note the position of iron and copper in Table 15-3.

$$Cu^{2+} + 2e^- \longrightarrow Cu$$
$$Fe^{2+} + 2e^- \longleftarrow Fe$$

The equation for the reaction that occurs between copper(II) ions and iron is found by reversing the iron half-reaction and adding the two half-reactions.

$$Cu^{2+} + Fe \longrightarrow Fe^{2+} + Cu$$

Over a period of many years, this reaction caused the corrosion and decay of the iron frame inside the statue. That is why extensive repairs were needed in 1986.

EXAMPLE 15-8

What reaction occurs when a piece of solid nickel metal is added to a solution of mercury(II) nitrate?

First, we write the formulas of the species mixed. Remember that we represent samples of metals by their symbols and we represent solutions of ionic compounds as the separate cation and anion.

$$Ni(s) + Hg^{2+} + NO_3^-$$

The mercury(II) ion is above nickel in the redox table, so we predict the following reaction obtained by adding the half-equations in the table:

$$Ni(s) + Hg^{2+} \longrightarrow Hg\ (\ell) + Ni^{2+}$$

Table 15-3 also shows why some metals dissolve in acidic solutions and some do not. The hydrogen ion occurs in the table as an oxidizing agent. Note the location of the hydrogen ion in Table 15-3. When any of the metals below H^+ in the table are added to an acidic solution, a reaction occurs, producing H_2 gas and the ion of the metal. These metals dissolve in acidic solutions. For example, note the relative positions of the hydrogen ion and magnesium metal in the table.

$$2H^+ + 2e^- \longrightarrow H_2(g)$$
$$Mg^{2+} + 2e^- \longleftarrow Mg$$

When magnesium metal is added to a solution of hydrochloric acid, a reaction occurs. Hydrochloric acid is a strong acid containing H_3O^+. Instead of H_3O^+, we use H^+ when writing redox equations. To write the overall equation, reverse the half-equation for the reducing agent and add the two half-equations.

$$2H^+ + Mg \longrightarrow Mg^{2+} + H_2(g)$$

DANGER: Never mix metals at the bottom of the table with acid solutions. They can react violently causing an explosion.

EXAMPLE 15-9

What reaction occurs when a piece of aluminum metal is added to a hydrochloric acid solution?

First, write the formulas of the species mixed. Use H^+ and Cl^- for hydrochloric acid.

$$Al(s) + H^+ + Cl^-$$

Hydrogen ion is above aluminum in the redox table. Thus, we expect the following reaction to take place. The equation is obtained by adding the two half-equations after each is multiplied by numbers to balance the electrons lost and gained. For practice you can do the multiplications and addition.

$$2Al(s) + 6H^+ \longrightarrow 2Al^{3+} + 3H_2(g)$$

Notice that H_2O also appears in Table 15-3 as an oxidizing agent. Some more active metals readily react with water. We expect that any metal below

ACTIVITY 15-6

(a) Which metals in Table 15-3 do you think are sometimes found in nature as the free metals? (b) Which metals in Table 15-3 cannot exist as metals in water?

ACTIVITY 15-7

What reaction occurs when solid copper metal is added to a solution of silver nitrate, $AgNO_3$?

H_2O in the table reacts with H_2O to form H_2 and the metal ion. But, Mg reacts only in very hot water, and Al does not react at normal temperatures because an impervious layer of aluminum oxide coats aluminum samples. This also explains why aluminum is corrosion-resistant and keeps its shine. The other three metals below H_2O in the table react vigorously with water. For instance, when potassium metal is placed in water, a very exothermic reaction occurs. We get the equation for the reaction by reversing the potassium half-reaction, multiplying it by 2, and adding it to the water half-reaction.

$$2H_2O + 2e^- \longrightarrow 2OH^- + H_2(g)$$
$$2K \longrightarrow 2K^+ + 2e^-$$
$$\overline{2K + 2H_2O \longrightarrow 2K^+ \quad + \quad 2OH^- + H_2(g)}$$

Rust

Corrosion refers to the reaction of metals with atmospheric oxygen. Iron metal corrodes to form rust. Rusting of iron occurs when iron combines with oxygen in a moist environment.

$$4Fe(s) + 3O_2 \xrightarrow{\quad H_2O \quad} 2Fe_2O_3(s) \quad \text{(rust)}$$

Many common metals corrode to form a protective surface layer of metal oxide. The layer prevents further reaction with oxygen, and the metals retain their metallic luster. Examples of such metals are chromium, nickel, tin, and aluminum. In a moist environment, however, such as in the presence of sea spray or acidic chemicals, most metals are consumed by corrosion. In contrast to oxides of other metals, iron oxides easily flake from the surface of iron, exposing fresh metal surfaces to further oxidation. Stainless steels are alloys containing nickel and chromium that help to protect the metal surfaces and make them corrosion-resistant. Rain, made acidic when acidic air pollutants dissolve, is responsible for significant rusting of steel bridges, buildings, and railroad tracks. Sea sprays and the salts used in winter deicing of roads promote the rusting of steel in automobiles.

One way to protect iron and steel structures is to cover them with paint. Of course, if the paint peels, rusting can occur. Some large bridges, such as the Golden Gate Bridge in California, are painted repeatedly to inhibit corrosion. Iron coated with a layer of zinc metal is called galvanized iron. Consult Table 15-3 and explain how zinc in contact with the iron prevents the formation of Fe^{2+} and, therefore, rusting.

Cathodic protection is a technique used to protect steel in water heaters, buried fuel tanks, underground pipes, and hulls of ships. A metal that is a stronger reducing agent than iron is attached, by wire, to the iron object. Pieces of magnesium metal are used for corrosion protection of tanks, pipes, and water heaters. The magnesium metal is more easily oxidized than the iron so it corrodes instead of the iron. Since the magnesium metal is consumed by oxidation, however, it is replaced occasionally. Nevertheless, cathodic protection saves money by preventing corrosion of expensive metal installations.

(a) Give the equation for the reaction that occurs when a piece of solid cobalt metal is mixed with a solution of hydrochloric acid. (b) Give the equation for the reaction that occurs when a piece of calcium metal is added to water. [Hint: One of the products is solid $Ca(OH)_2$.]

15-11 BATTERIES AND FUEL CELLS

In Section 15-4, we saw that it is possible to get useful electrical energy from a spontaneous redox reaction by separating the oxidizing and reducing agents in half-cell compartments. The compartments are connected by a barrier containing a conducting electrolyte to provide electrical contact between the compartments. The redox reaction occurs spontaneously if we mixed the contents of the two compartments. Such a reaction, however, does not give any useful electrical work. To create a potential source of useful electrical energy, a conducting electrode is placed in each half-cell. When the electrodes are connected by a wire, the half-reactions occur in each compartment and electrons flow through the connecting wire. As oxidation and reduction occur, electrons flow from the anode, where oxidation takes place, through the wire to the cathode, where reduction occurs. Of course, the flow of electrical current is used to do electrical work (light a bulb, run a motor, etc.). Much of the energy released in the reaction is converted to electrical energy, but some is lost as heat. Hence, batteries sometimes get hot as they are used.

Batteries

A device in which the chemical energy of a redox reaction is converted to electrical energy is a **battery** or **electrical cell.** The terminals of a battery are the electrode contacts. The oxidizing agent and reducing agent are contained in the cell and are separated by an electrolyte solution. The cell has the potential of doing electrical work. When we connect the electrodes by an external conductor, we get useful electrical work. Opening the circuit stops the electron flow and the reaction.

The strength of the cell depends on the relative strengths of the oxidizing and reducing agents. The strength is expressed as the voltage or potential difference between the electrodes. Higher voltages are obtained with stronger oxidizing and reducing agents or by connecting cells in series.

A common example of a battery is the **lead storage battery** used in automobiles (see Figure 15-5). A lead storage battery consists of cells using an anode coated with lead metal and a cathode coated with lead(IV) oxide, PbO_2. The electrodes are immersed in a sulfuric acid electrolyte solution. A 6-volt battery has three of these cells in series, and a 12-volt battery has six. The discharge reaction for such a battery is

$$Pb(s) + PbO_2(s) + 2SO_4^{2-}(aq) + 4H^+(aq) \longrightarrow 2PbSO_4 + 2H_2O + energy$$

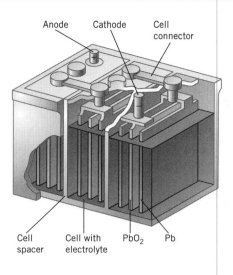

FIGURE 15-5 A three-cell lead storage battery.

Anode   Cathode   Cell connector

Cell spacer   Cell with electrolyte PbO_2 Pb

During the discharge, the lead anode is oxidized. This oxidation half-reaction is

$$Pb(s) + SO_4^{2-}(aq) \longrightarrow PbSO_4(s) + 2e^-$$
anode

Simultaneously, the lead(IV) oxide cathode is reduced as shown by the reduction half-reaction:

$$PbO_2 + 4H^+(aq) + SO_4^{2-}(aq) + 2e^- \longrightarrow PbSO_4(s) + 2H_2O$$
cathode

A lead storage battery runs down when the lead sulfate accumulates at the electrodes. A lead storage battery is recharged by pumping electrons into the battery in the direction opposite to the discharge direction. In a car, the battery is recharged by the alternator using energy from the engine. Reversing the electrode reactions in the battery changes the electrodes back into lead and lead(IV) oxide. Reversing the discharge reaction gives the charging reaction.

Ideally, the alternator on a car should keep a battery charged, and it should last indefinitely. In time, however, the electrode materials begin to flake off and the internal parts loosen, causing the battery to fail. Common dry-cell batteries used in flashlights and other electrical devices are not rechargeable. A dry cell runs down when the oxidizing and reducing agents are consumed. Figure 15-6 shows the components of a typical **dry cell.** They are called dry cells because the electrolyte material is damp but is not a liquid like the electrolytes used in lead batteries. Common dry cells use zinc as a reducing agent and manganese(IV) oxide as an oxidizing agent. Nickel–cadmium dry cells are similar, but they are rechargeable. A mercury cell, also shown in Figure 15-6, is a small battery used in watches and calculators. Mercury cells use zinc as an oxidizing agent and mer-

A C T I V I T Y

15-9

Give the equation for the reaction that occurs when a lead storage battery is charged.

cury(II) oxide as a reducing agent. All common kinds of batteries contain the heavy metals lead, manganese, cadmium, or mercury. These heavy metals are toxic. As a consequence, you should recycle batteries and not throw them in the trash.

A battery stops working when the reactants are consumed. A cell in which the oxidizing and reducing agents are continually supplied and the products continually removed is a **fuel cell**. Fuel cells are the subject of a great deal of research and may in the future be available as sources of industrial and domestic electricity. Fuel cells for use in electric cars are currently being developed. Special fuel cells are used as electrical sources in spacecraft. As shown in Figure 15-7, a typical fuel cell in a spacecraft involves the reaction of hydrogen gas and oxygen gas. The electrodes are nickel metal in contact with a potassium hydroxide solution. The hydrogen and oxygen fuels are fed into the cells continuously to produce electricity. Hydrogen gas reacts at the anode.

$$\text{oxidation} \quad H_2(g) + 2OH^-(aq) \longrightarrow 2H_2O + 2e^-$$

Oxygen gas reacts at the cathode.

$$\text{reduction} \quad 4e^- + O_2(g) + 2H_2O \longrightarrow 4OH^-(aq)$$

The net discharge reaction, obtained by combining the half-reactions, is the combination of hydrogen and oxygen to form water.

$$2H_2(g) + O_2(g) \longrightarrow 2H_2O + \text{electrical energy}$$

Fuel cells are used in spacecraft to run electrical equipment and instruments; the water produced is used as needed.

A typical dry-cell battery

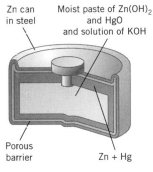

A typical mercury cell

FIGURE 15-6 Typical dry-cell battery and mercury-cell battery.

FIGURE 15-7 A hydrogen–oxygen fuel cell.

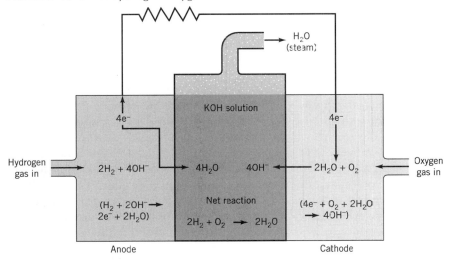

15-12 ELECTROLYSIS

The normal spontaneous discharge reaction of a lead battery is reversed by pumping electrons into the cell in the reverse direction of discharge. In a car, this process is done by the alternator, but another electron source such as a generator or even a stronger battery could be used. The electron source can supply enough potential difference or voltage to cause the electrode reactions to occur. In other words, electrical energy is used as a source of energy to cause a nonspontaneous redox reaction to occur.

Imagine placing a piece of metal, carrying a negative electrical charge, into a solution containing cations and anions. The negatively charged metal attracts the positive ions in the solution. Thus, the cations migrate toward the metal and form a layer of positive ions around the metal. If a positively charged piece of metal is placed in the solution, the anions are attracted to the metal and form a layer of negative ions around the metal.

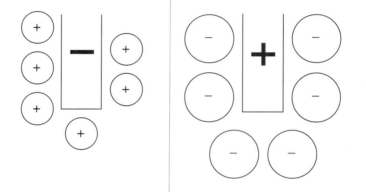

How is it possible to obtain the two charged pieces of metal? It can be done by connecting the two pieces of metal with wires to the two terminals of a battery or a generator, as shown in Figure 15-8. The battery or generator serves as a device for pumping electrons from one piece of metal to the other. This ease of movement of electrons in metals comes from the unique structure of metals. Since one piece of metal connected to the battery has an excess of electrons, it is negatively charged. In contrast, the piece of metal that is deficient in electrons is positively charged.

The electron pump or battery serves to keep a voltage or potential difference between the two pieces of metal. If the two pieces of metal connected to the electron pump are immersed in a solution containing ions, the cations move toward the negative piece of metal and the anions toward the positive piece of metal. The pieces of metal act as electrodes. The negative electrode that attracts the cations is called the **cathode,** and the positive electrode that attracts the anions is called the **anode.** That is why positive ions are called cations and negative ions are called anions.

An interesting phenomenon occurs if the potential difference between the electrodes is great enough. The battery or generator provides a driving

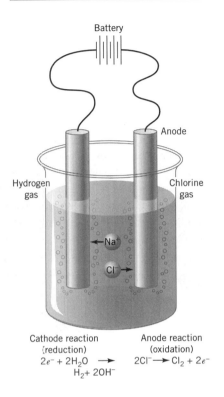

Cathode reaction
(reduction)

$2e^- + 2H_2O \longrightarrow$
$H_2 + 2OH^-$

Anode reaction
(oxidation)

$2Cl^- \longrightarrow Cl_2 + 2e^-$

FIGURE 15-8 The electrolysis of sodium chloride solution in an electrolytic cell.

force that causes electron loss and gain. At the cathode surface, some species gain electrons. At the anode, some species lose electrons. A redox reaction is induced using an external source of electrical energy. That is, since the battery or generator pumps electrons, a situation arises in which one species in solution loses electrons at the anode and another gains electrons at the cathode. In this manner, a complete electrical circuit is set up in which electrons move to the cathode, where they are gained by one species in solution. Simultaneously, another species in solution loses electrons to the anode to provide more electrons that are pumped to the cathode. This situation is essentially the reverse of a battery, which supplies electrical energy by a spontaneous redox reaction. The process in which electricity passes through a solution and a redox reaction occurs is called **electrolysis.** The term *electrolysis* means breaking down by electricity. Electrolysis can also occur when electricity passes through a molten sample of an ionic compound that contains separate cations and anions.

15-13 ELECTROLYSIS OF SODIUM CHLORIDE

The electrolysis of a water solution of sodium chloride, called a brine solution, serves as a good example of industrial electrolysis. In this process, which is illustrated in Figure 15-8, an external source of electricity serves as

Chlorine and Sodium Hydroxide

an electron pump. When the electrodes are connected to the source of electricity, the anode attracts chloride ions, where they lose electrons and form molecular chlorine. This half-reaction is

$$2Cl^-(aq) \longrightarrow Cl_2(g) + 2e^- \qquad \text{occurs at the anode}$$

Since molecular chlorine is diatomic, two chloride ions must react to produce one chlorine molecule. A half-reaction in which a species loses or gains electrons at an electrode and changes to a new species is called an **electrode reaction.** An electrode reaction, such as that given above, occurs only when a simultaneous reaction involving the gain of electrons takes place at the other electrode.

In this example, we expect the sodium ions to migrate toward the cathode. Sodium ions, however, do not gain electrons in aqueous solution

Table 15-4   *Some Examples of Electrolysis*

Electrolytic Production of Aluminum	
Aluminum oxide from bauxite ore is dissolved in the molten mineral cryolite. The metal is subjected to electrolysis to form aluminum metal.	$2Al_2O_3 + 3C \xrightarrow[\text{cryolite}]{\text{molten}} 3CO_2 + 4Al$ $Al^{3+} + 3e^- \longrightarrow Al \qquad$ cathode $C + 2O^{2-} \longrightarrow CO_2 + 4e^- \qquad$ anode
Electrolytic Production of Magnesium	
Magnesium chloride obtained from seawater is melted and subjected to electrolysis to give magnesium metal.	$MgCl_2 \xrightarrow{\text{molten}} Mg + Cl_2$ $Mg^{2+} + 2e^- \longrightarrow Mg \qquad$ cathode $2Cl^- \longrightarrow Cl_2 + 2e^- \qquad$ anode
Electrolysis of Water	
The electrolysis of water gives hydrogen and oxygen gases. Some nonelectrolyzable ions must be present in water for electrolysis.	$2H_2O \longrightarrow 2H_2 + O_2$ $2H_2O + 2e^- \longrightarrow H_2 + 2OH^- \qquad$ cathode $2H_2O \longrightarrow 4e^- + O_2 + 4H^+ \qquad$ anode
Electroplating of Silver	
In silver plating by electrolysis, a pure silver anode is used with a base metal cathode. During electrolysis, silver is plated on the cathode.	$Ag(\text{pure}) \longrightarrow Ag(\text{plate})$ $Ag^+ + 1e^- \longrightarrow Ag(\text{plate}) \qquad$ cathode $Ag \longrightarrow Ag^+ + 1e^- \qquad$ anode
Electroplating of Tin	
Tin plating of "tin" cans involves the electrolytic transfer of tin from a pure tin anode to a steel sheet. The tin-plated steel is used to make cans.	$Sn(\text{pure}) \longrightarrow Sn(\text{plate})$ $Sn^{2+} + 2e^- \longrightarrow Sn(\text{plate}) \qquad$ cathode $Sn \longrightarrow Sn^{2+} + 2e^- \qquad$ anode

to form metal because water is more easily reduced. Refer to Table 15-3 and note that having solid sodium metal in water is not possible. Water gains electrons according to the electrode reaction

occurs at cathode $2e^- + 2H_2O \longrightarrow H_2(g) + 2OH^-(aq)$

Notice that this electrode reaction produces the hydroxide ion, a base. The electrode reactions are actual chemical reactions involving electron transfer. The electrode reactions occur simultaneously when the solution of sodium chloride is subjected to electrolysis. For every two electrons gained at the cathode, two electrons are lost at the anode. The overall reaction corresponding to the electron transfer that occurs during the electrolysis of the sodium chloride is

$$2Cl^-(aq) + 2H_2O \xrightarrow{\text{electrolysis}} Cl_2(g) + H_2(g) + 2OH^-(aq)$$

The sodium ion present in the original brine is a spectator ion in the reaction. When a rather concentrated solution of sodium chloride is subjected to electrolysis, chlorine gas is produced at the anode and hydrogen gas at the cathode. Furthermore, for every two chloride ions that react, two hydroxide ions form; this step balances the charge. The result is that a solution of sodium chloride is transformed into a solution of sodium hydroxide and chlorine gas and hydrogen gas are released.

In the United States, more than 5 million tons of chlorine are manufactured each year by the chlor-alkali process using an electrolysis reaction similar to that described above. The production of some metals, such as aluminum, and electroplating, such as silver plating and chromium plating, also uses electrolysis. Table 15-4 lists some common examples of industrial electrolysis.

Aluminum

Aluminum

The largest-scale industrial electrolysis in the world is used in the making of aluminum metal from its ore. Aluminum is the most abundant metal in nature and is found in various aluminum silicate minerals, rocks, and soils. Aluminum is a low-density, corrosion-resistant metal used in a variety of structural materials and products, including cooking utensils, high-tension electrical lines, aluminum cans, and aluminum foil. Structural alloys of aluminum, containing small amounts of other metals and silicon, are used in buildings, window frames, airplanes, trailers, trucks, and automobiles.

Aluminum is found as a compound in the aluminum ore called bauxite. Once it is mined, bauxite is treated to form aluminum oxide. The industrial Hall process uses aluminum oxide as a raw material. More than six million tons of aluminum are produced in the United States each year by this process, developed in 1886 by Charles Hall, an American chemist. In the Hall process, aluminum oxide is dissolved in a compound called cryolite, Na_3AlF_6, at 800°C to

1000°C. Carbon electrodes are immersed in the molten mixture and electricity is passed through the melt. See Table 15-4 for the electrode reactions. Electrolysis reduces the aluminum ions to aluminum metal at the cathode. The carbon is oxidized to carbon dioxide at the anode. The overall reaction that takes place is electrolysis:

$$2Al_2O_3 + 3C \xrightarrow{\text{electrolysis}} 4Al + 3CO_2$$

At the high temperature of the process, the aluminum forms as a liquid that is tapped from the electrolysis vessel and then cast as a solid. See the figure in Section 7-8.

The refining of aluminum requires large amounts of electricity. As a result, aluminum-producing plants are situated near sources of cheap electricity such as hydroelectric dams. Not only is energy used in the electrolysis of aluminum ore, but energy is also expended in the mining, transporting, and purifying of the ore. More energy is needed to forge the metal into various materials. The energy used to make one aluminum can and use it once is approximately 7000 kJ or 7000 Btu. In contrast, 65% less energy is used to make a new can from recycled aluminum metal. Recycling aluminum not only saves bauxite but also saves energy.

How many tons of aluminum oxide are needed to make six million tons of aluminum by the Hall process? How many tons of carbon are needed to make six million tons of aluminum? (Hint: You can use tons in the mass-to-mass stoichiometric calculation without changing to grams. The gram units in the molar masses will cancel, leaving tons for the answer.)

FOCUS 15

Oxidation and Reduction

Oxidation number: The charge an element has in a monatomic ion or the hypothetical charge it would have if the shared electrons in covalent bonds are assigned to the more electronegative elements.

Oxidation: Loss of electrons is oxidation (LOX), which corresponds to the increase in oxidation number of an element (ION).

Reduction: Gain of electrons is reduction (GER), which corresponds to the decrease in oxidation number of an element (DON).

Oxidation–reduction reactions or redox reactions: Electron transfer reactions are characterized by changes in the oxidation numbers of some elements involved in the reaction. A redox reaction involves an increase in the oxidation number of one element and a decrease for another element.

Reducing agent: An electron donor that reduces another species. A species that reduces another species and is itself oxidized in the process. It is that species having an element that undergoes an increase in oxidation number, changing from a lower to a higher oxidation number.

Oxidizing agent: An electron acceptor that oxidizes another species. A species that oxidizes another species and is itself reduced in the process. It is that species having an element that undergoes a decrease in oxidation number, changing from a higher to a lower oxidation number.

Redox table: A list of oxidizing agents ranked according to decreasing strength and the corresponding reducing agents ranked according to increasing strength. Generally, any oxidizing agent reacts with any reducing agent below it in a redox table. In contrast, no oxidizing agent reacts with a reducing agent lying above it in the table.

In a reaction predicted from a redox table, the oxidizing agent reacts with the reducing agent to form the products given in the table. To write a complete net-ionic equation for a predicted redox reaction:

1. Copy the half-reaction of the oxidizing agent from the table as it appears.
2. Copy the half-reaction of the reducing agent but reverse it from the way it appears in the table.
3. Multiply each half-reaction by the proper numbers to give the same number of electrons lost as are gained, and add them to give the final equation.

Electromotive series of metals or the activity series of metals: A redox table that lists common metals according to their strength as reducing agents and metal ions according to their strengths as oxidizing agents. Any oxidizing agent reacts with any reducing agent below it in the table and does not react with a reducing agent above it. The series reveals which metals and metal ions react with one another. More active metals are those that are strong reducing agents and readily form metal ions.

Battery or electrical cell: A device in which the chemical energy of a redox reaction is converted to electrical energy. The terminals of a battery are the electrode contacts, and the oxidizing agent and reducing agent are contained in the cell and are separated by an electrolyte.

Fuel cell: An electrical cell in which the oxidizing and reducing agents are continuously supplied and the products are continuously removed.

Electrolysis: The process in which electricity is passed through a solution and causes a redox reaction.

QUESTIONS

1. Match the following terms with the appropriate lettered definitions.

oxidation number

a. loss of electrons or increase in oxidation number

oxidation

b. gain of electrons or decrease in oxidation number

reduction

c. a table including metals and metal ions as

reducing agents and oxidizing agents ranked according to relative strengths

oxidation–reduction reaction

  d. an electron transfer or redox reaction

voltaic cell

  e. a device in which a redox reaction is used as a source of electrical energy

half-reaction

  f. an electrical cell in which oxidizing agent and reducing agent are supplied continually as they are used in a redox reaction

reducing agent

  g. an electron donor that reduces another species in a redox reaction

oxidizing agent

  h. an electron acceptor that oxidizes another species in a redox reaction

electromotive or activity series

  i. a device in which the oxidation half and the reduction half of a redox reaction are physically separated

battery

  j. the oxidation part or reduction part of a redox reaction

fuel cell

  k. the electrode at which reduction occurs

cathode

  l. a process in which electricity is passed through a solution and causes a redox reaction

anode

  m. a number associated with an element to indicate its state of chemical combination

electrolysis

  n. the electrode at which oxidation occurs

Section 15-1

2. What is an oxidation number?

3. If a compound contains two elements, which element would be expected to have a positive oxidation number and which would be expected to have a negative oxidation number?

4. When an element is in the form of a monatomic ion, what is the oxidation number of the element? What is the oxidation number of an element when it is uncombined with other elements? What are the expected oxidation numbers of oxygen and hydrogen in their compounds?

5. What is true about the sum of the oxidation numbers of the elements in a compound?

6. What is true about the sum of the oxidation numbers of the elements that compose a polyatomic ion?

7. *Determine the oxidation number of the element other than hydrogen and oxygen in the following. Use the expected oxidation numbers for hydrogen and oxygen.

    (a) NO_3^- (b) NO (c) NO_2
    (d) SO_2 (e) SO_3 (f) CO
    (g) HSO_4^- (h) HCO_3^- (i) PO_4^{3-}
    (j) CaH_2 (k) N_2H_4 (l) P_4O_6

8. Determine the oxidation number of the element other than hydrogen and oxygen in the following. Use the expected oxidation numbers for hydrogen and oxygen.

    (a) NH_4^+ (b) IO_3^- (c) CrO_4^{2-}
    (d) H_2SO_4 (e) Cl_2O (f) CH_4
    (g) $CaCO_3$ (h) MgO (i) I_2
    (j) HNO_2 (k) N_2O_4 (l) P_4O_{10}

Section 15-2

9. What is an electron transfer reaction?

10. What is a voltaic or galvanic cell?

Sections 15-5 to 15-7

11. Give definitions for oxidation and reduction.

12. Why do oxidation and reduction always occur together?

13. Describe the following terms:

    (a) half-reactions

    (b) oxidizing agent

    (c) reducing agent

    (d) redox reaction

14. *Tell whether or not each of the following reactions is an oxidation–reduction reaction. For each of the reactions that is oxidation–reduction, identify the

oxidizing agent and the reducing agent. (Hint: If no elements change oxidation number in a reaction, the reaction is not a redox reaction.)

(a) $CH_4 + 2O_2 \longrightarrow 2H_2O + CO_2$

(b) $Ag^+ + Cl^- \longrightarrow AgCl$

(c) $I_2O_5 + 5CO \longrightarrow I_2 + 5CO_2$

(d) $NH_3 + HCl \longrightarrow NH_4Cl$

(e) $H_3O^+ + F^- \longrightarrow HF + H_2O$

(f) $SO_2 + NO_2 \longrightarrow SO_3 + NO$

(g) $2Cl_2 + CH_4 \longrightarrow 4HCl + C$

(h) $Zn^{2+} + 2OH^- \longrightarrow Zn(OH)_2$

(i) $Ni + 2H^+ \longrightarrow Ni^{2+} + H_2$

15. *For each of the redox reactions in Question 14, give the oxidation numbers of the elements that change oxidation number. For example, in part (a), C changes from -4 to $+4$ and O changes from 0 to -2.

16. Tell whether or not each of the following reactions is an oxidation–reduction reaction. For each of the reactions that is oxidation–reduction, identify the oxidizing agent and the reducing agent. (Hint: If no elements change oxidation number in a reaction, the reaction is not a redox reaction.)

(a) $Ca + 2H_2O \longrightarrow Ca(OH)_2 + H_2$

(b) $SO_3 + H_2O \longrightarrow H_2SO_4$

(c) $2CO + O_2 \longrightarrow 2CO_2$

(d) $Pb^{2+} + CrO_4^{2-} \longrightarrow PbCrO_4$

(e) $Cl_2 + 2I^- \longrightarrow I_2 + 2Cl^-$

(f) $Al^{3+} + 3OH^- \longrightarrow Al(OH)_3$

(g) $4Cl_2 + CH_4 \longrightarrow CCl_4 + 4HCl$

(h) $Zn + 2H^+ \longrightarrow Zn^{2+} + H_2$

(i) $2H_2 + O_2 \longrightarrow 2H_2O$

17. For each of the redox reactions in Question 16, give the oxidation numbers of the elements that change oxidation number. For example, in part (a), Ca changes from 0 to $+2$ and two of the H's change from $+1$ to 0.

18. In each of the following industrial or biological redox reactions, identify the oxidizing agent and the reducing agent.

(a) $N_2 + 3H_2 \longrightarrow 2NH_3$

(the synthesis of ammonia)

(b) $2H_2 + CO \longrightarrow CH_3OH$

(one step in the synthesis of methanol)

(c) $CH_4 + 2H_2O \longrightarrow CO_2 + 4H_2$

(the production of hydrogen)

(d) $Fe_2O_3 + 3CO \longrightarrow 2Fe + 3CO_2$

(one step in the refining of iron)

(e) $6CO_2 + 6H_2O \longrightarrow C_6H_{12}O_6 + 6O_2$

(photosynthesis of glucose)

(f) $C_6H_{12}O_6 - 2O_2 \longrightarrow$
$2HC_2H_3O_2 + 2CO_2 + 2H_2O$

(aerobic fermentation of glucose)

Section 15-8

19. What is a redox couple?

20. *In the reaction

$$Zn + 2H^+ \longrightarrow Zn^{2+} + H_2$$

Zn and Zn^{2+} are a redox couple and H^+ and H_2 are a redox couple. In each of the following redox reactions, identify two sets of redox couples.

(a) $CH_4 + 2H_2O \longrightarrow CO_2 + 4H_2$

(b) $Fe_2O_3 + 3CO \longrightarrow 2Fe + 3CO_2$

(c) $Ca + 2H_2O \longrightarrow Ca(OH)_2 + H_2$

(d) $Cl_2 + 2I^- \longrightarrow I_2 + 2Cl^-$

(e) $2Ag^+ + Zn \longrightarrow 2Ag - Zn^{2+}$

21. How is it possible to measure the strength of an oxidizing agent?

22. In a table of oxidizing and reducing agents in which the oxidizing agents are listed according to decreasing strengths, why are the corresponding reducing agents listed according to increasing strengths?

Section 15-9

23. *Using Table 15-2 as a guide, give balanced net-ionic equations for any redox reactions that occur when the following solutions or substances are mixed. The term *acidified* means that acid has been added to a solution to supply hydrogen ions.

(a) An NaBr solution is added to an acidified solution of hydrogen peroxide, H_2O_2.

(b) An NaI solution is added to an acidified solution of $Na_2Cr_2O_7$.

(c) An FeSO_4 solution is added to an acidified solution of $KMnO_4$.

(d) An oxalic acid solution, $H_2C_2O_4$, is added to an acidified solution of H_2O_2.

(e) An $Fe(NO_3)_3$ solution is added to a KI solution.

24. Using Table 15-2 as a guide, give balanced net-ionic equations for any redox reactions that occur when the

following solutions or substances are mixed. The term *acidified* means that acid has been added to a solution to supply hydrogen ions.

(a) Chlorine, Cl_2, is added to an NaBr solution.

(b) Iodine, I_2, is added to an NaCl solution.

(c) An oxalic acid solution, $H_2C_2O_4$, is added to an acidified solution of $Na_2Cr_2O_7$.

(d) A nitric acid solution is added to a solution of $Fe(NO_3)_2$.

25. The elemental form of a halogen will replace the halide ion of any halogen that occurs below it in the periodic table. Use Table 15-2 to explain this statement. Make a short redox table that includes all the halogens as oxidizing agents and all the halide ions as reducing agents. Where in the table would you fit fluorine and the fluoride ion?

26. Ocean water contains low concentrations of bromide ions, Br^-. Elemental bromine is obtained from ocean water by adding chlorine, Cl_2. Refer to Table 15-2 and give a balanced net-ionic equation to show the reaction for this process.

Section 15-10

27. *Using Table 15-3 as a guide, give balanced net-ionic equations for any redox reactions that occur when the following solutions or substances are mixed.

(a) A small piece of potassium metal is added to water.

(b) A piece of zinc is added to water.

(c) A copper penny is added to an $Hg(NO_3)_2$ solution.

(d) A piece of zinc is added to a $Pb(C_2H_3O_2)_2$ solution.

(e) A solution of $NiCl_2$ is mixed with a solution of $Zn(NO_3)_2$.

28. Using Table 15-3 as a guide, give balanced net-ionic equations for any redox reactions that occur when the following solutions or substances are mixed.

(a) A small piece of calcium metal is added to a hydrochloric acid solution. **CAUTION: Do not try this experiment.**

(b) A piece of iron metal is added to a solution of $Cu(NO_3)_2$.

(c) A piece of aluminum metal is added to a solution of $AgNO_3$.

(d) A piece of silver metal is added to a hydrochloric acid solution.

29. Corrosion or rusting of iron occurs when oxygen oxidizes iron in moist environments to form ionic iron compounds. Use Table 15-3 to explain the following statements.

(a) Galvanized iron is zinc-coated iron; coating iron with zinc inhibits rusting.

(b) Iron pipes buried in the ground can be protected from corrosion by connecting a sheet of magnesium metal to the pipeline.

(c) Aluminum screws should not be used with iron materials.

30. Copper metal is obtained from low-grade ore by dissolving any copper compounds found in the ores and then adding scrap iron. Use Table 15-3 to explain how this works and give an equation.

31. Use Table 15-3 to explain why highly acidic foods should not be wrapped in aluminum foil.

32. When silver metal becomes tarnished, it is coated with a compound of silver. The tarnish can be removed by wrapping a silver object in an aluminum foil and placing it in a container of salty water. Use Table 15-3 to explain this. (Hint: The silver compound in the tarnish contains Ag^+ ions, which react; the salt is not involved in any chemical reaction.)

33. An alloy of silver that contains tin and some mercury can be used for tooth fillings. This alloy does not dissolve in the foods we eat. Use Table 15-3 to explain why silver and mercury are good metals for tooth fillings.

34. Use Table 15-3 to explain why the metals silver, mercury, and copper are sometimes found in nature as uncombined metals, whereas other metals, especially, sodium, calcium and potassium, are never found in nature as uncombined metals.

35. Acidic air pollutants are absorbed in rain to make acid rain. Use Table 15-3 to explain why acid rains encourage the decomposition and rusting of iron bridges and other iron structures.

36. Use Table 15-3 to explain why it can be dangerous for sodium or potassium metal to come in contact with water.

37. It is sometimes said that metals "dissolve" in acids. Give an explanation of this statement using Table 15-3.

38. What is rust and how does it occur?

39. How can rust be prevented or inhibited?

Section 15-11

40. What is a battery and how does a battery function? What are the anode and cathode of a battery?

41. Describe a lead storage battery and give the overall discharge reaction. How is a lead storage battery recharged and what happens when it is recharged?

42. Describe a fuel cell. What advantage does a fuel cell have over a battery?

Sections 15-12 and 15-13

43. What is electrolysis?

44. What reaction occurs when a concentrated solution of sodium chloride is subjected to electrolysis?

45. List some examples of the use of electrolysis.

46. Describe the industrial process for making aluminum metal.

47. Why is the recycling of aluminum important?

Questions to Ponder

48. Since some batteries contain toxic metals, do you believe batteries should have deposits that can be redeemed when they are recycled? Argue your case.

49. Aluminum refining plants use relatively large amounts of electricity. Often, such plants pay discount prices for electricity. State whether or not you believe aluminum refining should be subsidized with low electricity rates. Explain your views.

CHAPTER
16

NUCLEAR

CHEMISTRY

16-1 THE NUCLEAR AGE

The nuclear age with all its perils and potential promise began just over 100 years ago. It is interesting that the story starts with the discovery of X rays by Wilhelm Roentgen, a German physicist, in 1895. Roentgen was studying the phenomenon of luminescence by which certain chemicals, when exposed to light, emit light of a different wavelength. You are probably most familiar with a type of luminescence called phosphorescence in which a material glows after light is shined on it. In any case, Roentgen was investigating the effect of cathode rays on chemicals that had luminescent properties. He covered the cathode-ray tube in black paper and observed that the luminescence still occurred even using the covered tube. Some strange radiation or ray was coming from the cathode-ray tube that penetrated thick paper and even thin sheets of metal. He called the radiation X rays, using X for the unknown. Today, of course, we know that X rays are one, although a high energy and penetrating, form of electromagnetic radiation. The original name is still used. In short time, it was learned that X rays essentially passed through soft tissue but were absorbed by bone and, thus, could be used to "look" inside the body. As a result, X radiation was soon used for medical purposes; it was some years later that scientists came to realize how potentially dangerous it is. So momentous was Roentgen's discovery that in 1901 he was awarded the very first Nobel Prize in physics.

Only one year later, Antoine Becquerel, a French physicist, was studying fluorescence, a type of luminescence in which certain chemicals emit light when energized. He thought that such chemicals could possibly emit X rays. He wrapped some photographic film in black paper, placed fluorescing minerals on top of the film, and planned to expose the minerals to sunlight. Before doing so, he stored the film and minerals in a dark drawer. For some reason, he decided to develop the film after taking it from the dark drawer to see if the film had been exposed. To his surprise, the film was quite ex-

511

posed. He realized that he had discovered some unknown source of penetrating radiation, different from X rays, that came from the mineral and passed through the paper to expose the film. He had discovered a phenomenon in which some material in the mineral, without an outside source of energy, spontaneously produced energetic and penetrating radiation.

Marie Curie

Soon after Becquerel announced his discovery, it caught the attention of Marie Sklodowski Curie, a recent graduate of the Sorbonne University in Paris. She had just graduated at the top of her class in the field of physical chemistry. Soon she realized that Becquerel's discovery was a result of the spontaneous, uncontrolled decay of uranium atoms that produced radiation. She coined the term *radioactivity* to describe the process. In this way, Curie began an illustrious career as one of the most significant pioneers in the field of radioactivity. She and her husband, Pierre, investigated uranium ores, and in 1898 discovered two previously unknown elements, both of which were radioactive. They named these elements polonium, after Marie's country of birth Poland, and radium, after the term *radiation* or *radioactivity.* Pierre died in 1906 and Marie, as a single parent of two daughters, carried on her scientific work, which brought her two Nobel Prizes, one in physics and one in chemistry. In addition, she is the only prize winner to have a daughter who also won the Nobel Prize. In 1934, at the age of 67, she died of leukemia, which most likely was caused by her many years of working with radioactive materials. In the early days of research in radioactivity, the dangers were not known.

16-2 ATOMIC NUCLEI

Chemistry deals with the structure and behavior of atoms. Nuclear chemistry deals with the structures and behaviors of atomic nuclei, the tiny, massive centers of atoms. For purposes of discussion, we can view an atomic nucleus as a clump or aggregate of protons and neutrons. **Protons** and **neutrons** as nuclear particles are called **nucleons.** As we will see, other particles known as beta particles, alpha particles, and gamma rays are associated with the properties of nuclei. Table 16-1 has a summary of the symbols and properties of these nuclear particles.

Table 16-1   *Common Nuclear Particles*

NAME	MASS (U)	CHARGE	SYMBOLS
Proton	1.007825	1+	^1_1H, p
Neutron	1.008665	0	$^1_0 n$, n
Electron	0.000549	1−	$^0_{-1}e$, e^-
Alpha particle	4.00260	2+	^4_2He, α
Gamma ray	0	0	γ, $h\nu$

All atoms contain nuclei surrounded by electrons. When discussing the properties of nuclei, we will view them without consideration of these electrons. Keep in mind, however, that except a few important cases, nuclei are always parts of atoms, not separate particles. The nuclei of the atoms of a given element always have the same number of protons; this number is the **atomic number** of the element. Recall that **isotopes** are atoms of a given element that contain different numbers of neutrons. Most elements have atoms with more than one combination of neutrons and protons. The types of nuclei found in the isotopes of the various elements are called **nuclides.** Nuclide is a general term that refers to the various types of known nuclei. Many nuclides occur in nature, and others are made by methods discussed in Section 16-12.

A nuclide is represented by a special symbol. The atomic number gives the number of protons in a nuclide. By definition, the **nucleon number** or **mass number** of a nuclide is the sum of the number of protons and the number of neutrons. It is found by adding the number of neutrons to the number of protons. A specific nuclide is represented by a symbol like

$$_Z^A \text{X}$$

where X is the symbol of the element corresponding to the nuclide, A is the nucleon number, and Z is the atomic number. A few examples of nuclides are

$$_1^1\text{H} \qquad _2^1\text{H} \qquad _6^{12}\text{C} \qquad _8^{16}\text{O} \qquad _{92}^{238}\text{U}$$

Read a nuclide symbol as the element name followed by the nucleon number (i.e., $_1^1\text{H}$, hydrogen one; $_1^2\text{H}$, hydrogen two; $_6^{12}\text{C}$, carbon twelve; $_8^{16}\text{O}$, oxygen sixteen, $_{92}^{238}\text{U}$, uranium two-thirty-eight) A nuclide is sometimes referred to by the element name followed by the nucleon number. For instance, the above nuclides are hydrogen-1, hydrogen-2, carbon-12, oxygen-16 and uranium-238.

Hundreds of nuclides are known, and descriptions of them are available on the Web or in reference books. Table 16-2 lists the common nuclides of a few elements. Notice that if the nucleon number of a nuclide is known, the number of neutrons in a nuclide is easily found by subtracting the atomic number from the nucleon number: $A - Z$.

ACTIVITY 16-1

An atomic nucleus is very small compared with the diameter of an atom. Place a dime on the ground and imagine it is the nucleus of a hydrogen atom. Walk about one-half a kilometer or three-tenths of a mile from the dime and imagine that you are at the outer limits of the electron space of the hydrogen atom.

ACTIVITY 16-2

Make a list of all the isotopes in Table 16-2 and indicate the number of protons and neutrons in each isotope. In addition, include the neutron-to-proton ratio for each isotope, found by dividing the number of neutrons by the number of protons.

Table 16-2 *The Common Nuclides of Some Elements*

Hydrogen	$_1^1\text{H}$	$_1^2\text{H}$	$_1^3\text{H}$
Helium	$_2^3\text{He}$	$_2^4\text{He}$	
Carbon	$_6^{12}\text{C}$	$_6^{13}\text{C}$	$_6^{14}\text{C}$
Oxygen	$_8^{16}\text{O}$	$_8^{17}\text{O}$	$_8^{18}\text{O}$
Uranium	$_{92}^{234}\text{U}$	$_{92}^{235}\text{U}$	$_{92}^{238}\text{U}$

Nuclear

16-3 RADIOACTIVITY

Most naturally occurring nuclides are stable and retain their structure indefinitely. Some nuclides, however, are not stable and spontaneously come apart or decay over time. Unstable nuclides are radioactive. For example, the nucleus of a uranium-238 atom can spontaneously break apart to form a helium-4 nucleus (two protons and two neutrons) and a new nucleus containing all the remaining protons and neutrons.

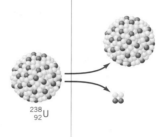

$^{238}_{92}U$

In a collection of radioactive nuclei, such spontaneous decay does not occur all at once. Instead, it continues over time at a characteristic rate until eventually all the nuclei in the collection decay.

By definition, **radioactivity** is the spontaneous decay of a nucleus to form another nucleus and a nuclear particle. Some nuclides are radioactive and some are not. Radioactivity relates to the ratio of the number of neutrons to protons in a nucleus. Generally, the ratio of neutrons to protons in stable nuclides ranges from a minimum of 1 to 1 to a maximum of about 1.6 to 1. A nuclide that has too low of a **neutron-to-proton ratio** is likely radioactive. Furthermore, a nuclide with too high a neutron-to-proton ratio is also likely radioactive. In fact, all nuclides with more than 83 protons are radioactive. Hence, all isotopes of elements beyond bismuth are radioactive. Apparently all nuclei with more than 83 protons spontaneously decay no matter how many neutrons are present. Look at a periodic table to see which elements follow bismuth, atomic number 83.

16-4 TYPES OF RADIOACTIVE DECAY

A radioactive nucleus decays by the splitting off of a nuclear particle and forming a new nucleus. In other words, the nucleus decays by breaking into a small nuclear particle and a different nucleus. The nuclear particles produced during radioactive decay are ejected from the original nucleus with large amounts of kinetic energy. Such particles radiate or move out in all directions from the collection of nuclei. This is why they are called nuclear radiation. These energetic nuclear particles make radioactive substances dangerous, but sometimes, in certain controlled situations, they are useful.

Radioactive decay occurs in specific ways called types or modes of decay. The three most common types of decay are described below. Other less common types are known, but they are not relevant to our discussion.

Alpha Particle Decay

A nucleus, called the **parent,** decays by emitting a high-speed helium-4 nucleus known as an **alpha particle (α).** The loss of an alpha particle forms a new nucleus, called the daughter. The **daughter nucleus** has a nucleon number that is four less than the parent nucleus and an atomic number that is two less. Alpha decay is represented by the general nuclear equation

$$\underset{\substack{\text{parent} \\ \text{nucleus}}}{^{A}_{Z}\text{X}} \longrightarrow \underset{\substack{\text{daughter} \\ \text{nucleus}}}{^{A-4}_{Z-2}\text{Y}} + \underset{\substack{\text{alpha} \\ \text{particle}}}{^{4}_{2}\text{He}}$$

where X is the parent element and Y is the newly formed daughter element. Some examples of alpha decay are

Uranium-238 decays to form thorium-234:

$$^{238}_{92}\text{U} \longrightarrow {}^{234}_{90}\text{Th} + {}^{4}_{2}\text{He}$$

Radium-226 decays to form radon-222:

$$^{226}_{88}\text{Ra} \longrightarrow {}^{222}_{86}\text{Rn} + {}^{4}_{2}\text{He}$$

The numerous alpha particles emitted from a sample of a radioactive substance are called **alpha radiation.** The alpha particles radiate from the sample in all directions.

A nuclear equation is different from a chemical equation but does represent the balance or conservation of nucleons characteristic of that nuclear process. In a **nuclear equation,** the sums of the nucleon numbers on each side of the arrow are equal and the sums of the atomic numbers are equal. Confirm this by summing the nucleon numbers in the superscripts and the atomic numbers in the subscripts.

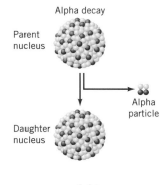

Alpha decay

Parent nucleus

Alpha particle

Daughter nucleus

ACTIVITY
16-3

In each of the two nuclear equations given as examples of alpha decay, confirm that the number of nucleons is conserved by adding the mass numbers and atomic numbers on each side of the arrow. What is the difference between the parent and daughter in each case? Find each parent and daughter in the periodic table.

Beta Particle Decay

Some nuclei decay by emitting a high-speed electron called a **beta particle** **(β).** The daughter nucleus formed by beta decay has the same nucleon number as the original nucleus and an atomic number that is one greater.

$$\underset{\substack{\text{parent} \\ \text{nucleus}}}{^{A}_{Z}\text{X}} \longrightarrow \underset{\substack{\text{daughter} \\ \text{nucleus}}}{^{A}_{Z+1}\text{Y}} + \underset{\substack{\text{beta} \\ \text{particle}}}{^{0}_{-1}e}$$

Beta particles are electrons that come from nuclei. During beta decay, it appears that a neutron in the nucleus decays to form a proton and an elec-

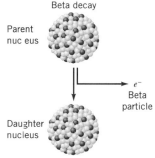

Beta decay

Parent nucleus

Beta particle

Daughter nucleus

tron ($n \rightarrow p + e^-$). The electron is ejected as a beta particle and the proton remains in the nucleus. Thus, beta decay of a nucleus decreases the number of neutrons by one and increases the number of protons by one. Some examples of beta decay are

Plutonium-241 decays to give americium-241:

$$^{241}_{94}\text{Pu} \longrightarrow {}^{241}_{95}\text{Am} + {}^{0}_{-1}e$$

Bismuth-210 decays to give polonium-210:

$$^{210}_{83}\text{Bi} \longrightarrow {}^{210}_{84}\text{Po} + {}^{0}_{-1}e$$

The numerous beta particles emitted from a sample of a radioactive substance are called **beta radiation.** The beta particles radiate from the sample in all directions.

Gamma Ray Emission

Sometimes, the daughter nucleus formed in an alpha decay or a beta decay is produced in an energetically excited state. It is in a higher than normal energy state. This excited energy state is not the normal state of the nucleus, and it quickly drops to a lower energy state. In the process, the nucleus emits a form of radiant energy or electromagnetic radiation called a **gamma ray (γ).** This process is similar to the emission of radiant energy by excited electrons in energized atoms as discussed in Chapter 8. The difference is that excited nuclei emit much-higher-energy radiation than excited atoms. Gamma rays, like X rays, are a form of radiant energy, but gamma rays normally have higher energies than X rays. The emission of gamma radiation by a radioactive substance can accompany the alpha radiation or beta radiation produced by the decay of nuclides of the substance.

In summary, the three major types of radiation are alpha radiation, beta radiation, and gamma radiation. Nuclides that are **alpha emitters** produce alpha radiation (α). Some alpha emitters also produce gamma radiation (α, γ). Nuclides that are **beta emitters** produce beta radiation (β). Some beta emitters also involve gamma radiation (β, γ). Some examples are

Thorium-230 decays by alpha and gamma emission:

$$^{230}_{90}\text{Th} \longrightarrow {}^{226}_{88}\text{Ra} + {}^{4}_{2}\text{He} + \gamma$$

thorium-230   alpha   gamma

Lead-210 decays by beta and gamma emission:

$$^{210}_{82}\text{Pb} \longrightarrow {}^{210}_{83}\text{Bi} + {}^{0}_{-1}e + \gamma$$

lead-210   beta   gamma

16-5 COMPLETING NUCLEAR EQUATIONS

If a parent nucleus is known to decay by a specific decay mode, we can predict the daughter by considering that the nucleon numbers and atomic numbers are equal in a nuclear equation. Note, however, that a beta particle has a negative charge. When a beta particle occurs in a nuclear equation, its atomic number is treated as though it is -1. For example, hydrogen-3, the radioactive isotope of hydrogen, decays by beta particle, $_{-1}^{0}e$, emission. What is the daughter produced? The incomplete nuclear equation is

$$_{1}^{3}\text{H} \longrightarrow _{-1}^{0}e + ?$$

The product must have a nucleon number of 3 to equal the 3 on the left and its atomic number must be 2, since $2 - 1$ on the right equals 1 on the left. The equation becomes

$$_{1}^{3}\text{H} \longrightarrow _{-1}^{0}e + _{2}^{3}?$$

By consulting a list of elements for atomic number 2, we see that the product must be an isotope of helium having the symbol $_{2}^{3}\text{He}$.

$$_{1}^{3}\text{H} \longrightarrow _{-1}^{0}e + _{2}^{3}\text{He}$$

EXAMPLE 16-1

Uranium-235 decays by alpha emission.

Consulting a list of elements for the atomic number of uranium and using $_{2}^{4}\text{He}$ for an alpha particle gives the incomplete nuclear equation

$$_{92}^{235}\text{U} \longrightarrow _{2}^{4}\text{He} + ?$$

Since the sum of the nucleon numbers of the products must be equal to the nucleon number of the parent, the daughter product has a nucleon number of 231 ($235 - 4$). Furthermore, since the sum of the atomic numbers must be equal on each side, the daughter has an atomic number of 90 ($92 - 2$). By consulting a list of elements for atomic number 90, it is found that the daughter product must be an isotope of thorium having the symbol $_{90}^{231}\text{Th}$. Thus, the complete nuclear equation is

$$_{92}^{235}\text{U} \longrightarrow _{2}^{4}\text{He} + _{90}^{231}\text{Th}$$

ACTIVITY
16-4

Complete the following incomplete nuclear equations:

(a) $_{94}^{241}\text{Pu} \longrightarrow _{95}^{241}\text{Am} + ?$

(b) $_{88}^{226}\text{Ra} \longrightarrow ? + _{2}^{4}\text{He}$

16-6 HALF-LIFE

Radioactive decay is random and spontaneous. It is not possible to stop it or prevent it. Such decay does follow a pattern that depends on the total number of radioactive nuclei in the sample. Radioactive decay is somewhat similar to the popping of corn. Imagine a collection of hot corn kernels. They begin to pop and continue to pop over time until the last few kernels pop. Obviously, the nuclei of a radioactive isotope are not heated like popcorn. They spontaneously decay and continue to decay over a time until they all decay. When popping corn, you may have noticed that at first many kernels pop, then the frequency of pops decreases as the number of kernels decreases. In a radioactive sample, the frequency of decays is great at first, then the activity decreases as the number of nuclei decreases. The rate of decay is proportional to the number of nuclei, so the frequency is great at the beginning and then tapers off.

A sample of a radioactive element continues to decay over time until all the nuclei decay. For some nuclides, this happens within seconds, minutes, hours, or days. For other nuclides, it may require months, years, or thousands of years for total decay. The rate of decay is a characteristic of a given radioactive nuclide. A common way to express the rate of decay of a radioactive element is in terms of its half-life.

Half-life is the time required for the decay of one-half of a sample of a radioactive substance. For instance, if we start with 10 g of radon-222, after 4 days 5 g remain, after 8 days 2.5 g remain, and so on. The amount of radon-222 decreases by one-half every 4 days, so the half-life is 4 days. Figure 16-1 illustrates the idea of a half-life. The half-lives of nuclides vary widely, ranging from fractions of seconds to billions of years. Each radioactive nuclide has a characteristic half-life. Table 16-3 lists a few typical half-lives.

FIGURE 16-1 To understand half-life, start with a given amount of radioactive nuclide, *N*. One-half the amount remains after one half-life has passed. One-fourth remains after two half-lives pass. The halving continues until the sample has completely decayed.

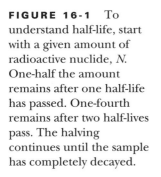
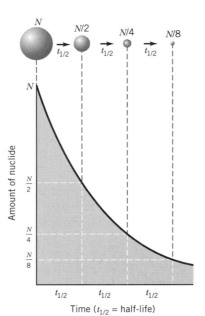

Table 16-3   *Some Typical Half-Lives*

NUCLIDE	SYMBOL	HALF-LIFE
Uranium-238	$^{238}_{92}U$	4.5 billion years
Plutonium-239	$^{239}_{94}Pu$	24,100 years
Carbon-14	$^{14}_{6}C$	5730 years
Americium-241	$^{241}_{95}Am$	432 years
Cobalt-60	$^{60}_{27}Co$	5.3 years
Iodine-125	$^{125}_{53}I$	60 days
Radon-222	$^{222}_{86}Rn$	3.8 days
Iodine-123	$^{123}_{53}I$	13.1 hours
Californium-242	$^{242}_{98}Cf$	3.5 minutes
Lawrencium-256	$^{256}_{103}Lr$	28 seconds
Polonium-214	$^{214}_{84}Po$	0.00016 second

The half-lives of nuclides are measured by experiment. The radioactivity of a sample of an element is found by periodically measuring the activity (in terms of the number of decays per unit time) of a sample over an appropriate period. As illustrated in Figure 16-2, the half-life is found from a plot of the activity versus time. As a sample decays, the activity is great when there are numerous radioactive nuclei and the activity decreases as the number of radioactive nuclei decreases. The decrease in activity with time follows a pattern, as illustrated in Figure 16-2. The half-life is deduced from the pattern. Incidentally, special experimental methods are needed to measure half-lives that are very long or very short.

Carbon-14

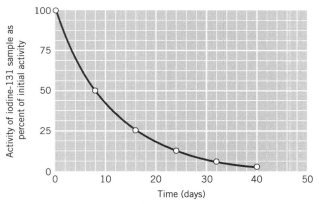

FIGURE 16-2   A plot of the activity of an iodine-131 sample versus time.

Carbon-14 Dating

Carbon-14 is a radioactive isotope of carbon having a half-life of 5730 years. It is formed in the atmosphere by a nuclear process involving atmospheric nitrogen. This process is a consequence of cosmic radiation striking nitrogen in Earth's atmosphere. Cosmic radiation contains high-energy alpha particles and protons that continually enter Earth's atmosphere from outer space (from the cosmos). Carbon-14 is continually formed as a result of cosmic radiation, and since it decays over time, there is a very low but constant steady-state concentration of carbon-14 in the atmosphere. Carbon-14 combines with oxygen to make radioactive carbon dioxide in the atmosphere. It appears that the concentration of radioactive carbon dioxide in the atmosphere has been essentially constant for thousands of years. Since plants use carbon dioxide in photosynthesis, all plants contain small amounts of carbon-14 distributed among the carbon-containing compounds. Moreover, since animals eat plants or plant eaters, all animals also contain small amounts of carbon-14 incorporated in carbon-containing compounds. In fact, all living plants and animals have a steady-state amount of carbon-14 since as some decays, more is assimilated by photosynthesis or the metabolism of foods. This fact has been experimentally confirmed by measuring the amount of carbon-14 per gram of recently living plant and animal tissue.

Once a plant or animal dies, no additional carbon-14 is added by life processes. The carbon-14 present at death decays and slowly decreases the amount of carbon-14 in the dead organism. Carbon-14 dating is a method of estimating the age of an ancient organic object by measuring the amount of carbon-14 per gram of the object and comparing this with the amount present in living organisms. For instance, if the measurement reveals that the carbon-14 content of an object is one-half the content in living organisms, the object must be about 5730 years old since this is one half-life of carbon-14. The carbon-14 content of a carbon-containing object is found by carefully measuring the activity of the sample due to this radioactive isotope. The age of the object is then estimated in reference to living systems.

Carbon-14 dating has been compared to dates obtained from tree rings and other objects of known age and is found to be accurate. Dating in this way has been a useful tool for estimating the age of many samples of organic matter found in archaeological sites and objects of historical interest. Some examples of the age of objects found by carbon-14 dating are tabulated here.

OBJECT	AGE (YEARS)
Charcoal from Lascaux cave, France	15,516
Hair from ancient Egyptian burial site	5,744
"Ice man" found in the Alps	5,300
Wood from an Egyptian tomb	2,190
Dead Sea Scrolls (Book of Isaiah)	1,917
Ancient corncobs from Texas site	1,553
Native American House Pit in Oregon	430

The carbon-14 activity in a piece of charcoal found at an archaeological site was three-fourths of the activity in living trees. What is the estimated age of the charcoal?

16-7 RADIOACTIVITY LEVELS

When working with a radioactive sample, it is important to know how radioactive it is or how "hot" it is. The **becquerel, Bq,** is the unit of radioactivity corresponding to 1 radioactive decay per second. This unit is named in honor of Antoine Becquerel, the French physicist who first discovered nuclear radiation. The curie is another unit used to express intensity of radioactivity. A **curie, Ci,** is a unit of activity equal to 37 billion decays per second, or 3.7×10^{10} Bq. It is named in honor of Marie Curie.

The actual amount in grams of a radioactive material that has an activity of 1 curie is related to the rate of decay of a radioactive element. For example, 1 g of radium-226 produces 1 curie of radiation. A curie is a relatively large amount of radioactivity. Normally, for safety, samples with **millicurie** (10^{-3} Ci) and **microcurie** (10^{-6} Ci) activities are used in laboratory work. Radioactive materials must be used with care and appropriate shielding.

As discussed in Section 16-11, nuclear radiation is dangerous to living systems. Special units are used to refer to radiation doses as they relate to humans. One **rad** (radiation-absorbed dose) is the radiation dose that transfers 1×10^{-5} J of energy per gram of tissue. The effect of radiation depends on the type of radiation and also its energy. The rem (radiation equivalent human) is used as a unit of radiation dose. A **rem** value is determined by measuring the dose in rad and multiplying by factors related to the kind of radiation (i.e., alpha, beta, gamma, or X rays, or neutrons). It is estimated that the minimum lethal whole-body dose for humans (50% die in 30 days) is 590 rem. Fortunately, such high doses of radiation are rare. Unfortunately, many victims of the use of atomic bombs in the Japanese cities of Hiroshima and Nagasaki and victims of the Chernobyl nuclear accident (see Section 16-22) received lethal rem doses and died of radiation exposure. A typical American is exposed to less than 0.2 rem in an entire year. This exposure is mainly from low-level environmental sources. All humans are constantly exposed to nuclear radiation from cosmic radiation, naturally occurring isotopes, radioactive materials that enter the environment from nuclear power plants, and fallout from nuclear weapons testing. This low-level but relatively constant environmental radiation is called **background radiation.**

Carefully take the cover off a battery-operated smoke detector. Read the label and record any information about the amount of radioactive material used in the detector. The isotope americium-241, used in smoke detectors, is an alpha emitter. Write a balanced nuclear equation showing its alpha decay and identify the daughter isotope. Caution: Never dismantle or take apart the pieces of a smoke detector.

Radiation

16-8 Radioactive Elements

Nearly 330 different nuclides are found in nature. These nuclides represent the naturally occurring isotopes of the elements. Some elements have only one naturally occurring nuclide (e.g., fluorine: $^{19}_{9}F$), whereas other elements have many natural nuclides (e.g., tin: $^{112}_{50}Sn$, $^{114}_{50}Sn$, $^{115}_{50}Sn$, $^{116}_{50}Sn$, $^{117}_{50}Sn$, $^{118}_{50}Sn$, $^{119}_{50}Sn$, $^{120}_{50}Sn$, $^{122}_{50}Sn$, $^{124}_{50}Sn$). Most natural nuclides are stable and are not radioactive. Recall that all the isotopes of the 29 elements of atomic number greater than bismuth are radioactive. A few naturally occurring nuclides of elements with atomic numbers lower than bismuth, such as $^{204}_{82}Pb$ and $^{40}_{19}K$, are radioactive.

Radioactive nuclides are unstable and undergo decay to form other radioactive nuclides or stable nuclides. Thus, the daughter of a specific decay is often radioactive, although sometimes the daughter is stable. Studies of the naturally occurring elements show that all radioactive nuclides of atomic number greater than 82 belong to one of three **decay chains or series.** In other words, these elements decay by alpha or beta emission according to definite patterns. Figure 16-3 shows the radioactive decay series that starts with uranium-238, which is the most common uranium isotope. In this series, uranium-238 decays into thorium-234, which, in time, decays into another radioactive daughter, and so on, until stable lead-206 forms. Other natural radioactive nuclides belong to one of the other series.

FIGURE 16-3
Uranium-238 radioactive decay series. The half-lives of the nuclides are shown along with the mode of decay (y = years, d = days, h = hours, m = minutes, and s = seconds).

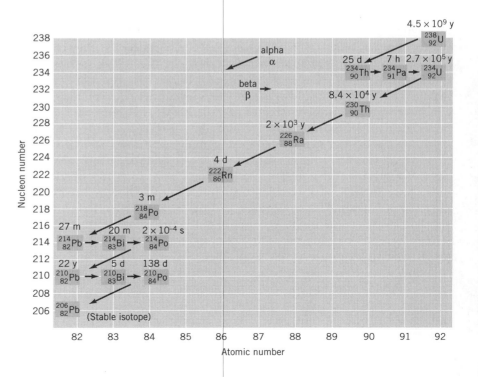

Knowledge of a radioactive decay series provides a way to estimate the age of Earth rocks and meteorites. If uranium-238 was deposited when the crust of Earth formed, a sample of undisturbed rock containing uranium-238 is expected to contain a certain amount of lead-206, the end product of the uranium-238 decay chain. The amount of lead-206 depends on how long the uranium had been decaying. Since the half-life of uranium-238 is 4.5×10^9 years, a measurement of the relative amounts of uranium-238 and lead-206 suggests the age of the rock. In one half-life, 1 g of $^{238}_{92}U$ would decay to form about 0.4 g of $^{206}_{82}Pb$, and 0.5 g of $^{238}_{92}U$ remains. By using this and similar methods of rock dating, involving other isotopes, the age of Earth is estimated at 4.6 billion years.

16-9 IONIZING RADIATION

The emission of radiation is a characteristic of radioactivity. The emitted radiation is used to detect and identify radioactive substances; the radiation, however, may be very dangerous to living organisms. The particles given off during alpha or beta decay are high-speed, high-energy particles. These particles are ejected from the decaying nucleus and travel outward into the material surrounding the radioactive substance. Alpha or beta particles lose their energy by interacting with the atoms and molecules of matter. They collide with atoms or molecules, causing them to lose electrons, as illustrated in Figure 16-4. In other words, these particles lose their energies by causing atoms and molecules to ionize. The electron and positive ion produced by such an interaction are called an **ion pair.**

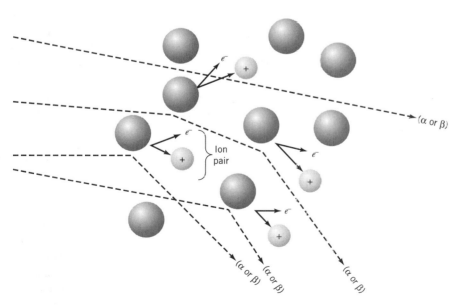

FIGURE 16-4 The formation of ion pairs occurs when ionizing radiation passes through matter. Some of the atoms and molecules are ionized, forming ion pairs. One ion pair consists of a separate electron and positive ion.

An alpha or beta particle passing through matter causes the formation of numerous ion pairs. The more massive alpha particles ($_2^4$He) are less penetrating than beta particles ($_{-0}^{-1}e$), but they produce more ion pairs. A typical alpha particle travels about 6 cm in the air and produces about 40,000 ion pairs, whereas a typical beta particle travels 1000 cm in the air and produces only about 2000 ion pairs. When these particles pass through materials more dense than air, they still produce ion pairs but travel shorter distances. Gamma radiation consists of electromagnetic radiation similar to X rays. Gamma rays are not massive like alpha and beta particles, but they also interact with matter to form ion pairs. Gamma rays are much more penetrating than alpha or beta radiation and can pass through dense samples of matter.

Because radiation causes the formation of ion pairs in the matter through which it passes, it is called **ionizing radiation.** Other less common types of nuclear decay processes produce radiation as X rays, and some produce high-energy, free-moving neutrons. These types of radiation also form ion pairs when they penetrate matter. Typical **X rays** are less penetrating than gamma rays, and **free neutrons** are an extremely penetrating form of radiation.

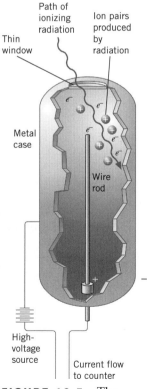

FIGURE 16-5 The operating principle of a Geiger–Müller tube.

16-10 DETECTION OF RADIATION

We cannot see, hear, or feel nuclear radiation. It is interesting, however, that some very radioactive samples have elevated temperatures because of the energy released by radioactive decay. Used fuel rods from nuclear reactors, for example, are stored in water to keep them from getting too hot. Methods of detecting radiation are based on the ionizing effects of the radiation. A typical device used to detect radiation is the **Geiger–Müller tube,** as illustrated in Figure 16-5. The tube consists of a metal cylinder with a thin plastic window on the end. The tube is filled with a special gas and has a thin metal rod in the center. Using an external source of power, a voltage difference is maintained between the rod and the cylinder. When a particle of ionizing radiation enters the tube, they cause the formation of ion pairs. The electrons formed by the ionization are strongly attracted to the positively charged center rod. They move toward the rod with great speed, and in the process, they act essentially like beta particles and produce more ion pairs. A relatively large number of electrons are formed and quickly flow toward the center rod.

The flow of electrons is equivalent to a small current passing through the tube. It stimulates an electronic counter, connected to the tube, that recognizes the current flow and registers it as an ionizing event caused by a particle of radiation. Each time a particle enters the tube and causes the electron "avalanche," the counter tallies the event. In this way, the Geiger–Müller tube and the counter provide a way to measure the **activity** of a radioactive substance. Silicon-based solid-state detectors are used as alternatives to Geiger–Müller tubes. These solid-state detectors are also used to detect the presence of nuclear radiation and to measure the activity of radioactive materials.

The number of decays per second or the number of decays per minute associated with a radioactive sample is measured with a radiation counter. This measurement is one way in which the half-life of a radioactive nuclide is found. Radiation counters are also used to detect the presence of radioactive elements and to find the amount of a radioactive element in a sample. Geiger counters are very useful, since nuclear radiation cannot be seen or felt. Geiger counters provide a way to "see and hear" radioactivity. In 1993, for instance, a small sample of radioactive cesium-137 was lost on a California freeway. The sample was found using a radiation detector.

16-11 RADIATION DANGER

Why is radiation dangerous? Gamma rays, X rays, and high-speed alpha and beta particles can penetrate living tissue and cause damage in varying degrees. Billions upon billions of high-speed particles and the accompanying gamma radiation are produced by a few milligrams of a radioactive substance. The radiation moves outward in all directions and penetrates any matter surrounding the substance. The radiation interacts with the matter and loses its energy by forming numerous ion pairs. Ionizing radiation can cause damage to biological materials since ion pairs within tissue cells can form altered and often highly reactive chemicals that react with other cell chemicals. These reactions disrupt normal cell chemistry and can cause permanent cell damage. Cells can, however, utilize chemical mechanisms to repair minor damage. Sometimes ionizing radiation damages cell chromosomes. Chromosome damage may cause the cell to malfunction or die. Occasionally, damage to chromosomes can cause cells to become cancerous, which results in tumor formation. Too much radiation is known to induce cancer.

Ionizing radiation can damage developing embryos and cause mutations in cells of the gonads (i.e., egg or sperm cells). Mutations result from damage to the genes of the chromosomes. The abnormal chromosomes are passed on to any children born from the cells. **Gene mutation** in these children can result in sickness, cancer, and even death. The extent of damage to cells caused by radiation depends on the amount of exposure that occurs. Obviously, **low doses** of radiation are not as dangerous as **high doses**, and **short-term exposure** is less dangerous than **long-term exposure.** When you get a medical X ray, this is a short-term exposure. X-ray technicians, on the other hand, have to protect themselves from long-term exposure. Obviously, you should avoid any unnecessary exposure to sources of radiation. Exposure to high levels of radiation may cause **radiation sickness** with symptoms of nausea, vomiting, and weakness. These symptoms are followed by a period characterized by fatigue, weight loss, fever, diarrhea, internal bleeding, and loss of hair. Cancer patients who are given radiation treatments suffer varying degrees of radiation sickness. If too much damage

CAUTION

RADIOACTIVE
MATERIAL

The
International
symbol for
radiation

ACTIVITY
16-7

Why should pregnant women avoid X rays, especially in the early stages of pregnancy? Call your local hospital or medical clinic and ask about their X-ray policy for pregnant women.

ACTIVITY 16-8

Suppose you have 1000 g of a radioactive element. Calculate the number of grams that remain after each half-life for 10 half-lives. Calculate the activity of any sample of radioactive material after 10 half-lives in terms of a percent of the initial activity. (Hint: Divide 100 by 2, ten times.)

Radon

is done to the body, a person can die of radiation sickness. Many people exposed to radiation during the World War II nuclear bomb attacks on the Japanese cities of Hiroshima and Nagasaki died of radiation sickness. Many more, however, died from the explosions and fires caused by the bombs.

Alpha particles cannot penetrate skin and are not dangerous externally. If you eat or inhale an alpha emitter, however, it can cause damage inside the body. Beta particles can penetrate the skin to some extent and may cause skin burns. Normal clothing serves to shield against exposure to beta particles. Gamma radiation, X rays, and neutrons can penetrate the body and thus are the most harmful type of external radiation. It is important to guard against undue exposure to these forms of penetrating radiation.

Since radioactive substances can be dangerous, they must be handled with special equipment and stored in protective, shielded containers. For example, radioactive wastes at nuclear power plants are stored in thick concrete-lined pools of water. How long a radioactive sample needs storage depends on the half-lives of the nuclides present. After about **10 half-lives** pass, the radioactivity of most nuclides decreases to a negligibly low level. Consequently, a general guideline is that after 10 half-lives, disposing of a sample of a radioactive element is possible. For example, the half-life of iodine-131 is 8.1 days, so a sample of this element needs storage for more than 81 days before disposal. Naturally, the safe disposal of a radioactive sample also depends on the type of element involved and the amounts of these elements. As discussed in Section 16-20, radioactive wastes from nuclear reactors must be stored for thousands of years before they are safe.

Radon in the Home

Rocks and soils contain small amounts of uranium and other radioactive elements that are daughters in the uranium decay series. These isotopes are partly responsible for background radiation, and they normally represent no hazard. One of the daughters of the uranium decay series radon-222, however, is a gas, and as a gas it can diffuse through soils and porous or cracked rocks. Furthermore, it can dissolve somewhat in water and accumulate in water wells. As a result, radon can accumulate in homes and buildings by diffusing through cracks and openings in foundations and through water pipes. Of course, this situation is only important in certain geographical regions where the uranium content of rocks and soils is relatively high.

ACTIVITY 16-9

Radon-222 is an alpha emitter. Write a nuclear equation to show the alpha decay of radon and identify the daughter. Write a nuclear equation for the alpha decay of the daughter and identify its daughter.

Radon-222 is an alpha emitter with a half-life of 3.8 days. It can potentially be harmful if inhaled. More significantly, it decays to give daughters that are various radioactive isotopes of polonium, bismuth, and lead. These radon daughters are solids, and small amounts can settle out in the environment and become attached to dust and smoke. When inhaled, these isotopes can lodge in the lungs and decay to produce alpha, beta, and gamma radiation. In this way, radon represents a lung cancer risk and is considered the leading cause of lung cancer among nonsmoking Americans.

Dangerous levels of radon are not common in most homes but may be significant in some geographical regions. Radon levels are highest in geological areas with granite and black shale covered by porous soils and where there are relatively high levels of uranium. Areas found of particular risk by the U.S. Environmental Protection Agency are eastern Pennsylvania, eastern Washington, Colorado, New Mexico, western Florida, and parts of Maine, New Jersey, and New York.

Radon levels are measured using commercially available test kits or with special instruments. Typically, indoor levels of radon are less than 1 picocurie per liter of air. A picocurie is 10^{-12} curie or a trillionth of a curie. If radon levels exceed 4 picocuries per liter, some remedial action is appropriate, such as patching foundation cracks, capping sump pumps, and increasing ventilation with fresh air.

16-12 NUCLEAR TRANSMUTATIONS

A high-speed nuclear particle can, under certain conditions, collide with a nucleus and cause a nuclear reaction that produces a different nucleus. **Transmutation** is the name given to such a nuclear reaction. In a transmutation, starting nuclei are changed to new nuclei. Some transmutations occur naturally, but most are done in nuclear laboratories using special techniques and equipment. An example of a transmutation is

$$^{14}_{7}\text{N} \quad + \quad ^{1}_{0}\text{n} \quad \longrightarrow \quad ^{14}_{6}\text{C} \quad + \quad ^{1}_{1}\text{H}$$

| target | projectile | product nucleus | particle |

In this reaction, the neutron is the projectile and the $^{14}_{7}\text{N}$ is the target nucleus. A transmutation involves a collision of a **projectile particle** with the **target nucleus** and results in new combinations of neutrons and protons as products. Incidentally, the above transmutation is the source of carbon-14 in the atmosphere. It forms when neutrons produced by cosmic radiation collide with nitrogen in the atmosphere.

Nuclear scientists use nuclear transmutations to prepare artificial nuclides. These are **humanmade isotopes.** Nearly all the naturally occurring nuclides have been used as targets, and a variety of nuclear particles, such as protons ($^{1}_{1}\text{H}$), deuterons ($^{2}_{1}\text{H}$), neutrons ($^{1}_{0}n$), alpha particles ($^{4}_{2}\text{He}$), and beta particles ($^{0}_{-1}e$), have been used as projectiles. The projectiles are given relatively high kinetic energies before being projected at the target nuclei. Particle accelerating devices such as cyclotrons, linear accelerators, synchrotrons, and nuclear reactors produce these high-energy projectiles. Special equipment is needed to produce humanmade isotopes. All the elements having atomic numbers greater than uranium (92) do not occur in nature. These transuranium elements (93 to 112) have been synthesized by nuclear scientists using specific transmutations. The following transmutation sequence is an example of the synthesis of the transuranium elements neptunium and plutonium.

$$\,^{238}_{92}U + \,^{1}_{0}n \longrightarrow \,^{239}_{92}U \longrightarrow \,^{239}_{93}Np \longrightarrow \,^{239}_{94}Pu$$

In this process, uranium-238 is the target and a neutron is the projectile. Uranium-238 captures a neutron to form uranium-239. Uranium-239 decays by beta emission to give the transuranium element named neptunium. Neptunium-239 decays by beta emission to give the transuranium element plutonium, which is an alpha emitter. The half-life of uranium-239 is 23.5 minutes, the half-life of neptunium-239 is 2.35 days, and the half-life of plutonium-239 is 24,100 years.

16-13 NUCLEAR FISSION

Some nuclear decay processes produce neutrons as products. The neutrons produced by these processes serve as projectiles in other nuclear transmutations. In 1938, German scientists Otto Hahn and Lise Meitner made a startling report about their research on the use of neutrons as projectiles with uranium target nuclei. This research showed that the isotope uranium-235 undergoes a wholly different kind of transmutation. Upon the capture of a neutron, the uranium-235 nucleus forms an unstable nucleus that splits into two nuclei of smaller size plus a few neutrons.

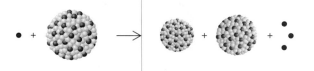

Nuclear fission is the name given to this unique type of transmutation. Fission means to break apart. A typical fission of uranium-235 is represented by the nuclear equation

$$\,^{235}_{92}U + \,^{1}_{0}n \longrightarrow \,^{85}_{36}Kr + \,^{148}_{56}Ba + 3\,^{1}_{0}n + \text{energy}$$

This example is just one of many possible fission reactions of uranium-235. When uranium-235 undergoes fission, the unstable nucleus can split in a variety of ways. Therefore, fission produces a mixture of **fission product nuclei.** As a simple analogy, if you try to break many pieces of chalk in half, not all pieces end up the same size. You would have an assortment of chalk "fission products." A specific fission event forms two fission product nuclei and some neutrons. The possible product nuclei of the fission of uranium-235 range in nucleon number from about 70 to 165. Most nuclei formed by fission are radioactive and some are stable. Thus, the collection of fission products includes a variety of elements that are highly radioactive.

Only a few kinds of isotopes undergo fission. These are called **fissile isotopes.** The most important fissile isotopes are **uranium-235** and **plutonium-239.** Uranium-235 is a rare natural isotope of uranium. It makes up about 0.7% of natural uranium. The most common isotope of uranium is uranium-238. Plutonium-239 does not occur in nature, but it is made by the transmutation described in Section 16-12. Plutonium is also formed from uranium in nuclear reactors as discussed in Section 16-23. Some isotopes of neptunium and americium are also fissile isotopes.

There are two important characteristics of fission. First, as indicated in the example of uranium-235, fission always gives neutrons as products. In fact, normally two or more neutrons form for every one neutron used to initiate a fission. Second, when fission occurs, more stable product nuclei form from the less stable parent nucleus. This change from less energetically stable to more stable nuclei produces a relatively large amount of energy. This energy represents **nuclear energy** that accompanies fission. The complete fission of 1 g of uranium-235 can produce about 80 billion joules of energy. The complete fission of 1 kg of uranium-235 produces as much energy as the burning of about 2600 tons of coal or 14,000 barrels of crude oil. The energy of fission is the basis for the production of atomic energy or nuclear energy.

16-14 ENERGY FROM FISSION

During fission, smaller, more stable nuclei form when energy is released. What is the source of this nuclear energy? Consider a typical fission of uranium-235 and the masses of the nuclei and particles involved.

$$^{235}_{92}\text{U} \quad + \quad ^{1}_{0}n \quad \longrightarrow \quad ^{94}_{38}\text{Sr} \; - \; ^{139}_{54}\text{Xe} \quad + \quad 3^{1}_{0}n$$

relative
masses   235.0439   1.0087   93.9154 138.9188   3(1.0087)

Adding the masses on each side of the equation gives 236.0526 on the left and 235.8603 on the right. In nuclear fission, loss of mass occurs as it changes to energy. This loss is the basis of nuclear energy. In the above fission process, when 235.0439 g of uranium-235 undergo fission, the loss in mass is the difference between the initial and final masses (236.0526 g − 235.8603 g = 0.1923 g). In a sense, mass is a form of energy. Energy is released in nuclear transmutations and nuclear decays because mass is changed to energy. The laws of conservation of matter and energy state that mass and energy are conserved in normal chemical processes. Nuclear decay and transmutations are not normal chemical processes. Mass is lost during these nuclear processes, and the loss changes to a corresponding amount of energy.

The relation between mass and energy is given by the famous formula first stated by Albert Einstein:

$$E = mc^2$$

where m is the mass in kilograms and c is the speed of light in meters per second. The speed of light refers to the speed at which radiant energy passes through empty space. The formula is used to calculate amounts of energy produced by nuclear processes.

EXAMPLE 16-2

In the fission of a 0.235 kg sample of uranium-235, 1.9×10^{-4} kg of mass is converted to energy. If the speed of light is 3.00×10^8 m/s and 1 J = 1 kg m²/s², how many joules of energy are produced by the fission?

The relationship between mass and energy is $E = mc^2$. To obtain the energy in joules, we use the mass in kilograms and the speed of light in meters per second.

$$E = mc^2 = (1.9 \times 10^{-4} \text{ kg})(3.00 \times 10^8 \text{ m/s})^2$$
$$= (1.9 \times 10^{-4} \text{ kg})(9.00 \times 10^{16} \text{ m}^2/\text{s}^2)$$
$$= 1.7 \times 10^{13} \text{ kg m}^2/\text{s}^2 \quad \text{or} \quad 1.7 \times 10^{13} \text{ J}$$

To put this number into perspective note, that it is equivalent to the energy released by the burning of about 20 million moles of methane gas.

ACTIVITY 16-10

What is the mass in grams of 20 million moles of methane? How many tons is this amount of methane?

ACTIVITY 16-11

Considering the nature of a chain reaction, why do you suppose the striking surface of paper matchbook is placed where you typically find it?

16-15 FISSION CHAIN REACTIONS

Neutrons are always products of the fission of uranium-235. The neutrons formed during fission can cause other uranium-235 nuclei to undergo fission. Each fission produces more neutrons, which in turn can cause further fission. In small samples of uranium-235, many neutrons are ejected into the surroundings. The fission that does occur in small samples is under control. As shown in Figure 16-6, if a sample contains enough uranium-235 nuclei, a **fission chain reaction** can occur. In a chain reaction, one fission step gives neutrons that cause further fission steps, these can cause still others, and so on. If, on the average, one neutron formed by a specific fission causes another fission, a fission chain reaction becomes self-propagating and self-sustaining. As an analogy, if we strike a match and use it to light another match, and so on, the process is self-propagating and we get a continuous supply of energy.

The minimum amount of fissionable material in which a fission chain reaction is self-sustaining is the **critical mass.** The critical mass of pure uranium-235 is less than 1 kg when the uranium sample is spherical for maximum fission efficiency. It is also important to note that not all neutrons produced in fission cause further fission. Some very high energy neutrons

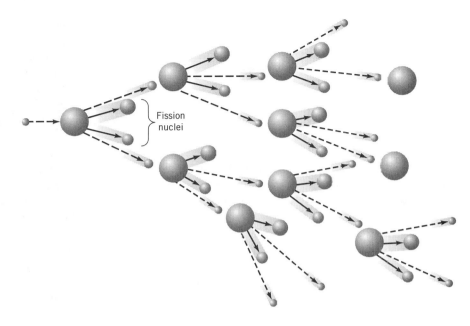

Fission
nuclei

FIGURE 16-6 In a nuclear chain reaction, the neutrons produced by a fission event can cause fission of other nuclei. Since each fission event typically produces twice the number of neutrons needed for one fission, a nuclear chain reaction can occur within a large collection of fissionable nuclei.

escape from the sample and others in the sample do not cause fission until they lose some kinetic energy. Neutrons of lower kinetic energy, called **thermal neutrons,** have a greater chance of causing further fission events.

When a fission chain reaction becomes self-sustaining, it is said to be **critical** and the reaction continues if fissionable nuclei are present. Under certain conditions, a fission chain reaction becomes uncontrollable and produces a tremendous number of fissions in a short period. This type of **supercritical** condition can result in an explosion and a release of large amounts of energy. This process is the basis for the construction of **atomic bombs** or **fission bombs** (see the margin). As an analogy, if we strike a match and it ignites two more matches that in turn ignite four more and the process continues as it multiplies into the millions and billions of ignitions, the result is the explosive release of energy.

Fission chain reactions are also used in nuclear power plants. In a nuclear reactor, the fission occurs at a subcritical or critical level. It is controlled by use of special neutron-absorbing materials within the reactor. The nuclear fuel components in a nuclear reactor are designed so that fission in the reactor cannot become supercritical; a reactor cannot explode like a nuclear bomb. The components of a reactor, however, can rise to very high temperatures as they absorb fission energy.

16-16 NUCLEAR REACTORS

A **nuclear reactor** is a device in which nuclear fission is carried out in a controlled fashion. A **power reactor** converts the energy of fission into electrical energy. Figure 16-7 shows the fundamental components of a nu-

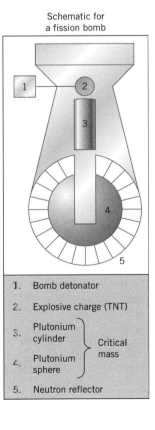

Schematic for
a fission bomb

1. Bomb detonator
2. Explosive charge (TNT)
3. Plutonium cylinder   } Critical
4. Plutonium sphere   } mass
5. Neutron reflector

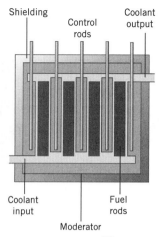

FIGURE 16-7 The components of a nuclear reactor.

Reactors

clear reactor. The **core** of the reactor contains **nuclear fuel rods.** The fuel rods contain fissionable isotopes as enriched uranium.

Recall that natural uranium contains about 0.7% uranium-235. Thus, there are about 7 uranium-235 atoms for every 993 uranium-238 atoms. Special enrichment processes increase the percentage of uranium-235 in uranium samples to a level of about 3%. At this level, self-sustaining fission in the uranium is possible. In contrast, nuclear weapons need uranium enriched to more than 90% uranium-235.

Nuclear fuel pellets are made from the 3% enriched uranium and are encased in steel or zirconium alloy rods. The pellets, made of a compound of uranium called uranium oxide, have cylindrical shapes about 2.5 cm long and 1.3 cm in diameter. The pellets are stacked on one another to make a fuel rod about 1.5 m long. Fuel rods are suspended in the center or core of the reactor. A typical reactor has around 40,000 fuel rods.

The **reactor core** also contains a substance called a **moderator** that serves to slow neutrons produced by fission. Fission neutrons normally have such high energies they do not cause enough further fission events. The moderator slows the neutrons to produce more thermal neutrons for further fission. Also in the core of the reactor are **control rods.** These rods, made of cadmium or boron steel, absorb neutrons. By moving the rods within the core, control of the number of neutrons is possible. When the control rods are partially removed, the number of neutrons increases to a level at which fission becomes self-sustaining. The reactor becomes critical. Lowering the rods into the core decreases fission to a subcritical level.

A **coolant** circulates within the core to absorb the energy of fission as heat energy. The heated coolant is the energy source used to make electricity. A simple analogy is the coolant in an automobile engine. The coolant absorbs the heat in the engine, is cooled by a radiator, and is returned to the engine. In a reactor, the hot coolant is used to make electricity and is cooled before returning to the reactor. Typical power reactors in the United States use water as a coolant and as a moderator. These water-using reactors are sometimes called **light water reactors (LWR).**

Shielding material surrounds the core of the reactor to prevent radiation leakage. Typical **shielding** consists of steel plates and thick layers of concrete. In a power reactor, a thick steel pressure vessel contains the reactor core. A power reactor is housed in a containment structure for safety. The containment structure has concrete walls 3 to 4 ft thick and a steel liner. Typically, an **emergency coolant** supply is available in a power reactor. If the primary coolant supply is interrupted in any way, the emergency core cooling system (ECCS) is started.

16-17 REACTORS IN USE

Around 350 nuclear power reactors are in use worldwide. In the United States, over 100 commercially licensed power reactors are in use. Nuclear

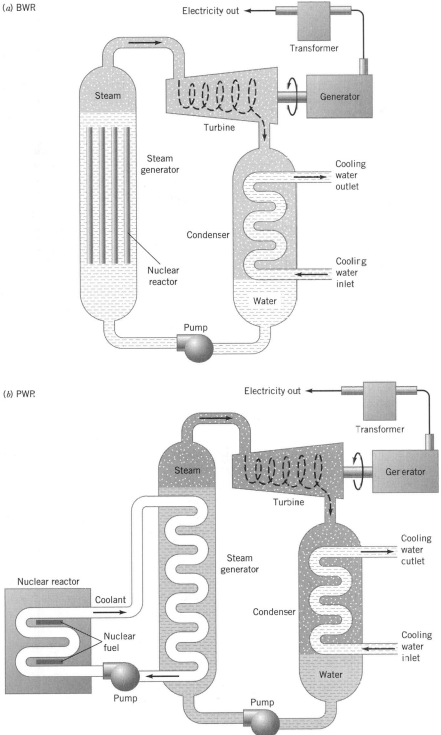

(a) BWR

Electricity out

Transformer

Generator

Turbine

Steam

Steam
generator

Nuclear
reactor

Condenser

Cooling
water
outlet

Cooling
water
inlet

Water

Pump

(b) PWR

Electricity out

Transformer

Generator

Turbine

Steam

Steam
generator

Condenser

Nuclear reactor

Coolant

Nuclear
fuel

Cooling
water
outlet

Cooling
water
inlet

Water

Pump

Pump

FIGURE 16-8 Light water nuclear power plants: (a) boiling water reactor, BWR; (b) pressurized water reactor, PWR.

power provides about 16% of the electricity generated in the United States. The useful lifetime of a power reactor is around 30 years, so many of the reactors in use today will likely be shut down by 2010.

Two types of power reactors are used in the United States. The **boiling water reactor (BWR)** design uses enriched uranium fuel. The core is in a pressurized vessel in which steam forms by the heat of fission (see Figure 16-8). The steam drives a turbine that generates electricity. In this design, the primary coolant converts to steam to drive the turbine and is condensed to liquid water using external cooling water.

The **pressurized water reactor** or **PWR** also uses enriched uranium fuel and water as the primary coolant (see Figure 16-8). These reactors run under high pressure so that the primary coolant water does not form steam. The water becomes superheated water with a temperature of about 300°C. The superheated water heats a secondary supply of water so that it becomes steam. The steam runs the turbine. Steam in the secondary water loop is condensed back to liquid water using external cooling water.

Incidentally, a typical nuclear power plant is about 30% efficient. Therefore, 30% of the energy of fission is converted to electricity and the remainder is waste heat. This waste heat is released to the atmosphere and warms the external cooling water. Conventional fossil fuel power plants have greater efficiencies than nuclear power plants.

16-18 FUSION

Fusion is another type of transmutation process that can produce energy. At high temperatures, it is possible for some lighter nuclei to fuse to form heavier nuclei. Fusion means to merge together. Fusion can produce enormous amounts of energy. In fact, the energy produced by the sun comes from fusion.

Nuclear scientists found that the temperatures produced during fission, known as **thermonuclear temperatures,** could cause certain fusion reactions. Nuclei carry electrical charges and strongly repel one another. The higher temperatures of fission give nuclei enough kinetic energy so that lighter nuclei can overcome the repulsive forces and undergo fusion. Typical fusion reactions involve hydrogen nuclides. The destructive power of a hydrogen bomb is a result of a very rapid self-sustaining fusion reaction. A hydrogen bomb uses the heat of a uranium or plutonium fission bomb to provide the thermonuclear temperatures needed for fusion. The fusionable isotopes hydrogen-2, deuterium, and hydrogen-3, tritium, are contained in a bomb along with lithium-6. The thermonuclear temperature induces the fusion of the hydrogen isotopes as shown here.

$$\,^2_1\text{H} + \,^3_1\text{H} \longrightarrow \,^4_2\text{He} + \,^1_0n + \text{energy}$$

The neutron formed in this process interacts with lithium-6 to give more tritium.

$$^{6}_{3}\text{Li} + ^{1}_{0}n \longrightarrow ^{4}_{2}\text{He} + ^{3}_{1}\text{H}$$

The tritium can then fuse with more deuterium to give a self-sustaining and explosively fast chain reaction that occurs in a fraction of a second.

The fusion process can produce larger amounts of energy for a given amount of material than the fission process. Uncontrolled thermonuclear fusion releases a large amount of energy and accounts for the tremendous destructive power of thermonuclear bombs. Since these **thermonuclear bombs** use hydrogen as fuel, they are sometimes called **hydrogen bombs** or **H-bombs.** Because they require a fission bomb to trigger the fusion process, they are also called **fission–fusion bombs.** No way has yet been found to carry out a controlled fusion process so that the energy released is available to make electricity. Controlled fusion, however, is the subject of much research, and one day fusion reactors may be developed to provide vast amounts of low-cost energy using common fusionable isotopes.

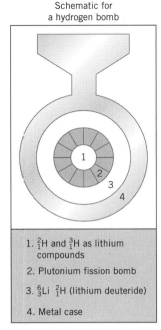

Schematic for a hydrogen bomb

1. $^{2}_{1}\text{H}$ and $^{3}_{1}\text{H}$ as lithium compounds
2. Plutonium fission bomb
3. $^{6}_{3}\text{Li}\ ^{2}_{1}\text{H}$ (lithium deuteride)
4. Metal case

16-19 PROS AND CONS OF NUCLEAR ENERGY

There are several advantages and disadvantages to the wide-scale use of fission reactors for electrical power. Power reactors can provide large amounts of energy and thus can replace conventional fossil fuel sources. At this time, power reactors are feasible alternatives to diminishing supplies of fossil fuels.

Since the fission process is not a combustion process, power reactors do not give off large amounts of greenhouse gases and waste gases that become air pollutants. Some highly radioactive gases, such as krypton-85, however, form in relatively small amounts during fission. These gases are different from normal air pollutants but are an especially dangerous kind of pollutant with which we have to contend. In fact, the use of power reactors introduces several new kinds of pollution problems.

Impurities in the reactor coolant can become radioactive when exposed to the reactor core. These are **low-level radioactive wastes** and must be disposed of by sealing in special containers placed in semipermanent storage. Some low-level radioactive wastes leak into the secondary cooling waters. When this water returns to the source, the radioactivity enters the environment. This kind of radioactive leakage must be monitored.

Figure 16-9 illustrates how radioactive substances enter the food chain and are transmitted to humans. The inhalation of radioactive materials can induce lung cancer. Some radioactive isotopes remain in the body because of their chemical similarity to common body chemicals. **Cesium-137,** a beta emitter of a 30-year half-life, is chemically similar to potassium and can become incorporated in all cells of the body. **Strontium-90,** a beta emitter of a 28-year half-life, is chemically similar to calcium. Since calcium

FIGURE 16-9

Transmission of radioactive substances to humans.

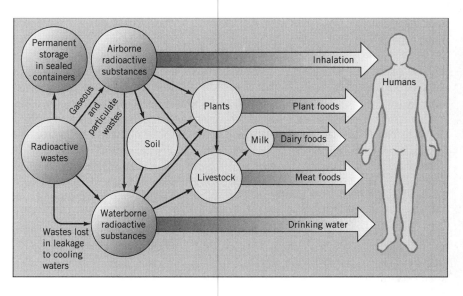

is a main constituent of bones and teeth, strontium-90 can accumulate in these regions of the body. The result is that the body is continually exposed to radiation from this nuclide.

16-20 RADIOACTIVE WASTES

When fission occurs, the fission products produced accumulate in the used or spent fuel rods of the reactor. Periodically, **spent rods** must be removed from a reactor and replaced. The fission products in the rods are extremely radioactive and hazardous. They include a variety of nuclides produced as fission products and some plutonium-239 produced by transmutation of the uranium-238 in the fuel rods. Materials from spent fuel rods are called **high-level radioactive wastes.** Such wastes must be kept in special containers for safety reasons. In 1980, an estimated 6700 metric tons (1 metric ton = 1000 kg) were in storage in the United States, and currently about 73,000 metric tons are in storage. These wastes represent around 40 billion curies of radiation. Spent fuel rod wastes are currently stored on nuclear reactor sites in continually cooled pools of water. Ideally, such wastes could be processed to remove the uranium and plutonium. Today, no spent fuel rod processing facilities are in use in the United States. Even if these elements are removed, the remaining wastes must be safely stored in isolated locations for thousands of years. In 1999, no national high-level storage sites were available in the United States. How high-level nuclear wastes are stored and the locations of storage sites are two of the most challenging dilemmas associated with the use of nuclear energy.

Another aspect of nuclear energy is the problem of **waste heat.** Nuclear power plants are currently less efficient in energy conversion than fossil fuel plants. A conventional nuclear power plant (meeting current safety standards) produces 50% more waste heat than a comparable fossil fuel plant. More efficient nuclear plants are in development. Nevertheless, if nuclear energy begins substantially to replace fossil fuel energy, waste heat is of greater concern.

16-21 REACTOR SAFETY

The fuel rods in the reactor core are the source of heat energy in a power reactor. The primary coolant has the important function of absorbing this heat. If the **primary coolant** supply is disrupted, the temperature of the core can rise dramatically. Theoretically, the temperature of the core could reach the level of **core meltdown.** At meltdown, the fuel rods melt to form a high-temperature mass that would maintain high temperatures because of continued fission. This mass would melt anything that tried to contain it. Within days it would melt through the core vessel and the concrete containment shielding below it. Then it would sink into the ground and probably be dispersed in a steam explosion when it reached the water table. The resulting explosion could scatter fission products over the environment near the plant site and downwind from the site. The possible penetration into the ground by a melted core is sometimes called the **China syndrome.** The China syndrome did not occur during the Chernobyl accident in 1986, but very hot masses of highly radioactive core material did accumulate beneath the destroyed reactor. See Section 16-22 for a discussion of the Chernobyl accident.

Reactor safety features help prevent a core meltdown and the accidental escape of fission products from the reactor. Nuclear power plants have a series of barriers between the nuclear fuel and the outside environment. A core meltdown is a highly unlikely event and no such disaster has occurred. The probability of such a meltdown accident is extremely low, but if one occurred, the results could be disastrous.

Core meltdown in a reactor could release radioactive materials into water and air in the surrounding environment. Winds or waterways could spread this material. The major problem with this type of pollution is that it is extremely dangerous to life and cannot be seen, tasted, or smelled. It takes special instruments to detect radiation. Another problem is psychological. Since radiation cannot be seen, people in contaminated areas often suffer extreme stress that can cause physical illnesses. Furthermore, some radioactive isotopes have long half-lives, which means that a contaminated area would need evacuation and would be unusable for habitation or agriculture for many years. A partial core melting could result in the permanent shutdown of the plant. Furthermore, cleanup and safety operations are very costly.

16-22 REACTOR ACCIDENTS

The most serious nuclear reactor accident in the United States occurred in the **Three Mile Island** reactor in Pennsylvania only 12 miles from the metropolitan area of Harrisburg. On March 28, 1979, a combination of equipment failure and human operator error resulted in the partial meltdown of the reactor. Fortunately, total meltdown did not occur, and the various containment features of the reactor worked to prevent the release of significant amounts of radioactive materials into the environment. Nonetheless, some radioactive material was released. The Three Mile Island incident made it clear that serious nuclear reactor accidents can occur. The damaged nuclear reactor was abandoned, and the cleanup cost about one billion dollars.

On April 26, 1986, a nuclear power reactor located near a city called Chernobyl in the Ukraine exploded, killing and hospitalizing numerous people and spreading radioactive materials in the local areas of the former USSR and other parts of Europe. The **Chernobyl reactor** used graphite, not water, as a moderator. The fuel in the reactor was about 75% used as it was being shut down for maintenance. During the shutdown, plant operators planned an experiment with the electrical generator associated with the reactor. To carry out the experiment they unwisely turned off the **ECCS** and withdrew nearly all the control rods as the reactor was being powered down. They also switched the primary coolant pumps from external electricity to electricity still being produced by the reactor and disabled the automatic shutdown mechanism of the reactor. All this was done on the assumption that since the reactor was being shut down, none of these safety features would be needed. Unfortunately, because of a design flaw in this type of reactor and the lack of operational safety features, the reactor experienced an increased rate of fission as it was being shut down. The high temperature caused by this power surge resulted in the destruction of the reactor.

The fuel rods reached a very high temperature and ruptured, spilling very hot material into the cooling water. The steam pressure became so great that it blew off the 1000-ton lid of the reactor, destroying the reactor and the reactor building. It is thought that a second explosion quickly followed the first. This explosion occurred from the hydrogen gas formed when steam reacted with the zirconium metal used in the reactor. In any case, the explosion resulted in a plume of debris that rose some 1200 m or three-quarters of a mile into the atmosphere. Following this, the hot graphite of the reactor began to burn producing a fire that lasted 11 days.

The explosion and fire spread radioactive materials from the fission products in the fuel rods. The area in a 30 km radius around the reactor was highly contaminated. In fact, 135,000 people left this area and will never be allowed to return. Lesser amounts of radioactive materials from the explosion and fire deposited as fallout in various regions of the republics of the former USSR and Europe. This disaster represents the single largest release of radioactive materials into the environment. It is estimated that the accident released from 50 million to 180 million curies of radiation. Because the isotopes had short half-lives, much of the radioactive material has dissipated. Among the most dangerous isotopes released was iodine-131 with an 8-day half-life. This isotope was absorbed into the thyroid glands of many people and has caused an increase in thyroid cancer. The other significant isotope is cesium-137 with a half-life of 30 years. This cesium deposited in the soil can become part of the food chain. Furthermore, the area around the reactor site is contaminated with plutonium-239.

The accident caused the direct death of 31 people, and more than 200 became hospitalized with radiation poisoning. Low levels of radiation affected millions of people, and the government moved thousands of people from the contaminated regions. Predicting the effect of the accident is difficult since nothing of this size has occurred in the past, but it is expected that radiation from the accident will result in at least 40,000 excess cancer deaths in the future. These cancers will occur mainly in the republics of the former USSR and neighboring regions of Europe. One hard lesson of Chernobyl is that serious nuclear accidents can happen.

16-23 PLUTONIUM

A final concern over the large-scale use of nuclear energy involves plutonium. **Plutonium-239,** produced in nuclear reactors, has a half-life of 24,100 years. A nuclear accident involving plutonium could contaminate an area for a very long time. The chance of such a disaster means the safety factors in nuclear reactors must be carefully developed and continuously reviewed. Earthquakes, human error of plant operators, and even the chance of sabotage, however, are unpredictable factors.

Plutonium is present in spent fuel rods. It may be extracted from spent fuel in special fuel-processing plants. In the year 1999, the only operating processing plant is in the United Kingdom. In addition, it is important to realize that the current dismantling of nuclear weapons in the United States and Russia is resulting in many tons of excess plutonium that must be safely stored for an indefinite time. The critical mass of plutonium-239 is about 5 kg. It is conceivable that a crude atomic bomb could be constructed from this amount of plutonium. Furthermore, plutonium can be used to make atomic bombs and hydrogen bombs by any country having the necessary technology.

In the United States, it has been a tradition to have a separation of nuclear power reactors and the military use of plutonium. That is, the plutonium produced in commercial nuclear reactors is not used to construct nuclear weapons. Fortunately, most countries do not have the technology to separate plutonium from nuclear reactor wastes. Any country that develops this technology, however, could make significant amounts of plutonium for the manufacture of nuclear weapons.

The desirable aspects and possible necessity, contrasted with the many disadvantages of nuclear power, present a significant environmental and social dilemma. We will continue to have to make decisions about the use of nuclear power use in the future.

The Genesis of the Universe

The universe began with a big bang. The big bang theory, even with its playful name, is an important scientific model about the origin of the universe. Scientific evidence shows that the universe is expanding, getting larger and larger. If it is expanding, it follows that the expansion started at some cosmic point or center. According to the theory, the universe began an estimated 10 billion to 15 billion years ago in a huge explosion. The result was the formation of energy and matter that began expanding. At first, the new universe was so hot even simple nuclei did not form, but soon protons, neutrons, and electrons formed. After the universe was about 3 minutes old, simple nuclei, like those of hydrogen, helium, and eventually lithium, formed from protons and neutrons. Hydrogen, with the simplest nucleus, was the major material of the early universe, and even today hydrogen is the most abundant element in the universe. Five hundred thousand years after the big bang, electrons combined with nuclei to form the first atoms.

These atoms followed the expansion of the universe and spread about. Gravitational forces of attraction resulted in clumping of hydrogen to form stars. Within the stars some hydrogen fused to form helium. Under extreme heat and pressure conditions that occur in stars, certain atomic nuclei can merge or fuse to make more complex nuclei.

$$\ _1^1\text{H} + \ _0^1n \longrightarrow \ _1^2\text{H}$$

$$\ _1^2\text{H} + \ _1^2\text{H} \longrightarrow \ _1^3\text{H} + \ _1^1\text{H}$$

$$\ _1^2\text{H} + \ _1^2\text{H} \longrightarrow \ _2^3\text{He} + \ _0^1n$$

$$\ _2^3\text{He} + \ _0^1n \longrightarrow \ _2^4\text{He}$$

$$\ _1^3\text{H} + \ _1^1\text{H} \longrightarrow \ _2^4\text{He}$$

Gravitational condensation of cosmic matter and fusion gave birth to early stars. Stars formed, then clumped into galaxies, and galaxies arranged into formations called groups, clusters, and superclusters. Early stars were vast accumulations of hydrogen, some of which fused to make helium and release energy. Over time, as these stars became smaller and hotter due to gravitational forces, helium fused to make beryllium, and hydrogen and beryllium fused to make boron.

$$\ _2^3\text{He} + \ _2^4\text{He} \longrightarrow \ _4^7\text{Be}$$

$$\ _4^7\text{Be} + \ _1^1\text{H} \longrightarrow \ _5^8\text{B}$$

In some larger stars, intense gravitational forces caused further fusion to form elements like carbon, oxygen, neon, magnesium, potassium, and silicon and eventually elements up to iron and some nickel. Iron is among the most stable of nuclei, along with nickel, so such stars did not produce elements of higher atomic numbers. Interestingly, many meteorites are composed of relatively pure iron mixed with some nickel, and the core at the center of Earth is iron.

Larger stars typically cool down as their hydrogen is used up by fusion. The cooling and intense gravitational forces cause the matter in the stars to be drawn inward, which makes the star suddenly collapse and then explode. The result is a supernova explosion producing temperatures so extreme that elements of lower atomic number undergo fusion to make nuclides of elements of atomic number greater than iron, up to and including uranium and beyond. These elements and the elements of lower atomic number are thrown into space by the explosion and become cosmic dust and debris. Some of this material condenses by gravitational forces along with cosmic hydrogen to form second-generation stars like our sun. Energy from our sun comes mostly from the fusion of hydrogen into helium. Other samples of cosmic material in the dust swirl about the sun condense to form planets like those of our solar system and Earth itself. In this way, the variety of cosmic processes have resulted in the mix of elements found on Earth. Humans, of course, are composed of the very elements that were made during the big bang and created in stars and supernova explosions.

FOCUS
16

Nuclear Chemistry

The general symbol for a nuclide is

$$^{A}_{Z}X$$

where X is the symbol of the element corresponding to the nuclide, A is the nucleon number, and Z is the atomic number.

Radioactivity: The spontaneous decay of a nucleus to form another nucleus and a nuclear particle.

Alpha particle decay: A nucleus, called the parent, decays by emitting a high-speed helium-4 nucleus known as an alpha particle (α). The loss of an alpha particle forms a new nucleus, called the daughter. The daughter nucleus has a nucleon number that is four less than the parent nucleus and an atomic number that is two less.

$$^{A}_{Z}X \longrightarrow {}^{A-4}_{Z-2}Y + {}^{4}_{2}He$$

parent	daughter	alpha
nucleus	nucleus	particle

Beta particle decay: The decay of a nucleus by emission of a high-speed electron called a beta particle (β). The daughter nucleus formed by beta particle decay has the same nucleon number as the parent nucleus and an atomic number that is one greater.

$$^{A}_{Z}X \longrightarrow {}^{A}_{Z+1}Y + {}^{0}_{-1}e$$

parent	daughter	beta
nucleus	nucleus	particle

Gamma ray emission: Sometimes the daughter nucleus formed in an alpha decay or a beta decay is produced in an energetically excited state or a higher than normal energy state. This excited energy state is not the normal state of the nucleus, and it quickly drops to a lower energy state. In the process, the nucleus emits a form of radiant energy called a gamma radiation (γ). Gamma rays have energies similar to X rays.

Half-life: The time required for the decay of one-half of a sample of a radioactive substance.

Becquerel, Bq: The unit of radioactivity corresponding to 1 radioactive decay per second.

Curie, Ci: A unit of activity equal to 37 billion decays per second, or 3.7×10^{10} Bq.

Rad (radiation-absorbed dose): The radiation dose that transfers 1×10^{-5} J of energy per gram of tissue. The effect of radiation depends on the type of radiation and its energy.

Rem (radiation equivalent human): A unit of radiation dose. A rem value is determined by measuring the dose in rads and multiplying by factors related to the kind of radiation (i.e., alpha, beta, gamma, or X rays, or neutrons).

Background radiation: Nuclear radiation from cosmic radiation, naturally occurring isotopes, radioactive materials that enter the environment from nuclear power plants, and fallout from nuclear weapons testing.

Nuclear transmutation: A process in which a high-speed nuclear particle collides with a nucleus, resulting in a nuclear reaction that produces a different nucleus. The high-speed particle is called the projectile, and the nucleus it collides with is called the target.

Nuclear fission: A nuclear transmutation in which a neutron projectile is captured by target nucleus like uranium-235 or plutonium-239. Upon the capture of a neutron, an unstable nucleus is formed that splits into two nuclei plus a few neutrons accompanied by the release of energy.

Energy released in nuclear processes comes from the conversion of mass to energy. The relation between mass and energy is given by the famous formula first stated by Albert Einstein: $E = mc^2$, where m is the mass in kilograms and c is the speed of light in meters per second. The speed of light refers to the speed at which radiant energy passes through empty space.

Nuclear power reactor: A device in which nuclear fission is carried out in a controlled fashion and the energy of fission is converted to electrical energy.

Nuclear fusion: A type of high-temperature nuclear transmutation in which some lighter nuclei fuse to form heavier nuclei accompanied by the release of energy.

QUESTIONS

1. Match the following terms with the appropriate lettered definitions.

nucleons

a. the sum of the number of protons and neutrons in a nucleus

isotopes

b. the time needed for the decay of one-half of a sample of a radioactive substance

nuclides

c. all naturally occurring and humanmade nuclei

nucleon number

radioactivity

alpha decay

beta decay

d. a unit of activity defined as 1 decay per second

e. a unit of activity defined as 37 billion decays per second

f. the spontaneous decay of a nucleus to form another nucleus and a nuclear particle

g. BWR or PWR

gamma radiation

half-life

becquerel

curie

rad

rem

ionizing radiation

nuclear transmutation

nuclear fission

critical mass

nuclear reactor

LWR

nuclear fusion

plutonium 239

h. atoms of an element with differing numbers of neutrons

i. protons and neutrons that compose nuclei

j. radioactive decay in which a nucleus loses an He-4 nucleus

k. radioactive decay in which a nucleus loses a high-speed electron

l. high energy electromagnetic radiation that is released during some radioactive decay processes

m. radiation absorbed close

n. radiation equivalent human

o. a device in which controlled nuclear fission is carried out

p. a humanmade isotope used to make nuclear weapons

q. the merging of lighter nuclei to make heavier nuclei accompanied by the release of energy

r. a process in which a projectile particle interacts with a target nucleus to form a new nucleus and a nuclear particle

s. the breaking apart of a nucleus to form smaller nuclei accompanied by the release of energy

t. the minimum amount of a fissionable material in which a fission chain reaction is self-sustaining

u. nuclear radiation that causes the formation of ion pairs as it passes through matter

Sections 16-1 and 16-2

2. Describe the structure of a typical atomic nucleus.

3. What are isotopes? What are nuclides?

4. Define the terms *atomic number* and *nucleon number*. How can the nucleon number and atomic number of a nuclide be used to find the number of neutrons in the nuclide?

5. Give the nuclide symbols used to represent the following nuclides. Refer to a list of elements for the necessary atomic numbers.

(a) carbon-14　　　　(b) oxygen-16

(c) uranium-238　　　(d) radium-226

(e) potassium-40　　 (f) plutonium-239

(g) iodine-131　　　 (h) cobalt-60

6. Determine the number of neutrons in the nuclei of each of the nuclides in Question 5.

7. Give the names of the following nuclides and determine the number of neutrons in nuclei of each nuclide.

(a) $^{235}_{92}U$　　　　(b) $^{51}_{24}Cr$

(c) $^{3}_{1}H$　　　　　(d) $^{249}_{98}Cf$

(e) $^{241}_{95}Am$　　　(f) $^{218}_{84}Po$

(g) $^{222}_{86}Rn$　　　(h) $^{90}_{38}Sr$

Section 16-3

8. The isotopes of some elements are radioactive. What does the term *radioactive* mean?

Section 16-4

9. Answer the following questions.

(a) What is alpha decay? What happens to the nucleon number and atomic number of a nuclide that undergoes alpha decay?

(b) What is beta decay? What happens to the nucleon number and atomic number of a nuclide that undergoes beta decay?

(c) What is gamma ray emission?

10. What is the structure and charge of an alpha particle? What is the structure and charge of a beta particle? How is it thought that a negatively charged beta particle can come from a nucleus?

11. What does the term *radiation* mean concerning radioactive decay? What are the three common types of nuclear radiation?

Section 16-5

12. *Write nuclear equations for the alpha decay of the following nuclides.

(a) $^{239}_{94}Pu$　　(b) $^{214}_{84}Po$　　(c) $^{238}_{92}U$

(d) $^{241}_{95}Am$　　(e) $^{207}_{84}Po$　　(f) $^{210}_{82}Pb$

13. Write nuclear equations for the alpha decay of the following nuclides.

    (a) $^{208}_{87}\text{Fr}$ (b) $^{242}_{94}\text{Pu}$ (c) $^{251}_{98}\text{Cf}$

    (d) $^{252}_{99}\text{Es}$ (e) $^{243}_{96}\text{Cm}$ (f) $^{230}_{90}\text{Th}$

14. *Write nuclear equations for the beta decay of the following nuclides.

    (a) $^{197}_{78}\text{Pt}$ (b) $^{14}_{6}\text{C}$ (c) $^{82}_{35}\text{Br}$

    (d) $^{214}_{82}\text{Pb}$ (e) $^{239}_{93}\text{Np}$ (f) $^{60}_{27}\text{Co}$

15. Write nuclear equations for the beta decay of the following nuclides.

    (a) $^{211}_{82}\text{Pb}$ (b) $^{28}_{12}\text{Mg}$ (c) $^{140}_{56}\text{Ba}$

    (d) $^{32}_{15}\text{P}$ (e) $^{42}_{19}\text{K}$ (f) $^{20}_{8}\text{O}$

16. *Complete the following nuclear equations by supplying the missing nuclide or particle.

    (a) _____ $\longrightarrow$ $^{28}_{14}\text{Si}$ + $^{0}_{-1}e$

    (b) $^{226}_{88}\text{Ra}$ $\longrightarrow$ _____ + $^{4}_{2}\text{He}$

    (c) $^{234}_{91}\text{Pa}$ $\longrightarrow$ $^{234}_{92}\text{U}$ + _____

    (d) _____ $\longrightarrow$ $^{234}_{90}\text{Th}$ + $^{4}_{2}\text{He}$

17. Complete the following nuclear equations by supplying the missing nuclide or particle.

    (a) $^{237}_{93}\text{Np}$ $\longrightarrow$ $^{4}_{2}\text{He}$ + _____

    (b) $^{210}_{84}\text{Po}$ $\longrightarrow$ _____ + $^{206}_{82}\text{Pb}$

    (c) $^{210}_{82}\text{Pb}$ $\longrightarrow$ _____ + $^{210}_{83}\text{Bi}$

    (d) $^{60}_{27}\text{Co}$ $\longrightarrow$ $^{0}_{-1}e$ + _____

Section 16-6

18. What is meant by the term *half-life*?

19. *Radon-222, an alpha emitter, has a half-life of 3.82 days.

    (a) Give a nuclear equation for the alpha decay of radon-222.

    (b) Radon is a gas. Why is it a dangerous alpha emitter when it is in the atmosphere?

    (c) If a radon-222 sample had a mass of 0.50 g, what is the mass of the sample after 11.5 days?

    (d) If we started with 1,000,000 radon-222 atoms, how many atoms remain after 10 half-lives had passed? (Hint: Use a calculator and divide by 2 for each half-life.)

20. Plutonium-239 is an alpha emitter with a half-life of 24,100 years. It is absorbed into the bones when it enters the body. It does not occur naturally but is produced as a by-product of nuclear reactors.

    (a) Give a nuclear equation for the alpha decay of plutonium-239.

    (b) Why is plutonium an especially dangerous nuclide if it enters the environment in significant amounts?

    (c) A dose of 1 μg (1 μg = 10^{-6} g) of plutonium-239 in the human body can cause serious radiation poisoning or death. How many plutonium atoms are in a 1.0 μg sample? (Hint: Use the nucleon number as the molar mass.)

    (d) A general guideline for radioactive materials is that 10 half-lives are needed to decrease the amount of a radioactive material to an "acceptable" level. If a part of the environment became contaminated with plutonium-239, how many years must pass before the acceptable level of contamination had been reached?

21. *Iodine-131, a beta emitter, has a half-life of about 8 days. It is used medically to diagnose thyroid disease and treat thyroid cancer. The body uses iodine to form thyroxine, a thyroid hormone, so most iodine in the body concentrates in the thyroid.

    (a) Give a nuclear equation for the beta decay of iodine-131.

    (b) If we start with 10 μg (1 μg = 10^{-6} g) of iodine-131, how many grams are left after 24 days? How many after 80 days? (Hint: Divide the amount by 2 for each half-life. Use a calculator.)

    (c) Why is iodine-131 dangerous when it is released by a nuclear power plant during an accident?

    (d) A precautionary measure for people in the vicinity of a nuclear accident is to take massive doses of nonradioactive iodine (in the form of iodide ion). How does this help prevent the dangerous effects of iodine-131?

Section 16-7

22. What is a becquerel? What is a curie? Some home smoke detectors contain 1 μCi of americium-241, an alpha emitter. How many nuclear disintegrations occur per second in this amount of the nuclide? **CAUTION: Never dismantle a smoke detector or let children touch or play with a smoke detector. For information about the disposal of an old detector, contact the manufacturer or your local fire department.**

23. What is a rad? What is a rem? What is background radiation?

Section 16-8

24. Which of the elements are naturally radioactive and have no stable isotopes?

25. What is a radioactive decay series? The half-life of radon-222 is 3.82 days, yet small amounts are found in nature. Why?

26. Why are uranium ores not only a source of uranium but also radium? How is the uranium and lead content of certain rocks used to estimate their age?

27. Helium gas is found trapped in porous rock deposits. How do you suppose this helium was formed?

Section 16-9

28. Describe what happens when radiation passes through matter. Why is it called ionizing radiation?

29. Why is nuclear radiation dangerous to humans and other living organisms? Which type of radiation is most dangerous?

Section 16-10

30. How does a Geiger–Müller tube function?

31. How can the activity of a radioactive nuclide be used to determine its half-life?

Section 16-11

32. What is radiation sickness?

33. How are cancer, genetic damage, and mutation associated with nuclear radiation?

Section 16-12

34. Define the following terms.

    (a) nuclear transmutation

    (b) target nuclei

    (c) projectile

    (d) humanmade isotopes

35. *Complete the following equations representing nuclear transmutations.

    (a) $^{59}_{26}\text{Fe} + {}^{1}_{0}n \longrightarrow {}^{60}_{27}\text{Co} + \underline{\hspace{1cm}}$

    (b) $^{246}_{96}\text{Cm} + {}^{12}_{6}\text{C} \longrightarrow \underline{\hspace{1cm}} + 4{}^{1}_{0}n$

    (c) $^{27}_{13}\text{Al} + {}^{4}_{2}\text{He} \longrightarrow \underline{\hspace{1cm}} + {}^{1}_{0}n$

    (d) $\underline{\hspace{1cm}} + {}^{0}_{-1}e \longrightarrow {}^{32}_{15}\text{P}$

    (e) $^{2}_{1}\text{H} + \underline{\hspace{1cm}} \longrightarrow {}^{4}_{2}\text{He} + {}^{1}_{0}n$

    (f) $\underline{\hspace{1cm}} + {}^{1}_{0}n \longrightarrow {}^{3}_{1}\text{H} + {}^{4}_{2}\text{He}$

    (g) $^{16}_{8}\text{O} + {}^{1}_{0}n \longrightarrow \underline{\hspace{1cm}} + {}^{4}_{2}\text{He}$

36. Complete the following equations representing nuclear transmutations.

    (a) $^{15}_{7}\text{N} + \underline{\hspace{1cm}} \longrightarrow {}^{12}_{6}\text{C} + {}^{4}_{2}\text{He}$

    (b) $^{24}_{12}\text{Mg} + {}^{2}_{1}\text{H} \longrightarrow {}^{25}_{12}\text{Mg} + \underline{\hspace{1cm}}$

    (c) $^{6}_{3}\text{Li} + {}^{1}_{0}n \longrightarrow {}^{4}_{2}\text{He} + \underline{\hspace{1cm}}$

    (d) $^{9}_{4}\text{Be} + {}^{4}_{2}\text{He} \longrightarrow {}^{12}_{6}\text{C} + \underline{\hspace{1cm}}$

    (e) $^{238}_{92}\text{U} + {}^{16}_{8}\text{O} \longrightarrow \underline{\hspace{1cm}} + 5{}^{1}_{0}n$

    (f) $^{3}_{1}\text{H} + {}^{2}_{1}\text{H} \longrightarrow {}^{4}_{2}\text{He} + \underline{\hspace{1cm}}$

    (g) $\underline{\hspace{1cm}} + {}^{1}_{1}\text{H} \longrightarrow {}^{15}_{7}\text{N} + {}^{4}_{2}\text{He}$

Section 16-13

37. Describe nuclear fission and a nuclear chain reaction.

38. Which two common nuclides are fissionable? Which of these is naturally occurring and which is humanmade?

39. What is meant by the critical mass of a fissionable nuclide?

Section 16-14

40. *Using the equation $E = mc^2$, calculate the energy released when 7.0160 g of lithium-7 and 1.0078 g of hydrogen-1 undergo the following fusion reaction:

$$^{7}_{3}\text{Li} + {}^{1}_{1}\text{H} \longrightarrow 2{}^{4}_{2}\text{He}$$

relative
masses   7.0160   1.0078   2(4.0026)

41. Using the equation $E = mc^2$, calculate the energy released when 226.0254 g of radium-226 undergo alpha decay.

$$^{226}_{88}\text{Ra} \longrightarrow {}^{222}_{36}\text{Rn} + {}^{4}_{2}\text{He}$$

relative
masses 226.0254   222.0176   4.0026

Section 16-15

42. Describe a fission chain reaction. What is a critical mass?

Section 16-16

43. List and describe the five basic components of a nuclear reactor. What is a LWR? What is the composition of the fuel rods used in a nuclear power reactor?

Section 16-17

44. Describe a nuclear power plant designed to produce electricity. Explain the differences between a boil-

ing water reactor (BWR) and a pressurized water reactor (PWR).

45. Why is it not possible to have a nuclear explosion in a nuclear power reactor?

Section 16-18

46. Describe nuclear fusion. How do fusion and fission differ?

Sections 16-19 to 16-23

47. Explain or give a description of each of the following.

(a) low-level radioactive wastes

(b) fission product wastes

(c) waste heat

(d) the "China syndrome" or core meltdown

48. List some pros and cons of nuclear power plants.

49. What is plutonium-239? How can this isotope be used and why is it dangerous?

50. Suggest some solutions to long-term disposal of high-level nuclear wastes.

51. Do you feel that nuclear energy should be an important source in the United States? Argue your case.

52. Describe the big bang theory and explain how the elements of the universe were formed.

SOME

USEFUL

CONVERSION

FACTORS

LENGTH

1 inch (in.) = 2.54 centimeters (cm) $\left(\dfrac{2.54\ \text{cm}}{1\ \text{in.}}\right)$

1 foot (ft) = 0.3048 meter (m) $\left(\dfrac{0.3048\ \text{m}}{1\ \text{ft}}\right)$

1 yard (yd) = 0.9144 meter (m) $\left(\dfrac{0.9144\ \text{m}}{1\ \text{yd}}\right)$

1 mile (mi) = 1.609 kilometers (km) $\left(\dfrac{1.609\ \text{km}}{1\ \text{mi}}\right)$

VOLUME

1 cup = 236.6 mL $\left(\dfrac{236.6\ \text{mL}}{1\ \text{cup}}\right)$

1 pint = 0.4731 liter (L) $\left(\dfrac{0.4731\ \text{L}}{1\ \text{pint}}\right)$

1 quart = 0.9461 liter (L) $\left(\dfrac{0.9461\ \text{L}}{1\ \text{quart}}\right)$

1 gallon (gal) = 3.785 liters (L) $\left(\dfrac{3.785\ \text{L}}{1\ \text{gal}}\right)$

MASS OR WEIGHT

1 pound (lb) = 453.6 grams (g) $\left(\dfrac{453.6 \text{ g}}{1 \text{ lb}}\right)$

1 ounce (oz) = 28.35 grams (g) $\left(\dfrac{28.35 \text{ g}}{1 \text{ oz}}\right)$

1 pound (lb) = 0.4536 kilogram (kg) $\left(\dfrac{0.4536 \text{ kg}}{1 \text{ lb}}\right)$

SPECIAL FACTORS

1 nanometer (nm) = 10^{-9} meters $\left(\dfrac{10^{-9} \text{ m}}{1 \text{ nm}}\right)$

1 liter (L) = 10^3 cubic centimeters (cm³) $\left(\dfrac{10^3 \text{ cm}^3}{1 \text{ L}}\right)$

1 milliliter (mL) = 1 cubic centimeter (cm³) $\left(\dfrac{1 \text{ cm}^3}{1 \text{ mL}}\right)$

1 atmosphere (atm) = 760 torr $\left(\dfrac{760 \text{ torr}}{1 \text{ atm}}\right)$

1 torr = 1 millimeter Hg (1 mm Hg) $\left(\dfrac{1 \text{ mm Hg}}{1 \text{ torr}}\right)$

1 atmosphere (atm) = 1.013 bar $\left(\dfrac{1.013 \text{ bar}}{1 \text{ atm}}\right)$

1 calorie = 4.184 joules (J) $\left(\dfrac{4.184 \text{ J}}{1 \text{ cal}}\right)$

Any of the above factors can be used in the inverted form if desired. For example, the factor relating kilograms to pounds is

$$\left(\frac{0.4536 \text{ kg}}{1 \text{ lb}}\right)$$

whereas the factor relating pounds to kilograms is

$$\left(\frac{1 \text{ lb}}{0.4536 \text{ kg}}\right)$$

Alternatively, the inverse or reciprocal of a factor is calculated by dividing the denominator into 1. Thus, the factor relating pounds to kilograms is found by dividing 0.4536 into 1.

$$\left(\frac{1 \text{ lb}}{0.4536 \text{ kg}}\right) = \left(\frac{2.205 \text{ lb}}{1 \text{ kg}}\right)$$

Another example is the relation of kilometers to miles. The factor is expressed as

$$\left(\frac{1 \text{ mi}}{1.609 \text{ km}}\right)$$

Dividing 1.609 into 1 the factor becomes

$$\left(\frac{0.6215 \text{ mi}}{1 \text{ km}}\right)$$

To relate areas that involve square length and volumes that involve cubic length, use the various length factors. For example, to relate square centimeters (cm^2) to square inches (in.^{2}), we square the centimeter-to-inch factor.

$$\left(\frac{2.54 \text{ cm}}{1 \text{ in.}}\right)^2 = \left(\frac{2.54^2 \text{ cm}^2}{1^2 \text{ in.}^2}\right) = \left(\frac{6.452 \text{ cm}^2}{1 \text{ in.}^2}\right)$$

Another example is the factor relating cubic centimeters to cubic meters, which is found by cubing the centimeter-to-meter factor.

$$\left(\frac{100 \text{ cm}}{1 \text{ m}}\right)^3 = \left(\frac{10^2 \text{ cm}}{1 \text{ m}}\right)^3 = \left(\frac{10^6 \text{ cm}^3}{1 \text{ m}^3}\right) \qquad \text{Note that } 10^2 \text{ cubed is } 10^6.$$

PRACTICE A1-1

(a) Calculate the inverse of all the factors given in this appendix. Express to the same number of digits given in the original factors and include the units in your results.

(b) Calculate a factor that relates square kilometers to square miles.

(c) Calculate a factor that relates cubic centimeters to cubic inches.

(d) Calculate a factor that relates cubic millimeters to cubic meters.

USING A HAND

CALCULATOR

IN CHEMISTRY

A hand calculator can be quite useful in chemical computations. A calculator will not solve problems for you! It is convenient, however, to use a calculator to carry out the arithmetic needed to obtain a numerical answer. When solving a problem, it is normally a good idea to work out the complete setup before actually doing any numerical calculations. Of course, you might have to do a few preliminary calculations to obtain the values of some factors involved in the setup. Once the problem has been set up, the calculator serves as a convenient and efficient tool to help with the arithmetical part of the calculation.

In addition to the $+$, $-$, $\times$, $\div$, and $=$ or enter keys, the most useful keys on a scientific calculator are

$+/-$ or CHS	Used to change the sign of a number. It makes a positive number negative or a negative number positive.
$(-)$	Used on some Texas Instrument (TI) calculators to make a number negative.
EXP or EE	Used to input exponential numbers. The key represents the $\times 10$ part of an exponential number. For practice, input 2.68×10^{-5} into your calculator.
log and 10^x	Used to calculate logarithms and inverse logarithms. Some calculators use the key sequence of INV log to find inverse log, which is equivalent to 10^x. See the discussion given below for practice using these keys.
x^2 and $\sqrt{}$	Used to square numbers and find square roots.
SCI or F↔E	Used to switch the operation to scientific notation or to change from fixed-point to exponential notation or vice versa.

Practice using your calculator by checking the worked-out examples given in the text. Some examples are given in this appendix for further practice. If the setup of a problem involves a series of multiplications and/or divisions, it is done as a sequence on the calculator. This sequence is called a chain calculation. For example,

$$285 \text{ mL} \left(\frac{780 \text{ torr}}{760 \text{ torr}}\right)\left(\frac{583 \text{ K}}{273 \text{ K}}\right) = ? \text{ mL}$$

Using an algebraic calculator having an equal key or enter key, a possible calculating sequence is

$$285 \otimes 780 \oslash 760 \otimes 583 \oslash 273 = \text{ (or enter)}$$

Other calculation sequences are possible since any series of multiplications or divisions would suffice. The advantage of the above sequence is that each number is used as it is encountered, working left to right. This process helps avoid the accidental omission of a number.

Calculators usually do not make mistakes (unless the battery is low), but fingers can make mistakes. It is good practice to double check a calculation to be sure that a keying error was not made. It is also good practice to check the value of a calculated answer to see if its value makes sense compared with the numbers used in the calculation. That is, if a calculated answer seems too small or too large, you may have made a calculation error. Of course, to judge whether your calculated answer is the correct size, you can estimate the answer to the calculation.

Try the above calculation on your calculator. Using a calculator, the display reads 624.64286 on an 8-digit model or 624.6428571 on a 10-digit model. The number of significant digits in a calculation involving measurements depends on the number of digits in the measurements. (See Sections 2-7 and 2-8.) A calculator does not make decisions on the number of significant digits. With some calculators, you have to set the number of digits you want after the decimal point. Be sure that you set the number at a value high enough so that answers are not truncated. The user of a calculator has to judge the number of digits needed and round the result from the calculator display. (See Chapter 2 for rounding rules.) For instance, suppose we drove a car 350 miles and used 21 gallons of gas. The mileage in miles per gallon can be found using a calculator.

$$\frac{350 \text{ mi}}{21 \text{ gal}} = 16.666667 \text{ mi/gal}$$

This answer is not correct since we knew the gallons to only two digits. The calculator gave us far too many digits, but it does not understand significant digits. So, we round the answer to two digits: 17 mi/gal.

Imagine that we carefully measured our mileage and found that 21.0 gallons were used in traveling 357.0 miles. Here the three digits in the gallons should give us a three-digit mileage. On many calculator models, the answer given by the calculator is

$$\frac{357.0 \text{ mi}}{21.0 \text{ gal}} = 17. \text{ mi/gal}$$

What has happened? Has the calculator deprived us of our additional significant digit? Again, the calculator does not deal with significant digits.

Anytime a string of zeros occurs after the decimal point, most calculators drop them and give only the remaining nonzero digits. A zero can be a significant digit, as it is in this calculation. Since we know that the answer should have three significant digits, we just add a zero to the result obtained from the display: 17.0 mi/gal. This situation arises only when a calculated answer has a string of zeros to the right of the decimal point. Some calculators will actually display the string of zeros, but we still have to decide on the correct number of significant digits.

MOLAR MASS COMPUTATION (SEE SECTION 4-9)

To find the molar mass of a compound like sucrose, $C_{12}H_{22}O_{11}$, from the formula, the molar masses of the elements are multiplied by the corresponding subscripts and the products are added. If we want a three- or four-digit molar mass, we usually carry more digits and then round after we have found the answer. The calculation of the molar mass of sucrose to three digits is

C	12(12.01)	
H	22(1.008)	The answer to three digits is 342.
O	11(16.00)	

The following are examples of the calculation sequence on various types of calculators. Try the calculation with your calculator.

Some calculators have algebraic operating systems in which multiplication and division operations are done left to right before any addition and subtraction operations left to right. The key sequence using a scientific calculator with algebraic operating system is

$$12 \otimes 12.01 \oplus 22 \otimes 1.008 \oplus 11 \otimes 16.00 = \text{ or enter}$$

An algebraic calculator with memory gives

$$12 \otimes 12.01 \ominus \boxed{M+} \ 22 \otimes 1.008 \ominus \boxed{M+} \ 11 \ominus 16.00 = \boxed{M+}\boxed{MR}$$

Be sure that memory is cleared or set to zero before calculating. If you have no memory on your calculator, write down each product as it is determined and then add them up.

pH AND pOH CALCULATIONS (SEE SECTION 13-16)

The pH and pOH of a solution are defined as (see Section 13-15)

$$pH = -\log[H_3O^+] \qquad pOH = -\log[OH^-]$$

To find the pH from the hydronium ion concentration or the pOH from the hydroxide ion concentration, a calculator with a log key is needed. The calculation sequence on a typical algebraic calculator using the numerical values of the concentrations is

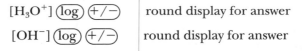

$+/-$ refers to change sign key, which is the CHS key on some calculators.
The key sequence on some TI calculators is

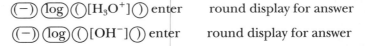

Some calculators supply the opening parenthesis after the log key and some imply the closed parenthesis. Test your own calculator. Practice by finding the pH corresponding to $[H_3O^+] = 2 \times 10^{-3}$ and the pOH corresponding to $[OH^-] = 5 \times 10^{-12}$. The answers are 2.7 and 11.3.

The hydronium ion concentration corresponding to a given pH or the hydroxide ion corresponding to a given pOH is found by changing the sign of the pH or pOH value and then finding 10 raised to the power of the negated pH or negated pOH.

$$[H_3O^+] = 10^{-pH} \qquad [OH^-] = 10^{-pOH}$$

Calculation sequences on various types of calculators are shown below (substitute the numerical value for pH or pOH as needed).

An algebraic calculator with a 10^x key gives

pH $\boxed{+/-}$ $\boxed{10^x}$ round display for answer

pOH $\boxed{+/-}$ $\boxed{10^x}$ round display for answer

A scientific calculator with algebraic operating system and no 10^x key gives

pH $\boxed{+/-}$ $\boxed{INV}\boxed{\log}$ round display for answer

pOH $\boxed{+/-}$ $\boxed{INV}\boxed{\log}$ round display for answer

The key sequence on some TI calculators is

$\boxed{10^x}\boxed{(}\boxed{(-)}$ pH $\boxed{)}$ enter   or $\boxed{10^x}\boxed{(}\boxed{(-)}$ pOH $\boxed{)}$ enter

Some calculators supply the opening parenthesis after the log key and some imply the closed parenthesis. For practice, find the $[H_3O^+]$ corresponding to a pH of 2.7 and find the $[OH^-]$ corresponding to a pOH of 11.3.

CALCULATIONS WITH EXPONENTIAL NUMBERS

Sometimes calculations involve exponential numbers. If you have a calculator that accepts exponential numbers, just key in the numbers in sequence as a chain calculation. On such calculators the exponent key is usually marked EE, EEX, or EXP and represents $\times$ 10. You may need to set the calculator in the scientific notation mode (SCI key) or switch from fixed-point to exponential notation (F↔E key).

As an example, suppose we want to find the number of gold atoms in 1.00 pound or 4.54×10^2 g of gold. The setup is

$$4.54 \times 10^2\, g \left(\frac{1\ mol\ Au}{197.0\ g}\right)\left(\frac{6.022 \times 10^{23}\ Au\ atoms}{1\ mol\ Au}\right)$$

If you have a calculator that accepts exponential numbers, just key in the numbers in sequence as a chain calculation. Thus, the setup above is keyed in as

4.54 (EE) 2 (÷) 197. (×) 6.022 (EE) 23 (=) or enter   round the answer as
                                                            you read it

Thus, the final answer is 1.39×10^{24} Au atoms.

As a final example, the following setup is used to find the mass in kilograms of one iron atom (to three digits):

$$1\ Fe\ atom \left(\frac{1\ mol\ Fe}{6.022 \times 10^{23}\ Fe\ atoms}\right)\left(\frac{55.85\ g}{1\ mol\ Fe}\right)\left(\frac{1\ kg}{10^3\ g}\right)$$

Using a calculator that allows exponential numbers, just key in the factors in sequence. Incidentally, for 10^3, use 1 EE 3 or simply key in 1000. On calculators that do not allow exponential numbers separate the exponential and non-exponential parts.

$$\frac{55.85}{6.022} \times \frac{1}{10^{23} \times 10^3}$$

Calculate the nonexponential part on the calculator and the exponent by hand.

55.85 (÷) 6.022 (=) 9.274   round the answer as you read it

The complete answer is 9.27×10^{-26} kg.

PRACTICE A2-1

Try the following calculations to see if you obtain the answer given.

$$(3.62 \times 10^{-5}) \frac{5.23 \times 10^4}{8.47 \times 10^8} = 2.24 \times 10^{-9}$$

POWER RULES

Sometimes we calculate with numbers written in exponential notation. Consequently, it is important to know how to calculate with powers of 10.

1. The product of factors involving 10 raised to a power is 10 raised to the sum of the powers.

    $$(10^a)(10^b) = 10^{a+b} \qquad (10^2)(10^3) = 10^{2+3} = 10^5$$

    If one power is negative, algebraic addition results in a subtraction.

    $$(10^a)(10^{-b}) = 10^{a-b} \qquad (10^3)(10^{-2}) = 10^{3-2} = 10^1$$

    Any number of factors may be involved.

    $$(10^4)(10^3)(10^{-2})(10^1) = 10^{4+3-2+1} = 10^6$$

2. The quotient of factors involving 10 raised to a power is 10 raised to the difference of the powers.

    $$\frac{10^a}{10^b} = 10^{a-b} \qquad \frac{10^3}{10^2} = 10^{3-2} = 10^1$$

    A negative power in the denominator results in the sum of the powers.

    $$\frac{10^3}{10^{-2}} = 10^{3-(-2)} = 10^{3+2} = 10^5$$

3. When a factor involving 10 raised to a power is raised to a power, the result is 10 raised to the product of the powers.

    $$(10^a)^b = 10^{ab} \qquad (10^3)^2 = 10^{3\times2} = 10^6$$

    For example,

    $$\left(\frac{10^3 \text{m}}{1 \text{ km}}\right)^2 = \frac{10^6 \text{m}^2}{1 \text{ km}^2}$$

    The product of a negative and positive power is negative.

    $$(10^{-3})^3 = 10^{-3\times3} = 10^{-9}$$

    For example,

    $$\left(\frac{10^{-3} \text{m}}{1 \text{ mm}}\right)^3 = \frac{10^{-9} \text{m}^3}{1 \text{ mm}^3}$$

4. When two factors involving the same power are divided or when two factors involving the same numerical power with opposite signs are multiplied, the result is 1, since 10^0 equals 1.

    $$\frac{10^a}{10^a} = 10^{a-a} = 10^0 = 1$$

    $$\frac{10^2}{10^2} = 10^{2-2} = 10^0 = 1$$

    $$(10^a)(10^{-a}) = 10^{a-a} = 10^0 = 1$$

    $$(10^3)(10^{-3}) = 10^{3-3} = 10^0 = 1$$

USING

THE VSEPR

THEORY

It is possible to predict the shape of any molecule in which the central atom is surrounded by four pairs of valence electrons. Do this by using the Lewis structure and deciding how many bonding electron pairs are involved. Some molecules have more than four pairs of electrons around the central atom, and some have fewer than four pairs. The shapes discussed here apply only to those molecules with four pairs of electrons, which is the most common situation. Other cases are not part of this discussion. To predict a shape:

1. Write the Lewis structure of the molecule or polyatomic ion.
2. Count the number of bonding pairs and nonbonding pairs (lone pairs) around the central atom.
3. Decide which case the structure fits:
   a. Four bonding pairs: tetrahedral
   b. Three bonding pairs and one nonbonding pair: triangular pyramid
   c. Two bonding pairs and two nonbonding pairs: angular or bent
   d. Two atom or diatomic molecule: always linear
4. If the structure does not fit any of these cases, is it one of the special cases involving multiple bonds?
   a. Two double bonds or one single and a triple bond: linear
   b. One double bond and two single bonds: triangular planar
   c. One double bond, one single bond, and one nonbonding pair: angular

EXAMPLE A3-1

What is the shape of sulfur dichloride, SCl_2, that has the following Lewis structure?

The central sulfur has four pairs of valence electrons. Since there are two bonding pairs and two nonbonding pairs, we predict its shape to be angular.

EXAMPLE A3-2

What a likely shape for a molecule of chloroform, $CHCl_3$? The Lewis structure is

Thus, a likely shape would be tetrahedral since there are four bonding pairs.

EXAMPLE A3-3

What is the likely shape of a hydrogen chloride, HCl, molecule?

H—C̈l: Any two-atom molecule is linear.

The VSEPR theory is also used to rationalize the shapes of some polyatomic ions. For instance, the Lewis structure of hydronium ion is

$$\left(H-\overset{..}{O}-H \atop \overset{|}{H} \right)^{+}$$

The shape would be triangular pyramidal since there are three bonding pairs and one nonbonding pair.

EXAMPLE A3-4

What is the likely shape of an ammonium ion, NH_4^+? The Lewis structure is

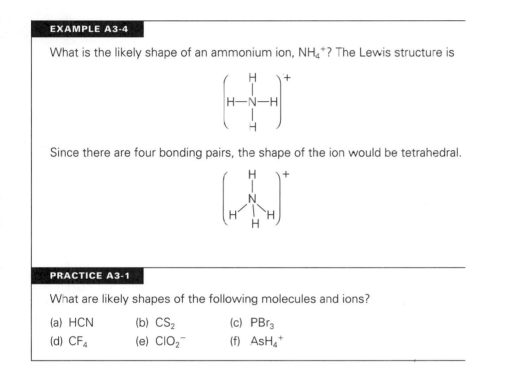

Since there are four bonding pairs, the shape of the ion would be tetrahedral.

PRACTICE A3-1

What are likely shapes of the following molecules and ions?

(a) HCN   (b) CS_2 (c) PBr_3

(d) CF_4 (e) ClO_2^- (f) AsH_4^+

The VSEPR theory is also used to rationalize the shapes of molecules or ions having more than one central atom. Do this by applying the theory to each atom that has a central position to which other atoms bond. For example, a molecule of methyl alcohol has a carbon bonded to three hydrogens and one oxygen. The oxygen is also bonded to another hydrogen.

$$
\begin{array}{ccc}
 & H & H \\
 & | & | \\
H- & C- & O: \\
 & | & \\
 & H &
\end{array}
$$

The carbon and the oxygen are both viewed as central atoms to which other atoms bond. To apply the VSEPR theory, we just count the number of electron pairs around each center and describe the shape around each center. Count the shared pair between the central atoms as a pair for both of them. In methyl alcohol, the atoms are arranged around the carbon in a tetrahedral shape since it has four bonding pairs of electrons. The atoms are arranged around the oxygen in an angular shape since it has two bonding pairs and two nonbonding pairs of electrons. Thus, the shape of methyl alcohol is an angular atomic arrangement attached to a tetrahedral arrangement:

Note that since methyl alcohol has one hydrogen bonded to oxygen, it is sometimes represented by the formula CH_3OH rather than CH_4O.

EXAMPLE A3-5

What is the shape of a molecule of acetaldehyde that has the Lewis structure

$$
\begin{array}{ccc}
\text{H} & \text{:O:} \\
| & || \\
\text{H—C—C} \\
| & | \\
\text{H} & \text{H}
\end{array}
$$

Both carbons are central atoms. The first carbon has four bonding pairs that correspond to a tetrahedral shape. The second carbon has two single bonds and a double bond that correspond to a triangular planar shape. Thus, the molecule has a tetrahedral portion attached to a triangular planar portion.

PRACTICE A3-2

What is the likely shape of a molecule of acetic acid that has the following Lewis structure? (Hint: There are three central atoms.)

The first carbon has four bonding pairs and a tetrahedral shape. The second carbon has two single bonds and a double bond, so it has a triangular shape. The oxygen has two bonding pairs and two lone pairs and has an angular shape. The overall shape of the molecule is a tetrahedron linked to a triangle, and the triangle is, in turn, linked to the angular oxygen.

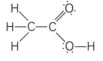

PRACTICE A3-3

What are the shapes of the following molecules? The bonding sequence is shown by the formulas.

(a) Ethyl alcohol, CH_3CH_2OH. (Hint: The molecule has three central atoms.)

(b) Acetone, CH_3COCH_3. (The three carbons are bonded in sequence and the oxygen is double bonded to the middle carbon.)

(c) Urea, NH_2CONH_2. (Two nitrogens are bonded to the carbon and the oxygen is double bonded to the carbon.)

(d) Butane, $CH_3CH_2CH_2CH_3$. (The four carbons are bonded in sequence.)

BALANCING

REDOX

EQUATIONS

The equations of some redox reactions can be balanced using the simple methods discussed in Chapter 5. Many redox reactions, however, cannot be easily balanced using these methods. Redox reactions involve electron transfer, and the same number of electrons lost must be gained during the reaction. This aspect of a redox reaction can serve as the basis of a pattern for balancing redox equations.

Several different methods are used to balance redox equations. A common and useful method is known as the oxidation number method. (Review the topic of oxidation numbers as discussed in Section 15-1.) This method is especially useful for redox reactions occurring in aqueous solutions. The balancing of an equation by this method illustrates how it works.

As an example, we'll use the reaction of dichromate ion, $Cr_2O_7^{2-}$, and iron(II) ion, Fe^{2+}, in aqueous acidic solution. When a solution containing the pale green iron(II) ion is added to a solution containing the bright orange dichromate ion and a small amount of some strong acid, a reaction occurs and the color of the mixture changes to a shade of green. The products of the redox reaction are the green chromium(III) ion, Cr^{3+}, and the yellow iron(III) ion, Fe^{3+}.

Initially, the reaction can be represented by an unbalanced equation showing the reactants and products.

$$Cr_2O_7^{2-} + Fe^{2+} \longrightarrow Fe^{3+} + Cr^{3-}$$

Because this reaction takes place in an acidic solution hydronium ion or simply hydrogen ion, H^+, is present as a charge carrier. Some redox reactions occur only in aqueous acidic solutions. In contrast, some redox reactions require the presence of hydroxide ion, so they occur only in basic solutions. We will see how these ions are involved as we learn to balance equations for redox reactions.

Here are the steps involved in balancing a redox equation.

1. Separate the reactants and products into two half-reactions involving the elements that are changing oxidation number. Write the skeletal equations for each half-reaction.

$$Cr_2O_7^{2-} \longrightarrow Cr^{3+}$$
$$Fe^{2+} \longrightarrow Fe^{3+}$$

These equations are called skeletal equations since the major reactant and product are given and other species may be added as the equations are balanced. Note that the skeletal equation for each half includes related species. One half includes the chromium-containing species, and the other half contains the iron-containing species.

2. Work with one half-reaction at a time. Deduce the oxidation numbers of the element that is changing oxidation number. Here, the oxidation number rules are used to deduce the oxidation numbers of the elements that are changing. See Chapter 15 to review the rules.

   In this example, the dichromate ion half is balanced first, and then the other half is balanced. Using the oxidation number rules, the oxidation numbers of chromium are deduced. Remember to find the oxidation number per atom.

$$\overset{+6}{Cr_2O_7^{2-}} \longrightarrow \overset{+3}{Cr^{3+}}$$

3. Temporarily balance the number of atoms of the element changing oxidation number.

   Since the dichromate ion on the left has two chromium atoms, we need a coefficient of 2 for Cr^{3+}. This value is called a temporary coefficient since it may become larger in the completely balanced equation.

$$\overset{+6}{Cr_2O_7^{2-}} \longrightarrow \overset{+3}{2Cr^{3+}}$$

4. Find the total oxidation number change by determining the change per atom and multiplying by the total number of atoms that change oxidation number. Also, decide whether electrons are lost or gained. An increase in oxidation number is loss of electrons, and a decrease in oxidation number is gain of electrons.

   The chromium changes from $+6$ to $+3$, which is a three-electron change. A decrease in oxidation number is reduction or a gain of electrons. Since there are two chromiums, changing the total oxidation number change corresponds to six electrons gained.

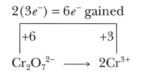

$$2(3e^-) = 6e^- \text{ gained}$$

5. Add the electrons lost or gained to the half-equation. Lost electrons go on the product side and gained electrons on the reactant side. It is important to put the electrons on the correct side of the equation.

   The six electrons gained are written on the reactant side of the half-equation.

$$6e^- + Cr_2O_7^{2-} \longrightarrow 2Cr^{3+}$$

6. Count the charges on both sides and balance the charges in the equation by adding H^+ or OH^- to either side. If the reaction occurs in acidic solution, use H^+, and if it occurs in basic solution, use OH^-. If the reaction occurs in a neutral solution, use H^+ or OH^- on the right as needed. That is, in a neutral solution, if negative charges are needed for balancing, use OH^-, and if positive charges are needed, use H^+. Whether the reaction solution is acidic, basic, or neutral must be known to balance the equation.

   In the example, the solution is acidic, so H^+ can be used to balance the charge. To balance the charge, count up the total charge on both sides. Electrons carry negative charges and ions carry positive or negative charges. Look for the charges in the superscript positions in ions. The idea of a charge balance comes from a reaction being unable to consume or produce charge. The same amount of positive and/or negative charge must occur on both sides of the arrow.

$$\begin{array}{ccc} 6e^- + & Cr_2O_7^{2-} & \longrightarrow 2Cr^{3+} \\ 6- & 2- & 2(3+) \\ & 8- & 6+ \end{array}$$

Note that the total charge is found on each side by multiplying the charge of any ion by its coefficient and adding any separate charges. In the example, there is a total of 8− on the left and 6+ on the right. The only way to balance the charge with hydrogen ion, H^+, is to add 14 hydrogen ions on the left. Since negative and positive charges on the same side cancel, this gives 6+ on both sides. In other words, since we are adding positive H^+ to balance the charge, we need to add enough on the left to cancel the 8− charges and have 6+ charges to balance the 6+ on the right.

$$14H^+ + 6e^- + Cr_2O_7^{2-} \longrightarrow 2Cr^{3+}$$

7. In the final step, balance the hydrogens and oxygens by adding the appropriate number of H_2O to either side. Count the number of hydrogens on each side, and then add the necessary number of waters.

In the example, the equation has 14 H^+, which contain 14 combined hydrogens on the left and no hydrogens on the right. Thus, 7 waters are needed to supply 14 hydrogens (2×7) on the right. Adding 7 waters on the right gives the completely balanced half-equation.

$$14H^+ + Cr_2O_7^{2-} + 6e^- \longrightarrow 2Cr^{3+} + 7H_2O$$

To be sure that the half-equation is balanced, check the chemical and charge balance. As can be seen from the equation, there are 14 hydrogens, 7 oxygens, and 2 chromiums on each side. There is a total of 6+ on each side. When tallying charge, remember that positive and negative charges on the same side of the arrow cancel. Thus, the 8− charge on the left cancels 8 of the + charges on the left, leaving a total of 6+ charges on each side.

The steps for balancing a redox half-reaction are summarized in Table A4-1. Any half-equation can be balanced by following these steps. For success, always follow the order of steps as listed. Of course, not all steps may be required in the balancing of the equations of some half-reactions.

Table A4-1   *Steps for Balancing Redox Half-Equations*

1.	**S**	*Skeletal half-equations*
		Separate the reactants and products into two skeletal half-reactions involving the elements that are changing oxidation number.
2.	**O**	*Oxidation numbers*
		Find the oxidation numbers of the elements that are changing.
3.	**T**	*Temporary coefficients*
		Balance the number of atoms of the elements that are changing oxidation number.
4.	**T**	*Total oxidation number changes*
		Find the total oxidation number change for the elements that are changing oxidation number.
5.	**E**	*Electrons added*
		Add electrons lost or gained.
6.	**C**	*Charge balance*
		Balance charge by adding H^+ in acidic solutions or OH^- in basic solutions or either on the right if neutral.
7.	**H**	*Hydrogens and oxygen balance*
		Balance the hydrogens and oxygens by adding water on either side.

Let's apply the same balancing steps to the other half of the example reaction.

Skeletal equation: $Fe^{2+} \longrightarrow Fe^{3+}$

The oxidation numbers for iron are

$$\overset{+2}{Fe^{2+}} \longrightarrow \overset{+3}{Fe^{3+}}$$

Oxidation numbers:

Temporary coefficients: No temporary coefficients are needed, and the total oxidation number change is

$$\text{1}e^{-} \text{ lost}$$

Total Oxidation $\quad \overset{+2}{Fe^{2+}} \quad \overset{+3}{Fe^{3+}}$

Number Change: $\quad Fe^{2+} \longrightarrow Fe^{3+}$

The lost electron is shown as a product.

Electrons added: $Fe^{2+} \longrightarrow Fe^{3+} + 1e^{-}$

Charge balance: Checking the charge, we see that there are $2+$ on the left and $3+$ and $1-$, for a total of $2+$, on the right. So the half-equation is balanced with a total of $2+$ on each side. Since no hydrogens or oxygens are involved in this half-reaction, it is completely balanced. In this case, the half-equation was found to be balanced after step 5. Of course, once a half-equation is balanced, there is no need to continue with additional steps.

The completely balanced redox equation is obtained by adding the two half-equations. First, the half-equations are multiplied by appropriate numbers to be sure that the total number of electrons lost equals the total number gained. Look at the two half-reactions in the example.

$$14H^{+} + Cr_2O_7^{2-} + 6e^{-} \longrightarrow 2Cr^{3+} + 7H_2O$$
$$Fe^{2+} \longrightarrow Fe^{3+} + 1e^{-}$$

All the terms in the iron half-equation are multiplied by 6 to have 6 electrons lost to balance the 6 electrons gained in the other half. Then, the half-equations are added and the electrons canceled.

$$14H^{+} + Cr_2O_7^{2-} + 6e^{-} \longrightarrow 2Cr^{3+} + 7H_2O$$
$$6Fe^{2+} \longrightarrow 6Fe^{3+} + 6e^{-}$$
$$\overline{14H^{+} + Cr_2O_7^{2-} + 6Fe^{2+} \longrightarrow 6Fe^{3+} + 2Cr^{3+} + 7H_2O}$$

To be sure the equation is balanced, check the chemical and charge balance. As can be seen from the equation, there are 14 hydrogens, 7 oxygens, 2 chromiums, and 6 irons on each side. There is a total of $24+$ on each side:

$$(14+) + (2-) + 6(2+) = 24+ \quad \text{and} \quad 6(3+) + 2(3+) = 24+$$

Sometimes, redox reactions are written using hydronium ion rather than hydrogen ion. An H^+ occurs in water as H_3O^+. A hydronium is viewed as hydrogen ion bonded to water. To change a balanced redox equation to include hydronium, change the hydrogen ions to hydronium ions and add the corresponding number of waters to the other side for balance. In the above example, the 14 H^+ are changed to 14 H_3O^+ and 14 H_2O are added to the right to give a total of 21 H_2O.

$$14H_3O^+ + Cr_2O_7^{2-} + 6Fe^{2+} \longrightarrow 6Fe^{3+} + 2Cr^{3+} + 21H_2O$$

EXAMPLE A4-1

Sodium metal reacts with water to form sodium ions and hydrogen gas. Give the balanced equation for this reaction if the skeletal equation is

$$Na + H_2O \longrightarrow Na^+ + H_2$$

Separate the reaction into two half-equations and balance each half. Separate by noting the species that are obviously related.

$$Na \longrightarrow Na^+$$

$$H_2O \longrightarrow H_2$$

First, balance the sodium half.

$$\overset{\text{1}e^- \text{ lost}}{\overset{\displaystyle 0 \qquad\qquad +1}{Na \longrightarrow Na^+}}$$

Adding the lost electron on the product side gives

$$Na \longrightarrow Na^+ + 1e^-$$

Checking the charge, we see that the half-equation is balanced.

Next, work with the second half-equation.

$$\overset{2(1e^-) = 2e^- \text{ gained}}{\overset{\displaystyle +1 \qquad\qquad 0}{H_2O \longrightarrow H_2}}$$

No temporary coefficients are needed, and since two hydrogens are changing from +1 to 0, 2 electrons are gained.

$$H_2O + 2e^- \longrightarrow H_2$$

Since the reaction solution begins as neutral, the charge is balanced by adding OH^- or H^+ as a product. In other words, the solution is neither acidic nor basic at the beginning. Two hydroxide ions on the right balance the charge of 2− on the left.

$$H_2O + 2e^- \longrightarrow H_2 + 2CH^-$$

The hydrogens and oxygens are balanced by adding water. Since there are four hydrogens on the right and only two on the left, another water is needed on the left to give a total of four hydrogens on the left.

$$2H_2O + 2e^- \longrightarrow H_2 + 2OH^-$$

Checking the chemical balance, we see that there are four H and two O on each side. The charge is balanced since there are 2− on each side.

The complete equation is obtained from the two half equations after multiplying the sodium half by 2 to balance the electrons lost with the electrons gained.

$$2Na \longrightarrow 2Na^+ + 2e^-$$
$$\underline{2H_2O + 2e^- \longrightarrow H_2 + 2OH^-}$$
$$2Na + 2H_2O \longrightarrow 2Na^+ + 2OH^- + H_2$$

EXAMPLE A4-2

Give the balanced equation for the reaction between zinc metal and hydrochloric acid that produces zinc ion and hydrogen gas. The skeletal equation is

$$Zn + H^+ \longrightarrow Zn^{2+} + H_2 \qquad \text{(acidic solution)}$$

Separate into two half-equations including related species.

$$Zn \longrightarrow Zn^{2+}$$

$$H^+ \longrightarrow H_2$$

First, balance the zinc half.

$$\overset{2e^- \text{ lost}}{\overbrace{\underset{Zn}{0} \longrightarrow \underset{Zn^{2+}}{+2}}}$$

Adding the lost electrons as a product gives

$$Zn \longrightarrow Zn^{2+} + 2e^-$$

A check of the charges shows the half-equation to be balanced.

Next, work with the skeletal equation for the hydrogen half.

$$\underset{H^+}{+1} \longrightarrow \underset{H_2}{0}$$

Since there is one hydrogen on the left and two on the right, a temporary coefficient of 2 on the left is needed.

$$\overset{+1}{2H^+} \longrightarrow \overset{0}{H_2}$$

The total oxidation number change is

$$2(1e^-) = 2e^- \text{ gained}$$

$$\boxed{\overset{+1}{2H^+} \quad\quad \overset{0}{H_2}} \longrightarrow$$

The electrons gained are added as reactants.

$$2H^+ + 2e^- \longrightarrow H_2$$

A check of the charges reveals that there are 2+ and 2− on the left and no charge on the right. Thus, the charge is balanced.

Since there are no oxygens involved in the half-equation, no waters are needed. The half-equation is balanced.

$$2H^+ + 2e^- \longrightarrow H_2$$

The completely balanced equation is obtained by adding the two half-equations. Neither half has to be multiplied by a factor since two electrons are involved in each half.

$$
\begin{array}{l}
2H^+ + 2e^- \longrightarrow H_2 \\
\underline{\quad\quad Zn \longrightarrow Zn^{2+} + 2e^-} \\
2H^+ + Zn \longrightarrow Zn^{2+} + H_2
\end{array}
$$

EXAMPLE A4-3

Give the balanced equation for the reaction corresponding to the skeletal equation given below.

$$MnO_4^- + CHO_2^- \longrightarrow MnO_2 + CO_3^{2-} \quad\quad \text{(basic solution)}$$

Separate into two half-equations, one including the manganese-containing species and one including the carbon-containing species.

$$MnO_4^- \longrightarrow MnO_2$$
$$CHO_2^- \longrightarrow CO_3^{2-}$$

First, balance the permanganate ion half.

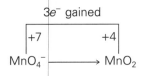

$$3e^- \text{ gained}$$

$$\boxed{\overset{+7}{MnO_4^-} \quad\quad \overset{+4}{MnO_2}} \longrightarrow$$

Adding the electrons gained gives

$$MnO_4^- + 3e^- \longrightarrow MnO_2$$

To balance the charge, we note that the solution is basic, so hydroxide ion, OH^-, is used. Four hydroxide ions are needed on the right to balance the $4-$ charges $[1- + 3(-) = 4-]$ on the left:

$$MnO_4^- + 3e^- \longrightarrow MnO_2 + 4OH^-$$

The hydrogen and oxygen are balanced by adding water. We need two H_2O on the left to balance the four H in the OH^- on the right.

$$2H_2O + MnO_4^- + 3e^- \longrightarrow MnO_2 + 4OH^-$$

Next, balance the other half-equation.

Skeletal equation: $CHO_2^- \longrightarrow CO_3^{2-}$

No temporary coefficients, Oxidation number change and electrons lost:

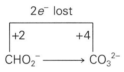

$$\overset{+2}{CHO_2^-} \longrightarrow \overset{+4}{CO_3^{2-}}$$

Include the lost electrons as a product:

$$CHO_2^- \longrightarrow CO_3^{2-} + 2e^-$$

Balance the charge with hydroxide ion. Remember that it is a basic solution. Three OH^- are needed on the left to give a total of four negative charges on each side:

$$3OH^- + CHO_2^- \longrightarrow CO_3^{2-} + 2e^-$$

Balance the hydrogens and oxygens by adding water. Two H_2O are needed on the right to balance the four H on the left. Note that three H are in the $3OH^-$ and one H is in the CHO_2^-:

$$3OH^- + CHO_2^- \longrightarrow CO_3^{2-} + 2e^- + 2H_2O$$

The completely balanced equation is obtained by adding the two half-equations after each is multiplied by a factor to adjust the electrons lost to equal the electrons gained. The permanganate half is multiplied by two and the other half by three as revealed by the half-equations.

$$2(2H_2O + MnO_4^- + 3e^- \longrightarrow MnO_2 + 4OH^-)$$
$$3(3OH^- + CHO_2^- \longrightarrow CO_3^{2-} + 2e^- + 2H_2O)$$

Multiplication and adding the two half-equations give

$$4H_2O + 2MnO_4^- + 6e^- \longrightarrow 2MnO_2 + 8OH^-$$
$$9OH^- + 3CHO_2^- \longrightarrow 3CO_3^{2-} + 6e^- + 6H_2O$$

$$4H_2O + 2MnO_4^- + 9OH^- + 3CHO_2^- \longrightarrow 3CO_3^{2-} + 2MnO_2 + 8OH^- + 6H_2O$$

Notice that water and hydroxide ion occur on both sides of the equation. When balancing a redox equation, you may note that some species appear on both sides. Just cancel the appropriate number of species to give a properly balanced equation. In this case, eight OH^- on the right cancel eight on the left, leaving one OH^-. Four H_2O on the left cancel four H_2O on the right, leaving two H_2O on the right. Thus, the completely balanced equation is

$$2MnO_4^- + 3CHO_2^- + OH^- \longrightarrow 3CO_3^{2-} + 2MnO_2 + 2H_2O$$

A check of the chemical and charge balance shows the equation to be balanced.

BALANCING WHOLE REDOX EQUATIONS

Many redox equations can be balanced by incorporating both the oxidation and reduction steps as a whole equation. A slight variation of the oxidation number method is used.

Here are the steps involved in balancing a whole redox equation using the first example given in the previous section of dichromate ion and iron(II) ion reacting in acid solution.

1. Start with the skeletal equation including the reactants and products.

$$Cr_2O_7^{2-} + Fe^{2+} \longrightarrow Fe^{3+} + Cr^{3+}$$

2. Deduce the oxidation numbers of the elements that are changing oxidation number.

$$\overset{+6}{Cr_2}O_7^{2-} + \overset{+2}{Fe^{2+}} \longrightarrow \overset{+3}{Fe^{3+}} + \overset{+3}{Cr^{3+}}$$

3. Temporarily balance the number of atoms of any element changing oxidation number. Since the dichromate ion on the left has two chromiums, we need a coefficient of 2 for Cr^{3+}.

$$\overset{+6}{Cr_2}O_7^{2-} + \overset{+2}{Fe^{2+}} \longrightarrow \overset{+3}{Fe^{3+}} + 2\overset{+3}{Cr^{3+}}$$

4. Find the total oxidation number change for each element changing by determining the change per atom and multiplying by the total number of atoms that change oxidation number.

$$2(3e^-) = 6e^- \text{ gained}$$

$$\overset{+6}{Cr_2}O_7^{2-} + \overset{+2}{Fe^{2+}} \longrightarrow \overset{+3}{Fe^{3+}} + 2\overset{+3}{Cr^{3+}}$$

$$1e^- \text{ lost}$$

5. Balance the electrons lost and gained by multiplying each half by the appropriate number. This part is the major difference between this

method and the half-equation method. No electrons need to be written in the equation. To balance the six electrons gained, multiply the iron half by six to make six electrons gained.

$$Cr_2O_7{}^{2-} + 6Fe^{2+} \longrightarrow 6Fe^{3+} + 2Cr^{3+}$$

6. Balance the charge by adding H^+ if acidic, OH^- if basic, or, if neutral, either H^+ or OH^- on the right.

In the example, the solution is acidic, so H^+ can be used to balance the charge.

$$
\begin{array}{llll}
Cr_2O_7{}^{2-} + 6Fe^{2+} & \longrightarrow & 6Fe^{3+} + 2Cr^{3+} \\
(2-) \qquad\quad 6(2+) & & 6(3+) \quad 2(3+) \\
(2-) + (12+) = 10+ & & (18+) + (6+) = 24+
\end{array}
$$

There is a total of 10+ on the left and 24+ on the right. The only way to balance the charge with hydrogen ion, H^+, is to add 14 hydrogen ions on the left.

$$14H^+ + Cr_2O_7{}^{2-} + 6Fe^{2+} \longrightarrow 6Fe^{3+} + 2Cr^{3+}$$

7. In the final step, balance the hydrogens and oxygens by adding the appropriate number of H_2O to either side.

In the example, the equation has $14H^+$, which contain 14 combined hydrogens on the left and no hydrogens on the right. Thus, 7 waters are needed on the right to give 14 hydrogens.

$$14H^+ + Cr_2O_7{}^{2-} + 6Fe^{2+} \longrightarrow 6Fe^{3+} + 2Cr^{3+} + 7H_2O$$

To be sure that the equation is balanced, check the chemical and charge balance. As can be seen from the equation, there are 14 hydrogens, 7 oxygens, 6 irons, and 2 chromiums on each side. There is a total of 24+ on each side.

EXAMPLE A4-4

Balance the following redox skeletal equation:

Skeletal: $MnO_4{}^- + CHO_2{}^- \longrightarrow CO_3{}^{2-} + MnO_2$ (basic)

Oxidation numbers: $\underset{+7}{MnO_4{}^-} + \underset{+2}{CHO_2{}^-} \longrightarrow \underset{+4}{CO_3{}^{2-}} + \underset{+4}{MnO_2}$

No temporary coefficients needed.

Total oxidation number change for each half:

$$
\begin{array}{c}
\overset{3e^-}{\overbrace{}} \\
\underset{+7}{MnO_4{}^-} + \underset{+2}{CHO_2{}^-} \longrightarrow \underset{+4}{CO_3{}^{2-}} + \underset{+4}{MnO_2} \\
\underbrace{}_{2e^-}
\end{array}
$$

Multiply each half to balance the electrons lost and gained. Multiply the Mn half by 2 and C half by 3.

$$2MnO_4^- + 3CHO_2^- \longrightarrow 3CO_3^{2-} + 2MnO_2$$

Tally the charges.

left: $(2-) + (3-) = 5-$ right: $3(2-) = 6-$

Balance the charge using H^+ or OH^-. Since the reaction conditions are basic, balance the charge by adding one OH^- on the left.

$$2MnO_4^- + 3CHO_2^- + OH^- \longrightarrow 3CO_3^{2-} + 2MnO_2$$

Tally the H on each side.

left: $3H + 1H = 4H$ right: $0H$

Balance the H and O by adding H_2O. Two H_2O are needed on the right.

$$2MnO_4^- + 3CHO_2^- + OH^- \longrightarrow 3CO_3^{2-} + 2MnO_2 + 2H_2O$$

Check the chemical and charge balance. The equation is completely balanced.

PRACTICE A4-1

Give balanced equations for the following redox reactions that take place in acidic solutions. Use the half-equation or whole-equation method for balancing.

(a) $ClO_3^- + I_3^- \longrightarrow Cl^- + I_2$ (Hint: Each I changes oxidation number by $\frac{2}{3}$, from $-\frac{1}{3}$ to 0.)

(b) $Ba + H^+ \longrightarrow Ba^{2+} + H_2$

(c) $Cr^{3+} + BiO_3^- \longrightarrow Cr_2O_7^{2-} + Bi^{3+}$

(d) $Ag + NO_3^- \longrightarrow Ag^+ + NO_2$

(e) $H_2C_2O_4 + HNO_2 \longrightarrow CO_2 + NO$

(f) $I_2 + ClO^- \longrightarrow IO_3^- + Cl^-$

(g) $Br^- + H_2O_2 \longrightarrow Br_2 + H_2O$ (Hint: The oxidation number of O in H_2O_2 is -1.)

(h) $XeO_3 + I_3^- \longrightarrow Xe + I_2$ (Hint: Each I changes oxidation number by $\frac{2}{3}$, from $-\frac{1}{3}$ to 0.)

(i) $MnO_4^- + H_2C_2O_4 \longrightarrow Mn^{2+} + CO_2$

Give balanced equations for the following redox reactions that take place in basic solutions. Use the half-equation or whole-equation method for balancing.

(a) $BrO_3^- + F_2 \longrightarrow BrO_4^- + F^-$

(b) $N_2H_4 + Cu(OH)_2 \longrightarrow N_2 + Cu$ (Hint: H in N_2H_4 has an oxidation number of $+1$.)

(c) $SO_3^{2-} + ClO_3^- \longrightarrow SO_4^{2-} + Cl^-$

(d) $NO_3^- + PbO \longrightarrow PbO_2 + NO_2^-$

(e) $2Cl_2 \longrightarrow Cl^- + ClO_3^-$ (Hint: Use one Cl_2 in each half reaction.)

(f) $IO_3^- + Cr(OH)_4^- \longrightarrow I^- + CrO_4^{2-}$ ([Hint: The oxidation number of Cr in $Cr(OH)_4^-$ is $+3$.])

(g) $Bi_2O_3 + ClO^- \longrightarrow BiO_3^- + Cl^-$

(h) $ClO_3^- + N_2H_4 \longrightarrow NO + Cl^-$ (Hint: H in N_2H_4 has an oxidation number of $+1$.)

(i) $Fe(OH)_2 + O_2 \longrightarrow Fe(OH)_3 + OH^-$

GLOSSARY
OF TERMS

Absolute Zero: The lowest possible temperature corresponding to zero on the Kelvin scale. $-273°C$

Acid (Brønsted–Lowry): A species that can lose a proton (H^+) in a chemical reaction. A proton donor.

$$H—A \longrightarrow H^+ + A^-$$

Acid–Base Indicator: A chemical species that changes color at the end point in an acid–base titration.

Acid–Base Reaction: A reaction in which a proton is transferred from an acid to a base to give the corresponding conjugate acid and base.

Acid Constant or Acid Ionization Constant, K_a: The equilibrium constant for a solution of a weak acid, HA, corresponding to the equilibrium constant expression for a weak acid.

$$K_a = \frac{[H_3O^+][A^-]}{[HA]}$$

Acid Rain: Rain that has a lower pH than normal rain due to absorption of acidic air pollutants.

Acidic Solution: A solution that has a hydronium ion concentration that is greater than pure water. An acid dissolved in water forms an acidic solution.

Acids: A class of chemical compounds that display specific chemical properties called acidic properties. Normally, the formula of an acid begins with hydrogen (e.g., HF, H_2S).

Activity Series of Metals: A redox table of metals and metal ions that ranks the metals according to increasing strength as reducing agents.

Actual Formula: The formula of a compound that reflects the actual atomic combination in the compound. The actual formula is the empirical formula or some small whole-number multiple of the empirical formula. For molecular compounds the actual formula is the molecular formula.

Alcohols: A series of organic compounds involving a hydroxy group, $-OH$, bonded to groups of atoms containing carbon and hydrogen (e.g., methyl alcohol, CH_3OH, and ethyl alcohol, CH_3CH_2OH).

Alkali Metals: The group 1 or group IA elements not including hydrogen.

Alkaline Earth Metals: The group 2 or group IIA elements.

Alpha Particle: A nuclear particle containing two protons and two neutrons and is equivalent to a helium-4 nucleus.

Alpha Particle Decay: A mode of radioactive decay in which a nucleus decays by emitting a high-speed helium-4 nucleus, ^4_2He, called an alpha particle, and a new nucleus is formed with a nucleon number that is four less than the original nucleus and an atomic number that is two less.

Amphiprotic: The ability of a chemical species to behave as an acid or a base.

Anion: A negatively charged ion.

Anode: The electrode in an electrical cell that attracts anions and at which oxidation occurs.

Aqueous Solution: A solution of a solute dissolved in water.

Atmosphere, atm: A unit of pressure defined as 1 atm = 760 torr. Also, the name used for the layer of gases that surrounds Earth.

Atmospheric Pressure: The pressure exerted by the gas particles in the air on all objects exposed to the atmosphere.

Atom: The smallest representative particle of an element.

Atomic Mass Unit (u):   A special unit of mass used to express isotope masses and atomic weights. An atom of the isotope carbon-12 is defined as having a mass of exactly 12 u. The masses of all other isotopes of elements are measured relative to carbon-12 and expressed in atomic mass units.

Atomic Number: The number of protons in the nuclei of the atoms of an element. The number of protons in an atom of an element equals the number of electrons in the atom. Each element has its own unique atomic number and is listed in the periodic table according to this number.

Atomic Theory: (1) Elements are composed of tiny, fundamental particles called atoms. (2) Atoms of a particular element are the same but differ from atoms of all other elements. (3) Atoms enter into combinations to form compounds.

Atomic Weight: The average mass of an atom of an element determined using the contribution of each natural isotope.

Avogadro's Law: Equal volumes of gases at the same temperature and pressure contain the same number of moles.

Avogadro's Number: The number of atoms in 1 mole of an element or the number of formula units in a mole of a compound: 6.022×10^{23}.

Barometer: A pressure-measuring device consisting of a mercury-filled glass tube sealed on one end and inverted in a bowl of mercury. Barometric pressures are measured in terms of millimeters of mercury (mm Hg). *See* torr.

Base (Brønsted–Lowry):   A species that can gain a proton (H^+) in a chemical reaction. A proton acceptor.

$$H^+ + B \longrightarrow B\!-\!H^+$$

Basic Solution: A solution that has a hydroxide ion concentration that is greater than pure water. A base dissolved in water forms a basic solution.

Battery: A device in which oxidation and reduction are carried out in separate compartments so that useful electrical work is obtained from a redox reaction.

Beta Particle: A nuclear particle that is a high-speed electron emanating from a nucleus.

Beta Particle Decay: A mode of radioactive decay in which a nucleus decays by emitting a high-speed electron, $_{0}^{-1}e$, called a beta particle, and a new nucleus is formed with the same nucleon number as the original nucleus and an atomic number one greater.

Binary Solution: A solution consisting of two components; a solvent and a solute.

Biochemistry: The study of compounds involved in biological processes and the changes they undergo during life processes.

Bohr's Atomic Model: The concept of the atom developed by Niels Bohr in which the atom is described as a nucleus around which electrons are in motion and occupy circular orbits.

Boiling Point: The temperature at which a liquid boils or rapidly changes from a liquid to a vapor. The temperature at which the vapor pressure of a liquid equals the prevailing pressure.

Boiling Point Elevation: The phenomenon in which the presence of a solute causes the boiling point of the solution to be higher than the boiling point of the pure solvent.

Bonding Pairs: Pairs of valence electrons around the central atom of a molecule or polyatomic ion that are involved in bonding to other atoms.

Boyle's Law: The volume of a gas is inversely proportional to the pressure at a constant temperature.

$$V = \frac{k}{P}$$

Brønsted–Lowry Acid: A species that can lose a proton (H^+) in a chemical reaction. A proton donor.

Brønsted–Lowry Base: A species that can gain a proton (H^+) in a chemical reaction. A proton acceptor.

Buffer Solution: A solution that resists a change in pH on the addition of small amounts of an acid or base. A simple buffer solution contains a weak acid in equilibrium with its conjugate base.

Calorie (cal):   The amount of heat needed to change the temperature of 1 g of water by 1°C. 1 cal = 4.184 J.

Carbohydrates or Sugars: Carbon-, hydrogen-, and oxygen-containing compounds produced mainly by plants and ranging in size from simple sugars like glucose to complex sugars called complex carbohydrates like starch. Carbohydrates serve as food energy sources for humans.

Catalyst: A chemical that increases the speed or rate of a chemical reaction without being chemically changed. A catalyst provides an alternate path for a reaction that is faster and gives the same products.

Cathode: The electrode in an electrical cell that attracts cations and at which reduction occurs.

Cathode-Ray Tube: A glass tube that has the air pumped out and metal electrodes sealed to opposite ends. When a high-voltage source is attached to the electrodes, electricity flows through the tube. The electricity flow occurs as a cathode-ray beam, which is a beam of free electrons.

Cation: A positively charged ion.

Celsius, °C: The unit of temperature used on the Celsius temperature scale.

Chain or Sequential Nuclear Fission: A series of subsequent fission events that results when the neutrons of fission cause the fission of more nuclei. Such a sequential reaction is possible in a large collection of fissionable nuclei since a single fission event produces two or more neutrons that can enter into further fission events.

Charles' Law: The volume of a gas is directly proportional to the temperature at a constant pressure.

$$V = kT$$

Chemical Bond: A force that holds atoms in combinations.

Chemical Equation: The symbolic representation of a chemical reaction showing the formulas of the reactants and products separated by an arrow.

Chemical Equilibrium: A dynamic equilibrium between reactants and products in which the rate that the reactants form the products equals the rate that the products form the reactants. A state of chemical equilibrium can exist between reactants and products in a reversible chemical reaction.

Chemical Formula: A representation of a compound that gives the symbol of each element in the compound followed by a subscript indicating the relative number of combined atoms.

Chemical Properties: Properties that relate to the chemical behavior of matter. Such properties relate

how a chemical reacts with other chemicals: does it burn in air, does it rust or corrode, does it react with water, and so on.

Chemical Reaction: A process in which one set of chemicals called reactants is converted into another set of chemicals called products.

Chemical Symbol: The international symbol used to represent a chemical element.

Chemistry: The science of the properties, composition, and behavior of materials.

Coefficient: A number placed in front of a formula when balancing an equation.

Colligative Properties: Properties of solutions that depend only on the presence of solute particles and not on the identity of the solute.

Columns: The vertical (up and down) groupings of elements within the periodic table. Also called families or groups.

Combined Gas Law: The gas law used to calculate changes in conditions of a gas.

$$\frac{V_2 P_2}{T_2} = \frac{V_1 P_1}{T_1}$$

Combustion: The reaction of a carbon-containing compound with oxygen to give carbon dioxide and water.

Common or Trivial Names: Names for compounds that have developed historically and give no information concerning chemical composition.

Compounds: Pure chemicals made up of combinations of chemical elements. A given compound always contains the same elements in definite proportions by mass.

Concentration: An expression of the amount of solute per unit amount of solution or unit amount of solvent.

Condensation: The change from a vapor to a liquid.

Conjugate Acid–Base Pair: An acid and a base related by the loss or gain of a proton.

Conversion Factor: A factor relating units or properties that is used to convert a measurement made in terms of one unit to another unit.

Core Melting: The melting of the core of a nuclear reactor resulting from the interruption of coolant flow and the subsequent dramatic increase in temperature from the energy of fission. Core melting could result in destruction of the reactor and radioactive contamination of the environment.

Covalent Bond: The force of attraction between atoms arising from the sharing of an electron pair.

Covalent or Molecular Compound: A compound consisting of a collection of molecules. Compounds involving two or more nonmetals or those that include metalloids and nonmetals are molecular compounds.

Critical Mass: The minimum amount of fissionable material in which a fission chain reaction is self-sustaining and can become supercritical.

Crystal or Crystal Lattice: The three-dimensional pattern or arrangement of the chemical particles making up a solid.

Crystal Lattice Sites: The specific positions or locations that particles occupy in a three-dimensional crystalline solid.

Dalton's Law of Partial Pressures: The total pressure of a mixture of gases is the sum of the partial pressures of the components of the mixture.

$$P_T = P_A + P_B + \cdots$$

d Block: The elements of the periodic table that correspond to the filling of the d sublevels.

Decay Chains or Series: The isotopes of the elements with atomic numbers greater than 82 belong to one or another of several series of elements that decay, by alpha or beta emission. The elements in the series are formed in sequence according to the pattern of decay, and the series ends in the formation of a stable isotope.

Density: The mass per unit volume of a substance: $D = m/V$. A property that is used to describe a substance that is determined by measuring the mass and volume of a sample of a substance and then dividing the mass by the volume.

Diatomic Elements: Elements that occur in the form of molecules composed of two atoms of the element. The seven diatomic elements are H_2, O_2, N_2, F_2, Cl_2, Br_2, and I_2.

Dietetic Calorie: A unit used for food energy equal to 1 kilocalorie and often represented using a capital C on Calorie. 1 Cal = 1000 cal = 1 kcal.

Dilution: The process of preparing a solution of lower concentration by mixing more solvent with a solution of higher concentration.

Dipole Molecular Attraction: Attractions between polar molecules caused by the electrostatic forces of attraction between oppositely charged regions of molecules.

Distillation: The process of boiling a liquid and condensing the resulting vapors.

Double Bond or Double Covalent Bond: The sharing of two pairs of electrons between atoms so that they are bonded by two covalent bonds. Carbon atoms are capable of double-bond formation in certain organic compounds.

Ductile: A property of metals corresponding to the ability to be stretched into wires.

Dynamic Equilibrium: A phenomenon in which the rates of two opposing processes are equal. For a solute dissolved in a solvent, a state of dynamic equilibrium exists when the rate at which the solute dissolves equals the rate at which the solute comes out of solution ("undissolves"). Once a state of dynamic equilibrium exists, there is balance between the two processes and no net change occurs.

Electrical Cell: A device in which oxidation and reduction are carried out in separate compartments so that useful electrical work is obtained from a redox reaction.

Electrolysis: A process in which an external source of electricity is used to cause a nonspontaneous oxidation–reduction reaction to occur.

Electrolyte: A substance that forms an aqueous solution that conducts electricity. Electrolytes dissolve in water to form cations and anions.

Electron Dot Structure: *See* Lewis electron dot structure.

Electron Dot Symbol: A symbol for an element using the normal element symbol surrounded by dots that indicate the number of valence electrons.

Electron Orbitals: Each energy sublevel is made up of one or more electron orbitals in which electrons are located within an atom. Orbitals are visualized as three-dimensional electron clouds. The various orbitals in sublevels are referred to as s orbitals, p orbitals, d orbitals, and f orbitals.

Electron Transfer Reaction: A chemical reaction in which one species loses electrons and another species gains electrons. An oxidation–reduction or redox reaction.

Electronegativity: The tendency of an atom of an element to attract electrons in a chemical bond. Except for the noble gases, the electronegativities of elements increase from bottom to top in columns and from left to right in rows. Fluorine is the most electronegative element followed by oxygen and then chlorine and nitrogen.

Electronic Configuration: The pattern in which the electrons in an atom of a given element are distributed often shown as the distribution of the electrons in an atom in terms of the sublevels occupied.

Electrons: Negatively charged subatomic particles found in atoms and as particles of electricity.

Elements: Pure chemicals that cannot be separated into simpler chemicals. The fundamental forms of matter.

Emergency-Core Cooling: A reactor safety measure in which a reserve coolant is available to be used in case the primary coolant flow is interrupted.

Empirical Formula: The simplest formula of a compound that is deduced from the experimentally determined composition of the compound.

Endothermic Reaction: A chemical reaction in which heat is absorbed or required for the reaction to occur.

End Point: The point in a titration at which the indicator changes color.

Energy: The capacity for doing work that takes various forms, such as radiant energy, heat, potential energy, kinetic energy, chemical energy, and electrical energy.

Energy Levels: Various energy positions in the atom in which electrons are located. Sometimes called energy states or energy shells. They are referred to by the numbers 1, 2, 3, 4, 5, 6, and 7.

Energy States: The possible quantized energy positions of electrons in an atom. Sometimes called energy levels.

Energy Sublevels: Each energy level is made up of one or more sublevels in which electrons of varying energies can reside. Sublevels are sometimes called energy subshells. An energy level can have s, p, d, and f sublevels.

Enzymes: Biological catalysts involved in reactions of metabolism and cell function.

Equilibrium Constant Expression: An expression involving the ratio of the molar concentrations of the products to the reactants that represents that the ratio of molar concentrations is equal to a constant value for an equilibrium reaction. For the general reaction,

$$aA + bB \rightleftharpoons cC + dD$$

The equilibrium constant expression is

$$K_{eq} = \frac{[C]^c[D]^d}{[A]^a[B]^b}$$

The numerical value of an equilibrium constant is determined by substituting experimental values of equilibrium concentrations into the equilibrium constant expression for an equilibrium system.

Exothermic Reaction: A chemical reaction in which heat is released or produced.

f Block: The elements of the periodic table that correspond to the filling of the f sublevels.

Fissile Isotopes: The isotopes uranium-235 and plutonium-239 that can undergo fission.

Fission: A special kind of nuclear transmutation in which a neutron interacts with a nucleus, causing the splitting of the nucleus into two new nuclei, a few neutrons, and the release of energy.

Fission Products (High-Level Radioactive Wastes): A variety of radioactive substances that accumulate in nuclear reactor fuel rods when they form as products of fission. Fission products have to be collected periodically and disposed of by special methods. Fission products include many isotopes. The most notable are plutonium-239, cesium-137, and strontium-90.

Force: An action that can cause some effect (i.e., force of gravity, magnetic force, electrostatic force).

Formula Unit: An amount of a compound that contains the number of atoms of each element given in the formula. The smallest representative unit of a compound.

Freezing: The change from the liquid to the solid state. Also called solidification or crystallization.

Freezing Point: The temperature at which a liquid changes to a solid. For a given chemical, the melting point and freezing point are the same.

Freezing Point Depression: The phenomenon in which the presence of a solute causes the freezing point of the solution to be lower than the freezing point of the pure solvent.

Fuel Cell: A device in which oxidation and reduction are carried out in separate compartments to obtain useful electrical work. The reactants are continually supplied and the products are continually removed to give a continuous source of electricity.

Fusion: A special type of nuclear transmutation that occurs at very high temperatures (thermonuclear temperatures) and in which certain lighter nuclei fuse together to form heavier nuclei and in the process release energy.

Gamma Ray Emission: The daughter nuclei formed in alpha or beta decay are in some cases energetically excited. When such an excited nuclei fall to lower energy states, they emit electromagnetic radiation called gamma rays.

Gas: A physical state in which a sample of matter has neither a defined volume nor a defined shape and occupies the entire container in which it is placed.

Gas Constant, R: The proportionality constant in the ideal gas law, $PV = nRT$, which applies to any gas behaving according to the kinetic molecular theory. The common value of R is used for gases with volume expressed in liters, pressure in atmospheres and temperature in Kelvin is

$$\left(\frac{0.0821 \; L \, atm}{K \, mol}\right)$$

Gas Discharge Tube: A cathode-ray tube that contains a small amount of some gas. The flow of electricity through the tube results in the emission of colored light that is characteristic of the gas in the tube.

Geiger–Müller Tube: An electronic tube designed to detect the presence of ionizing radiation. Radiation passing through the tube causes ionization of a gas contained in the tube. This ionizing event is amplified by the tube as a flow of electrical current that is detected and registered by electronic circuits attached to the tube.

Groups or Families of Elements: Vertical columns of elements in the periodic table. Elements within a group or family have similar properties.

Half-Life: A characteristic property of a radioactive substance defined as the time in which one-half of a sample of a radioactive substance decays.

Heat: A mode of energy transfer between objects. Heating an object can increase the heat content and the temperature, whereas cooling can decrease the heat content and temperature. Heat is a measure of the total energy content of a sample of matter.

Heat of Condensation: The amount of heat released when a specific amount of vapor condenses at a given temperature.

Heat of Crystallization: The amount of heat released when a specific amount of liquid solidifies or freezes.

Heat of Fusion: The amount of heat needed to melt a specific amount of a solid at a given temperature.

Heat of Reaction or Enthalpy Change: The energy exchanged in a chemical reaction. Heats of reaction are usually expressed in units of kilocalories per mole or kilojoules per mole.

Heat of Solution: The energy exchange that is associated with the dissolving of a solute in water. Heats of solution are usually expressed in units of kilocalories or kilojoules per mole of solute.

Heat of Vaporization: The amount of heat required to change a specific amount of liquid to the vapor state at a given temperature.

Humanmade Isotopes: Nuclides formed by nuclear transmutations induced in the laboratory or a nuclear reactor.

Hund's Rule: Electrons in a set of orbitals within a sublevel tend to occupy empty orbitals and pair only when all orbitals have one electron.

Hydrocarbons: Organic compounds containing only carbon and hydrogen.

Hydrogen Bond: Electrostatic attraction between hydrogen atoms of one polar molecule and a highly electronegative atom (F, O, N) of another polar molecule.

Hydronium Ion: The ion formed when water gains a proton, H_3O^+.

Ideal Gas: An idealized or theoretical gas that obeys the kinetic molecular theory and the ideal gas law.

Ideal Gas Law: A general gas law relating the pressure, volume, number of moles, and temperature. The product of the pressure, P, and the volume, V, of an ideal gas equals the product of the number of moles, n, the gas constant, R, and the Kelvin temperature, T, of the gas.

$$PV = nRT$$

Inner Transition Elements: Those elements comprising the f block of the periodic table (the actinide series and the lanthanide series). They are listed in the lower portion of the periodic table.

Intermolecular Attractions: The attractive forces that exist between molecules, sometimes called Van der Waals forces.

Ion: A chemical particle that carries a negative or positive charge. A monatomic ion is a charged atom, and a polyatomic ion is a group of covalently bonded atoms that carries a charge.

Ion Combination Reaction: A chemical reaction that forms a precipitate when solutions containing cations ions and anions of an insoluble substance are mixed. Also called a precipitation reaction.

Ionic Bond: The electrostatic force of attraction between ions that binds ions in ionic compounds.

Ionic Compound: A compound consisting of a collection of ions held together by ionic bonds. Compounds of metals and nonmetals are typically ionic compounds. Ionic compounds are normally crystalline solids.

Ionization Energy: The amount of energy that is required to remove one highest energy level electron from an atom of an element.

Ionizing Radiation: Radiation that causes ion pair formation or ionization of molecules or atoms in the material through which it passes.

Ion Pairs: Electrons and positive ions formed by ionization of molecules or atoms of material through which radiation passes. The high energy of the particles of radiation causes the ionization, and a given particle of radiation forms numerous ion pairs.

Inhibitor: A chemical that slows down or decreases the rate of a chemical reaction.

Isotopes: Atoms of the same element that contain the same number of protons but different numbers of neutrons. Atoms of an element that have different masses due to differing numbers of neutrons.

Joule: The metric or SI unit for energy.

1 cal = 4.184 J.

K_a: The equilibrium constant for a solution of a weak acid, HA, corresponding to the acid ionization constant expression for a weak acid.

$$K_a = \frac{[H_3O^+][A^-]}{[HA]}$$

Kinetic Energy: Energy of motion. The kinetic energy of a moving object is given by the expression K.E. = $\frac{1}{2} mv^2$, where m is the mass and v is the speed.

Kinetic Molecular Theory: The view of a gas as a diffuse collection of molecules in rapid, random, straight-line motion continually colliding with one another and any objects in their vicinity.

Law of Conservation of Energy: Energy cannot be created or destroyed in chemical processes.

Law of Conservation of Matter: Matter cannot be created or destroyed in chemical processes.

Law of Constant Composition: A compound always contains the same elements, and they are present in definite proportions by mass.

Le Châtelier's Principle: When a factor affecting an equilibrium system is changed, the equilibrium shifts in a direction that tends to counteract the change.

Lewis Electron Dot Structure: A symbolic representation of a molecule showing the shared pairs of electrons between atoms and any other pairs of outer-energy-level electrons associated with each atom.

Limiting Reactant: The reactant in a chemical reaction that is present in an amount that limits the amount of product formed.

Liter, L: A unit of volume used in chemistry defined as the volume equal to 1000 cubic centimeters or 1000 cm^3.

Liquefaction: The conversion of a gas to a liquid.

Liquid: A physical state in which a sample of matter occupies a definite volume and takes on the shape of the portion of the container it occupies.

Lone Pairs or Nonbonding Pairs: Pairs of valence electrons around the central atom of a molecule or polyatomic ion that are not involved in bonding to other atoms.

Low-Level Radioactive Wastes: Impurities in the coolant of the reactor that become radioactive when exposed to the reactor core or radioactive materials from a reactor that dissolve in the coolant.

Macromolecule: A relatively large molecule containing hundreds or thousands of covalently bonded atoms.

Malleable: A metallic property related to the ability to be bent, cast, and formed.

Mass: A property of matter that determines its resistance to being set in motion or resistance to any change in motion. Mass relates to the amount of material in a sample of a substance.

Mass Number: The sum of the number of protons and neutrons in a nucleus of an atom. Also called the nucleon number. Each different isotope of an element has its own unique mass number.

Mass Spectrometer: An instrument used to measure the atomic masses of isotopes relative to carbon-12.

Matter: Anything that occupies space and has mass.

Maxwell–Boltzmann Distribution: At a given temperature, the kinetic energies of the particles in a sample of a substance fit a pattern of distribution shown as a plot of the percent of particles with a particular value of kinetic energy versus possible kinetic energy values.

Metabolism: The numerous chemical reactions that occur in the body, including digestion of food, utilization of food molecules for energy, and the breakdown and formation of cells.

Melting: The change of a solid to a liquid. Also called fusion.

Melting Point: The characteristic temperature at which a solid changes to a liquid.

Metalloids: Those elements (Si, Ge, As, Sb, Te, Po) that display both metallic and nonmetallic properties. They are located between the metals and non-metals in the periodic table.

Metals: Those elements that have metallic properties. They are good electrical and heat conductors, are flexible enough to be deformed, and possess metallic luster. Of the 112 elements, 87 are metals. Most of the elements are metals, and they are located to the left of the metalloids in the periodic table.

Metric System: An international system of measurement in which defined units of mass, length, time, and temperature have been established.

Molality, m: The number of moles of solute per kilogram of solvent.

Molarity, M: The number of moles of solute per liter of solution. $M = n/V$

Molar Mass of a Compound: The number of grams per mole of a compound found by multiplying the molar mass of each element by the corresponding subscript in the formula and adding up the products.

Molar Mass or Number of Grams per Mole of an Element: The number of grams of an element corresponding to 1 mole. For a given element, the molar mass is numerically equal to the atomic weight of the element and has units of grams in the numerator and 1 mole in the denominator.

Molar Ratio: A numerical ratio expressing the number of moles of one element to the number of moles of another element or the number of moles of an element per mole of a compound. The subscripts in the formula reveal the molar ratios. For example, in water, H_2O, there are

$$\left(\frac{2 \text{ mol H}}{1 \text{ mol O}}\right), \left(\frac{2 \text{ mol H}}{1 \text{ mol H}_2\text{O}}\right), \text{ and } \left(\frac{1 \text{ mol O}}{1 \text{ mol H}_2\text{O}}\right)$$

Molar Volume: The volume occupied by 1 mole of an ideal gas at STP. The value of the molar volume is

$$\left(\frac{22.4 \text{ L}}{1 \text{ mol}}\right)_{\text{STP}}$$

Mole of a Compound: The amount of a compound that contains Avogadro's number of formula units. The amount of a compound that contains the number of moles of each element given by the subscripts in the formula of the compound.

Mole of an Element: The amount of an element that contains as many atoms as there are carbon atoms in 12 grams of carbon-12. The number of atoms is Avogadro's number.

Molecular Formula: A formula of a molecular compound that indicates the type and number of atoms comprising a molecule. A molecular formula is a conventional formula and does not convey the bonding sequence in the molecule.

Molecule: A group of two or more covalently bonded atoms.

Monatomic Ion: A simple ion that is an atom with a negative or positive charge.

Multiple Bond: The sharing of two pairs (double bond) or three pairs (triple bond) of electrons between two atoms.

Nanometer: A metric unit of length used to describe atomic, molecular, or ionic sizes. One nanometer equals 10^{-9} m.

Net-Ionic Equation: An equation for a reaction occurring in aqueous solution showing only the molecules and ions that react and are produced.

Neutral Solution: A solution in which the concentrations of hydronium ion and hydroxide ion are equal, as in pure water. A solution that is neither acidic nor basic.

Neutralization Reaction: A reaction involving a solution of a strong acid and a solution of strong base that produces a neutral solution.

Neutrons: Subatomic particles that have no charge and are found in atomic nuclei.

Noble Gases: The elements comprising the column to the far right of the periodic table. These elements have completely filled s and p sublevels in the outer energy level and form few compounds. The noble gases are He, Ne, Ar, Kr, Xe, and Rn.

Nonbonding Electrons: *See* lone pairs.

Nonelectrolyte: A substance that forms an aqueous solution that does not conduct electricity.

Nonmetals: Those elements that do not possess metallic properties. They are gases or brittle solids, are poor or nonconductors, and have no metallic luster. There are 19 nonmetals, and they are located in the upper right portion of the periodic table.

Nonpolar Molecular Attractions: Attractions between nonpolar molecules caused by the attractions of electrons of nonpolar molecules for the nuclei of other nonpolar molecules.

Nonpolar Molecule: A molecule with no net separation of positive and negative charge centers. A nonpolar molecule has either no polar bonds or polar bonds that are distributed in a symmetrical manner about the center of the molecule.

Normal Boiling Point: The temperature at which a liquid boils when exposed to a pressure of 1 atmosphere. The temperature at which the vapor pressure of a liquid equals one atomosphere.

Nuclear Fission: A special kind of nuclear transmutation in which a neutron interacts with a nucleus, causing the splitting of the nucleus into two new nuclei, a few neutrons, and the release of energy.

Nuclear Fusion: A nuclear transmutation in which two smaller nuclei combine or merge to form a larger nucleus and release energy.

Nuclear Reactor: A device in which self-sustaining fission is carried out under controlled conditions. The basic components of a nuclear reactor include (1) nuclear fuel rods that contain fissionable material, (2) a moderator used to slow down neutrons to induce fission, (3) control rods that absorb neutrons and are manipulated to control or stop fission, (4) a coolant circulating about the fuel rods to carry away the energy of fission as heat, and (5) a thick shielding around the reactor to protect people from the radioactive core of the reactor.

Nuclear Transmutation: A phenomenon in which a high-speed nuclear particle (projectile) collides with a nucleus (target) causing a nuclear reaction in which a new nucleus and particle are formed.

Nucleon Number: The sum of the number of protons and neutrons in a nucleus of an atom. Also called the mass number. Each different isotope of an element has its own unique nucleon number.

Nucleons: Basic nuclear particles; protons and neutrons.

Nucleus: The small, highly dense, positively charged center of atoms made up of clusters of protons and neutrons.

Nuclides: A general term used to refer to the nuclei of the various isotopes of all the elements.

Octet Rule: Atoms tend to lose, gain, or share electrons so that they attain a total of eight outer-energy-level electrons.

Orbital: *See* electron orbitals.

Organic Compounds: A special group of carbon-containing compounds, the study of which is the concern of organic chemistry. Most organic compounds include carbon and hydrogen combined with oxygen, nitrogen, phosphorus, and/or sulfur.

Organic Chemistry: The field of chemistry devoted to the study of those compounds of carbon considered to be organic compounds.

Osmosis: The phenomenon in which the water passes from the solution of lower concentration of solute to the solution of higher concentration of solute through a semipermeable membrane separating the two solutions.

Outer-Energy-Level Electrons: Those electrons in an atom that are located in the highest numbered energy level occupied by electrons.

Oxidation: Loss of electrons; the increase in oxidation number of an element.

Oxidation Half-Reaction: The portion of an electron transfer reaction involving oxidation or loss of electrons.

Oxidation Number: The charge an element has in a monatomic ion or the hypothetical charge it would have if the shared electrons in covalent bonds are assigned to the more electronegative elements.

Oxidation–Reduction Reaction: An electron transfer reaction in which one species loses electrons (is oxidized) and another species gains electrons (is reduced). Also called a redox reaction.

Oxidizing Agent: A species that causes another species to be oxidized and is reduced in the process; an electron acceptor.

Oxoacid or Oxyacid: An acid that contains hydrogen, oxygen, and some other element (e.g., H_2SO_4, HNO_3).

Oxoanions or oxyanions: Negative polyatomic ions containing oxygen and another element. Some oxoanions are related to or derived from the common oxoacids (e.g., SO_4^{2-}, NO_3^-).

Parts per Million (ppm): An expression of the number of parts a given component is per million parts of the whole.

Pauli Exclusion Principle: An orbital can contain a maximum of two electrons, and those two electrons must have opposite electron spin states.

p Block: The elements of the periodic table that correspond to the filling of the outer p sublevels.

Percent by Mass Concentration: The amount of a component in a solution expressed as a percent of the total mass of solution.

Percent-by-Mass Composition of a Compound: The expression of the composition of a compound in terms of the mass of each element present with respect to the mass of the compound; the number of grams of each element in 100 g of the compound.

Period: Horizontal (left to right) row in the periodic table.

Periodic Table: A table of the elements in which the various elements are listed according to increasing atomic number and are grouped in columns according to similarities in properties and electronic configurations.

pH: A term used to express the relative hydronium ion concentration in a solution. pH is defined as the negative logarithm of the hydronium ion concentration.

$$pH = -\log[H_3O^+]$$

For typical solutions, the pH values range from about 0 to 14. A solution of pH 7 is neutral. An acidic solution has a pH of less than 7, and a basic solution has a pH greater than 7.

Photochemical Smog or Smog: The variety of chemicals that accumulate in the immobile air mass during a temperature inversion. These chemicals include primary air pollutants and numerous secondary pollutants. The secondary pollutants are formed by sunlight-induced (photochemical) reactions between primary pollutants and atmospheric gases.

Physical Law: A general statement concerning an important and consistent pattern in nature.

Physical Properties: Properties that relate to the physical nature of matter. Examples are appearance, color, state, density, conductivity, melting point, and boiling point.

Plastics: A group of synthetic organic compounds containing macromolecules. Plastics are molded, laminated, extruded, and so forth, into various forms such as films, fibers, bottles, and other consumer products.

Plutonium or Plutonium-239: The fissionable isotope of plutonium produced in nuclear reactors by the transmutation of uranium-238. It may be a major fuel used in nuclear reactors in the future, but it can also be used to make nuclear weapons.

pOH: A term used to express the relative hydroxide ion concentration in a solution. pOH is defined as the negative logarithm of the hydroxide ion concentration.

$$pOH = -\log[OH^-]$$

For typical solutions, the pOH values range from about 0 to 14. A solution of pOH 7, is neutral. A basic solution has a pOH of less than 7, and an acidic solution has a pOH greater than 7.

Polar Bond: A covalent bond in which the more electronegative atom attracts the electrons to a greater extent than the other atom, producing an uneven distribution of charge.

Polar Molecular Attractions: Attractions between polar molecules caused by the electrostatic forces of attraction between oppositely charged regions of molecules.

Polar Molecule: A molecule in which there is a net separation of positive and negative charge centers. A polar molecule has polar bonds and an unsymmetrical shape that results in a separation of positive and negative charge centers.

Polyatomic Ion: A group of two or more covalently bonded atoms that carries an electrical charge as a unit.

Power Reactor: A nuclear reactor in which the heat absorbed by the coolant is used to generate steam from which electricity is produced.

Precipitate: An insoluble solid formed in a precipitation or ion combination reaction.

Precipitation Reaction: *See* ion combination reaction.

Prefix Nomenclature: A system of nomenclature for binary nonmetal–nonmetal or metalloid nonmetal compounds in which the number of combined atoms of each element is indicated by a prefix preceding the name of the element. Such names always end in *-ide*.

Pressure: Force per unit area. The pressure exerted by a gas results from the force of the rapidly moving particles of the gas per unit area of the container.

Pressure–Temperature Law: The pressure of a gas is directly proportional to the temperature at a constant volume.

$$P = kT$$

Primary Air Pollutants: Waste gases produced by activities of an industrial society. The five primary air pollutants as defined by the U.S. Environmental Protection Agency (EPA) are carbon monoxide, sulfur dioxide, nitrogen oxides, volatile organic compounds, and particulates.

Products: The chemicals produced in a chemical reaction.

Protons: Positively charged subatomic particles found in atomic nuclei. Also used to refer to hydrogen ions involved in acid–base chemistry. In acid–base chemistry, a proton is a hydrogen ion (H^+) formed by the breaking of a covalent bond involving hydrogen and some other element.

$$H\!-\!A \longrightarrow H^+ + A^-$$

Proton-Transfer Reaction: An acid–base reaction in which protons transfer from an acid to a base.

Qualitative Analysis: A chemical analysis in which only the identity or presence of one or more components in a mixture or a compound is determined.

Quantitative Analysis: A chemical analysis in which the amount, concentration or percent of one or more components in a mixture or a compound is determined.

Quantum Jump: A change in the energy level of an electron within an atom. Energy is required for a jump from a lower to a higher level, and energy is released in a jump from a higher to a lower level.

Quantum Mechanical Model of Atom: The concept of the atom in which the atom is described as a nucleus around which the electrons exist in the form of electron clouds called electron orbitals. The electrons are located in energy levels. These energy levels are made up of energy sublevels, which in turn are composed of electron orbitals in which the electrons reside.

R: *See* gas constant.

Radiation: The high-speed particles emitted in radioactive decay. The common forms of nuclear radiation are alpha (α) radiation, beta (β) radiation, and gamma (γ) radiation.

Radiation Sickness: An illness resulting from malfunction of bodily processes caused by exposure to high levels of ionizing radiation. The illness is characterized by nausea, vomiting, weakness, weight loss, fever, diarrhea, internal bleeding, loss of hair, and, can cause death.

Radioactivity: Spontaneous decay of a nucleus to form another nucleus and a nuclear particle.

Rate of Reaction: A measure of how fast a chemical reaction occurs. Normally, the rate of a reaction is expressed in terms of how fast a reactant is used up or how fast a product is produced using units of moles per second or moles per minute.

Reactants: The initial or starting chemicals in a chemical reaction.

Reactor: *See* nuclear reactor.

Redox Reaction: An electron transfer reaction in which one species loses electrons (is oxidized) and another species gains electrons (is reduced).

Reducing Agent: A species that causes another species to be reduced and is oxidized in the process; an electron donor.

Reduction: Gain of electrons; the decrease in oxidation number of an element.

Reduction Half-Reaction: The portion of an electron transfer reaction involving reduction or gain of electrons.

Rem, or Roentgen Equivalent Human: An absorbed dose unit used to express the amount of radiation absorbed by humans upon exposure to radiation.

Representative Elements: The elements comprising the s and p blocks of the periodic table that make up the following groups:

Group 1 or IA	Alkali metals
Group 2 or IIA	Alkaline-earth metals
Group 13 or IIIA	Boron group
Group 14 or IVA	Carbon group
Group 15 or VA	Nitrogen group
Group 16 or VIA	Oxygen group
Group 17 or VIIA	Halogens
Group 18 or VIIIA	Noble gases

Reversible Reaction: A chemical reaction in which the reactants form the products and the products form the reactants. A double arrow is used to represent a reversible reaction.

Rows: The horizontal (left to right) sequences or periods of elements in the periodic table.

Salts: A general term sometimes used to refer to ionic compounds.

Saturated Solution: A solution in which the dissolved and undissolved solute are in dynamic equilibrium. A solution containing as much solute as possible at a specific temperature.

s Block: The elements of the periodic table that correspond to the filling of the outer s sublevels.

Semimetals: *See* metalloids.

Shielding Effect: The decrease in the attraction of the nucleus for outer-level electrons resulting from the presence of the inner core of the lower energy level electrons.

SI Units: An international system of measurement units (*Système International d'Unités*) that is essentially the same as the metric system.

Significant Digits: All the numerical digits that are part of a measurement. They reflect how accurately the measurement was made and do not include zeros that are used as placeholders to indicate the position of the decimal point. Sometimes called significant figures.

Simple Ion: An atom carrying a positive or negative charge resulting from the loss or gain of electrons; a monatomic ion.

Single Bond: A covalent bond involving the sharing of one pair of electrons.

Smog: The variety of chemicals that accumulate in the immobile air mass during a temperature inversion. These chemicals include primary air pollutants and numerous secondary pollutants. The secondary pollutants are formed by sunlight-induced (photochemical) reactions between primary pollutants and atmospheric gases.

Solid: A physical state in which a sample of matter occupies a definite volume and has a defined, rigid shape.

Solidification, Crystallization, or Freezing: The change from a liquid to a solid.

Solubility: The amount of a chemical that can be dissolved in a specific amount of solvent to form a saturated solution.

Solute: The component of a solution that has been dissolved by the solvent.

Solution: An intimate mixture of two or more substances in which the components intermingle on an atomic, molecular, or ionic basis. A homogeneous mixture of a solute in a solvent.

Solvent: The component of a solution that is the dissolver. The component of a solution that is of the same physical state as the solution or is present in the greatest amount if all components are of the same state.

Specific Heat: The number of joules or calories needed to change the temperature of 1 g by 1°C (sometimes represented by the symbol c_p). The units of specific heat are (J/g °C) or (cal/g °C).

Spectator Ions: Ions that are present in solution during a reaction but do not actually react.

Standard Solution: A solution of known concentration.

Standard Temperature and Pressure (STP): A commonly used set of reference conditions for gases defined as

0°C or 273 K and 760 torr or 1 atm

States of Matter: The physical forms—solid, liquid, or gas—in which matter occurs.

Stoichiometry: Calculations involving mass relations in chemical reactions.

STP: *See* standard temperature and pressure.

Strong Acids: Those acids that completely react with water to form hydronium ion (H_3O^+) and the conjugate base of the acid. Common strong acids are hydrochloric acid, sulfuric acid, and nitric acid.

Strong Bases: Substances like NaOH, KOH, and $Ba(OH)_2$ that are soluble in water and dissolve to give hydroxide ion and the metal cation. The most common strong bases are NaOH and KOH.

Structural Formula: A formula for a molecular compound that indicates the atoms present and the bonding sequence of the atoms. The covalent bonds between atoms are shown as lines connecting the symbols of the bonded atoms.

Subatomic Particles and Nuclear Particles: Particles including protons, neutrons, electrons, and alpha particles that structure atoms and atomic nuclei.

Sublimation: The change from the solid state to the vapor or gaseous state.

Supercritical Fission: A sequential nuclear fission process in a sample of fissile isotope that can result in a nuclear explosion.

Supersaturated Solution: An unstable solution that contains more solute than a saturated solution at the same temperature.

Surface Tension: In a liquid, the surface molecules experience unbalanced attractive forces that cause a net inward pull that tends to draw surface molecules into the body of the liquid, producing a strain or tension on the surface.

Synthesis: A chemical process by which a specific chemical is made from other chemicals.

Synthetic Organic Compounds: Substances synthesized or prepared from naturally occurring organic compounds (usually obtained from petroleum) by industrial chemical processes that convert the natural products to more useful compounds. Synthetics include such consumer products as plastics, plasticizers, paints, pesticides, preservatives, and pharmaceuticals.

Systematic Names: Names for compounds based on a system of nomenclature designed to indicate the chemical nature of the compounds.

Temperature: An expression of the degree of hotness or coldness of an object that reflects the average kinetic energy of its particles. Temperatures are measured relative to specific reference temperatures used to establish temperature scales. The Celsius and Kelvin scales are commonly used in science.

Temperature Inversion: The atmospheric situation that occurs when a cooler air mass moves in at lower altitude and underlies a warmer surface air mass. This movement results in the trapping of a warm air mass between the cool air mass below and the cooler air of high altitude. The cool-warm-cool layers of air differ from the normal warm-cool layers, and the condition is called an inversion. A temperature inversion can result in an immobile air mass in which air pollutants accumulate. Inversions normally occur during warm, clear weather.

Tetravalent: The tendency of the atoms of an element to form four covalent bonds in compounds. Carbon is tetravalent.

Theory or Model: A concept that serves as a description of why objects behave as they do or why certain natural processes occur.

Titration: The process of adding a measured amount of a solution of known concentration to a sample of another solution for purposes of determining the concentration of the solution or the amount of some species in the solution.

Torr: A unit of pressure defined as 1 torr = 1 mm Hg.

Transition Metals or Elements: Those elements comprising the *d* block of the periodic table. They are located in the table between the *s* block and the *p* block.

Transmutation: *See* nuclear transmutation.

Transuranium Elements: Those elements having atomic numbers higher than uranium. They do not occur naturally but rather are formed by special transmutations accomplished by nuclear scientists.

Triple Bond or Triple Covalent Bond: The sharing of three pairs of electrons between atoms so that they are bonded by three covalent bonds. Carbon atoms are capable of triple-bond formation in certain organic compounds.

Unit Factor: A ratio of two equivalent terms used to convert from one unit to another. A type of conversion factor.

Universal Gas Constant: *See* gas constant or *R*.

Valence Electrons: The outer-energy-level electrons of an atom available for bond formation.

Van der Waals Forces: The intermolecular attractive forces that exist in collections of molecules.

Vapor Pressure: The pressure of a vapor associated with a liquid or solid.

Viscosity: A property of liquids characterized by an internal resistance to flow arising from the attractive forces between molecules.

Volatility: The readiness or ability of a liquid to evaporate.

Voltaic Cell: *See* electrical cell.

Volume Ratio: The ratio of the number of liters of one gas to that of another gas involved in a chemical reaction. These ratios are used in volume-to-volume stoichiometric computations.

Weak Acids: Those acids that react only slightly with water when dissolved in water.

Weight: The product of the mass of an object and the gravitational attraction of the object by Earth. Sometimes weight is used to mean mass.

Chapter 2

7. (a) 6.959500×10^5 km (b) $568,400,000,000,000,000,000,000,000$ kg
   (c) $29,980,000,000$ cm/s (d) 1.672×10^{-24} g (e) $31,536,000$ s
   (f) 2.8841×10^{-8} cm (g) 1.252×10^5 cm/s (h) $345,000$ cm/s

10. $0.017 \text{ m}\left(\dfrac{1000 \text{ mm}}{1 \text{ m}}\right) = 17$ mm; 71 mm is longer

    $0.5 \text{ g}\left(\dfrac{1000 \text{ mg}}{1 \text{ g}}\right) = 500$ mg; 500 mg is greater

12. (a) $\left(\dfrac{1 \text{ L}}{1000 \text{ ml}}\right)$ (b) $\left(\dfrac{1000 \text{ g}}{1 \text{ kg}}\right)$ (c) $\left(\dfrac{100 \text{ mL}}{1 \text{ dL}}\right)$ (d) $\left(\dfrac{1 \text{ m}}{100 \text{ cm}}\right)$

    (e) $\left(\dfrac{1000 \text{ mg}}{1 \text{ g}}\right)$ (f) $\left(\dfrac{0.1 \text{ cm}}{1 \text{ mm}}\right)$ (g) $\left(\dfrac{1000 \text{ ms}}{1 \text{ s}}\right)$ (h) $\left(\dfrac{1000 \text{ m}}{1 \text{ km}}\right)$

13. (a) $4.37 \text{ cm}\left(\dfrac{1 \text{ m}}{100 \text{ cm}}\right) = 0.0437$ m (b) $0.024 \text{ kg}\left(\dfrac{1000 \text{ g}}{1 \text{ kg}}\right)\left(\dfrac{1000 \text{ mg}}{1 \text{ g}}\right) = 2.4 \times 10^4$ mg

    (c) $2.8 \text{ cm}\left(\dfrac{1 \text{ m}}{100 \text{ cm}}\right)\left(\dfrac{1,000,000,000 \text{ nm}}{1 \text{ m}}\right) = 2.8 \times 10^7$ nm (d) $30\text{s}\left(\dfrac{1000 \text{ ms}}{1 \text{ s}}\right) = 3.0 \times 10^4$ ms

    (e) $537 \text{ mm}\left(\dfrac{1 \text{ m}}{1000 \text{ mm}}\right) = 0.537$ m (f) $750 \text{ mL}\left(\dfrac{1 \text{ L}}{1000 \text{ mL}}\right) = 0.750$ L

    (g) $16,890 \text{ mg}\left(\dfrac{1 \text{ g}}{1000 \text{ mg}}\right)\left(\dfrac{1 \text{ kg}}{1000 \text{ g}}\right) = 0.01689$ kg (h) $0.0821 \text{ L}\left(\dfrac{1000 \text{ ml}}{1 \text{ L}}\right) = 82.1$ mL

15. (a) 2 (b) 4 (c) 2 (d) 1 (e) 5 (f) 4 (g) 3 (h) 3

17. (a) 3 (b) 2 (c) 3 (d) 2 (e) 4 (2 past decimal) (f) 5 (2 past decimal)
    (g) 5 (2 past decimal) (h) 2

20. $1 \text{ yd}\left(\dfrac{0.9144 \text{ m}}{1 \text{ yd}}\right)\left(\dfrac{1000 \text{ mm}}{1 \text{ m}}\right) = 914 \text{ mm}$ **22.** $403 \text{ ft}\left(\dfrac{1 \text{ yd}}{3 \text{ ft}}\right)\left(\dfrac{0.9144 \text{ m}}{1 \text{ yd}}\right) = 123 \text{ m}$

27. $\left(\dfrac{55 \text{ mi}}{1 \text{ hr}}\right)\left(\dfrac{1.609 \text{ km}}{1 \text{ mi}}\right) = \left(\dfrac{88 \text{ km}}{1 \text{ hr}}\right)$

29. $\left(\dfrac{0.0018 \text{ oz}}{\text{trip}}\right)\left(\dfrac{28.35 \text{ g}}{1 \text{ oz}}\right)\left(\dfrac{1000 \text{ mg}}{1 \text{ g}}\right) = \left(\dfrac{51 \text{ mg}}{\text{trip}}\right)$

$1.00 \text{ kg}\left(\dfrac{1000 \text{ g}}{1 \text{ kg}}\right)\left(\dfrac{1000 \text{ mg}}{1 \text{ g}}\right)\left(\dfrac{1 \text{ trip}}{51 \text{ mg}}\right) = 20{,}000 \text{ trips (2 significant digits)}$

32. $1.00 \text{ L}\left(\dfrac{1.057 \text{ qt}}{1 \text{ L}}\right)\left(\dfrac{1 \text{ gal}}{4 \text{ qt}}\right) = 0.264 \text{ gal}$ **37.** $34.0 \text{ L}\left(\dfrac{1 \text{ gal}}{3.785 \text{ L}}\right)\left(\dfrac{41.5 \text{ mi}}{1 \text{ gal}}\right) = 373 \text{ mi}$

40. **(a)** $\left(\dfrac{34.2 \text{ g}}{34.1 \text{ mL}}\right) = 1.00 \text{ g/mL water}$ **(b)** $\left(\dfrac{2.568 \text{ kg}}{133 \text{ mL}}\right)\left(\dfrac{1000 \text{ g}}{1 \text{ kg}}\right) = 19.3 \text{ g/mL gold}$

(c) $\left(\dfrac{43.24 \text{ g}}{47 \text{ mL}}\right) = 0.92 \text{ g/mL ice}$ **(d)** $\left(\dfrac{49.1 \text{ g}}{61.9 \text{ mL}}\right) = 0.793 \text{ g/mL ethyl alcohol}$

(e) $\left(\dfrac{158.5 \text{ g}}{58.7 \text{ cm}^3}\right) = 2.70 \text{ g/cm}^3 \text{ aluminum}$ **(f)** $\left(\dfrac{1.35 \text{ g}}{1.06 \text{ L}}\right) = 1.27 \text{ g/L air}$

43. $\text{density} = \dfrac{\text{mass}}{\text{volume}}$

(a) $10.0 \text{ L}\left(\dfrac{1000 \text{ cm}^3}{1 \text{ mL}}\right)\left(\dfrac{1.00 \text{ g}}{1 \text{ cm}^3}\right) = 1.00 \times 10^4 \text{ g}$ **(b)** $6.72 \text{ g}\left(\dfrac{1 \text{ cm}^3}{2.07 \text{ g}}\right) = 3.25 \text{ cm}^3$

(c) $31.3 \text{ g}\left(\dfrac{1 \text{ cm}^3}{2.70 \text{ g}}\right) = 11.6 \text{ cm}^3$ **(d)** $456 \text{ cm}^3\left(\dfrac{7.87 \text{ g}}{1 \text{ cm}^3}\right) = 3.59 \times 10^3 \text{ g}$

(e) $237 \text{ mL}\left(\dfrac{1.59 \text{ g}}{1 \text{ mL}}\right) = 377 \text{ g}$ **(f)** $5.9 \times 10^3 \text{ mL}\left(\dfrac{1 \text{ L}}{1000 \text{ mL}}\right)\left(\dfrac{1.29 \text{ g}}{1 \text{ L}}\right) = 7.6 \text{ g}$

(g) $15 \text{ gal}\left(\dfrac{3.785 \text{ L}}{1 \text{ gal}}\right)\left(\dfrac{1000 \text{ mL}}{1 \text{ L}}\right)\left(\dfrac{0.70 \text{ g}}{1 \text{ mL}}\right) = 40{,}000 \text{ g or 40 kg}$ $40 \text{ kg}\left(\dfrac{1 \text{ lb}}{0.4536 \text{ kg}}\right) = 88 \text{ lb}$

45. $6.98 \times 10^4 \text{ g}, 154 \text{ lb}$ **47.** $\left(\dfrac{1.00 \text{ g}}{1 \text{ cm}^3}\right)\left(\dfrac{1 \text{ lb}}{453.6 \text{ g}}\right)\left(\dfrac{28317 \text{ cm}^3}{1 \text{ ft}^3}\right) = \left(\dfrac{62.4 \text{ lb}}{1 \text{ ft}^3}\right)$

53. $\left(\dfrac{31.6 \text{ g}}{1.59 \text{ mL}}\right) = 19.9 \text{ g/mL}$; the ring is not pure gold. It must contain some denser metal.

56. $\left(\dfrac{1.44 \times 10^6 \text{ kg}}{72 \text{ mL}}\right)\left(\dfrac{1000 \text{ g}}{1 \text{ kg}}\right) = \dfrac{2.0 \times 10^7 \text{ g}}{1 \text{ mL}}$ **59.** $\left(\dfrac{550 \text{ g}}{578.3 \text{ mL}}\right) = \left(\dfrac{0.951 \text{ g}}{1 \text{ mL}}\right)$; it will float.

61. $V = \dfrac{3.142 \, d^3}{6} = \dfrac{(3.142)(0.3742 \text{ cm})^3}{6} = 0.02744 \text{ cm}^3$ $\left(\dfrac{0.216 \text{ g}}{0.02744 \text{ cm}^3}\right) = 7.87 \text{ g/cm}^3$

68. **(a)** $37.0°C$ **(b)** $77°F$ **(c)** $373K$

Chapter 3

13. **(a)** $\left(\dfrac{8.00 \text{ g C}}{56.0 \text{ g}}\right) \times 100 = 14.3\% \text{ C}$ **(b)** $\left(\dfrac{26.3 \text{ g H}}{235 \text{ g}}\right) \times 100 = 11.2\% \text{ H}$ **(c)** $\left(\dfrac{11.9 \text{ g C}}{28.25 \text{ g}}\right) \times 100 = 42.1\% \text{ C}$

15. **(a)** $\left(\dfrac{40.0 \text{ g C}}{100 \text{ g acetic acid}}\right)$ **(b)** $\left(\dfrac{42.1 \text{ g C}}{100 \text{ g sugar}}\right)$ **(c)** $\left(\dfrac{14.3 \text{ g C}}{100 \text{ g baking soda}}\right)$

29. These data support the law of constant composition. The same compound obtained from four different sources contains the same elements in definite proportions by mass.

52.

		Protons	Neutrons	Electrons
(a)	carbon-14	6	8	6
(b)	chlorine-37	17	20	17
(c)	uranium-234	92	142	92
(d)	oxygen-18	8	10	8
(e)	phosphorus-32	15	17	15
(f)	calcium-45	20	25	20
(g)	strontium-90	38	52	38
(h)	iron-56	26	30	26
(i)	iodine-125	53	72	53
(j)	barium-137	56	81	56

63. silicon-28 (27.98 u) $92.21/100 = 25.800$ u
silicon-29 (28.98 u) $4.70/100 = 1.362$ u
silicon-30 (29.97 u) $3.09/100 = 0.926$ u
Atomic weight 28.09 u

Chapter 4

7. (a) $24.8 \text{ g}\left(\dfrac{1 \text{ mol Ag}}{107.868 \text{ g}}\right) = 2.30 \times 10^{-1}$ mol Ag (b) 958 mol Fe

(c) $6.71 \text{ g}\left(\dfrac{1 \text{ mol S}}{32.07 \text{ g}}\right) = 2.09 \times 10^{-1}$ mol S (d) 3.80×10^{-4} mol W

(e) $235 \text{ tons}\left(\dfrac{1000 \text{ kg}}{1 \text{ ton}}\right)\left(\dfrac{1000 \text{ g}}{1 \text{ kg}}\right)\left(\dfrac{1 \text{ mol Al}}{26.98154 \text{ g}}\right) = 8.71 \times 10^{6}$ mol Al (f) 1.20 mol He

(g) $131 \text{ mg}\left(\dfrac{1 \text{ g}}{1000 \text{ mg}}\right)\left(\dfrac{1 \text{ mol Si}}{28.0855 \text{ g}}\right) = 4.66 \times 10^{-3}$ mol Si (h) 18.7 mol Mg

(i) $2.76 \text{ kg}\left(\dfrac{1000 \text{ g}}{1 \text{ kg}}\right)\left(\dfrac{1 \text{ mol Cu}}{63.546 \text{ g}}\right) = 43.4$ mol Cu (j) 4.22 mol Pb

(k) $0.30 \text{ g}\left(\dfrac{1 \text{ mol Be}}{9.01218 \text{ g}}\right) = 3.3 \times 10^{-2}$ mol Be

9. (a) $124 \text{ mol Pt}\left(\dfrac{195.08 \text{ g}}{1 \text{ mol Pt}}\right) = 2.42 \times 10^{4}$ g (b) 3.05×10^{-4} g

(c) $6.04 \times 10^{-7} \text{ mol Si}\left(\dfrac{28.086 \text{ g}}{1 \text{ mol Si}}\right) = 1.70 \times 10^{-5}$ g (d) 9.75×10^{3} g

(e) $11.9 \text{ mol Ar}\left(\dfrac{39.948 \text{ g}}{1 \text{ mol Ar}}\right) = 475$ g (f) 5.28×10^{-8} g

11. (a) $84.8 \text{ g}\left(\dfrac{1 \text{ mol Hg}}{200.59 \text{ g}}\right)\left(\dfrac{6.022 \times 10^{23} \text{ atoms Hg}}{1 \text{ mol Hg}}\right) = 2.55 \times 10^{23}$ atoms Hg

(b) $0.249 \text{ g}\left(\dfrac{1 \text{ mol Ti}}{47.90 \text{ g}}\right)\left(\dfrac{6.022 \times 10^{23} \text{ atoms Ti}}{1 \text{ mol Ti}}\right) = 3.13 \times 10^{21}$ atoms Ti

(c) 3.77×10^{22} atoms Kr (d) 3.805×10^{25} atoms Ni

(e) $8.23 \text{ mg} \left(\dfrac{1 \text{ g}}{1000 \text{ mg}}\right)\left(\dfrac{1 \text{ mol Ag}}{107.868 \text{ g}}\right)\left(\dfrac{6.022 \times 10^{23} \text{ atoms Ag}}{1 \text{ mol Ag}}\right) = \begin{matrix} 4.59 \times 10^{19} \\ \text{atoms Ag} \end{matrix}$

(f) $1.00 \text{ oz} \left(\dfrac{31.1 \text{ g}}{1 \text{ oz}}\right)\left(\dfrac{1 \text{ mol Au}}{196.9666 \text{ g}}\right)\left(\dfrac{6.022 \times 10^{23}}{1 \text{ mol Au}}\right) = 9.51 \times 10^{22} \text{ atoms Au}$

13. $\left(\dfrac{1 \text{ mol}}{6.022 \times 10^{23} \text{ atoms}}\right)\left(\dfrac{1.0079 \text{ g}}{1 \text{ mol H}}\right) = \left(\dfrac{1.674 \times 10^{-24} \text{ g}}{1 \text{ atom H}}\right)$

$\left(\dfrac{1.674 \times 10^{-24} \text{ g}}{1 \text{ atom H}}\right)\left(\dfrac{15.9994 \text{ u/1 atom O}}{1.0079 \text{ u/1 atom H}}\right) = \left(\dfrac{2.657 \times 10^{-23} \text{ g}}{1 \text{ atom O}}\right)$

15. (a)

	N 19.1 g	**H** 9.6 g	**O** 22.0 g	**C** 49.3 g
g/mol	14.007	1.0079	15.9994	12.011
Moles	1.364 mol N	9.525 mol H	1.375 mol O	4.105 mol C
Ratios	1	7	1	3
Formula	NH_7OC_3			

(b)

	K 20.8 g	**H** 2.7 g	**C** 25.5 g	**O** 51.0 g
g/mol	39.10	1.0079	12.011	15.9994
Moles	0.532 mol K	2.68 mol H	2.12 mol C	3.19 mol O
Ratios	1	5	4	6
Formula	$KH_5C_4O_6$			

(c) CaF_2 **(d)** C_3H_3O

(e)

	Na 12.1 g	**B** 11.3 g	**O** 71.3 g	**H** 5.29 g
g/mol	22.99	10.81	15.9994	1.0079
Moles	0.526 mol Na	1.045 mol B	4.456 mol O	5.228 mol H
Ratios	1	2	8.5	10
Formula	$NaB_2O_{8.5}H_{10}$ or $Na_2B_4O_{17}H_{20}$			

(f) $C_7H_5O_3NS$

(g)

	Ba 68.4 g	**P** 10.3 g	**O** 21.3 g	**(h)** $C_3H_4O_3$
g/mol	137.3	30.97	15.9994	
Moles	0.498 mol Ba	0.333 mol P	1.33 mol O	
Ratios	1.5	1	4	
Formula	$Ba_{1.5}PO_4$ or $Ba_3P_2O_8$			

17. g O = 250.0 g − 142.85 g − 15.40 − 23.80 g = 68.0 g

	C 142.85 g	**H** 15.40 g	**N** 23.80 g	**O** 68.0 g
g/mol	12.011	1.0079	14.007	15.9994
Moles	11.89 mol C	15.28 mol H	1.699 mol N	4.252 mol O
Ratios	7	9	1	2.5
Formula	$C_7H_9NO_{2.5}$ or $C_{14}H_{18}N_2O_5$			

19. g O = 61.68 g − 32.38 g = 29.30 g Al_2O_3

25. (a)   2K = (39.0983×2) (b) 72.15 g/mol C_5H_{12}
        2C = (12.011×2)
      + 4O = (15.9994×4)
          166.2 g/mol $K_2C_2O_4$

    (c)   C = (12.011) (d) 58.12 g/mol C_4H_{10}
       3H = (1.0079×3)
      + Br = (79.904)
        94.94 g/mol CH_3Br

    (e)   Ca = 40.078   (f) 146.2 g/mol $C_7H_{16}NO_2$
      + 2F = (18.998×2)
        78.07 g/mol CaF_2

    (g)   7C = (12.011×7) (h) 84.31 g/mol $MgCO_3$
      5H = (1.0079×5)
      3N = (14.0067×3)
      + 6O = (15.9994×6)
        227.1 g/mol $C_7H_5N_3O_6$

27. (a) $819 \text{ g}\left(\dfrac{1 \text{ mol NH}_3}{17.03 \text{ g}}\right) = 48.1 \text{ mol NH}_3$ (b) $0.0418 \text{ mol K}_2Cr_2O_7$

    (c) $0.1086 \text{ g}\left(\dfrac{1 \text{ mol NaCN}}{49.008 \text{ g}}\right) = 2.216 \times 10^{-3} \text{ mol NaCN}$ (d) $8.27 \times 10^{-2} \text{ mol C}_{12}H_{22}O_{11}$

    (e) $2.19 \times 10^6 \text{ g}\left(\dfrac{1 \text{ mol NaCl}}{58.44 \text{ g}}\right) = 3.75 \times 10^4 \text{ mol NaCl}$ (f) 0.245 mol C_2H_2

    (g) $56.8 \text{ g}\left(\dfrac{1 \text{ mol C}_9H_{11}NO_2}{165.2 \text{ g}}\right) = 0.344 \text{ mol C}_9H_{11}NO_2$

29. (a) $3.7 \text{ mol K}_2C_2O_4\left(\dfrac{166.2 \text{ g}}{1 \text{ mol K}_2C_2O_4}\right) = 6.1 \times 10^2 \text{ g}$ (b) $1.89 \times 10^4 \text{ g}$

    (c) $5.87 \text{ mol CH}_3Br\left(\dfrac{94.94 \text{ g}}{1 \text{ mol CH}_3Br}\right) = 557 \text{ g}$ (d) 55.0 g

    (e) $7.46 \times 10^{-3} \text{ mol CaF}_2\left(\dfrac{78.07 \text{ g}}{1 \text{ mol CaF}_2}\right) = 0.582 \text{ g}$ (f) 31.4 g

    (g) $4.35 \times 10^5 \text{ mol C}_7H_5N_3O_6\left(\dfrac{227.1 \text{ g}}{1 \text{ mol C}_7H_5N_3O_6}\right) = 9.88 \times 10^7 \text{ g}$ (h) 5.50×10^2

31. $2.442 \text{ lb}\left(\dfrac{453.6 \text{ g}}{1 \text{ lb}}\right)\left(\dfrac{1 \text{ mol MgSO}_4}{120.4 \text{ g}}\right) = 9.20 \text{ mol MgSO}_4$

33. $16 \text{ oz}\left(\dfrac{28.3 \text{ g}}{1 \text{ oz}}\right)\left(\dfrac{1 \text{ mol NaHCO}_3}{84.01 \text{ g}}\right) = 5.4 \text{ mol NaHCO}_3$

35. $6.97 \times 10^{-2} \text{ mol C}_{12}H_{22}O_{11}$

37. (a)

	C	**H**	**N**	**O**
	32.35 g	**2.96 g**	**6.86 g**	**7.84 g**
g/mol	12.011	1.0079	14.007	15.9994
Moles	2.693 mol C	2.937 mol H	0.490 mol N	0.490 mol O
Ratios	5.5	6	1	1

Empirical $C_{5.5}H_6NO$ or $C_{11}H_{12}N_2O_2$
formula

Empirical molar mass = $11C = (12.011 \times 11)$
$12H = (1.0079 \times 12)$
$2N = (14.007 \times 2)$
$\underline{2O = (15.9994 \times 2)}$
204.2 g/mol

$$\frac{\text{actual molar mass}}{\text{empirical molar mass}} = \frac{204 \text{ g/mol}}{204.2 \text{ g/mol}} = \frac{1}{1}$$

The representative formula is empirical formula = $C_{11}H_{12}N_2O_2$.

(b) Empirical formula CCl_2F_2

empirical molar mass = $C = 12.011$

$Cl = (35.453 \times 2)$
$\underline{F = (18.998 \times 2)}$
120.9 g/mol

$$\frac{\text{actual molar mass}}{\text{empirical molar mass}} = \frac{120 \text{ g/mol}}{121 \text{ g/mol}} = \frac{1}{1}$$

(c) Empirical formula $C_{2.66}H_{2.65}O$ or $C_8H_8O_3$

The representative formula is empirical formula = $C_8H_8O_3$.

The representative formula is empirical formula = CCl_2F_2.

(d)

	Hg 12.75 g	Cl 2.25 g
g/mol	200.59	35.453
Moles	0.636 mol Hg	0.0635 mol Cl
Ratios	1	1

Empirical formula = HgCl

Empirical molar mass = $Hg = 200.59$
$\underline{Cl = 35.453}$
236.0 g/mol

$$\frac{\text{actual molar mass}}{\text{empirical molar mass}} = \frac{472 \text{ g/mol}}{236.0 \text{ g/mol}} = \frac{2}{1}$$

The representative formula = Hg_2Cl_2.

(e)

	C 14.8 g	H 2.49 g	O 19.8 g
g/mol	12.011	1.0079	15.9994
Moles	1.23 mol C	2.47 mol H	1.24 mol O
Ratios	1	2	1

Empirical formula = CH_2O

Empirical molar mass = $1C = 12.011$
$2H = (1.0079 \times 2)$
$\underline{O = 15.9994}$
30.03 g/mol

$$\frac{\text{actual molar mass}}{\text{empirical molar mass}} = \frac{150 \text{ g/mol}}{30 \text{ g/mol}} = \frac{5}{1}$$

The representative formula = $C_5H_{10}O_5$.

(f)

	C **84.1 g**	**H** **15.9 g**
g/mol	12.011	1.0079
Moles	7.00 mol C	15.77 mol H
Ratios	1	2.25

Empirical formula $= CH_{2.25}$ or C_4H_9

Empirical molar mass $= 4C = (12.011 \times 4)$

$\qquad\qquad\qquad\quad \underline{9H = (1.0079 \times 9)}$

$\qquad\qquad\qquad\qquad\quad 57.12$ g/mol

$$\frac{\text{actual molar mass}}{\text{empirical molar mass}} = \frac{115 \text{ g/mol}}{57 \text{ g/mol}} = \frac{2}{1}$$

The representative formula $= C_8H_{18}$.

39. **(a)** $\left(\dfrac{4 \text{ mol H}}{1 \text{ mol CH}_4}\right)$ **(b)** $\left(\dfrac{11 \text{ mol O}}{1 \text{ mol C}_{12}H_{22}O_{11}}\right)$ **(c)** $\left(\dfrac{9 \text{ mol C}}{1 \text{ mol C}_9H_8O_4}\right)$

(d) $\left(\dfrac{3 \text{ mol H}}{1 \text{ mol C}_3H_8}\right)$ **(e)** $\left(\dfrac{1 \text{ mol Ca}}{1 \text{ mol CaCO}_3}\right)$ **(f)** $\left(\dfrac{2 \text{ mol Na}}{1 \text{ mol Na}_2SO_4}\right)$

41. **(a)** $2.00 \text{ mol CH}_4 \left(\dfrac{4 \text{ mol H}}{1 \text{ mol CH}_4}\right) = 8.00 \text{ mol H}$

(b) $64.0 \text{ g}\left(\dfrac{1 \text{ mol CH}_4}{16.04 \text{ g}}\right)\left(\dfrac{4 \text{ mol H}}{1 \text{ mol CH}_4}\right) = 16.0 \text{ mol H}$

(c) $25.4 \text{ g}\left(\dfrac{1 \text{ mol CH}_4}{16.04 \text{ g}}\right)\left(\dfrac{4 \text{ mol H}}{1 \text{ mol CH}_4}\right)\left(\dfrac{1.008 \text{ g}}{1 \text{ mol H}}\right) = 6.38 \text{ g}$

(d) $278 \text{ g}\left(\dfrac{1 \text{ mol C}_{12}H_{22}O_{11}}{342.3 \text{ g}}\right)\left(\dfrac{11 \text{ mol O}}{1 \text{ mol C}_{12}H_{22}O_{11}}\right) = 8.93 \text{ mol O}$

(e) $454 \text{ g}\left(\dfrac{1 \text{ mol C}_{12}H_{22}O_{11}}{342.3 \text{ g}}\right)\left(\dfrac{11 \text{ mol O}}{1 \text{ mol C}_{12}H_{22}O_{11}}\right)\left(\dfrac{16.00 \text{ g}}{1 \text{ mol O}}\right) = 233 \text{ g}$

(f) $0.725 \text{ g}\left(\dfrac{1 \text{ mol CaCO}_3}{100.1 \text{ g}}\right)\left(\dfrac{1 \text{ mol Ca}}{1 \text{ mol CaCO}_3}\right) = 7.24 \times 10^{-3} \text{ mol Ca}$

(g) $48 \text{ g}\left(\dfrac{1 \text{ mol CaCO}_3}{100.1 \text{ g}}\right)\left(\dfrac{1 \text{ mol Ca}}{1 \text{ mol CaCO}_3}\right)\left(\dfrac{40.08 \text{ g}}{1 \text{ mol Ca}}\right) = 19 \text{ g}$

(h) $0.256 \text{ g}\left(\dfrac{1 \text{ mol C}_6H_{12}O_6}{180.2 \text{ g}}\right)\left(\dfrac{12 \text{ mol H}}{1 \text{ mol C}_6H_{12}O_6}\right) = 1.70 \times 10^{-2} \text{ mol H}$

(i) $125 \text{ g}\left(\dfrac{1 \text{ mol C}_6H_{12}O_6}{180.2 \text{ g}}\right)\left(\dfrac{12 \text{ mol H}}{1 \text{ mol C}_6H_{12}O_6}\right)\left(\dfrac{1.008 \text{ g}}{1 \text{ mol H}}\right) = 8.39 \text{ g}$

(j) $12.0 \text{ mol Na}_2SO_4\left(\dfrac{2 \text{ mol Na}}{1 \text{ mol Na}_2SO_4}\right) = 24 \text{ mol Na}$

(k) $5.00 \times 10^2 \text{ g}\left(\dfrac{1 \text{ mol Na}_2SO_4}{142.0 \text{ g}}\right)\left(\dfrac{2 \text{ mol Na}}{1 \text{ mol Na}_2SO_4}\right) = 7.04 \text{ mol Na}$

(l) $5.00 \times 10^2 \text{ g}\left(\dfrac{1 \text{ mol Na}_2SO_4}{142.0 \text{ g}}\right)\left(\dfrac{2 \text{ mol Na}}{1 \text{ mol Na}_2SO_4}\right)\left(\dfrac{22.99 \text{ g}}{1 \text{ mol Na}}\right) = 162 \text{ g}$

43. Au_2S_3 has 80.38% Au; Au_2S has 92.47% Au

45. **(a)** molar mass $C_7H_5SNO_3 =$ $7C = (12.011 \times 7)$ $\dfrac{14.0067}{183.18} \times 100 = 7.646\%$ by mass N

$5H = (1.0079 \times 5)$

$S = 32.066$

$N = 14.0067$

$\underline{3O = (15.9994 \times 3)}$

183.18 g/mol

(b) 65.34% by mass I

(c) molar mass $SO_3 =$ $S = 32.066$ $\dfrac{32.066}{80.064} \times 100 = 40.05\%$ by mass S   **(d)** 49.97% by mass C

$\underline{3O = (15.9994 \times 3)}$

80.064 g/mol

(e) molar mass $NaHCO_3 =$ $Na = 22.990$ $\dfrac{22.990}{84.007} \times 100 = 27.37\%$ by mass Na

$H = 1.0079$

$C = 12.011$

$\underline{3O = (15.9994 \times 3)}$

84.007 g/mol

48. **(a)** $\left(\dfrac{1 \text{ mol N}}{1 \text{ mol NO}}\right)$ $\dfrac{14.0067}{30.006} \times 100 = 46.68\%$ by mass N

$\%$ O $= 100.00 - 46.68\%$ N $= 53.32\%$

(b) 46.65% by mass N

6.713% by mass H

20.00% by mass C

26.64% by mass O

(c) $\left(\dfrac{1 \text{ mol K}}{1 \text{ mol KClO}_3}\right)$ $\dfrac{39.098}{122.55} \times 100 = 31.90\%$ by mass K

$\left(\dfrac{1 \text{ mol Cl}}{1 \text{ mol KClO}_3}\right)$ $\dfrac{35.453}{122.55} \times 100 = 28.93\%$ by mass Cl

$\left(\dfrac{3 \text{ mol O}}{1 \text{ mol KClO}_3}\right)$ $\dfrac{3(15.9994)}{122.55} \times 100 = 39.17\%$ by mass O

(d) 81.71% by mass C

$\%$ H $= 100.00 - 81.71\%$ C $= 18.29\%$ H

(e) $\left(\dfrac{1 \text{ mol Mg}}{1 \text{ mol MgCO}_3}\right)$ $\dfrac{24.305}{84.314} \times 100 = 28.83\%$ by mass Mg

$\left(\dfrac{1 \text{ mol C}}{1 \text{ mol MgCO}_3}\right)$ $\dfrac{12.011}{84.314} \times 100 = 14.25\%$ by mass C

$\left(\dfrac{3 \text{ mol O}}{1 \text{ mol MgCO}_3}\right)$ $\dfrac{3(15.9994)}{84.314} \times 100 = 56.93\%$ by mass O

(f) 79.95% by mass C

9.691% by mass H

10.36% by mass N

50. $\left(\dfrac{1 \text{ mol F}}{1 \text{ mol LiF}}\right)$ $\dfrac{18.998}{25.939} \times 100 = 73.24\%$ by mass F

$\left(\dfrac{2 \text{ mol F}}{1 \text{ mol BaF}_2}\right)$ $\dfrac{2(18.998)}{175.32} \times 100 = 21.67\%$ by mass F

$\left(\dfrac{2 \text{ mol F}}{1 \text{ mol CaF}_2}\right)$ $\dfrac{2(18.998)}{78.074} \times 100 = 48.67\%$ by mass F

$\left(\dfrac{1 \text{ mol F}}{1 \text{ mol NaF}}\right)$ $\dfrac{18.998}{41.988} \times 100 = 45.25\%$ mass F

LiF has the highest percentage by mass fluorine.

Chapter 5

16. (a) $2Zn + O_2 \longrightarrow 2ZnO$ (b) $4Cr + 3O_2 \longrightarrow 2Cr_2O_5$ (c) $2KClO_3 \longrightarrow 2KCl + 3O_2$

(d) $4Li + O_2 \longrightarrow 2Li_2O$ (e) $2B + 3F_2 \longrightarrow 2BF_3$ (f) $2C_3H_6 + 9O_2 \longrightarrow 6CO_2 + 6H_2O$

(g) $3Hg + N_2 \longrightarrow Hg_3N_2$ (h) $N_2 + 3H_2 \longrightarrow 2NH_3$ (i) $WO_3 + 3H_2 \longrightarrow W + 3H_2O$

(j) $2C_4H_{10} + 13O_2 \longrightarrow 8CO_2 + 10H_2O$ (k) $6Ca + 2NH_3 \longrightarrow 3CaH_2 + Ca_3N_2$ (l) $2P + 5S \longrightarrow P_2S_5$

(m) $2KNO_3 + 4C \longrightarrow K_2CO_3 + 3CO + N_2$

18. (a) 2,1,2   (b) 4,3,2   (c) 2,2,3   (d) 4,1,2   (e) 2,3,2   (f) 1,3,2,2

(g) 4,3,2   (h) 1,1,2   (i) 3,1,1,1   (j) 2,9,6,6   (k) 4,5,1

21. (a) Methane gas reacts with diatomic oxygen gas to form carbon dioxide gas and water.

(b) Solid sulfur reacts with diatomic oxygen gas to form sulfur dioxide gas.

(c) Solid calcium reacts with diatomic fluorine gas to form solid calcium fluoride.

(d) Solid mercury(II) oxide when heated reacts to form liquid mercury and diatomic oxygen gas.

(e) Liquid phosphorus trichloride reacts with diatomic chlorine gas to form solid phosphorus pentachloride.

Chapter 6

10. (a) PBr_5 nonmetal–nonmetal   phosphorus pentabromide

(b) ClO_2 nonmetal–nonmetal   chlorine dioxide

(c) Cu_2S metal–nonmetal   copper(I) sulfide

(d) AlN metal–nonmetal   aluminum nitride

(e) N_2O nonmetal–nonmetal   dinitrogen oxide

(f) $Ba(OH)_2$ metal–polyatomic ion   barium hydroxide

(g) $Ni(C_2H_3O_2)_2$ metal–polyatomic ion   nickel(II) acetate

(h) $SnCl_4$ metal–nonmetal   tin(IV) chloride

(i) BCl_3 nonmetal–nonmetal   boron trichloride

(j) Na_2CrO_4 metal–polyatomic ion   sodium chromate

(k) H_3PO_4 oxoacid   phosphoric acid

(l) $HgCl_2$ metal–nonmetal   mercury(II) chloride

(m) Hg_2Br_2 metal–nonmetal   mercury(I) bromide

(n) Cr_2O_3 metal–nonmetal   chromium(III) oxide

(o) $Fe(CN)_3$ metal–polyatomic ion   iron(III) cyanide

(p) Ca_3P_2 metal–nonmetal   calcium phosphide

(q) $KClO_3$ metal–polyatomic ion   potassium chlorate

(r) $HF(aq)$	acid	hydrofluoric acid
(s) $Sn(NO_2)_2$	metal–polyatomic ion	tin(II) nitrite
(t) H_2SeO_4	oxoacid	selenic acid
(u) $NaCN$	metal–polyatomic ion	sodium cyanide
(v) N_2O_3	nonmetal–nonmetal	dinitrogen trioxide
(w) SF_6	nonmetal–nonmetal	sulfur hexafluoride
(x) HNO_3	oxoacid	nitric acid
(y) CuO	metal–nonmetal	copper(II) oxide
(z) $NaOH$	metal–nonmetal	sodium hydroxide
(aa) CCl_4	nonmetal–nonmetal	carbon tetrachloride
(bb) I_2O_5	nonmetal–nonmetal	diiodine pentoxide
(cc) H_2S	nonmetal–nonmetal	hydrogen sulfide
		dihydrogen sulfide
(dd) H_2Se	nonmetal–nonmetal	hydrogen selenide
		dihydrogen selenide
(ee) Na_2CO_3	metal–polyatomic ion	sodium carbonate
(ff) $SbBr_5$	metalloid–nonmetal	antimony pentabromide
(gg) $LiNO_2$	metal–polyatomic ion	lithium nitrite

12.

(a) sodium nitrate, $NaNO_3$

(b) tin(II) acetate, $Sn(C_2H_3O_2)_2$

(c) manganese(II) sulfate, $MnSO_4$

(d) ammonium acetate, $NH_4C_2H_3O_2$

(e) chlorine trifluoride, ClF_3

(f) diarsenic pentasulfide, As_2S_5

(g) dinitrogen trioxide, N_2O_3

(h) calcium fluoride, CaF_2

(i) iron(II) nitrate, $Fe(NO_3)_2$

(j) magnesium carbonate, $MgCO_3$

(k) methane, CH_4

(l) copper(I) oxide, Cu_2O

(m) silver phosphate, Ag_3PO_4

(n) strontium cyanide, $Sr(CN)_2$

(o) nitric acid, HNO_3

(p) ammonium sulfide, $(NH_4)_2S$

(q) phosphoric acid, H_3PO_4

(r) magnesium oxalate, MgC_2O_4

(s) zinc chromate, $ZnCrO_4$

(t) mercury(II) iodide, HgI_2

(u) potassium sulfite, K_2SO_3

(v) sodium dichromate, $Na_2Cr_2O_7$

(w) carbon disulfide, CS_2

(x) dinitrogen pentoxide, N_2O_5

(y) magnesium chloride, $MgCl_2$

(z) hydrogen fluoride, HF

16.

(a) carbon tetrachloride	nonmetal–nonmetal
(b) iron(II) sulfate	metal–polyatomic ion
(c) magnesium oxide	metal–nonmetal
(d) mercury(II) bromide	metal–nonmetal
(e) potassium iodide	metal–nonmetal
(f) lithium arsenide	metal–metalloid
(g) sodium acetate	metal–polyatomic ion
(h) potassium permanganate	metal–polyatomic ion
(i) phosphorus trifluoride	nonmetal–nonmetal
(j) hydrogen chloride	nonmetal–nonmetal
(k) uranium(VI) fluoride	metal–nonmetal

(l) cobalt(II) chloride   metal–nonmetal

(m) barium sulfite   metal–polyatomic ion

(n) calcium fluoride   metal–nonmetal

(o) tin(II) bromide   metal–nonmetal

(p) zinc hydroxide   metal–polyatomic ion

(q) sodium bromide   metal–nonmetal

(r) aluminum sulfide   metal–nonmetal

(s) silver chloride   metal–nonmetal

(t) nickel(II) hydroxide   metal–polyatomic ion

(u) chromium(VI) oxide   metal–nonmetal

(v) acetic acid   oxoacid

(w) ammonium iodide   ammonium ion–nonmetal

(x) calcium oxalate   metal–polyatomic ion

19. (a) nitric acid   (b) nitrite ion   (c) phosphoric acid   (d) bromate ion

    (e) selenic acid   (f) arsenate ion   (g) sulfurous acid

Chapter 7

2. $0.055 \text{ mL Hg}\left(\dfrac{13.6 \text{ g}}{1 \text{ mL Hg}}\right)\left(\dfrac{1 \text{ mol Hg}}{200.59 \text{ g}}\right) = 3.7 \times 10^{-3} \text{ mol Hg}$

4. $37.5 \text{ cm}^3\left(\dfrac{2.75 \text{ g}}{1 \text{ cm}^3}\right)\left(\dfrac{1 \text{ mol SiO}_2}{60.084 \text{ g}}\right) = 1.72 \text{ mol SiO}_2$

6. $3.15 \times 10^{-6} \text{ g}\left(\dfrac{1 \text{ mol C}_{15}\text{H}_{11}\text{O}_4\text{I}_4\text{N}}{776.874 \text{ g}}\right)\left(\dfrac{4 \text{ mol I}}{1 \text{ mol C}_{15}\text{H}_{11}\text{O}_4\text{I}_4\text{N}}\right) = 1.62 \times 10^{-8} \text{ mol I}$

8. $280 \text{ mg}\left(\dfrac{1 \text{ g}}{10^3 \text{ mg}}\right)\left(\dfrac{1 \text{ mol Mg}}{24.31 \text{ g}}\right)\left(\dfrac{1 \text{ mol C}_{55}\text{H}_{72}\text{MgN}_4\text{O}_5}{1 \text{ mol Mg}}\right)\left(\dfrac{893.5 \text{ g}}{1 \text{ mol C}_{55}\text{H}_{72}\text{MgN}_4\text{O}_5}\right) = 10.3 \text{ g}$

10. $2 \text{ tsp}\left(\dfrac{1 \text{ oz}}{8 \text{ tsp}}\right)\left(\dfrac{28.35 \text{ g}}{1 \text{ oz}}\right)\left(\dfrac{1 \text{ mol NaHCO}_3}{84.007 \text{ g}}\right)\left(\dfrac{6.022 \times 10^{23} \text{ NaHCO}_3}{1 \text{ mol NaHCO}_3}\right) = 5.1 \times 10^{22} \text{ NaHCO}_3$

11. $2.25 \text{ kg rock}\left(\dfrac{1000 \text{ g}}{1 \text{ kg}}\right)\left(\dfrac{12.5 \text{ g Al}_2\text{O}_3}{100 \text{ g rock}}\right)\left(\dfrac{1 \text{ mol Al}_2\text{O}_3}{101.961 \text{ g}}\right) = 2.76 \text{ mol Al}_2\text{O}_3 \text{ and } 8.28 \text{ mol O}^{2-}$

16. $2\text{C}_2\text{H}_2(g) + 5\text{O}_2(g) \longrightarrow 4\text{CO}_2(g) + 2\text{H}_2\text{O}(g)$

    2 moles C_2H_2 and 5 moles O_2 react to form 4 moles CO_2 and 2 moles H_2O.

    (a) $\left(\dfrac{2 \text{ mol C}_2\text{H}_2}{5 \text{ mol O}_2}\right)$ (b) $\left(\dfrac{2 \text{ mol C}_2\text{H}_2}{4 \text{ mol CO}_2}\right)$ (c) $\left(\dfrac{2 \text{ mol C}_2\text{H}_2}{2 \text{ mol H}_2\text{O}}\right)$ (d) $\left(\dfrac{5 \text{ mol O}_2}{4 \text{ mol CO}_2}\right)$

17. (a) $0.570 \text{ mol C}_2\text{H}_2\left(\dfrac{5 \text{ mol O}_2}{2 \text{ mol C}_2\text{H}_2}\right) = 1.43 \text{ mol O}_2$ (b) $0.200 \text{ mol C}_2\text{H}_2$ (c) $17.6 \text{ mol H}_2\text{O}$

    (d) $3.89 \text{ mol O}_2\left(\dfrac{4 \text{ mol CO}_2}{5 \text{ mol O}_2}\right) = 3.11 \text{ mol CO}_2$

20. $\text{C}_2\text{H}_5\text{OH} + 3\text{O}_2 \longrightarrow 2\text{CO}_2 + 3\text{H}_2\text{O}$

    (a) $52.6 \text{ g C}_2\text{H}_5\text{OH}\left(\dfrac{1 \text{ mol C}_2\text{H}_5\text{OH}}{46.068 \text{ g}}\right)\left(\dfrac{3 \text{ mol O}_2}{1 \text{ mol C}_2\text{H}_5\text{OH}}\right) = 3.43 \text{ mol O}_2$

(b) $3.43 \text{ mol O}_2\left(\dfrac{31.998 \text{ g}}{1 \text{ mol O}_2}\right) = 110 \text{ g}$ (c) 1001 g

(d) $52.6 \text{ g}\left(\dfrac{1 \text{ mol C}_2\text{H}_5\text{OH}}{46.068 \text{ g}}\right) = 1.142 \text{ mol C}_2\text{H}_5\text{OH}$

$75.0 \text{ g}\left(\dfrac{1 \text{ mol O}_2}{31.999 \text{ g}}\right) = 2.344 \text{ mol O}_2$

68.8 g CO_2 are formed, and O_2 is the limiting reactant.

22. $CaCO_3(s) + 2HCl(aq) \longrightarrow CaCl_2(aq) + CO_2(g) + H_2O$

(a) $500 \text{ mg}\left(\dfrac{1 \text{ g}}{10^3 \text{ mg}}\right)\left(\dfrac{1 \text{ mol CaCO}_3}{100.089 \text{ g}}\right)\left(\dfrac{1 \text{ mol CO}_2}{1 \text{ mol CaCO}_3}\right)\left(\dfrac{44.009 \text{ g}}{1 \text{ mol CO}_2}\right) = 0.220 \text{ g}$

(b) $1.00 \text{ g}\left(\dfrac{1 \text{ mol CaCO}_3}{100.09 \text{ g}}\right)\left(\dfrac{2 \text{ mol HCl}}{1 \text{ mol CaCO}_3}\right) = 0.0200 \text{ mol HCl}$

(c)) $1200 \text{ mg}\left(\dfrac{1 \text{ g}}{10^3 \text{ mg}}\right)\left(\dfrac{1 \text{ mol Ca}^{2+}}{40.08 \text{ g}}\right)\left(\dfrac{1 \text{ mol CaCO}_3}{1 \text{ mol Ca}^{2+}}\right)\left(\dfrac{100.089 \text{ g}}{1 \text{ mol CaCO}_3}\right) = 2.997 \text{ g}$

(d) $9.45 \text{ g}\left(\dfrac{1 \text{ mol HCl}}{36.461 \text{ g}}\right) = 0.2592 \text{ mol HCl}$

$28.4 \text{ g}\left(\dfrac{1 \text{ mol CaCO}_3}{100.089 \text{ g}}\right) = 0.2837 \text{ mol CaCO}_3$

5.70 g CO_2 are formed, and HCl is the limiting reactant.

24. (a) $1.00 \text{ mol CaCN}_2\left(\dfrac{2 \text{ mol NH}_3}{1 \text{ mol CaCN}_2}\right)\left(\dfrac{17.0304 \text{ g}}{1 \text{ mol NH}_3}\right) = 34.1 \text{ g}$ (b) 992 g

(c) $500 \text{ kg}\left(\dfrac{10^3 \text{ g}}{1 \text{ kg}}\right)\left(\dfrac{1 \text{ mol NH}_3}{17.0304 \text{ g}}\right)\left(\dfrac{1 \text{ mol CaCN}_2}{2 \text{ mol NH}_3}\right)\left(\dfrac{80.104 \text{ g}}{1 \text{ mol CaCN}_2}\right) = 1.18 \times 10^6 \text{ g}$

(d) $500 \text{ kg}\left(\dfrac{70 \text{ g CaCN}_2}{100 \text{ g}}\right)\left(\dfrac{10^3 \text{ g}}{1 \text{ kg}}\right)\left(\dfrac{1 \text{ mol CaCN}_2}{80.104 \text{ g}}\right)\left(\dfrac{2 \text{ mol NH}_3}{1 \text{ mol CaCN}_2}\right)\left(\dfrac{17.0304 \text{ g}}{1 \text{ mol NH}_3}\right) = 1.5 \times 10^5 \text{ g or } 1.5 \times 10^2 \text{ kg}$

26. (a) $11.6 \text{ g}\left(\dfrac{1 \text{ mol Na}}{22.99 \text{ g}}\right)\left(\dfrac{1 \text{ mol Ti}}{4 \text{ mol Na}}\right) = 1.26 \times 10^{-1} \text{ mol Ti}$ (b) 97.2 g

(c) $5.0 \text{ kg}\left(\dfrac{10^3 \text{ g}}{1 \text{ kg}}\right)\left(\dfrac{1 \text{ mol Ti}}{47.90 \text{ g}}\right)\left(\dfrac{4 \text{ mol Na}}{1 \text{ mol Ti}}\right)\left(\dfrac{22.99 \text{ g}}{1 \text{ mol Na}}\right) = 9.6 \times 10^3 \text{ g}$ (d) 1.05 g

(e) $\left(\dfrac{4 \text{ mol Na}}{1 \text{ mol TiCl}_4}\right)$ $965 \text{ g}\left(\dfrac{1 \text{ mol TiCl}_4}{189.692 \text{ g}}\right) = 5.087 \text{ mol TiCl}_4$

$480 \text{ g}\left(\dfrac{1 \text{ mol Na}}{22.99 \text{ g}}\right) = 20.88 \text{ mol Na}$ $\left(\dfrac{20.88 \text{ mol Na}}{5.087 \text{ mol TiCl}_4}\right) = \left(\dfrac{4.1 \text{ mol Na}}{1 \text{ mol TiCl}_4}\right)$

$TiCl_4$ is the limiting reactant.

$5.087 \text{ mol TiCl}_4\left(\dfrac{1 \text{ mol Ti}}{1 \text{ mol TiCl}_4}\right)\left(\dfrac{47.88 \text{ g}}{1 \text{ mol Ti}}\right) = 244 \text{ g}$

28. (a) $1.36 \times 10^{-6} \text{ mol O}_2\left(\dfrac{2 \text{ mol H}_2\text{O}_2}{1 \text{ mol O}_2}\right)\left(\dfrac{34.0138 \text{ g}}{1 \text{ mol H}_2\text{O}_2}\right) = 9.25 \times 10^{-5} \text{ g}$

(b) $4.61 \times 10^{-12} \text{ g}\left(\dfrac{1 \text{ mol H}_2\text{O}_2}{34.0138 \text{ g}}\right)\left(\dfrac{2 \text{ mol H}_2\text{O}}{2 \text{ mol H}_2\text{O}_2}\right)\left(\dfrac{18.0152 \text{ g}}{1 \text{ mol H}_2\text{O}}\right) = 2.44 \times 10^{-12} \text{ g}$

(c) $1.00 \text{ mg} \left(\dfrac{1 \text{ g}}{10^3 \text{ mg}} \right) \left(\dfrac{1 \text{ mol H}_2\text{O}}{18.0152 \text{ g}} \right) \left(\dfrac{2 \text{ mol H}_2\text{O}_2}{2 \text{ mol H}_2\text{O}} \right) = 5.55 \times 10^{-5} \text{ mol H}_2\text{O}_2$

(d) $3.32 \times 10^{-8} \text{ g} \left(\dfrac{1 \text{ mol H}_2\text{O}_2}{34.0138 \text{ g}} \right) \left(\dfrac{1 \text{ mol O}_2}{2 \text{ mol H}_2\text{O}_2} \right) \left(\dfrac{31.998 \text{ g}}{1 \text{ mol O}_2} \right) = 1.56 \times 10^{-8} \text{ g}$

30. $C_6H_{12}O_6 \longrightarrow 2C_2H_5OH + 2CO_2$

(a) $525 \text{ g} \left(\dfrac{1 \text{ mol C}_6\text{H}_{12}\text{O}_6}{180.16 \text{ g}} \right) \left(\dfrac{2 \text{ mol C}_2\text{H}_5\text{OH}}{1 \text{ mol C}_6\text{H}_{12}\text{O}_6} \right) \left(\dfrac{46.07 \text{ g}}{1 \text{ mol C}_2\text{H}_5\text{OH}} \right) = 269 \text{ g}$ (b) 163 mol CO_2

(c) $255 \text{ kg} \left(\dfrac{10^3 \text{ g}}{1 \text{ kg}} \right) \left(\dfrac{1 \text{ mol C}_2\text{H}_5\text{OH}}{46.069 \text{ g}} \right) \left(\dfrac{2 \text{ mol CO}_2}{2 \text{ mol C}_2\text{H}_5\text{OH}} \right) \left(\dfrac{44.01 \text{ g}}{1 \text{ mol CO}_2} \right) \left(\dfrac{1 \text{ kg}}{10^3 \text{ g}} \right) = 244 \text{ kg}$

33. (a) $\left(\dfrac{1 \text{ mol HC}_2\text{H}_3\text{O}_2}{1 \text{ mol NaHCO}_3} \right)$

$49.5 \text{ g} \left(\dfrac{1 \text{ mol NaHCO}_3}{84.007 \text{ g}} \right) = 0.5892 \text{ mol NaHCO}_3$ $42.6 \text{ g} \left(\dfrac{1 \text{ mol HC}_2\text{H}_3\text{O}_2}{60.0524 \text{ g}} \right) = 0.7094 \text{ mol HC}_2\text{H}_3\text{O}_2$

$\left(\dfrac{0.7094 \text{ mol HC}_2\text{H}_3\text{O}_2}{0.5892 \text{ mol NaHCO}_3} \right) = \left(\dfrac{1.2 \text{ mol HC}_2\text{H}_3\text{O}_2}{1 \text{ mol NaHCO}_3} \right)$ NaHCO$_3$ is the limiting reactant.

$0.5892 \text{ mol NaHCO}_3 \left(\dfrac{1 \text{ mol CO}_2}{1 \text{ mol NaHCO}_3} \right) \left(\dfrac{44.01 \text{ g}}{1 \text{ mol CO}_2} \right) = 25.9 \text{ g}$

(b) $0.5892 \text{ mol NaHCO}_3 \left(\dfrac{1 \text{ mol HC}_2\text{H}_3\text{O}_2}{1 \text{ mol NaHCO}_3} \right) \left(\dfrac{60.0524 \text{ g}}{1 \text{ mol HC}_2\text{H}_3\text{O}_2} \right) = 35.4 \text{ g}$

$42.6 \text{ g} - 35.4 \text{ g} = 7.2 \text{ g HC}_2\text{H}_3\text{O}_2 \text{ in excess}$

35. (a) $\left(\dfrac{1 \text{ mol C}_4\text{H}_6\text{O}_3}{1 \text{ mol C}_7\text{H}_6\text{O}_3} \right)$

$210 \text{ g} \left(\dfrac{1 \text{ mol C}_4\text{H}_6\text{O}_3}{102.09 \text{ g}} \right) = 2.057 \text{ mol C}_4\text{H}_6\text{O}_3$ $219 \text{ g} \left(\dfrac{1 \text{ mol C}_7\text{H}_5\text{O}_3}{138.12 \text{ g}} \right) = 1.586 \text{ mol C}_7\text{H}_6\text{O}_3$

$\left(\dfrac{2.057 \text{ mol C}_4\text{H}_6\text{O}_3}{1.586 \text{ mol C}_7\text{H}_6\text{O}_3} \right) = \left(\dfrac{1.3 \text{ mol C}_4\text{H}_6\text{O}_3}{1 \text{ mol C}_7\text{H}_6\text{O}_3} \right)$ C$_7$H$_6$O$_3$ is the limiting reactant.

$1.586 \text{ mol C}_7\text{H}_6\text{O}_3 \left(\dfrac{1 \text{ mol C}_9\text{H}_8\text{O}_4}{1 \text{ mol C}_7\text{H}_6\text{O}_3} \right) \left(\dfrac{180.16 \text{ g}}{1 \text{ mol C}_9\text{H}_8\text{O}_4} \right) = 286 \text{ g}$

(b) C$_4$H$_6$O$_3$ is in excess.

$1.586 \text{ mol C}_7\text{H}_6\text{O}_3 \left(\dfrac{1 \text{ mol C}_4\text{H}_6\text{O}_3}{1 \text{ mol C}_7\text{H}_6\text{O}_3} \right) \left(\dfrac{102.098 \text{ g}}{1 \text{ mol C}_4\text{H}_6\text{O}_3} \right) = 162 \text{ g}$

$210 \text{ g} - 162 \text{ g} = 48 \text{ g C}_4\text{H}_6\text{O}_3 \text{ in excess}$

37. (a) $\left(\dfrac{1 \text{ mol H}_2\text{O}}{1 \text{ mol CaO}} \right)$

$25.0 \text{ g} \left(\dfrac{1 \text{ mol CaO}}{56.08 \text{ g}} \right) = 0.4458 \text{ mol CaO}$ $21.0 \text{ g} \left(\dfrac{1 \text{ mol H}_2\text{O}}{18.02 \text{ g}} \right) = 1.167 \text{ mol H}_2\text{O}$

$\left(\dfrac{1.167 \text{ mol H}_2\text{O}}{0.4458 \text{ mol CaO}} \right) = \left(\dfrac{2.6 \text{ mol H}_2\text{O}}{1 \text{ mol CaO}} \right)$ CaO is the limiting reactant.

$0.4458 \text{ mol CaO} \left(\dfrac{1 \text{ mol Ca(OH)}_2}{1 \text{ mol CaO}} \right) \left(\dfrac{74.095 \text{ g}}{1 \text{ mol Ca(OH)}_2} \right) = 33.0 \text{ g}$

(b) $0.4458 \text{ mol CaO}\left(\dfrac{1 \text{ mol } H_2O}{1 \text{ mol CaO}}\right)\left(\dfrac{18.02 \text{ g}}{1 \text{ mol } H_2O}\right) = 8.03 \text{ g}$

$21.0 \text{ g} - 8.03 \text{ g} = 13.0 \text{ g } H_2O$ in excess

40. $69.2 \text{ g}\left(\dfrac{1 \text{ mol } C_7H_6O_3}{138.12 \text{ g}}\right)\left(\dfrac{1 \text{ mol } C_9H_8O_4}{1 \text{ mol } C_7H_6O_3}\right)\left(\dfrac{180.16 \text{ g}}{1 \text{ mol } C_9H_8O_4}\right) = 90.3 \text{ g}$

percent yield $= \dfrac{82.5}{90.3} \times 100 = 91.4\%$

46. **(a)** $1.00 \text{ kcal}\left(\dfrac{4.184 \text{ kJ}}{1 \text{ kcal}}\right)\left(\dfrac{1 \text{ mol } C_{12}H_{22}O_{11}}{5649 \text{ kJ}}\right)\left(\dfrac{342.3 \text{ g}}{1 \text{ mol } C_{12}H_{22}O_{11}}\right) = 0.254 \text{ g}$

(b) $220 \text{ g}\left(\dfrac{1 \text{ mol } C_{12}H_{22}O_{11}}{342.3 \text{ g}}\right)\left(\dfrac{5649 \text{ kJ}}{1 \text{ mol } C_{12}H_{22}O_{11}}\right) = 3630 \text{ kJ}$ $3630 \text{ kJ}\left(\dfrac{1 \text{ kcal}}{4.184 \text{ kJ}}\right) = 868 \text{ kcal} = 868 \text{ Cal}$

(c) $1.000 \text{ g}\left(\dfrac{1 \text{ mol } C_{12}H_{22}O_{11}}{342.3 \text{ g}}\right)\left(\dfrac{5649 \text{ kJ}}{1 \text{ mol } C_{12}H_{22}O_{11}}\right)\left(\dfrac{1 \text{ kcal}}{4.184 \text{ kJ}}\right) = 3.944 \text{ kcal}$

48. $N_2(g) + O_2(g) + 181 \text{ kJ} \longrightarrow 2NO(g)$

$N_2(g) + O_2(g) \longrightarrow 2NO(g)$ $\Delta H = +90.5 \text{ kJ/mol NO}$

$175.2 \text{ g } N_2\left(\dfrac{1 \text{ mol } N_2}{28.0135 \text{ g}}\right)\left(\dfrac{181 \text{ kJ}}{1 \text{ mol } N_2}\right) = 1.13 \times 10^3 \text{ kJ}$

50. $\left(\dfrac{803 \text{ kJ}}{1 \text{ mol } CH_4}\right)\left(\dfrac{1 \text{ mol } CH_4}{16.04 \text{ g}}\right) = 50.1 \text{ kJ/g } CH_4$ $\left(\dfrac{2046 \text{ kJ}}{1 \text{ mol } C_3H_8}\right)\left(\dfrac{1 \text{ mol } C_3H_8}{44.10 \text{ g}}\right) = 46.39 \text{ kJ/g } C_3H_8$

$\left(\dfrac{2261 \text{ kJ}}{2 \text{ mol } C_4H_{10}}\right)\left(\dfrac{1 \text{ mol } C_4H_{10}}{58.12 \text{ g}}\right) = 38.90 \text{ kJ/g } C_4H_{10}$ CH_4 is the best fuel in terms of energy per gram.

51. **(a)** $1.00 \times 10^3 \text{ kJ}\left(\dfrac{1 \text{ g } CH_4}{50.1 \text{ kJ}}\right) = 20.0 \text{ g}$ **(b)** $65.0 \text{ g}\left(\dfrac{46.39 \text{ kJ}}{1 \text{ g } C_3H_8}\right)\left(\dfrac{1 \text{ Btu}}{1.05 \text{ kJ}}\right) = 2.87 \times 10^3 \text{ Btu}$

(c) $35.0 \text{ Btu}\left(\dfrac{1.05 \text{ kJ}}{1 \text{ btu}}\right)\left(\dfrac{1 \text{ g } C_4H_{10}}{45.78 \text{ kJ}}\right) = 0.803 \text{ g}$ **(d)** $\left(\dfrac{30 \times 10^3 \text{ Btu}}{1 \text{ hr}}\right)\left(\dfrac{1.05 \text{ kJ}}{1 \text{ btu}}\right)\left(\dfrac{1 \text{ g}}{50.1 \text{ kJ}}\right) = 6.3 \times 10^2 \text{ g/hr}$

Chapter 8

14. $2d, 3f, 1p$

19. **(a)** $1s^1$

(b) $1s^2 2s^2 2p^6 3s^2 3p^1$

(c) $1s^2 2s^2 2p^6 3s^2 3p^6 4s^2 3d^{10} 4p^1$

(d) $1s^2 2s^2 2p^6 3s^2 3p^6 4s^2 3d^{10} 4p^4$

(e) $1s^2 2s^2 2p^6 3s^2 3p^6 4s^2 3d^{10} 4p^6$

(f) $1s^2 2s^2 2p^6 3s^2 3p^6 4s^2 3d^{10} 4p^6 5s^2 4d^2$

(g) $1s^2 2s^2 2p^6 3s^2 3p^6 4s^2 3d^{10} 4p^6 5s^2 4d^{10} 5p^2$

(h) $1s^2 2s^2 2p^4$

(i) $1s^2 2s^2 2p^6 3s^2 3p^6 4s^2 3d^{10} 4p^6 5s^2 4d^{10} 5p^6 6s^2 4f^{14} 5d^8$ (actual configuration is an exception to this found by the counting rule)

(j) $1s^2 2s^2 2p^6 3s^2 3p^3$

(k) $1s^2 2s^2 2p^6 3s^2 3p^6 4s^2 3d^{10} 4p^6 5s^2 4d^{10} 5p^5$

38. **(a)** transition, metal   **(f)** representative, metalloid

(b) noble gas   **(g)** inner transition, metal

(c) representative, metal   (h) representative, nonmetal

(d) noble gas   (i) transition, metal

(e) transition, metal   (j) representative, metal

40. (a) $1s^1$ (b) [Ne] $3s^23p^1$ (c) [Ar] $4s^23d^{10}4p^1$ (d) [Ar] $4s^23d^{10}4p^5$

(e) [Ar] $4s^23d^{10}4p^6$ (f) [Kr] $5s^24d^2$ (g) [Kr] $5s^24d^{10}5p^2$ (h) [He] $2s^22p^4$

(i) [Xe] $6s^24f^{14}5d^8$ (actual configuration is exception to this found by the counting rule)

(j) [Ne] $3s^23p^3$ (k) [Kr] $5s^24d^{10}5p^5$

49. $0.208 \text{ nm} \left(\dfrac{1 \times 10^{-7} \text{ cm}}{1 \text{ nm}} \right) \left(\dfrac{1 \text{ in.}}{2.54 \text{ cm}} \right) = 8.19 \times 10^{-9} \text{ in.}$

51. $6.022 \times 10^{23} \text{ atoms} \left(\dfrac{0.154 \text{ nm}}{1 \text{ C atom}} \right) \left(\dfrac{1 \text{ m}}{1 \times 10^9 \text{ nm}} \right) = 9.27 \times 10^{13} \text{ m}$

54. $5 \times 10^{-6} \text{ in.} \left(\dfrac{2.54 \text{ cm}}{1 \text{ in.}} \right) \left(\dfrac{1 \times 10^7 \text{ nm}}{1 \text{ cm}} \right) \left(\dfrac{1 \text{ atom Au}}{0.29 \text{ nm}} \right) = 4 \times 10^2 \text{ atoms}$

55. (a) 7; 14.00674; p; representative nonmetal; [He] $2s^22p^3$; 5

(b) 11; 22.98977; s; representative metal; [Ne] $3s^1$; 1

(c) 26; 55.845; d; transition metal; [Ar] $4s^23d^6$; 2

(d) 28; 58.6934; d; transition metal; [Ar] $4s^23d^8$; 2

(e) 14; 28.0855; p; representative metalloid; [Ne] $3s^23p^2$; 4

(f) 102; 259.1009; f; inner transition metal; [Rn] $5f^{13}7s^26d^1$; 2

(g) 56; 137.327; s; representative metal; [Xe] $6s^2$; 2

(h) 10; 20.1797; p; noble gas; [He] $2s^22p^6$; 8

(i) 15; 30.97376; p; representative nonmetal; [Ne] $3s^23p^3$; 5

(j) 8; 15.9994; p; representative nonmetal; [He] $2s^22p^4$; 6

Chapter 9

11. (a) H^+ (b) K^+ (c) Be^{2+} (d) Sr^{2+} (e) Ga^{3+} (f) Fr^+ (g) Ca^{2+} (h) Rb^+

13. (a) S^{2-} (b) I^- (c) P^{3-} (d) Br^- (e) Se^{2-} (f) As^{3-}

19. (a) Na_2S (b) LiBr (c) CaO (d) MgF_2 (e) Al_2O_3 (f) $AlBr_3$

21. (a) LiCl (b) Mg_3N_2 (c) $BaCl_2$ (d) SrO (e) Na_2S (f) Al_2S_3 (g) Ca_3As_2 (h) Na_3P (i) K_2O

31. (a) K_2SO_3 (b) $Al(NO_2)_3$ (c) $Mg_3(PO_4)_2$ (d) $KC_2H_3O_2$ (e) $Ca(HCO_3)_2$

(f) AgI (g) $HgBr_2$ (h) $(NH_4)_2SO_4$

38. (a) $:\!\overset{\cdot\cdot}{\underset{\cdot\cdot}{Cl}}\!-\!\overset{\cdot\cdot}{\underset{\cdot\cdot}{Cl}}\!:$ (b) $H\!-\!\overset{\cdot\cdot}{\underset{\cdot\cdot}{I}}\!:$

(c)
$$\begin{array}{c} H \\ | \\ :\!\overset{\cdot\cdot}{\underset{\cdot\cdot}{Cl}}\!-\!C\!-\!\overset{\cdot\cdot}{\underset{\cdot\cdot}{Cl}}\!: \\ | \\ :\!\underset{\cdot\cdot}{\overset{\cdot\cdot}{Cl}}\!: \end{array}$$

(d)
$$\begin{array}{c} \overset{\cdot\cdot}{\underset{\cdot}{O}} \\ {\diagup} \quad {\diagdown} \\ H \qquad H \end{array}$$

(e)
$$\begin{array}{c} \overset{\cdot\cdot}{\underset{\cdot}{Te}} \\ {\diagup} \quad {\diagdown} \\ H \qquad H \end{array}$$

(f)
$$\begin{array}{c} \overset{\cdot\cdot}{\underset{\cdot}{O}} \\ {\diagup} \quad {\diagdown} \\ :\!\overset{\cdot\cdot}{\underset{\cdot\cdot}{Cl}} \qquad \overset{\cdot\cdot}{\underset{\cdot\cdot}{Cl}}\!: \end{array}$$

(g)
$$\begin{array}{c} :\!\overset{\cdot\cdot}{\underset{\cdot}{O}} \qquad \overset{\cdot\cdot}{\underset{\cdot}{O}}\!: \\ {\diagdown} \quad {\diagup} \\ Cl \\ {\diagup} \quad {\diagdown} \\ :\!\underset{\cdot\cdot}{O}. \qquad .\underset{\cdot\cdot}{O}.\!-\!H \end{array}$$

(h)
$$\begin{array}{c} H \quad H \\ | \qquad | \\ :\!N\!-\!N\!: \\ | \qquad | \\ H \quad H \end{array}$$

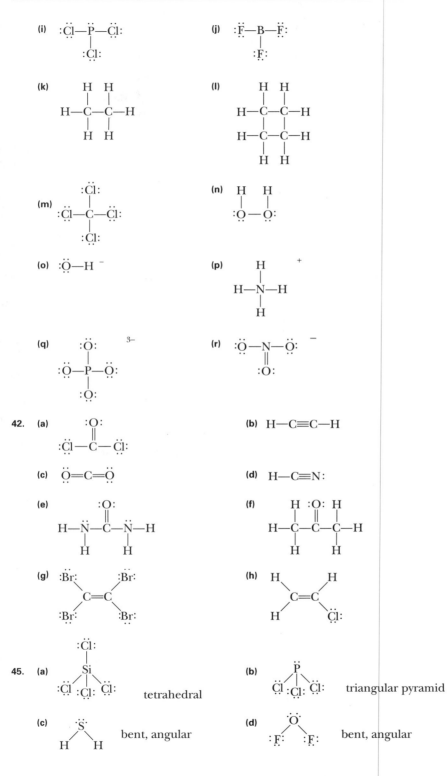

(i) $:\ddot{C}l-P-\ddot{C}l:$
  $:\ddot{C}l:$

(j) $:\ddot{F}-B-\ddot{F}:$
  $:\ddot{F}:$

(k)
```
    H   H
    |   |
H—C—C—H
    |   |
    H   H
```

(l)
```
    H   H
    |   |
H—C—C—H
    |   |
H—C—C—H
    |   |
    H   H
```

(m) $:\ddot{C}l-C-\ddot{C}l:$ with $:\ddot{C}l:$ above and $:\ddot{C}l:$ below

(n)
```
    H   H
    |   |
:O—O:
  ..  ..
```

(o) $:\ddot{O}-H^{-}$

(p)
```
      H          +
      |
H—N—H
      |
      H
```

(q)
```
      :O:       3-
      |
:O—P—O:
      |
      :O:
```

(r) $:\ddot{O}-N-\ddot{O}:$ ⁻ with $\|$ and $:O:$ below the N

42. (a)
```
      :O:
      ‖
:Cl—C—Cl:
```

(b) $H-C\equiv C-H$

(c) $\ddot{O}=C=\ddot{O}$

(d) $H-C\equiv N:$

(e)
```
         :O:
         ‖
H—N—C—N—H
    |       |
    H       H
```

(f)
```
    H  :O:  H
    |   ‖   |
H—C—C—C—H
    |       |
    H       H
```

(g)
```
:Br:      :Br:
    \      /
     C=C
    /      \
:Br:      :Br:
```

(h)
```
H          H
  \        /
   C=C
  /        \
H          Cl:
```

45. (a)
```
      :Cl:
      |
      Si
     /|\
:Cl :Cl: Cl:
```
tetrahedral

(b)
```
      ..
      P
     /|\
Cl :Cl: Cl:
```
triangular pyramid

(c)
```
     ..
     S
    / \
   H   H
```
bent, angular

(d)
```
     ..
     O
    / \
:F:   :F:
```
bent, angular

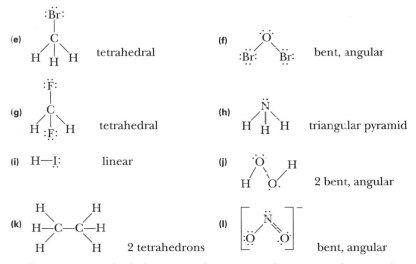

(e) tetrahedral

(f) bent, angular

(g) tetrahedral

(h) triangular pyramid

(i) H—I: linear

(j) 2 bent, angular

(k) 2 tetrahedrons

(l) bent, angular

47. **(b)** linear **(c)** tetrahedral **(d)** angular **(e)** angular **(f)** angular **(i)** triangular pyramid **(j)** triangular **(m)** tetrahedral **(o)** linear **(p)** tetrahedral **(r)** triangular

Chapter 10

23. $\left(\dfrac{39.948 \text{ g}}{1 \text{ mol Ar}}\right)\left(\dfrac{749 \text{ torr}}{1.00 \text{ g}}\right)\left(\dfrac{417 \text{ mL}}{200 \text{ K}}\right) = 6.24 \times 10^4 \text{ mL torr/K mol}$

25. $n = PV/RT = \dfrac{(2.25 \text{ atm})(30 \text{ gal})(3.785 \text{ L/gal})}{(0.0821 \text{ L atm/K mol})(303 \text{ K})} = 1.0 \times 10^1 \text{ mol C}_2\text{H}_2$

27. 1.40×10^3 L   **29.** 5.56 atm

31. $763 \text{ torr}\left(\dfrac{1 \text{ atm}}{760 \text{ torr}}\right)\left(\dfrac{5.91 \times 10^6 \text{ L}}{300 \text{ K}}\right)\left(\dfrac{\text{K mol He}}{0.0821 \text{ L atm}}\right)\left(\dfrac{4.002 \text{ g}}{1 \text{ mol He}}\right) = 9.64 \times 10^5 \text{ g}$

34. **(a)** 377 mL **(b)** 238 mL **(c)** 950 mL **(d)** 95.0 mL   **36.** 102 L   **37.** 116 torr

43. **(a)** $750 \text{ mL}\left(\dfrac{337 \text{ K}}{305 \text{ K}}\right) = 829 \text{ mL}$ **(b)** $750 \text{ mL}\left(\dfrac{610 \text{ K}}{305 \text{ K}}\right) = 1.50 \times 10^3 \text{ mL}$ **(c)** $750 \text{ mL}\left(\dfrac{152.5 \text{ K}}{305 \text{ K}}\right) = 375 \text{ mL}$

45. **(a)** 1.13 atm **(b)** 2.02 atm **(c)** 0.505 atm   **48.** 2.93 atm

54. **(a)** $1.00 \text{ L}\left(\dfrac{1.00 \text{ atm}}{0.675 \text{ atm}}\right)\left(\dfrac{393 \text{ K}}{273 \text{ K}}\right) = 2.13 \text{ L}$ **(b)** 1.00 L **(c)** 4.00 L   **56.** 1.26×10^4 L

70. $n = PV/RT = \dfrac{(8.45 \times 10^{-19} \text{ atm})(1.00 \text{ L})}{(0.0821 \text{ L atm/K mol})(291 \text{ K})} = 3.549 \times 10^{-20} \text{ mol}$

$3.549 \times 10^{-20} \text{ mol}\left(\dfrac{6.022 \times 10^{23} \text{ molecules}}{1 \text{ mol}}\right) = 2.14 \times 10^4 \text{ molecules}$

73. 197 g/mol

75. The empirical formula is C_2H_5 and the molecular formula is C_4H_{10}.

79. **(a)** 255 mL **(b)** 0.560 g **(c)** 15.0 mL

80. **(a)** 1.55×10^{-2} L **(b)** 32.9 g **(c)** 173 L

81. (a) 2.69 atm

(b) The pressure of CO_2 builds up as the gas is formed. When the pressure is great enough, the excess CO_2 escapes to the atmosphere, causing a burp.

(c) 0.592 g $NaHCO_3$

Chapter 11

9. (a) polar (b) nonpolar (c) nonpolar (d) polar (e) polar

50. 189 kcal   **52.** 242 kJ

53. $\left(\dfrac{1188 \text{ cal}}{25.0 \text{ g}}\right)\left(\dfrac{284.48 \text{ g}}{1 \text{ mol}}\right) = 1.35 \times 10^4 \text{ cal/mol}$

59. $0.500 \text{ gal}\left(\dfrac{3.785 \text{ L}}{1 \text{ gal}}\right)\left(\dfrac{1000 \text{ mL}}{1 \text{ L}}\right)\left(\dfrac{1.00 \text{ g}}{1 \text{ mL}}\right)\left(\dfrac{1.00 \text{ cal}}{\text{g }^\circ C}\right)\left(\dfrac{1 \text{ kcal}}{1000 \text{ cal}}\right)(37^\circ C) = 7.0 \times 10^1 \text{ kcal}$

61. $150 \text{ gal}\left(\dfrac{3.785 \text{ L}}{1 \text{ gal}}\right)\left(\dfrac{1000 \text{ mL}}{1 \text{ L}}\right)\left(\dfrac{1.00 \text{ g}}{1 \text{ mL}}\right)(35^\circ C - 10^\circ C)\left(\dfrac{1 \text{ cal}}{\text{g }^\circ C}\right) = 1.4 \times 10^7 \text{ cal}$

63. $(0.80 \text{ J/g }^\circ C)$ **65.** $(32.1 \text{ g})(58.5^\circ C - 27.2^\circ C)\left(\dfrac{0.129 \text{ J}}{\text{g }^\circ C}\right) = 130 \text{ J}$

67. $(500 \text{ g})(40^\circ C - 10^\circ C)\left(\dfrac{0.031 \text{ cal}}{\text{g }^\circ C}\right) = 4.7 \times 10^2 \text{ cal}$

$(500 \text{ g})(40^\circ C - 10^\circ C)\left(\dfrac{1 \text{ cal}}{\text{g }^\circ C}\right) = 1.5 \times 10^4 \text{ cal}$

69. $1.60 \times 10^2 \text{ kcal}$ **71.** $523 \text{ J} + 8.34 \times 10^3 \text{ J} + 1.05 \times 10^4 \text{ J} + 5.64 \times 10^4 \text{ J} = 7.58 \times 10^4 \text{ J}$

Chapter 12

18. $\left(\dfrac{18.9 \text{ g}}{0.341 \text{ L}}\right)\left(\dfrac{1 \text{ mol HC}_2\text{H}_3\text{O}_2}{60.0516 \text{ g}}\right) = 0.923 \ M \text{ HC}_2\text{H}_3\text{O}_2$

20. $\left(\dfrac{0.088 \text{ g}}{237 \text{ mL}}\right)\left(\dfrac{1 \text{ mol C}_8\text{H}_{10}\text{N}_4\text{O}_2}{194.2 \text{ g}}\right)\left(\dfrac{1000 \text{ mL}}{1 \text{ L}}\right) = 1.9 \times 10^{-3} \ M \text{ C}_8\text{H}_{10}\text{N}_4\text{O}_2$

24. $\left(\dfrac{20.5 \text{ mL}}{0.115 \text{ L}}\right)\left(\dfrac{1.049 \text{ g}}{1 \text{ mL}}\right)\left(\dfrac{1 \text{ mol HC}_2\text{H}_3\text{O}_2}{60.052 \text{ g}}\right) = 3.11 \ M \text{ HC}_2\text{H}_3\text{O}_2$

26. (a) $\left(\dfrac{19.5 \text{ g}}{0.785 \text{ L}}\right)\left(\dfrac{1 \text{ mol Na}_2\text{SO}_4}{142.037 \text{ g}}\right)\left(\dfrac{1 \text{ mol SO}_4^{2-}}{1 \text{ mol Na}_2\text{SO}_4}\right) = 0.175 \ M \text{ SO}_4^{2-}$

(b) $\left(\dfrac{19.5 \text{ g}}{0.785 \text{ L}}\right)\left(\dfrac{1 \text{ mol Na}_2\text{SO}_4}{142.037 \text{ g}}\right)\left(\dfrac{2 \text{ mol Na}^+}{1 \text{ mol Na}_2\text{SO}_4}\right) = 0.350 \ M \text{ Na}^+$

28. $0.0323 \text{ mol KNO}_3\left(\dfrac{1 \text{ L}}{0.215 \text{ mol KNO}_3}\right)\left(\dfrac{10^3 \text{ mL}}{1 \text{ L}}\right) = 150 \text{ mL}$

30. $1.75 \text{ g}\left(\dfrac{1 \text{ mol H}_2\text{O}_2}{34.0138 \text{ g}}\right)\left(\dfrac{1 \text{ L}}{0.363 \text{ mol H}_2\text{O}_2}\right)\left(\dfrac{10^3 \text{ mL}}{1 \text{ L}}\right) = 142 \text{ mL}$

33. (a) $0.300 \text{ L}\left(\dfrac{0.925 \text{ mol KOH}}{1 \text{ L}}\right)\left(\dfrac{56.11 \text{ g}}{1 \text{ mol KOH}}\right) = 15.6 \text{ g}$ (b) 21.5 g

(c) $1.17 \text{ L}\left(\dfrac{0.125 \text{ mol C}_2\text{H}_5\text{OH}}{1 \text{ L}}\right)\left(\dfrac{46.0684 \text{ g}}{1 \text{ mol C}_2\text{H}_5\text{OH}}\right) = 6.74 \text{ g}$ (d) 1.9 g

(e) $0.725 \text{ L}\left(\dfrac{2.12 \text{ mol Li}_2\text{CO}_3}{1 \text{ L}}\right)\left(\dfrac{73.89 \text{ g}}{1 \text{ mol Li}_2\text{CO}_3}\right) = 114 \text{ g}$

35. $5.85 \text{ L}\left(\dfrac{2.99 \times 10^{-4} \text{ mol C}_9\text{H}_8\text{O}_4}{1 \text{ L}}\right)\left(\dfrac{180.1582 \text{ g}}{1 \text{ mol C}_9\text{H}_8\text{O}_4}\right) = 0.315 \text{ g}$

37. $0.946 \text{ L}\left(\dfrac{9.75 \text{ mol NH}_3}{1 \text{ L}}\right)\left(\dfrac{17.031 \text{ g}}{1 \text{ mol NH}_3}\right) = 157 \text{ g}$ **38.** $25.0 \text{ mL}\left(\dfrac{1.37 \, M}{0.375 \, M}\right) = 91.3 \text{ mL}$

40. 4.17 mL **42.** $50 \text{ mL}\left(\dfrac{0.075 \, M}{3.00 \, M}\right) = 1.3 \text{ mL}$ **44.** $1.2 \text{ L}\left(\dfrac{0.050 \, M}{6.00 \, M}\right) = 1.0 \times 10^{-2} \text{ L or } 10 \text{ mL}$

46. (a) $2.50 \, M\left(\dfrac{10.0 \text{ mL}}{40.0 \text{ mL}}\right) = 0.625 \, M$ (b) $2.50 \, M\left(\dfrac{75.0 \text{ mL}}{40.0 \text{ mL}}\right) = 4.69 \, M$

(c) $\left(\dfrac{25.0 \text{ g}}{1.25 \text{ L}}\right)\left(\dfrac{1 \text{ mol FeCl}_3}{162.2 \text{ g}}\right) = 0.123 \, M \text{ FeCl}_3$ $0.123 \, M \text{ FeCl}_3 + 2.5 \, M \text{ FeCl}_3 = 2.6 \, M \text{ FeCl}_3$

48. $15.00 \text{ g}\left(\dfrac{1 \text{ mol Al}}{26.9815 \text{ g}}\right)\left(\dfrac{6 \text{ mol HCl}}{2 \text{ mol Al}}\right)\left(\dfrac{1000 \text{ mL}}{1.00 \text{ mol HCl}}\right) = 1.67 \times 10^3 \text{ mL}$

50. $40.0 \text{ mL}\left(\dfrac{12.00 \text{ mol}}{1000 \text{ mL}}\right)\left(\dfrac{1 \text{ mol CO}_2}{2 \text{ mol HCl}}\right)\left(\dfrac{22.4 \text{ L}}{1 \text{ mol}}\right) = 5.38 \text{ L CO}_2$

52. $300 \text{ g}\left(\dfrac{12.5 \text{ g C}_{12}\text{H}_{22}\text{O}_{11}}{100 \text{ g}}\right) = 37.5 \text{ g C}_{12}\text{H}_{22}\text{O}_{11}$ **54.** $448 \text{ g}\left(\dfrac{3.0 \text{ g H}_2\text{O}_2}{100 \text{ g}}\right) = 13 \text{ g H}_2\text{O}_2$

56. $\left(\dfrac{1.31 \text{ g}}{1 \text{ kg}}\right)\left(\dfrac{1 \text{ kg}}{10^3 \text{ g}}\right)10^6 = 1310 \text{ ppm Mg}^{2+} \text{ or } 1.31 \times 10^3 \text{ ppm}$

58. $1.5 \times 10^9 \text{ km}^3\left(\dfrac{10^{15} \text{ cm}^3}{1 \text{ km}^3}\right)\left(\dfrac{1.005 \text{ g}}{1 \text{ cm}^3}\right)\left(\dfrac{1.39 \times 10^{-4} \text{ g}}{10^6 \text{ g}}\right) = 2.1 \times 10^{14} \text{ g Ag}^+$

60. $\left(\dfrac{0.68 \text{ g}}{2.0 \text{ L}}\right)\left(\dfrac{1 \text{ L}}{10^3 \text{ g}}\right)10^6 = 340 \text{ ppm}$; it is likely to be lethal.

62. $\left(\dfrac{3.0 \text{ g}}{97.0 \text{ g H}_2\text{O}}\right)\left(\dfrac{1 \text{ mol H}_2\text{O}_2}{34.01 \text{ g}}\right)\left(\dfrac{10^3 \text{ g H}_2\text{O}}{1 \text{ kg H}_2\text{O}}\right) = 0.91 \text{ m H}_2\text{O}_2$

71. (a) $\text{Na}^+, \text{SO}_4^{2-}$ (b) $\text{K}^+, \text{Cr}_2\text{O}_7^{2-}$ (c) Na^+, F^- (d) $\text{Li}^+, \text{NO}_3^-$ (e) $\text{NH}_4^+, \text{Cl}^-$ (f) $\text{Ag}^+, \text{NO}_3^-$

73. (a) $\text{H}_3\text{O}^+, \text{NO}_3^-$ (b) $\text{C}_2\text{H}_5\text{OH}$ (c) H_2S (d) $\text{C}_{12}\text{H}_{22}\text{O}_{11}$ (e) HF
(f) $\text{Cu}^{2+}, \text{SO}_4^{2-}$ (g) H_2O_2 (h) $\text{Ca}^{2+}, \text{Cl}^-$ (i) NH_3

77. (a) $\text{Zn} + \text{Cu}^{2+} \longrightarrow \text{Zn}^{2+} + \text{Cu}$
(b) $\text{Cl}_2 + 2\text{Br}^- \longrightarrow 2\text{Cl}^- + \text{Br}_2$
(c) $\text{Zn} + 2\text{H}_3\text{O}^+ \longrightarrow \text{Zn}^{2+} + \text{H}_2 + 2\text{H}_2\text{O}$
(d) $\text{CrO}_4^{2-} + \text{Ba}^{2+} \longrightarrow \text{BaCrO}_4$
(e) $\text{OH}^- + \text{HPO}_4^{2-} \longrightarrow \text{H}_2\text{O} + \text{PO}_4^{3-}$
(f) $\text{CaCO}_3 + 2\text{H}_3\text{O}^+ \longrightarrow \text{Ca}^{2+} + \text{CO}_2 + 3\text{H}_2\text{O}$

82. (a) $\text{Hg}_2^{2+} + 2\text{Br}^- \longrightarrow \text{Hg}_2\text{Br}_2$
(b) $2\text{F}^- + \text{Ba}^{2+} \longrightarrow \text{BaF}_2$
(c) no reaction
(d) no reaction

(e) $Ca^{2+} + 2OH^- \longrightarrow Ca(OH)_2$

(f) $3Ba^{2+} + 2PO_4^{3-} \longrightarrow Ba_3(PO_4)_2$

(g) $Pb^{2+} + 2I^- \longrightarrow PbI_2$

84. (a) $Mg^{2+} + 2OH^- \longrightarrow Mg(OH)_2$

(b) $Ca^{2+} + 2F^- \longrightarrow CaF_2$

(c) $3Pb^{2+} + 2PO_4^{3-} \longrightarrow Pb_3(PO_4)_2$

(d) $Pb^{2+} + 2Cl^- \longrightarrow PbCl_2$

(e) no reaction

(f) $Ag^+ + I^- \longrightarrow AgI$

(g) $Fe^{3+} + 3OH^- \longrightarrow Fe(OH)_3$

(h) $Ba^{2+} + SO_4^{2-} \longrightarrow BaSO_4$

$Cd^{2+} + 2OH^- \longrightarrow Cd(OH)_2$

(i) no reaction

(j) $Pb^{2+} + CrO_4^{2-} \longrightarrow PbCrO_4$

(k) $Mg^{2+} + 2F^- \longrightarrow MgF_2$

90. $\left(\dfrac{3.0 \text{ g}}{97.0 \text{ g } H_2O}\right)\left(\dfrac{1 \text{ mol } H_2O_2}{34.01 \text{ g}}\right)\left(\dfrac{10^3 \text{ g } H_2O}{1 \text{ kg } H_2O}\right) = 0.909 \; m \; H_2O_2$

$1.86°C/m \times 0.909 \text{ m} = 1.7°C$ freezing point: $0°C - 1.7°C = -1.7°C$

92. $\left(\dfrac{50.0 \text{ g}}{200 \text{ g } H_2O}\right)\left(\dfrac{1 \text{ mol } C_{12}H_{22}O_{11}}{342.3 \text{ g}}\right)\left(\dfrac{10^3 \text{ g } H_2O}{1 \text{ kg } H_2O}\right) = 0.730 \; m \; C_{12}H_{22}O_{11}$

$0.52°C/m \times 0.730 \text{ m} = 0.38°C$ boiling point: $100°C + 0.38°C = 100.38°C$

Chapter 13

8.

ACID	BASE
HNO_2	NO_2^-
H_3O^+	H_2O
HSO_4^-	SO_4^{2-}
NH_4^+	NH_3
H_2O	OH^-

32. $\left(\dfrac{0.02115 \text{ L}}{0.0100 \text{ L}}\right)\left(\dfrac{0.515 \text{ mol NaOH}}{1 \text{ L}}\right)\left(\dfrac{1 \text{ mol } HC_2H_3O_2}{1 \text{ mol NaOH}}\right) = 1.09 \; M \, HC_2H_3O_2$

33. $\left(\dfrac{0.0275 \text{ L}}{0.0200 \text{ L}}\right)\left(\dfrac{0.675 \text{ mol } OH^-}{1 \text{ L}}\right)\left(\dfrac{1 \text{ mol } H_3O^+}{1 \text{ mol } OH^-}\right) = 0.928 \; M \, HCl$

34. 2.17 g

45. (a) pH 8.? or 8.05   basic   (b) pH 7.? or 7.57   basic

(c) pH 6.? or 6.16   acidic   (d) pH 3.00   acidic

(e) pH 14 or 14.00   basic   (f) pH 0.? or 0.30   acidic

(g) pH 1.? or 1.49   acidic   (h) pH 6.? or 6.96   acidic

(i) pH 3.? or 3.64   acidic   (j) 3.? or 3.55   acidic

(k) pH 7.? or 7.40   basic

47. (a) pOH 4.? or 4.64   basic   (b) pOH 5.? or 5.21   basic

(c) pOH 5.? or 5.16   basic   (d) pOH 11.00   acidic

(e) pOH 0.00   basic   (f) pOH 13.? or 13.70   acidic

(g) pOH 5.? or 5.49   basic   (h) pOH 4.? or 4.04   basic

(i) pOH 10.? or 10.11   acidic   (j) pOH 9.? or 9.68   acidic

(k) pOH 6.? or 6.22   basic

Chapter 14

9. (a) The finely divided starch particles from the flour may explode when ignited in air.

(b) The increase in temperature in the internal combustion engine increases the rate of the reaction.

(c) The increase in pressure of a reaction involving gases increases the rate.

(d) The microorganisms can make ammonia at low temperature by use of enzymes that serve as catalysts.

(e) The increase in temperature increases the rate of internal metabolic reactions of the fish.

(f) The increased concentration of oxygen will increase the rate of combustion, resulting in a fire hazard.

(g) The wood shavings represent an increased state or subdivision of the wood.

(h) As the concentration of the oxygen decreases, the candle will stop burning.

20. 1.8×10^{-5}

22. (a) $K = \dfrac{[H_2][Cl_2]}{[HCl]^2}$ (b) $K = \dfrac{[SO_3]^2}{[SO_2]^2[O_2]}$ (c) $K = \dfrac{[O_3]^2}{[O_2]^3}$

(d) $K = \dfrac{[H_3O^+][CN^-]}{[HCN]}$ (e) $K = \dfrac{[NH_4^+][OH^-]}{[NH_3]}$

25. (a) $C_2H_4 + H_2O \rightleftharpoons C_2H_5OH$ (b) $CO + H_2O \rightleftharpoons H_2 + CO_2$

(c) $CO_2 + H_2O \rightleftharpoons H_3O^+ + HCO_3^-$ (d) $HF + H_2O \rightleftharpoons H_3O^- + F^-$

29. (a) $K_a = \dfrac{[H_3O^+][HPO_4^{2-}]}{[H_2PO_4^-]}$ (b) $K_a = \dfrac{[H_3O^+][CO_3^{2-}]}{[HCO_3^-]}$ (c) $K_a = \dfrac{[H_3O^+][NH_3]}{[NH_4^+]}$

30. (a) 0.012   (b) 5.7×10^{-11} (c) 5.6×10^{-10}

36. (a) Cooling causes a shift toward the H_2–I_2.

(b) Adding H_2 causes a shift toward the HI side.

(c) A catalyst does not affect the equilibrium.

(d) A decrease in pressure has no effect since there are 2 moles of gas on each side.

(e) An increase in HI concentration causes a shift toward the H_2–I_2 side.

38. (a) Increasing the temperature causes a shift toward the N_2–H_2 side.

(b) Increasing the concentration of H_2 causes a shift toward the NH_3 side.

(c) Decreasing the concentration of N_2 causes a shift toward the N_2–H_2 side.

(d) Decreasing the temperature causes a shift toward the NH_3 side.

(e) Increasing the pressure causes a shift toward the NH_3 side since this side has fewer moles of gas.

40. (a) Cooling causes a shift toward the $H_2O(s)$ side.

(b) A catalyst does not affect the equilibrium.

(c) Adding extra OH^- causes a shift toward the $Mg(OH)_2(s)$ side.

(d) Adding H_3O^+ causes a shift toward the CO_2–H_2O side.

(e) Decreasing the pressure causes a shift toward the $CO_2(g)$ side.

(f) Both changes cause the equilibrium to shift toward the H_2O side.

(g) Adding Cl^- causes a shift toward the $AgCl(s)$ side.

46. pH = 1.46   48. $[H_3O^+] = 7.7 \times 10^{-3}\ M$, pH = 2.11   52. $[H_3O^+] = 3.0 \times 10^{-5}\ M$

53. $[H_3O^+] = 2.6 \times 10^{-4}\ M$ 54. pH = 7.51   55. pH = 3.69

Chapter 15

7. (a) $+5$ (b) $+2$ (c) $+4$ (d) $+4$ (e) $+6$ (f) $+2$ (g) $+6$ (h) $+4$ (i) $+5$ (j) $+2$ (k) -2 (l) $+3$

14. O.A. = oxidizing agent   R.A. = reducing agent

(a) redox O_2 = O.A. CH_4 = R.A.   (b) not redox

(c) redox I_2O_5 = O.A.   CO = R.A.   (d) not redox

(e) not redox   (f) redox NO_2 = O.A. | SO_2 = R.A.

(g) redox Cl_2 = O.A. CH_4 = R.A.   (h) not redox

(i) redox H^+ = O.A.   Ni = R.A.

15. (a) C -4 to $+4$ O   0 to -2 (c) I $+5$ to 0   C $+2$ to $+4$

(f) S $+4$ to $+6$ N $+4$ to $+2$ (g) Cl   0 to -1 C -4 to 0

(i) Ni   0 to $+2$ H $+1$ to 0

20. (a) CH_4 and CO_2; H_2O and H_2 (b) Fe_2O_3 and Fe; CO and CO_2

(c) Ca and $Ca(OH)_2$; H_2O and H_2 (d) Cl_2 and Cl^-; I^- and I_2

(e) Ag^+ and Ag; Zn and Zn^{2+}

23. (a) $Na^+ + Br^- + H_2O_2 + H^+$

$H_2O_2 + 2H^+ + 2Br^- \longrightarrow Br_2 + 2H_2O$

(b) $Na^+ + Cr_2O_7{}^{2-} + Na^+ + I^- + H^+$

$Cr_2O_7{}^{2-} + 6I^- + 14H^+ \longrightarrow 2Cr^{3+} + 3I_2 + 7H_2O$

(c) $K^+ + MnO_4{}^- + Fe^{2+} + Cl^- + H^+$

$MnO_4{}^- + 5Fe^{2+} + 8H^+ \longrightarrow 4H_2O + Mn^{2+} + 5Fe^{3+}$

(d) $H_2C_2O_4 + H_2O_2 + H^+$

$H_2C_2O_4 + H_2O_2 \longrightarrow 2CO_2 + 2H_2O$

(e) $Fe^{3+} + NO_3{}^- + K^+ + I^-$

$2Fe^{3+} + 2I^- \longrightarrow 2Fe^{2+} + I_2$

27. (a) $2K(s) + 2H_2O \longrightarrow 2K^+ + H_2 + 2OH^-$ (b) No reaction

(c) $Cu(s) + Hg^{2+} \longrightarrow Hg(\ell) + Cu^{2+}$ (d) $Zn(s) + Pb^{2+} \longrightarrow Zn^{2+} + Pb(s)$

(e) no reaction

Chapter 16

12. (a) $^{239}_{94}Pu \longrightarrow {}^{235}_{92}U + {}^4_2He$ (b) $^{214}_{84}Po \longrightarrow {}^{210}_{82}Pb + {}^4_2He$

(c) $^{238}_{92}U \longrightarrow {}^{234}_{90}Th + {}^4_2He$ (d) $^{241}_{95}Am \longrightarrow {}^{237}_{93}Np + {}^4_2He$

(e) $^{207}_{84}Po \longrightarrow {}^{203}_{82}Pb + {}^4_2He$ (f) $^{210}_{82}Pb \longrightarrow {}^{206}_{80}Hg + {}^4_2He$

14. (a) $^{197}_{78}Pt \longrightarrow {}^{197}_{79}Au + {}^0_{-1}e$ (b) $^{14}_6C \longrightarrow {}^{14}_7N + {}^0_{-1}e$

(c) $^{82}_{35}Br \longrightarrow {}^{82}_{36}Kr + {}^0_{-1}e$ (d) $^{214}_{82}Pb \longrightarrow {}^{214}_{83}Bi + {}^0_{-1}e$

(e) $^{239}_{93}Np \longrightarrow {}^{239}_{94}Pu + {}^0_{-1}e$ (f) $^{60}_{27}Co \longrightarrow {}^{60}_{28}Ni + {}^0_{-1}e$

16. (a) $^{28}_{13}Al$ (b) $^{222}_{86}Rn$ (c) $^0_{-1}e$ (d) $^{238}_{92}U$

19. (a) $^{222}_{86}Rn \longrightarrow {}^{218}_{84}Po + {}^4_2He$

(b) Radon in the atmosphere can be breathed with air.

(c) 0.063 g after 3 half-lives   (d) 977 atoms after 10 half-lives

21. (b) 1.3×10^{-6} g, 9.8×10^{-9} g

35. (a) $^0_{-1}e$ (b) $^{254}_{102}No$ (c) $^{30}_{15}P$ (d) $^{32}_{16}S$ (e) ^3_1H (f) ^6_3Li (g) $^{13}_6C$

40. 1.67×10^{12} J

INDEX

Alphabetical List of Elements with Atomic Numbers and Atomic Weights*

Element	Symbol	Atomic Number	Atomic Mass	Element	Symbol	Atomic Number	Atomic Mass
Actinium	Ac	89	227.0278	Mercury	Hg	80	200.59(2)
Aluminum	Al	13	26.981538(2)	Molybdenum	Mo	42	95.94(1)
Americium	Am	95	243.0614	Neodymium	Nd	60	144.234(3)
Antimony	Sb	51	121.760(1)	Neon	Ne	10	20.1797(6)
Argon	Ar	18	39.948(1)	Neptunium	Np	93	237.0482
Arsenic	As	33	74.92160(2)	Nickel	Ni	28	58.6934(2)
Astatine	At	85	209.9871	Niobium	Nb	41	92.90638(2)
Barium	Ba	56	137.327(7)	Nitrogen	N	7	14.00674(7)
Berkelium	Bk	97	247.0703	Nobelium	No	102	259.1009
Beryllium	Be	4	9.012182(3)	Osmium	Os	76	190.23(3)
Bismuth	Bi	83	208.98038(3)	Oxygen	O	8	15.9994(3)
Bohrium	Bh	107	262.12	Palladium	Pd	46	106.42(1)
Boron	B	5	10.811(7)	Phosphorus	P	15	30.973761(2)
Bromine	Br	35	79.904(1)	Platinum	Pt	78	195.078(2)
Cadmium	Cd	48	112.411(8)	Plutonium	Pu	94	244.0642
Calcium	Ca	20	40.078(4)	Polonium	Po	84	208.9824
Californium	Cf	98	251.0796	Potassium	K	19	39.0983(1)
Carbon	C	6	12.01107(8)	Praseodymium	Pr	59	140.90765(2)
Cerium	Ce	58	140.116(1)	Promethium	Pm	61	144.9127
Cesium	Cs	55	132.90545(2)	Protactinium	Pa	91	231.03588(2)
Chlorine	Cl	17	35.4527(9)	Radium	Ra	88	226.0254
Chromium	Cr	24	51.9961(6)	Radon	Rn	86	222.0176
Cobalt	Co	27	58.933200(9)	Rhenium	Re	75	186.207(1)
Copper	Cu	29	63.546(3)	Rhodium	Rh	45	102.90550(2)
Curium	Cm	96	247.0703	Rubidium	Rb	37	85.4678(3)
Dubnium	Db	105	262.114	Ruthenium	Ru	44	101.07(2)
Dysprosium	Dy	66	162.50(3)	Rutherfordium	Rf	104	261.11
Einsteinium	Es	99	252.083	Samarium	Sm	62	150.36(3)
Erbium	Er	68	167.26(3)	Scandium	Sc	21	44.955910(8)
Europium	Eu	63	151.964(1)	Seaborgium	Sg	106	263.118
Fermium	Fm	100	257.0951	Selenium	Se	34	78.96(3)
Fluorine	F	9	18.9984032(5)	Silicon	Si	14	28.0855(3)
Francium	Fr	87	223.0197	Silver	Ag	47	107.8682(2)
Gadolinium	Gd	64	157.25(3)	Sodium	Na	11	22.989770(2)
Gallium	Ga	31	69.723(1)	Strontium	Sr	38	87.62(1)
Germanium	Ge	32	72.61(2)	Sulfur	S	16	32.066(6)
Gold	Au	79	196.96655(2)	Tantalum	Ta	73	180.9479(1)
Hafnium	Hf	72	178.49(2)	Technetium	Tc	43	98.9072
Hassium	Hs	108	265	Tellurium	Te	52	127.60(3)
Helium	He	2	4.002602(2)	Terbium	Tb	65	158.92534(2)
Holmium	Ho	67	164.93032(2)	Thallium	Tl	81	204.3833(2)
Hydrogen	H	1	1.00794(7)	Thorium	Th	90	232.0381(1)
Indium	In	49	114.818(3)	Thulium	Tm	69	168.93421(2)
Iodine	I	53	126.90447(3)	Tin	Sn	50	118.710(7)
Iridium	Ir	77	192.217(3)	Titanium	Ti	22	47.867(1)
Iron	Fe	26	55.845(2)	Tungsten	W	74	183.84(1)
Krypton	Kr	36	83.80(1)	Ununnilium	Uun	110	269
Lanthanum	La	57	138.9055(2)	Unununium	Uuu	111	272
Lawrencium	Lr	103	262.11	Ununbium	Uub	112	277
Lead	Pb	82	207.2(1)	Uranium	U	92	238.0289(1)
Lithium	Li	3	6.941(2)	Vanadium	V	23	50.9415(1)
Lutetium	Lu	71	174.967(1)	Xenon	Xe	54	131.29(2)
Magnesium	Mg	12	24.3050(6)	Ytterbium	Yb	70	173.04(3)
Manganese	Mn	25	54.938049(9)	Yttrium	Y	39	88.90585(2)
Meitnerium	Mt	109	266	Zinc	Zn	30	65.39(2)
Mendelevium	Md	101	258.10	Zirconium	Zr	40	91.224(2)

*In some atomic weights, the uncertainties in the last digit between ±1 and ±9 are given in parentheses.